Handbook of MIMO Systems

Handbook of MIMO Systems

Edited by Jake Turner

STATES
ACADEMIC PRESS
www.statesacademicpress.com

States Academic Press,
109 South 5th Street,
Brooklyn, NY 11249, USA

ISBN: 978-1-63989-261-7

Cataloging-in-Publication Data

Handbook of MIMO systems / edited by Jake Turner.
p. cm.
Includes bibliographical references and index.
ISBN 978-1-63989-261-7
1. MIMO systems. 2. Wireless communication systems.
3. Telecommunication systems. I. Turner, Jake.
TK5103.4836 .H36 2022
621.384--dc23

For information on all States Academic Press publications visit our website at www.statesacademicpress.com

STATES
ACADEMIC PRESS

Contents

Preface

MIMO or multiple-input multiple-output is a wireless technology used to increases the data capacity of an RF radio by deploying multiple transmitting and receiving antennas. In a multiple-input multiple-output system, the data is transmitted from multiple antennas via the same path in the same bandwidth. This results in each signal reaching the receiving antenna through a different path leading to more reliable data. The receiver is designed to consider the slight time difference between reception of each signal, additional noise, interference, and even lost signals. Some of the typical MIMO configurations are 2X2 MIMO, 3X3 MIMO, 4X4 MIMO, and 8X8 MIMO. This book elucidates the concepts and innovative models around prospective developments with respect to MIMO systems. The ever-growing need for advanced technology is the reason that has fueled the research in this field in recent times. This book will provide comprehensive knowledge to the readers.

The information shared in this book is based on empirical researches made by veterans in this field of study. The elaborative information provided in this book will help the readers further their scope of knowledge leading to advancements in this field.

Finally, I would like to thank my fellow researchers who gave constructive feedback and my family members who supported me at every step of my research.

Editor

1

Beam-domain hybrid time-switching and power-splitting SWIPT in full-duplex massive MIMO system

Kui Xu, Zhexian Shen*, Yurong Wang and Xiaochen Xia

Abstract

In this paper, we consider the beam-domain hybrid time-switching (TS) and power-splitting (PS) simultaneous wireless information and power transfer (SWIPT) protocol design in full-duplex (FD) massive multiple-input multiple-output (MIMO) system, where the FD base station (BS) simultaneously serves a set of downlink half-duplex (HD) users (cellular users) and a set of fixed uplink HD users (fixed sensor nodes) which are uniformly distributed in its coverage area. In order to reduce the computational complexity, we investigate the beam-domain representation of massive MIMO channels based on the basis expansion model, and then the beam-domain SWIPT protocol which lies in intelligently scheduling the users and sensors based on the beam-domain distributions of their associated channels to mitigate SI and enhance transmission efficiency is designed for full-duplex massive MIMO system. The whole beam-domain hybrid TS and PS SWIPT protocol is divided into two phases based on the ideal of TS. The first phase is used for cellular users uplink training and sensor nodes energy harvesting as well as downlink training, wherein the cellular users transmit uplink pilots for uplink channel estimation at the BS, while the BS simultaneously transmits energy signals to the sensor nodes. Based on the idea of PS, the sensor nodes utilize the received energy signals for energy harvesting and downlink channel estimation. In the second phase, the BS forms the transmit beamformers for information transmission to the users as well as the receive beamformers for the sensors transmit their data to the BS simultaneously. By optimizing the TS ratio and transmit powers at the BS in two phases, the system achievable sum rate performance is maximized. Simulation results show the superiority of the proposed protocol on spectral efficiency compared with the existing massive MIMO SWIPT protocol.

Keywords: Full-duplex massive MIMO, Beam-domain channel representation, Simultaneous wireless information and power transfer, Beamforming, Achievable sum rate

1 Introduction

In the existing fourth generation (4G) system, multiple-input multiple-output (MIMO) technology has been used to improve the spectral efficiency (SE) performance. By deploying M antennas at transmitter, and N antennas at receiver, MIMO takes advantage of the spatial multiplexing gain of $\min(M,N)$ to improve the date rate. With the exponential growth of mobile Internet, the data traffic will increase by 1000 times beyond 2020. The existing 4G system cannot meet the increasing traffic demand of the future mobile communication, which put forward a great challenge to the fifth generation (5G) wireless communication systems [1].

As one of the key technologies in 5G, massive MIMO [1, 2], which uses large-scale antenna arrays (with 100 or more antennas) to replace the currently adopted multiple antennas, has been proved to be able to greatly improve the SE of wireless communications system [3–5]. As the number of antennas in massive MIMO grows larger, the simple linear precoder and detector tend to optimize, and the noise as well as the uncorrelated interference can be effectively eliminated [1, 2]. What is more, the spatial resolution of massive MIMO is remarkably improved when compared with the conventional MIMO, and the signal

*Correspondence: 18805198368@139.com
Institute of Communication Engineering, Army Engineering University of PLA, No. 2, Biaoying Street, Nanjing, China

beam can be focused within a narrow range; hence, various kinds of interference can be substantially suppressed [4, 6–10].

In the conventional half-duplex (HD) systems, i.e., time-division duplex (TDD) systems or frequency-division duplex (FDD) systems, the uplink and downlink are separated by orthogonal time or frequency resources, hence theoretically wastes half of the time-frequency resource [11]. As a potential SE enhancement technique in 5G systems, full-duplex (FD) wireless transmission scheme allows the transceiver to use the same frequency resource for simultaneous transmission and reception [12–16]. FD technology can potentially double the transmission capability of the wireless physical layer. In terms of practical FD implementation, the effect of self-interference (SI) due to the coupling of own high-powered transmit signals to the receiver must be reduced [11]. Thanks to the promising analog/digital and spatial domain SI cancelation techniques that can achieve high transmit-receive isolation which have been reported in the recent literatures [11, 17], now, FD is becoming feasible in the near future.

1.1 Motivation and related work

In addition to the SE, energy efficiency (EE) also has gained wide research attention for the design of wireless networks [18, 19]. Energy constraints introduce an upper limit on the power of transmission and associated signal processing in wireless devices, especially for battery-powered equipments [20]. As a sustainable solution to maintain the lifetime of energy constrained wireless networks, energy harvesting (EH) technique has recently received significant attention [21–23]. EH technology can effectively solve the charging problem in battery-powered wireless networks, wherein the wireless nodes are inaccessible or a large number of wireless nodes are widely distributed.

Various sources of renewable energy, such as ambient heat, wind, solar, and vibration, can be used to power wireless communication devices [24]. However, one common problem of collecting energy from such sources is its random nature. Energy availability variations with location, time, or environmental conditions make resource allocation in such systems very challenging. Radio frequency (RF) signals radiated by ambient transmitters can be identified as a viable new inspiration for EH. Compared with conventional energy sources, RF signals can carry both information and energy simultaneously. Wireless devices collecting energy of RF signals to power the signal transmission are named as simultaneous wireless information and power transfer (SWIPT), which has recently attracted great attention for various wireless channels [25–28]. Owing to its features, SWIPT can be used to remotely power a variety of applications such as wireless sensor networks, body area networks, wireless charging facilities, and future cellular networks. Recently, several important advances of SWIPT in the antenna and RF EH circuit designs, have greatly increased the feasibility of EH in practical wireless applications [29–31].

The SWIPT receiver was originally designed for simultaneous information decoding (ID) and EH from the same received signal [32, 33], which is found to be unrealistic in the later research [34]. Due to the non-ideal EH circuit design, part of the RF signal energy will be lost during the ID process [35]. Hence, practical circuits used for EH are not able to decode the information carried by the signal directly. In the following SWIPT receiver designs, antennas with different functionalities are equipped for the information transceiver and the energy harvester, respectively. In [36], antennas with different power sensitivities are designed for receivers of different uses, i.e., − 20 to − 10 dBm for EH and − 60 dBm for ID.

According to the signal partition method for EH and ID, there are mainly two types of practical SWIPT receiver designs, namely the time-switching (TS) and the power-splitting (PS) receiver architectures [18]. The TS receiver alternately switches between EH and ID according to the TS ratio [36]. The received RF signal is first sent to the EH receiver and then to the information receiver. In the PS receiver, the power of the incoming signal is split into two streams according to the PS ratio [26]. One portion of the received signal is sent to the EH receiver, and the remaining portion is sent to the information receiver.

The design of transceiver for MISO interference channel with EH is studied in [37], where the PS method is adopted and the transmit beamforming vectors as well as receive PS ratios are jointly optimized. The authors in [38, 39] consider a one-way single-antenna amplify-and-forward (AF) SWIPT relay networks, where both the TS and PS protocols are studied. The joint transceiver designs for multiuser MISO relay systems with EH is studied in [40], where the received signal is divided into two parts for ID and EH based on the PS principle. Since the two-way relaying (TWR) system can further improve the SE, many works also studied the SWIPT protocols in the TWR scenario. The authors in [41] considered a two-way AF relaying system with SWIPT, where an EH relay node is used to help two source nodes exchange information. In [42], the authors consider the sum-throughput maximization problem in TWR network, where all the nodes are wireless-powered. The authors in [43] consider a SWIPT AF TWR network, where two source nodes harvest energy from multiple relay nodes.

The above works on SWIPT relay network are studied in HD scenario, where devices can either transmit or receive on a single frequency band, but not simultaneously. Since FD can potentially double the wireless physical layer capacity, recently, the transceiver design problem in FD SWIPT systems has arised. In [44], the joint optimization problem of relay beamforming, the receiver PS ratio as well as the transmit power at the sources are investigated for SWIPT system with a FD MIMO AF relay to maximize the system achievable sum rate. The design of robust nonlinear transceivers for MIMO FD wireless-powered relay networks is studied in [45], where the effects of imperfect channel state information (CSI) is took into consideration. The joint transceiver design for a FD cloud radio access network with SWIPT is considered in [46]. The joint transceiver designs for FD K-Pair MIMO interference channel with SWIPT is considered in [47]. Robust secure beamforming scheme for wireless-powered FD MIMO systems with self-energy recycling is studied in [48].

Despite the previous works on EH in FD system or MIMO system, a few works have been done on the SWIPT protocol design in cellular system with massive MIMO base station (BS). In [21], the joint FD self-backhaul and EH protocol for small cell networks with massive MIMO is investigated. In [49], SWIPT protocol for three-dimensional (3D) massive MIMO system is designed, where the matched filter (MF) precoder is adopted at the BS. In [50], the SWIPT protocol is designed for multipair TWR system with massive MIMO, where the linear precoders, i.e., zero-forcing (ZF) and maximal ratio combining (MRC), are adopted at the relay. A low-complexity SWIPT scheme with retrodirective maximum ratio transmission (MRT) beamforming is studied in [51], where all energy receivers (ERs) send a common beacon signal simultaneously to the energy transmitter (ET) in the uplink and the ET simply conjugates and amplifies its received sum-signal and transmits to all ERs in the downlink for SWIPT. For SWIPT system with massive MIMO, instantaneous full dimensional CSI is needed to perform linear precoding. To obtain full dimensional CSI, the uplink training overhead scales linearly with the number of user equipments (UE) in TDD system, the downlink training overhead scales linearly with the number of antennas in FDD system, and the corresponding CSI feedback yield an unacceptably high overhead, and therefore poses a significant bottleneck on the achievable SE [52].

1.2 Contributions

In this paper, we consider the beam-domain hybrid TS and PS SWIPT protocol design in FD massive MIMO system, where the FD BS simultaneously serves a set of downlink HD users (cellular users) and a set of fixed uplink HD users (fixed sensor nodes) which are uniformly distributed in its coverage area.

The contributions of this paper are summarized as follows:

- We investigate the beam-domain (BD) representation of massive MIMO channels based on the basis expansion model (BEM). In this way, the full dimensional channel can be compressed into BD effective channel (BDEC), as a result, the computational and implementation complexity can be greatly reduced and the SE performance can be improved.
- The BD hybrid TS and PS SWIPT protocol which lies in intelligently scheduling the users and sensors based on the BD distributions of their associated BDECs to mitigate SI and enhance SE performance is designed for FD massive MIMO system. Specifically, the whole BD hybrid TS and PS SWIPT protocol is divided into two phases based on the ideal of TS. *Phase I* is used for cellular users uplink training and sensor nodes EH as well as downlink training. During this phase, the cellular users transmit pilots for uplink UE-BS channel estimation at the BS, while the BS simultaneously transmits energy signals to the sensor nodes. Based on the idea of PS, the sensor nodes utilize the received energy signals for EH and downlink Sensor-BS channel estimation. In *phase II*, the BS forms the transmit beams for information transmission to the users as well as the receive beams for the sensors transmit their data to the BS simultaneously.
- By optimizing the TS ratio and transmit powers at the BS in two phases, the system achievable sum rate performance is maximized. Simulation results shown the superiority of the proposed protocol on spectral efficiency compared with the conventional HD massive MIMO SWIPT protocol.

The rest of the paper is organized as follows. The system model is described in Section 2. Sections 3 and 4 consider the beam-domain channel representation and proposed BD massive MIMO FD SWIPT transmission scheme, respectively. Section 5 consider the practical beam-domain user and sensor grouping problem. Section 6 derives the system achievable sum rate and the optimum TS ratio as well as transmit powers at the BS in two phases. Section 7 presents the simulation results. Section 8 draws the conclusions.

Notations: In this paper, $\mathbb{E}(\cdot)$ denotes the expectation. $\mathbf{A}^{\{B_1,B_2\}}$ denotes the submatrix of $\mathbf{A}$ by keeping its rows indexed by set B_1 and columns indexed by set B_2. $\mathbf{A}^{\{B,:\}}$ and $\mathbf{A}^{\{:,B\}}$ denote the submatrix of $\mathbf{A}$ by keeping its rows and columns indexed by set B, respectively. $\delta(\cdot)$ denotes the dirac delta function. $(\cdot)^T$, $(\cdot)^*$, $(\cdot)^H$, $|\cdot|$, $\|\cdot\|$, and $tr(\cdot)$ denote transpose, conjugate, conjugate transpose, determinant, Frobenius norm, and trace of a matrix, respectively.

2 System model

We consider a FD massive MIMO SWIPT system that the BS simultaneously serves a set of downlink users (cellular users) $\mathbf{K}_D = \{1, 2, \cdots, K_d\}$ and a set of fixed uplink sensors $\mathbf{K}_U = \{1, 2, \cdots, K_u\}$ which are uniformly distributed in the coverage area of the BS, as shown in Fig. 1. The number of downlink users is K_d, and the number of sensors is K_u. The BS is equipped with a separate massive antenna array. The total number of antennas at the BS is $2N$ of which N antennas are dedicated to the downlink and N antennas are used for the uplink. We assume that all users and sensors are work in HD mode and have a single antenna. Each sensor employs a rectanna for EH and uses the harvested energy to power its subsequent data transmission.

We consider frame-based transmissions over Rayleigh fading channels. The length of one frame is fixed to T seconds, which is assumed to be less than the coherence interval of the channel. We use $\mathbf{h}_{k_u} \in \mathbb{C}^{N\times 1}$ and $\mathbf{h}_{k_d} \in \mathbb{C}^{N\times 1}$ to denote the uplink UE-BS channel vector from the user k_u to the receive antenna array of BS and the downlink UE-BS channel vector from the transmit antenna array of BS to the user k_d. Let $\mathbf{h}^e_{k_u} \in \mathbb{C}^{N\times 1}$ and $\mathbf{h}^e_{k_d} \in \mathbb{C}^{N\times 1}$ denote the uplink sensor-BS channel vector from the sensor node k_d to the receive antenna array of the BS and the downlink sensor-BS channel vector from the transmit antenna array of the BS to the sensor node k_d. We use $\mathbf{H}_{SI} \in \mathbb{C}^{N\times N}$ to denote the SI channel matrix from the transmit antenna array to the receive antenna array of the BS.

Note that the channel model between the user and the BS is the same as the channel model between the sensor and the BS; in the following sections, we only introduce the uplink and downlink channel model between the user and the BS.

2.1 Uplink/downlink channel model

We consider the general cluster-based channel model [53] where the received signal at the BS from the uplink user k_u is a sum of the contributions from M_u scattering clusters and the received signal at user k_u is a sum of the contributions from M_u scattering clusters. The direction of arrival (DOA) of signals resulting from the ith cluster to the BS is within the region $[\theta^{\min}_{k_{u,i}}, \theta^{\max}_{k_{u,i}}]$ and the direction of departure (DOD) region of signals resulting from the ith scattering cluster to user k_d is within the region $[\theta^{\min}_{k_{d,i}}, \theta^{\max}_{k_{d,i}}]$. Thus, the channel vector between the user k_u or k_d and the BS can be expressed as [53]

$$\mathbf{h}_{k_\xi} = \sum_{i=1}^{M_\xi} \int_{\theta^{\min}_{k_\xi,i}}^{\theta^{\max}_{k_\xi,i}} \mathbf{a}(\theta)\, r_{k_\xi,i}(\theta)\, d\theta \tag{1}$$

where $\xi \in \{u, d\}$, $\mathbf{a}(\theta) = \left[1, \exp\left(-j\frac{2\pi d \sin(\theta)}{\lambda}\right), \cdots, \exp\left(-j\frac{2\pi d (N-1)\sin(\theta)}{\lambda}\right)\right]^T$ is the array response vector with d and λ denoting the antenna spacing and carrier wavelength, respectively. $r_{k_\xi,i}(\theta)$ denotes the complex-valued response gain of uplink/downlink channel.

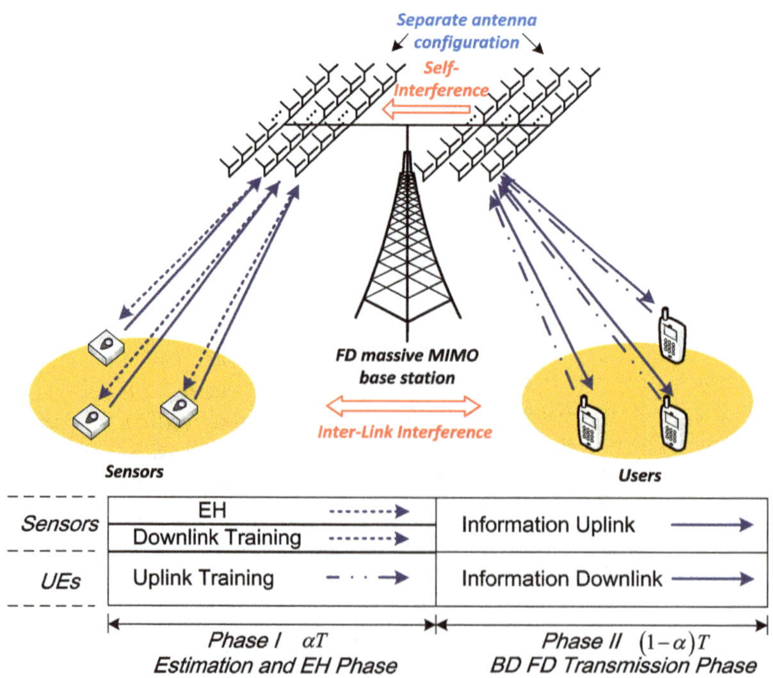

Fig. 1 Illustration of full-duplex massive MIMO SWIPT system and data frame structure

2.2 SI channel model

The SI signal can be viewed as the contributions of signals from M_{SI} scattering clusters with different DOA and DOD regions. Thus, the SI channel matrix $\mathbf{H}_{SI}$ can be expressed as

$$\mathbf{H}_{SI} = \sum_{i=1}^{M_{SI}} \int_{\theta_{R,i}^{\min}}^{\theta_{R,i}^{\max}} \int_{\theta_{T,i}^{\max}}^{\theta_{T,i}^{\max}} r_{SI,i}\left(\theta_R, \theta_T\right) \mathbf{a}\left(\theta_R\right) \mathbf{a}^H\left(\theta_T\right) d\theta_R d\theta_T \tag{2}$$

where $r_{SI,i}(\theta_R, \theta_T)$ denotes the complex-valued response gain.

In (1) and (2), we assume that the complex-valued response gains with different incidence angles are uncorrelated [54], that is

$$\begin{aligned} \mathbb{E}\left[r_{k_u,i}\left(\theta\right) r_{k_u,i}^*\left(\theta'\right)\right] &= S_{k_u,i}\left(\theta\right) \delta\left(\theta - \theta'\right) \\ \mathbb{E}\left[r_{k_d,i}\left(\theta\right) r_{k_d,i}^*\left(\theta'\right)\right] &= S_{k_d,i}\left(\theta\right) \delta\left(\theta - \theta'\right) \\ \mathbb{E}\left[r_{SI,i}\left(\theta_R, \theta_T\right) r_{SI,i}^*\left(\theta_R', \theta_T'\right)\right] &= S_{SI,i}\left(\theta\right) \delta\left(\theta_R - \theta_R'\right) \delta\left(\theta_T - \theta_T'\right) \end{aligned} \tag{3}$$

where $S_{\omega,i}(\cdot), \omega \in \{k_u, k_d, SI\}$, represents the product of the large-scale fading and channel power angle spectrum (PAS).

Note that the above studied channel model can be transformed into several well-known massive MIMO channels. For example, by setting $M_u = M_d = 1$, we can get the "one ring" channel model introduced in [7]. The one ring model is typically used in the macro-cell environment where the uplink/downlink received signals are scattered by the vicinity of the users [53]. Moreover, by setting

$$\begin{aligned} r_{k_u,i}\left(\theta\right) &= \sum_j r_{k_u,i,j} \delta\left(\theta - \theta_j\right) \\ r_{k_d,i}\left(\theta\right) &= \sum_j r_{k_d,i,j} \delta\left(\theta - \theta_j\right) \\ r_{SI,i}\left(\theta_R, \theta_T\right) &= \sum_{j,l} r_{k_{SI},i,j,l} \delta\left(\theta_R - \theta_{R,j}\right) \delta\left(\theta_T - \theta_{T,l}\right) \end{aligned} \tag{4}$$

we obtain the ray-cluster-based spatial channel model, which is usually used for millimeter wave massive MIMO channel [55].

3 Beam-domain channel representation

In order to reduce the number of dimensions of the channel to be estimated as well as the amount of feedback, we resort to the BEM [56]. Using the BEM, the BD channel representation can be obtained by projecting the channel vector (matrix) on common bases. In this way, the channel can be compressed in the BD under certain selected basis spaces and the channel dimension required to be estimated can be reduced greatly. Moreover, by exploiting the BD structure of SI channel, it is possible to eliminate the SI without using the instantaneous SI channel knowledge and hence realize efficient SWIPT transmission.

Under the BEM, the uplink/downlink UE-BS channel vector $\mathbf{h}_{k_\xi}, \xi \in \{u, d\}$ can be expanded from a set of uniform basis vectors $\{\mathbf{f}_1, \mathbf{f}_2, \cdots, \mathbf{f}_N\} \in \mathbb{C}^{N \times 1}$, that is

$$\mathbf{h}_{k_\xi} = \sum_{m=1}^{N} \tilde{h}_{k_\xi,m} \mathbf{f}_m = \mathbf{F} \tilde{\mathbf{h}}_{k_\xi} \tag{5}$$

where $\mathbf{F} = [\mathbf{f}_1, \mathbf{f}_2, \cdots, \mathbf{f}_N]$. In this paper, the basis vector $\mathbf{f}_i$ is also called a beam [8], and $\tilde{\mathbf{h}}_{k_\xi} = [\tilde{h}_{k_\xi,1}, \tilde{h}_{k_\xi,2}, \cdots, \tilde{h}_{k_\xi,N}]$ is called the BD channel vector, which is obtained by projecting the channel vector $\mathbf{h}_{k_\xi}$ onto the set of all the beams $\Gamma = \{\mathbf{f}_1, \mathbf{f}_2, \cdots, \mathbf{f}_N\}$. According to (1) and (5), we can obtain

$$\tilde{\mathbf{h}}_{k_\xi} = \mathbf{F}^H \mathbf{h}_{k_\xi} = \sum_{i=1}^{M_\xi} \int_{\theta_{k_\xi,i}^{\min}}^{\theta_{k_\xi,i}^{\max}} \mathbf{F}^H \mathbf{a}\left(\theta\right) r_{k_\xi,i}\left(\theta\right) d\theta \tag{6}$$

By projecting channel vectors onto certain selected basis spaces, we can approximate the channel vector from user k_ξ to BS as [52]

$$\mathbf{h}_{k_\xi} \approx \sum_{m \in B_{k_\xi}} \tilde{h}_{k_\xi,m} \mathbf{f}_m = \mathbf{F}^{\left\{B_{k_\xi}\right\}} \tilde{\mathbf{h}}_{k_\xi}^{\left\{B_{k_\xi}\right\}} \tag{7}$$

where B_{k_ξ} is the selected active beam set (ABS) of uplink/downlink which contains the indexes of beams with non-negligible BD channel gains. All the elements selected in the set B_{k_ξ} is related to the DOA/DOD, and the detailed method for determining all the elements in the set of B_{k_ξ} will be described in the following section. $\mathbf{F}^{\{B_{k_\xi}\}}$ is the corresponding active beam spaces, whose columns are consisted of the beams in B_{k_ξ}. The reduced-dimension BD channel vector $\tilde{\mathbf{h}}_{k_\xi}^{\left\{B_{k_\xi}\right\}} \in \mathbb{C}^{\left|B_{k_\xi}\right| \times 1}$ is called the BDEC. Note that (7) holds with equality as $N \to \infty$.

Based on (1) and (7), the BDEC vector can be expressed as

$$\begin{aligned} \tilde{\mathbf{h}}_{k_\xi}^{\left\{B_{k_\xi}\right\}} &= \left(\mathbf{F}^{\left\{B_{k_\xi}\right\}}\right)^H \mathbf{h}_{k_\xi} \\ &= \sum_{i=1}^{M_\xi} \int_{\theta_{k_\xi,i}^{\min}}^{\theta_{k_\xi,i}^{\max}} \left(\mathbf{F}^{\left\{B_{k_\xi}\right\}}\right)^H \mathbf{a}\left(\theta\right) r_{k_\xi,i}\left(\theta\right) d\theta \end{aligned} \tag{8}$$

From (8), the original channel vector can be recovered from the BDEC vector if the DOA/DOD (and hence the ABS) information is known. As a result, in order to obtain $\mathbf{h}_{k_\xi}$, it is enough to estimate the $\left|B_{k_\xi}\right|$- dimension BDEC during the training phase. Hence, the consumption of training resources will be significantly reduced. The ABSs $B_{k_\xi}, \xi \in \{d, u\}$ can be designed by solving the following problem

$$
\begin{aligned}
&\min_{B_{k_\xi}} \left|B_{k_\xi}\right| \\
&s.t. \frac{\mathbb{E}\left[\left\|\tilde{\mathbf{h}}_{k_\xi}^{\left\{B_{k_\xi}\right\}}\right\|^2\right]}{\mathbb{E}\left[\left\|\tilde{\mathbf{h}}_{k_\xi}\right\|^2\right]} = \frac{\sum_{i=1}^{M_\xi} \int_{\theta_{k_\xi,i}^{\min}}^{\theta_{k_\xi,i}^{\max}} \left\|\left(\mathbf{F}^{\left\{B_{k_\xi}\right\}}\right)^H \mathbf{a}(\theta)\right\|^2 s_{k_\xi,i}(\theta)d\theta}{\sum_{i=1}^{M_\xi} \int_{\theta_{k_\xi,i}^{\min}}^{\theta_{k_\xi,i}^{\max}} \left\|\mathbf{F}^H \mathbf{a}(\theta)\right\|^2 s_{k_\xi,i}(\theta)d\theta} \geq \varpi
\end{aligned} \tag{9}
$$

where $\varpi < 1$ is the threshold.

In the same way, we can decompose the SI channel in the BD form, that is $\tilde{\mathbf{H}}_{SI} = \mathbf{F}^H \mathbf{H}_{SI} \mathbf{F}$. Since the average BD SI channel gain has non-negligible value when the beam angle of receive side $\mathbf{f}_m$ lies in $\cup_{i=1}^{M_{SI}} \left(\sin\theta_{R,i}^{\min} - \varepsilon, \sin\theta_{R,i}^{\max} + \varepsilon\right)$, and the beam angle of transmit side $\mathbf{f}_n$ lies in $\cup_{i=1}^{M_{SI}} \left(\sin\theta_{T,i}^{\min} - \varepsilon, \sin\theta_{T,i}^{\max} + \varepsilon\right)$, where ε approaches to zero in the large N regime. Hence, the average BD SI channel gain $\mathbb{E}\left[\left|\left[\tilde{\mathbf{H}}_{SI}\right]_{m,n}\right|^2\right] = \mathbb{E}\left[\left|\mathbf{f}_m^H \mathbf{H}_{SI} \mathbf{f}_n\right|^2\right]$.x

Let $B_{\mathrm{SI,R}}$ and $B_{\mathrm{SI,T}}$ denote the ABS of the SI channel at the receive side and transmit side, respectively. $B_{\mathrm{SI,R}}$ and $B_{\mathrm{SI,T}}$ can be determined by solving problem in (10) on the top of this page.

$$
\begin{aligned}
&\min_{B_{\mathrm{SI,R}}, B_{\mathrm{SI,T}}} \max\left\{\left|B_{SI,R}\right|, \left|B_{SI,T}\right|\right\} \\
&s.t. \\
&\frac{\mathbb{E}\left[\left\|\tilde{\mathbf{H}}_{\mathrm{SI}}^{\left\{B_{\mathrm{SI,R}}, B_{\mathrm{SI,T}}\right\}}\right\|^2\right]}{\mathbb{E}\left[\left\|\tilde{\mathbf{H}}_{\mathrm{SI}}\right\|^2\right]} \\
&= \frac{\sum_{i=1}^{M_{\mathrm{SI}}} \int_{\theta_{R,i}^{\min}}^{\theta_{R,i}^{\max}} \int_{\theta_{T,i}^{\min}}^{\theta_{T,i}^{\max}} s_{\mathrm{SI,i}}(\theta_R, \theta_T) \left\|\left(\mathbf{F}^{\left\{B_{\mathrm{SI,R}}\right\}}\right)^H \mathbf{a}(\theta_R)\mathbf{a}^H(\theta_T)\mathbf{F}^{\left\{B_{\mathrm{SI,T}}\right\}}\right\|^2 d\theta_R d\theta_T}{\sum_{i=1}^{M_{\mathrm{SI}}} \int_{\theta_{R,i}^{\min}}^{\theta_{R,i}^{\max}} \int_{\theta_{T,i}^{\min}}^{\theta_{T,i}^{\max}} s_{\mathrm{SI,i}}(\theta_R, \theta_T) \left\|\mathbf{F}^H \mathbf{a}(\theta_R)\mathbf{a}^H(\theta_T)\mathbf{F}\right\|^2 d\theta_R d\theta_T} \\
&\geq \varpi
\end{aligned} \tag{10}
$$

Hence, we can approximately express the BD SI channel matrix as $\tilde{\mathbf{H}}_{\mathrm{SI}}^{\left\{B_{\mathrm{SI,R}}, B_{\mathrm{SI,T}}\right\}} \approx \left(\mathbf{F}^{\left\{B_{\mathrm{SI,R}}\right\}}\right)^H \mathbf{H}_{\mathrm{SI}} \mathbf{F}^{\left\{B_{\mathrm{SI,T}}\right\}}$, and $\tilde{\mathbf{H}}_{\mathrm{SI}}^{\left\{B_{\mathrm{SI,R}}, B_{\mathrm{SI,T}}\right\}} \in \mathbb{C}^{|B_{\mathrm{SI,R}}| \times |B_{\mathrm{SI,T}}|}$ can be expressed as

$$
\begin{aligned}
\tilde{\mathbf{H}}_{\mathrm{SI}}^{\left\{B_{\mathrm{SI,R}}, B_{\mathrm{SI,T}}\right\}} &= \left(\mathbf{F}^{\left\{B_{\mathrm{SI,R}}\right\}}\right)^H \mathbf{H}_{\mathrm{SI}} \mathbf{F}^{\left\{B_{\mathrm{SI,T}}\right\}} \\
&= \sum_{i=1}^{M_{\mathrm{SI}}} \int_{\theta_{R,i}^{\min}}^{\theta_{R,i}^{\max}} \int_{\theta_{T,i}^{\min}}^{\theta_{T,i}^{\max}} r_{\mathrm{SI,i}}\left(\theta_R, \theta_T\right) \left(\mathbf{F}^{\left\{B_{\mathrm{SI,R}}\right\}}\right)^H \\
&\quad \times \mathbf{a}\left(\theta_R\right) \mathbf{a}^H\left(\theta_T\right) \mathbf{F}^{\left\{B_{\mathrm{SI,T}}\right\}} d\theta_R d\theta_T
\end{aligned} \tag{11}
$$

4 BD hybrid TS and PS SWIPT protocol

In this section, we propose a BD scheme to realize SWIPT in the cellular system with FD massive MIMO BS. In order to accomplish the channel estimation, the transmit and receive antenna arrays of the BS are assumed to be shared antennas, i.e., the transmit and receive RF chains share the same antenna [10]. In the frame-based BD massive MIMO FD SWIPT system, each frame is divided into two phases, *phase I: estimation and EH phase* and *phase II: BD FD transmission phase*, as shown in Fig. 1.

In the *phase I* of time period αT $(0 \leq \alpha \leq 1)$, users transmit pilots for channel estimation at the BS, while the BS simultaneously transmits energy signals to the sensor nodes for EH and channel estimation. Due to the separate antenna configuration at the BS; therefore, the reciprocity of uplink and downlink channels does not hold. The BS uses the transmit antenna array to receive the pilot signals of users for uplink UE-BS channel estimation and uses the receive antenna array to transmit the energy signals. We assume that the sensor opens the receive circuit during the interval αT, and part of the received energy signal is used for channel estimation and part of the received energy signal is used for EH; hence, the sensors can obtain the uplink sensor-BS channel. Specifically, the BS uses the transmit antenna array to receive the pilot signals of UEs, that is y_t^U as shown in Fig. 2. Thus, the uplink UE-BS channel can be estimated during *phase I*. Simultaneously, the BS uses receive antenna array to transmit the energy signals to all the sensors. The signal received by sensor node can be expressed as y_r^S. Part of the received energy signal, $\sqrt{\beta} y_r^S$, is used for EH and part of the received signal, $\sqrt{1-\beta} y_r^S$, is used for downlink sensor-BS channel estimation, where β denotes the PS ratio. Note that the detailed training signals (pilot signals and energy signals) design will illustrated in the next subsection.

In the *phase II* of time period $(1-\alpha)T$, the BS obtains the downlink UE-BS CSI based on channel reciprocity and then forms the beams for information transmission to the users. Moreover, the sensors feedback the uplink sensor-BS CSI to the BS and transmit their data during the interval $(1-\alpha)T$ using the harvested energy. Specifically, the sensors transmit signal $\sqrt{\beta\eta} y_t^S$ to the BS, where $\eta < 1$ denotes the energy conversion efficiency. Note that the resource consumption for CSI feedback of sensors are omitted in this paper.

4.1 Training design

Let $\Phi_u \in \mathbb{C}^{\tau_u \times \max_{g_u \in G_u} K_{g_u}}$ be the orthogonal pilot sequence set for uplink UE-BS training, where τ_u denotes the length of pilot sequence which satisfies $\tau_u \geq \max_{g_u \in G_u} K_{g_u}$, g_u denotes the user group, K_{g_u} is the number of users in the group g_u, and G_u denotes the set of user groups. The pilot sequences allocated for group g_u can be given by $\Phi_{g_u} = \Phi_u^{\{:,1,:K_{g_u}\}}$ and $\left|\left[\Phi_{g_u}\right]_{i,j}\right|^2 = p_u$, where p_u denotes the power of each uplink pilot symbol. In this paper, we consider the fixed sensors, hence it is reasonable to assume that the

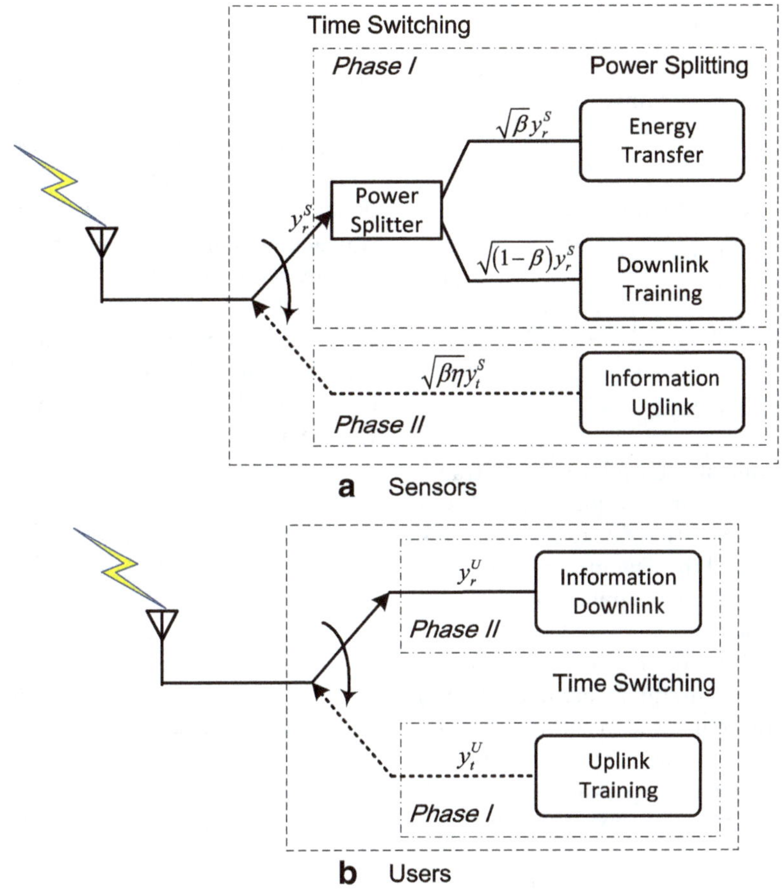

Fig. 2 Illustration of the signal flow for the proposed beam-domain hybrid TS and PS SWIPT protocol

downlink sensors group g_d^e and the active beam sets $B_{g_d^e}$ are known. Let $|B_{g_d^e}| = b_d^e$ and $\Phi^{\mathrm{E}}_{g_d} \in \mathbb{C}^{\tau_d \times b_d^e}$ be the orthogonal energy sequence set, where τ_d denotes the length of energy sequence which satisfies $\tau_d \geq b_d^e$. Let $K_{g_d^e}$ denotes the number of sensors in the group g_d^e, and G_d^e denotes the set of downlink sensor groups.

If $\tau_d \neq \tau_u$, we define

$$\begin{cases} \dot{\Phi}^{\mathrm{E}}_{g_d} = \Phi^{\mathrm{E}\{1:\tau_u\}}_{g_d} & \text{if } \tau_d \geq \tau_u \\ \dot{\Phi}^{\mathrm{E}}_{g_d} = \begin{bmatrix} \Phi^{\mathrm{E}}_{g_d} \\ 0_{\left(\tau_u - \tau_d\right) \times b_d} \end{bmatrix} & \text{if } \tau_d < \tau_u \end{cases} \tag{12}$$

and $\left|\left[\Phi^{\mathrm{E}}_{g_d}\right]_{i,j}\right|^2 = p_d^e$, the received signal at the BS is given by

$$\mathbf{Y}_u = \sum_{g_u \in G_u} \mathbf{H}_{g_u} \Phi^T_{g_u} + \mathbf{H}_{SI} \sum_{g_d^e \in G_d^e} \mathbf{F}^{\left\{B_{g_d^e}\right\}} \left(\dot{\Phi}^{\mathrm{E}}_{g_d}\right)^T + \mathbf{N}_u \tag{13}$$

where $\mathbf{H}_{g_u} = \left[\mathbf{h}_{g_u,1}, \cdots, \mathbf{h}_{g_u,K_{g_u}}\right]$ denotes the channel matrix from the user group g_u to the BS, $\mathbf{N}_u$ denotes the AWGN with variance σ. The received signal at the sensor group g_d^e can be expressed as

$$\mathbf{Y}_{g_d^e} = \mathbf{H}^H_{g_d^e} \sum_{g_d^e \in G_d^e} \mathbf{F}^{\left\{B_{g_d^e}\right\}} \left(\Phi^{\mathrm{E}}_{g_d}\right)^T + \mathbf{N}_{g_d^e} \tag{14}$$

where $\mathbf{H}_{g_d^e} = \left[\mathbf{h}_{g_d^e,1}, \cdots, \mathbf{h}_{g_d^e,K_{g_d^e}}\right]$ and $\mathbf{N}_{g_d^e}$ denotes the AWGN with variance σ.

4.2 UE-BS BDEC estimation

By multiplying both sides of (13) with $\left(\mathbf{F}^{\{B_{g_u}\}}\right)^H$, we arrive at the BD receive pilot signal from user group g_u

$$\begin{aligned} \mathbf{Y}_{g_u} &= \left(\mathbf{F}^{\{B_{g_u}\}}\right)^H \mathbf{Y}_u \\ &= \tilde{\mathbf{H}}^{\{B_{g_u}\}}_{g_u} \Phi^T_{g_u} + \sum_{g_u' \in G_u / \{g_u\}} \tilde{\mathbf{H}}^{\{B_{g_u}\}}_{g_u'} \Phi^T_{g_u'} \\ &\quad + \sum_{g_d^e \in G_d^e} \tilde{\mathbf{H}}^{\left\{B_{g_u}, B_{g_d^e}\right\}}_{SI} \left(\dot{\Phi}^{\mathrm{E}}_{g_d}\right)^T + \left(\mathbf{F}^{\{B_{g_u}\}}\right)^H \mathbf{N}_u \end{aligned} \tag{15}$$

where $\tilde{\mathbf{H}}_{SI}^{\{B_{g_u},B_{g_d^e}\}} = \left(\mathbf{F}^{\{B_{g_u}\}}\right)^H \mathbf{H}_{SI}\mathbf{F}^{\{B_{g_d^e}\}}$. With (15), the least squares (LS) estimator of the BDEC vector for the user $g_{u,k}$ within group g_u can be obtained as

$$\begin{aligned}\tilde{\mathbf{h}}_{g_{u,k},LS}^{\{B_{g_u}\}} &= \frac{1}{\tau_u p_u}\mathbf{Y}_{g_u}\Phi_u^*\mathbf{e}_k \\ &= \tilde{\mathbf{h}}_{g_{u,k}}^{\{B_{g_u}\}} + \sum_{g'_u\in G_u/\{g_u\}} \tilde{\mathbf{h}}_{g'_{u,k}}^{\{B_{g_u}\}} \\ &+ \frac{1}{\tau_u p_u}\sum_{g_d^e\in G_d^e} \tilde{\mathbf{H}}_{SI}^{\{B_{g_u},B_{g_d^e}\}}\left(\Phi_{g_d}^{\mathrm{E}}\right)^T\Phi_u^*\mathbf{e}_k \\ &+ \frac{1}{\tau_u p_u}\left(\mathbf{F}^{\{B_{g_u}\}}\right)^H\mathbf{N}_u\Phi_u^*\mathbf{e}_k\end{aligned} \tag{16}$$

where p_u denotes the power of each pilot symbol transmitted by users. The second term of right-hand sides of (16) indicates the pilot contamination due to the use of the same pilot sequence set over all the user groups. The third term is the SI due to the simultaneous uplink training and downlink energy transmission.

In the practical system with finite number of BS antennas, the linear minimum mean square error (LMMSE) estimator can be used to refine the results of the LS estimator in (16) to mitigate the residual pilot contamination and SI. Based on the general expression of LMMSE estimator, the refined estimates can be expressed as

$$\begin{aligned}\tilde{\mathbf{h}}_{g_{u,k},LM}^{\{B_{g_u}\}} &= \sum_{i=1}^{M_u}\int_{\theta_{g_{u,k,i}}^{\min}}^{\theta_{g_{u,k,i}}^{\max}}\left(\mathbf{F}^{\{B_{g_u}\}}\right)^H\mathbf{a}(\theta)\,\mathbf{a}^H(\theta)\,\mathbf{F}^{\{B_{g_u}\}}S_{g_{u,k,i}}(\theta)\,d\theta \\ &\times\left(\sum_{g'_u\in G_u}\sum_{i=1}^{M_u}\int_{\theta_{g'_{u,k,i}}^{\min}}^{\theta_{g'_{u,k,i}}^{\max}}\left(\mathbf{F}^{\{B_{g_u}\}}\right)^H\mathbf{a}(\theta)\,\mathbf{a}^H(\theta)\,\mathbf{F}^{\{B_{g_u}\}}S_{g'_{u,k,i}}(\theta)\,d\theta\right. \\ &+\left(\frac{1}{\tau_u p_u}\right)^2\sum_{B_{g_d^e},B_{g'^e_d}\in G_d^e}\sum_{i=1}^{M_{SI}}\int_{\theta_{R,i}^{\min}}^{\theta_{R,i}^{\max}}\int_{\theta_{T,i}^{\min}}^{\theta_{T,i}^{\max}}\mathbf{G}_{\theta_R,\theta_T}^{\{B_{g_u},B_{g_d^e}\}}\Psi_k \\ &\left.\times\left(\mathbf{G}_{\theta_R,\theta_T}^{\{B_{g_u},B_{g'^e_d}\}}\right)^H S_{SI,i}(\theta_R,\theta_T)\,d\theta_R d\theta_T + \frac{\sigma}{\tau_u p_u}\mathbf{I}_{b_u}\right)^{-1}\tilde{\mathbf{h}}_{g_{u,k},LS}^{\{B_{g_u}\}}\end{aligned} \tag{17}$$

where $\mathbf{G}_{\theta_R,\theta_T}^{\{B_{g_u},B_{g'^e_d}\}} \triangleq \left(\mathbf{F}^{\{B_{g_u}\}}\right)^H\mathbf{a}(\theta_R)\,\mathbf{a}^H(\theta_T)\,\mathbf{F}^{\{B_{g'^e_d}\}}$ and meanwhile $\Psi_k = \left(\dot{\Phi}_{g_d}^{\mathrm{E}}\right)^T\Phi_u^*\mathbf{e}_k\mathbf{e}_k^H\Phi_u^T\left(\dot{\Phi}_{g_d}^{\mathrm{E}}\right)^*$. The BS-to-users channel CSI $\tilde{\mathbf{h}}_{g_{d,k},LM}^{\{B_{g_d}\}}$ is obtained by using the reciprocity properties of the wireless channel.

4.3 Energy transfer for sensors

During *phase I*, the received signal at the $g_{d,k}^e$th sensor can be expressed by

$$\mathbf{y}_{g_{d,k}^e} = \mathbf{h}_{g_{d,k}^e}^H\sum_{g'^e_d\in G_d^e}\mathbf{F}^{\{B_{g'^e_d}\}}\left(\Phi_{g_d}^{\mathrm{E}}\right)^T + \sum_{k_u=1}^{K_d}h_{k_u,g_{d,k}^e}\Phi_{k_u}^T + \mathbf{n}_{g_{d,k}^e} \tag{18}$$

where $h_{k_u,g_{d,k}^e} \in \mathbb{C}^{1\times 1}$ denotes the inter-user interference (IUI) channel between the k_uth user and the $g_{d,k}^e$th sensor node, Φ_{k_u} is one row of Φ_u and denotes the pilot sequence transmit by user k_u, and $\mathbf{n}_{g_{d,k}^e} \sim \mathcal{CN}(0,\sigma^2)$ denotes the AWGN at the $g_{d,k}^e$th sensor node.

Generally, the EH receiver will harvest energy from part of the whole signal, that is $\sqrt{\beta}\mathbf{y}_{g_{d,k}^e}$. However, the noise is negligible compared with the high-power signal and is thereby omitted in the harvested energy [39]. We further assume that the amount of energy harvested from the cellular users' transmissions is negligible due to their low transmit power. Therefore, the $g_{d,k}^e$th sensor node transmit power during the remaining $(1-\alpha)T$ time can be written as

$$\begin{aligned}p_{g_{d,k}^e}^{\mathrm{H}} &= \frac{\eta\beta}{(1-\alpha)T}\left\|\mathbf{h}_{g_{d,k}^e}^H\sum_{g'^e_d\in G_d^e}\mathbf{F}^{\{B_{g'^e_d}\}}\left(\Phi_{g_d}^{\mathrm{E}}\right)^T\right\|^2 \\ &\approx \frac{\eta\beta}{(1-\alpha)T}\left\|\mathbf{h}_{g_{d,k}^e}^H\mathbf{F}^{\{B_{g_d^e}\}}\left(\Phi_{g_d}^{\mathrm{E}}\right)^T\right\|^2 \\ &= \frac{\eta\beta}{(1-\alpha)T}\left\|\left(\tilde{\mathbf{h}}_{g_{d,k}^e}^{\{B_{g_d^e}\}}\right)^H\left(\Phi_{g_d}^{\mathrm{E}}\right)^T\right\|^2 \\ &= \frac{\eta\beta}{(1-\alpha)T}\tau_d p_{g_d^e}\left\|\tilde{\mathbf{h}}_{g_{d,k}^e}^{\{B_{g_d^e}\}}\right\|^2 \triangleq \frac{\alpha}{(1-\alpha)}\hat{p}_{g_{d,k}^e}^{\mathrm{H}}\end{aligned} \tag{19}$$

where $\|\cdot\|$ denotes the Euclidean norm of a vector, $0 < \eta < 1$ denotes the energy conversion efficiency. $\hat{p}_{g_{d,k}^e}^{\mathrm{H}} = p_{g_d^e}q_{g_{d,k}^e}$, where $q_{g_{d,k}^e} = \eta\beta\left\|\tilde{\mathbf{h}}_{g_{d,k}^e}^{\{B_{g_d^e}\}}\right\|^2$. It is worth noting that for a fixed sensor node, the downlink sensor grouping and the ABS remain unchanged. Therefore, when the instantaneous channel information is not available, effective energy transmission can be achieved only by using the information of the ABS, regardless of how the energy sequences are allocated.

The left part of the whole signal, that is $\sqrt{1-\beta}\mathbf{y}_{g_{d,k}^e}$, is used for the channel estimation. Since the sensor knows the values of β and can accurate compensate it, hence we omit it in the following channel estimation. The LS channel estimator of the downlink sensor-BS channel can be expressed as

$$\begin{aligned}\tilde{\mathbf{h}}_{g_{d,k}^e,LS}^{\{B_{g_d^e}\}} &= \frac{1}{\tau_d p_d^e}\left(\mathbf{Y}_{g_d^e}\left(\Phi_{g_d}^{\mathrm{E}}\right)^*\right)^H\mathbf{e}^k \\ &= \tilde{\mathbf{h}}_{g_{d,k}^e}^{\{B_{g_d^e}\}} + \sum_{g'^e_d\in G_d^e/g_d^e}\tilde{\mathbf{h}}_{g_{d,k}^e}^{\{B_{g'^e_d}\}} + \frac{1}{\tau_d p_d^e}\left(\mathbf{N}_{g_d^e}\left(\Phi_{g_d}^{\mathrm{E}}\right)^*\right)^H\mathbf{e}^k\end{aligned} \tag{20}$$

where p^e_d denotes the power of each downlink energy sequence symbol, i.e., $\left|\left[\Phi^{\mathrm{E}}_{g_d}\right]_{i,j}\right|^2 = p_{g^e_d}$. Similarly, we can refine the estimates with the LMMSE procedure, which resulting in

$$\tilde{\mathbf{h}}^{\left\{B_{g^e_d}\right\}}_{g^e_{d,k},LM} = \sum_{g'^e_d \in G^e_d} \sum_{i=1}^{M_d} \int_{\theta^{\min}_{g^e_{d,k,i}}}^{\theta^{\max}_{g^e_{d,k,i}}} \left(\mathbf{F}^{\left\{B_{g^e_d}\right\}}\right)^H \mathbf{a}(\theta)\,\mathbf{a}^H(\theta)\mathbf{F}^{\left\{B_{g'^e_d}\right\}} S_{g^e_{d,k,i}}(\theta)\,d\theta \times \left(\sum_{g'^e_d, g''^e_d \in G^e_d} \sum_{i=1}^{M_d} \int_{\theta^{\min}_{g^e_{d,k,i}}}^{\theta^{\max}_{g^e_{d,k,i}}} \left(\mathbf{F}^{\left\{B_{g'^e_d}\right\}}\right)^H \mathbf{a}(\theta)\,\mathbf{a}^H(\theta)\mathbf{F}^{\left\{B_{g''^e_d}\right\}} S_{g^e_{d,k,i}}(\theta)\,d\theta + \frac{\sigma}{\tau_d p_d} \right)^{-1} \tilde{\mathbf{h}}^{\left\{B_{g^e_d}\right\}}_{g^e_{d,k},LS} \tag{21}$$

The uplink sensors-to-BS channel CSI $\tilde{\mathbf{h}}^{\left\{B_{g^e_u}\right\}}_{g^e_{u,k},LM}$ is obtained by using the reciprocity properties of the wireless channel.

4.4 Beam-domain FD information transmission

The BS uses the estimated BDEC to perform beamforming to transmit information to the K_d users. At the same time, it receives data from the K_u sensors.

4.4.1 BS-to-UE downlink transmission

In downlink, the received signal at user group g_d can be expressed as

$$\mathbf{y}_{g_d} = \mathbf{H}^H_{g_d} \sum_{g'_d \in G_d} \mathbf{x}_{g'_d} + \mathbf{H}^H_{K_u,K_d} \mathbf{x}_{u,K_u} + \mathbf{n}_{g_d} \tag{22}$$

where $\mathbf{n}_{g_d}$ denotes the AWGN vector with variance σ, $\mathbf{x}_{g_d} \in \mathbb{C}^{N\times 1}$ denotes the precoded transit signal of the BS, $\mathbf{H}_{g_d} = \left[\mathbf{h}_{g_d,1}, \cdots, \mathbf{h}_{g_d,K_{g_d}}\right]$ denotes the channel matrix from the user group g_d to the BS, $\mathbf{H}_{K_u g_d} \in \mathbb{C}^{K_u \times K_{g_d}}$ denotes the channel matrix from K_u sensors to the user group g_d, and $\mathbf{x}_{u,K_u} \in \mathbb{C}^{K_u \times 1}$ denotes the transmit signal vector of K_u sensors.

By using the definition $\mathbf{x}_{g_d} = \mathbf{F}^{\left\{B_{g_d}\right\}} \tilde{\mathbf{x}}_{g_d}$, and substituting it into (22), we can rewrite the BD received signal at user group g_d as

$$\tilde{\mathbf{y}}_{g_d} = \left(\tilde{\mathbf{H}}^{\left\{B_{g_d}\right\}}_{g_d}\right)^H \tilde{\mathbf{x}}_{g_d} + \sum_{g'_d \in G_d} \left(\tilde{\mathbf{H}}^{\left\{B_{g'_d}\right\}}_{g_d}\right)^H \tilde{\mathbf{x}}_{g'_d} + \mathbf{H}^H_{K_u,K_d} \mathbf{x}_{u,K_u} + \mathbf{n}_{g_d} \tag{23}$$

where $\tilde{\mathbf{H}}^{\left\{B_{g_d}\right\}}_{g_d} = \left(\mathbf{F}^{\left\{B_{g_d}\right\}}\right)^H \mathbf{H}_{g_d}$. Note that the second term on the right-hand sides of (23) indicate the inter-group interferences (IGIs). Using the interference control scheme, such as cell sectorization-based uplink-downlink user scheduling [52], the interference from sensors to users, that is, the third term on the right-hand sides of (23), can be mitigated. Hence, the BD received signal at user group g_d can be rewritten as

$$\tilde{\mathbf{y}}_{g_d} = \left(\tilde{\mathbf{H}}^{\left\{B_{g_d}\right\}}_{g_d}\right)^H \tilde{\mathbf{x}}_{g_d} + \sum_{g'_d \in G_d} \left(\tilde{\mathbf{H}}^{\left\{B_{g'_d}\right\}}_{g_d}\right)^H \tilde{\mathbf{x}}_{g'_d} + \mathbf{n}_{g_d} \tag{24}$$

To mitigate the IGI and IUI, the BS employs the BD beamforming. The intended signal of user group g_d, i.e., $\mathbf{s}_{g_d} \in \mathbb{C}^{K_{g_d}\times 1}$, is precoded by the beamforming matrix $\mathbf{W}_{g_d} = \left[\mathbf{w}_{g_d,1}, \mathbf{w}_{g_d,2}, \cdots, \mathbf{w}_{g_d,K_{g_d}}\right] \in \mathbb{C}^{b_d \times K_{g_d}}$, and b_d denotes the number of beams in the ABS B_{g_d}, that is $|B_{g_d}| = b_d$, which means that the users in the user group g_d detect signal only on their active beam spaces. Thus, the BD transmit signal vector for user group g_d can be expressed as $\tilde{\mathbf{x}}_{g_d} = \mathbf{W}_{g_d}\mathbf{s}_{g_d} \in \mathbb{C}^{b_d\times 1}$. Using these on (24), the BD received signal at user g_{d_k} can be expressed as

$$\tilde{y}_{g_{d,k}} = \left(\tilde{\mathbf{h}}^{\left\{B_{g_d}\right\}}_{g_{d,k}}\right)^H \mathbf{w}_{g_{d,k}} s_{g_{d,k}} + \left(\tilde{\mathbf{h}}^{\left\{B_{g_d}\right\}}_{g_{d,k}}\right)^H \sum_{k'=1,k'\neq k}^{K_{g_d}} \mathbf{w}_{g_{d,k'}} s_{g_{d,k'}} + \sum_{g'_d \in G_d / \{g_d\}} \left(\tilde{\mathbf{h}}^{\left\{B_{g'_d}\right\}}_{g_{d,k}}\right)^H \mathbf{W}_{g'_d}\mathbf{s}_{g'_d} + n_{g_{d,k}} \tag{25}$$

The optimal beamforming scheme to maximize the sum rate has been proved NP-hard [57]. Thus, in this paper, we consider the suboptimal linear precoding scheme, i.e., ZF beamforming, to approximate the system performance upper bound. Assuming the channel estimators in (19), the transmit beamforming matrices of the BS can be expressed as

$$\mathbf{W}_{g_d} = \tilde{\mathbf{H}}^{\left\{B_{g_d}\right\}}_{g_{d},LM} \left(\left(\tilde{\mathbf{H}}^{\left\{B_{g_d}\right\}}_{g_{d},LM}\right)^H \tilde{\mathbf{H}}^{\left\{B_{g_d}\right\}}_{g_{d},LM} \right)^{-1} \Upsilon^{-1/2}_{g_d} \tag{26}$$

where Υ_{g_d} is a diagonal normalized matrix with $\left[\Upsilon_{g_d}\right]_{l,l}$ can be expressed as $\left[\Upsilon_{g_d}\right]_{l,l} = \mathbf{e}^H_l \left(\left(\tilde{\mathbf{H}}^{\left\{B_{g_d}\right\}}_{g_{d},LM}\right)^H \tilde{\mathbf{H}}^{\left\{B_{g_d}\right\}}_{g_{d},LM} \right)^{-1} \mathbf{e}_l$.

4.4.2 Sensor-to-BS uplink transmission

In uplink, the sensor $g^e_{d,k}$ transmit signal using the harvested energy $p^{\mathrm{H}}_{g^e_{d,k}}$. In the following, the transmit power of sensor $g^e_{d,k}$ is set to $p_{g^e_{u,k}} = p^{\mathrm{H}}_{g^e_{d,k}} = \left|s_{g^e_{u,k}}\right|^2$, where $s_{g^e_{u,k}}$ denotes the transmit signal of sensor $g^e_{u,k}$. The received signal at the BS can be expressed as

$$\mathbf{y}_u^e = \sum_{g'^e_u \in G_u} \mathbf{H}_{g'^e_u} \mathbf{s}_{g'^e_u} + \mathbf{H}_{SI} \sum_{g'_d \in G_d} \mathbf{x}_{g'_d} + \mathbf{n}_u^e \tag{27}$$

where $\mathbf{n}_u^e$ denotes the AWGN vector with variance σ. $\mathbf{s}_{g_u^e} \in \mathbb{C}^{K_{g_u}^e \times 1}$ denotes the transmit signal of sensor group g_u^e. $\mathbf{x}_{g_d} \in \mathbb{C}^{N \times 1}$ denotes the precoded transit signal of the BS. By multiplying both sides of (27) with $\mathbf{F}^{\left\{B_{g_u^e}\right\}}$, we arrive at the BD received signal at the BS from sensor group g_u^e.

$$\begin{aligned}\tilde{\mathbf{y}}_{g_u^e} &= \mathbf{F}^{\left\{B_{g_u^e}\right\}} \mathbf{y}_u^e \\ &= \tilde{\mathbf{H}}_{g_u^e}^{\left\{B_{g_u^e}\right\}} \mathbf{s}_{g_u^e} + \sum_{g'^e_u \in G_u^e / \{g_u^e\}} \tilde{\mathbf{H}}_{g'^e_u}^{\left\{B_{g_u^e}\right\}} \mathbf{s}_{g'^e_u} + \sum_{g'_d \in G_d} \tilde{\mathbf{H}}_{SI}^{\left\{B_{g_u^e}, B_{g'_d}\right\}} \tilde{\mathbf{x}}_{g'_d} + \tilde{\mathbf{n}}_{g_u^e}\end{aligned} \tag{28}$$

where $\tilde{\mathbf{n}}_{g_u^e} = \mathbf{F}^{\left\{B_{g_u^e}\right\}} \mathbf{n}_u^e$. To detect the signals from sensor group g_u^e, the BS multiplies the BD received signal (28) with the receive beamforming matrix $\mathbf{W}_{g_u^e} = \left[\mathbf{w}_{g_{u,1}^e}, \mathbf{w}_{g_{u,2}^e}, \cdots, \mathbf{w}_{g_u^e, K_{g_u}^e}\right] \in \mathbb{C}^{b_u^e \times K_{g_u}^e}$, i.e., $\dot{\mathbf{y}}_{g_u^e} = \mathbf{W}_{g_u^e}^H \tilde{\mathbf{y}}_{g_u^e}$. The kth entry of $\dot{\mathbf{y}}_{g_u^e}$, that is $\dot{y}_{g_{u,k}^e}$, can be expressed as

$$\begin{aligned}\dot{y}_{g_{u,k}^e} =& \mathbf{w}_{g_{u,k}^e}^H \tilde{\mathbf{h}}_{g_{u,k}^e}^{\left\{B_{g_u^e}\right\}} s_{g_{u,k}^e} + \mathbf{w}_{g_u^e, k_{g_u}^e}^H \sum_{k'=1, k' \neq k}^{K_{g_u}^e} \tilde{\mathbf{h}}_{g_{u,k'}^e}^{\left\{B_{g_u^e}\right\}} s_{g_{u,k'}^e} \\ &+ \mathbf{w}_{g_{u,k}^e}^H \sum_{g'^e_u \in G_u^e / \{g_u^e\}} \tilde{\mathbf{H}}_{g'^e_u}^{\left\{B_{g_u^e}\right\}} \mathbf{s}_{g'^e_u} + \mathbf{w}_{g_{u,k}^e}^H \sum_{g'_d \in G_d} \tilde{\mathbf{H}}_{SI}^{\left\{B_{g_u^e}, B_{g'_d}\right\}} \tilde{\mathbf{x}}_{g'_d} + \tilde{n}_{g_u^e}\end{aligned} \tag{29}$$

The transmit symbol of sensor $g_{u,k}^e$ can be recovered by using signal $\dot{y}_{g_{u,k}^e}$. The signal-to-interference-and-noise ratio (SINR) at the BS of sensor $g_{u,k}^e$ can be expressed as

$$\mathrm{SINR}_{g_{u,k}^e} = \frac{p_{g_{u,k}^e} \left| \mathbf{w}_{g_{u,k}^e}^H \tilde{\mathbf{h}}_{g_{u,k}^e}^{\left\{B_{g_u^e}\right\}} \right|^2}{\sigma^2 \|\mathbf{w}_{g_{u,k}^e}\|^2 + \Theta_{g_{u,k}^e}} \tag{30}$$

where $\Theta_{g_{u,k}^e}$ denotes the sum of powers of channel estimation error. IUI within the group and IGI can be expressed as (31), where Λ_{g_d} denotes the diagonal input covariance matrix of the downlink group g_d. $\Lambda_{g_u^e} = \mathbb{E}\left[\mathbf{s}_{g_u^e} \mathbf{s}_{g_u^e}^H\right]$ denotes the diagonal input covariance matrix of the downlink group g_u^e.

$$\begin{aligned}\Theta_{g_{u,k}^e} =& p_{g_{u,k}^e} \left| \mathbf{w}_{g_{u,k}^e}^H \Delta\tilde{\mathbf{h}}_{g_{u,k}^e}^{\left\{B_{g_u^e}\right\}} \right|^2 + \sum_{k'=1, k' \neq k}^{K_{g_u}^e} p_{g_{u,k'}^e} \left| \mathbf{w}_{g_{u,k}^e}^H \Delta\tilde{\mathbf{h}}_{g_{u,k'}^e, LM}^{\left\{B_{g_u^e}\right\}} \right|^2 \\ &+ \sum_{g'^e_u \in G_u / \{g_u^e\}} \left\| \mathbf{w}_{g_{u,k}^e}^H \tilde{\mathbf{H}}_{g'^e_u}^{\left\{B_{g_u^e}\right\}} \Lambda_{g'^e_u}^{1/2} \right\|^2 \\ &+ \sum_{g'_d \in G_d} \left\| \mathbf{w}_{g_{u,k}^e}^H \tilde{\mathbf{H}}_{SI}^{\left\{B_{g_u^e}, B_{g'_d}\right\}} \mathbf{W}_{g'_d} \Lambda_{g'_d}^{1/2} \right\|^2\end{aligned} \tag{31}$$

Hence, the optimum $\mathbf{w}_{g_{u,k}^e}$ is the one that maximizes the $\mathrm{SINR}_{g_{u,k}}$ and, thus, is obtained by solving the following problem

$$\max_{\left\|\mathbf{w}_{g_{u,k}^e}\right\|=1} \frac{p_{g_{u,k}^e} \mathbf{w}_{g_{u,k}^e}^H \tilde{\mathbf{h}}_{g_{u,k}^e}^{\left\{B_{g_u^e}\right\}} \left(\tilde{\mathbf{h}}_{g_{u,k}^e}^{\left\{B_{g_u^e}\right\}}\right)^H \mathbf{w}_{g_{u,k}^e}}{\mathbf{w}_{g_{u,k}^e}^H \left(\Xi_{g_{u,k}^e}\right) \mathbf{w}_{g_{u,k}^e}} \tag{32}$$

where $\Xi_{g_{u,k}^e}$ can be expressed as

$$\begin{aligned}\Xi_{g_{u,k}^e} =& p_{g_{u,k}^e} \Delta\tilde{\mathbf{h}}_{g_{u,k}^e}^{\left\{B_{g_u^e}\right\}} \left(\Delta\tilde{\mathbf{h}}_{g_{u,k}^e}^{\left\{B_{g_u^e}\right\}}\right)^H \\ &+ \sum_{k'=1, k' \neq k}^{K_{g_u}^e} p_{g_{u,k'}^e} \Delta\mathbf{h}_{g_{u,k'}^e, LM}^{\left\{B_{g_u^e}\right\}} \left(\Delta\tilde{\mathbf{h}}_{g_{u,k'}^e, LM}^{\left\{B_{g_u^e}\right\}}\right)^H \\ &+ \sum_{g'^e_u \in G_u / \{g_u^e\}} \tilde{\mathbf{H}}_{g'^e_u}^{\left\{B_{g_u^e}\right\}} \Lambda_{g'^e_u}^{1/2} \left(\tilde{\mathbf{H}}_{g'^e_u}^{\left\{B_{g_u^e}\right\}} \Lambda_{g'^e_u}^{1/2}\right)^H \\ &+ \sum_{g'_d \in G_d} \tilde{\mathbf{H}}_{SI}^{\left\{B_{g_u^e}, B_{g'_d}\right\}} \mathbf{W}_{g'_d} \Lambda_{g'_d}^{1/2} \left(\tilde{\mathbf{H}}_{SI}^{\left\{B_{g_u^e}, B_{g'_d}\right\}} \mathbf{W}_{g'_d} \Lambda_{g'_d}^{1/2}\right)^H + \sigma^2 \mathbf{I}_{b_u^e}\end{aligned} \tag{33}$$

The problem in (32) is in generalized Rayleigh quotient form. It is well known that the maximum value in (32) is obtained when $\mathbf{w}_{g_{u,k}^e}$ equals to the generalized eigenvector corresponding to the maximum generalized eigenvalue.

5 Beam-domain user and sensor grouping

The key idea of the proposed BD FD massive MIMO SWIPT scheme lies in partitioning users and sensors according to their ABSs to realize efficient energy and information transmission. In particular, we divide the users and sensors into groups under the following two criterions:

- *Criterion 1*: The sensors or users with the same ABS are collected in the same sensor/user group. The ABSs of different sensor/user groups are non-overlapping with certain guard interval $\mho$. Let

B_{g_u} and $B_{g'_u}$ be the ABSs of two arbitrary user groups g_u and g'_u, and B_{g_u} and $B_{g'_u}$ should satisfy $\mathcal{D}(B_{g_u}, B_{g'_u}) \geq \mho$, as shown in Fig. 3. Let B_{g_d} and $B_{g'_d}$ be the ABSs of two arbitrary sensor groups g_d and g'_d, and B_{g_d} and $B_{g'_d}$ should satisfy $\mathcal{D}(B_{g_d}, B_{g'_d}) \geq \mho$.

- *Criterion 2*: The complete ABS of all the user groups or sensor groups is non-overlapping with the receive ABS or transmit ABS of the SI channel under certain guard interval $\mho$. Let G_u and G_d be the complete set of sensor groups and user groups, respectively. The ABSs B_{g_u} and B_{g_d} satisfy $\mathcal{D}(\bigcup_{g_u \in G_u} B_{g_u}, B_{SI,R}) > \mho$ or $\mathcal{D}(\bigcup_{g_d \in G_d} B_{g_d}, B_{SI,T}) > \mho$.

In the practical implementation, the users and sensors with different ABSs must be partitioned so that the above two conditions are satisfied as close as possible, as shown in Fig. 3. For all the users or sensors, the ABS of the SI channel at the receive side $B_{SI,R}$ and transmit side $B_{SI,T}$ should be took into consideration when carry out grouping. In the following, we proposed a BD user and sensor grouping scheme, which consists of the following two steps.

Step 1: Remove the ABS of the SI channel at the receive side $B_{SI,R}$ and transmit side $B_{SI,T}$ from the set of all the beams Γ with a certain guard interval $\mho$.

Step2: Remove all the beams that satisfy the adjacent $\mho$ beams including itself which are not occupied by any user or sensor. We obtain serval beam sets, each of which contains several continuous beams occupied by different users or sensors. For each beam set with several continuous beams, the corresponding users or sensors form a user and sensor group.

In this way, the self-interference at the BS can be mitigated effectively based on the statistical CSI of SI channel, that is, the ABS of SI channel, rather than the instantaneous CSI of SI channel. Note that users and sensors should be grouped independently. Meanwhile, to mitigate the interference between the uplink and downlink, the user group or sensor group with overlap beams cannot be allocated with the same frequency band for transmission.

6 Achievable rate analysis and optimization

We carry out the achievable rate analysis and optimization in this section. In this paper, the part of the time used for channel estimation at the BS and energy harvesting αT can be determined by $\alpha T = \tau_d \geq \tau_u$. It is worth noting that when $\alpha T \geq \tau_u$, more time is allocated to the sensors for EH which as a result will increase the harvested energy and hence increase the transmit power of sensors. But for the users, less time is allocated for transmitting and hence reduce the SE.

6.1 Downlink achievable rate

6.1.1 Downlink achievable rate with perfect CSI

With perfect CSI, the downlink achievable rate of user $g_{d,k}$ can be expressed as

$$\ddot{R}_{g_{d,k}}(\alpha) = (1-\alpha)\log_2\left(1+\ddot{\gamma}_{g_{d,k}}\right) \tag{34}$$

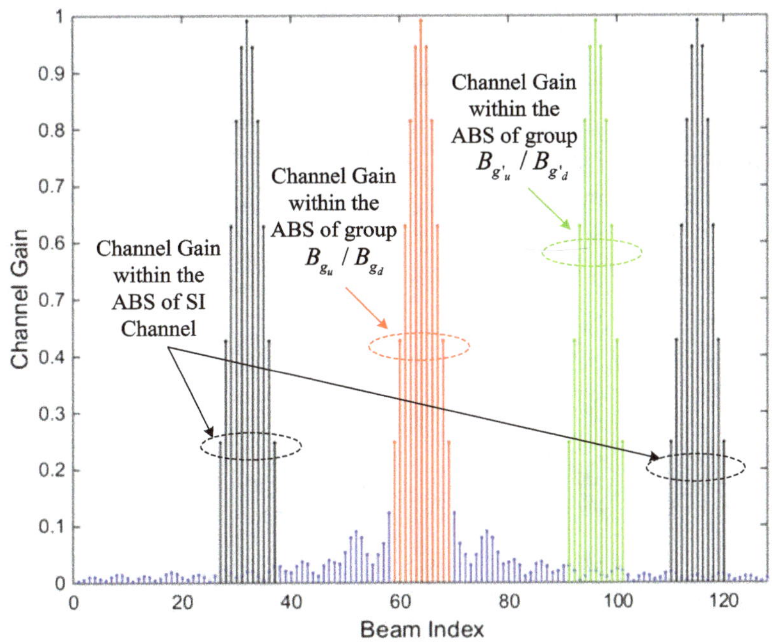

Fig. 3 Illustration of the ABSs of different sensor/user groups and the SI channel

where $\ddot{\gamma}_{g_{d,k}}$ denotes the SINR of user $g_{d,k}$ and can be written as

$$\ddot{\gamma}_{g_{d,k}} = \frac{p_{g_{d,k}} \mathbb{E}\left[\left[\Upsilon_{g_d}^{-1}\right]_{k,k}\right]}{\sum_{g'_d \in G_d / \{g_d\}} \left\| \left(\tilde{\mathbf{h}}_{g_{d,k}}^{\{B_{g'_d}\}}\right)^H \mathbf{W}_{g'_d} \Lambda_{g'_d} \right\|^2 + 1} \tag{35}$$

6.1.2 Downlink achievable rate with imperfect CSI

Since perfect channel estimation does not exist in real systems, hence we let $\Delta\tilde{\mathbf{h}}_{g_{d,k},LM}^{\{B_{g_d}\}} = \mathbf{h}_{g_{d,k},LM}^{\{B_{g_d}\}} - \tilde{\mathbf{h}}_{g_{d,k},LM}^{\{B_{g_d}\}}$ denote the channel estimation error of LMMSE estimator. According to (25) and using the bounding technique in [58], the average achievable rates at user $g_{d,k}$ can be expressed as (36) where $p_{g_{d,k}} = \mathbb{E}\left[|s_{g_{d,k}}|^2\right]$ denotes the transmit powers. The powers of channel estimation error (CEE), IUI within the group and IGI, are explained in the equation.

$$R_{g_{d,k}}(\alpha) = (1-\alpha)\log_2\left(1 + p_{g_{d,k}}\mathbb{E}\left[\left[\Upsilon_{g_d}^{-1}\right]_{k,k}\right]\left(\underbrace{p_{g_{d,k}}\left\|\left(\Delta\tilde{\mathbf{h}}_{g_{d,k},LM}^{\{B_{g_d}\}}\right)^H \mathbf{w}_{g_{d,k}}\right\|^2}_{\mathrm{CEE}_{g_{d,k}}} + \underbrace{\sum_{k'=1,k'\neq k}^{K_{g_d}} p_{g_{d,k'}}\left\|\left(\Delta\tilde{\mathbf{h}}_{g_{d,k},LM}^{\{B_{g_d}\}}\right)^H \mathbf{w}_{g_{d,k'}}\right\|^2}_{\mathrm{IUI}_{g_{d,k}}} + \underbrace{\sum_{g'_d \in G_d/\{g_d\}}\left\|\left(\tilde{\mathbf{h}}_{g_{d,k}}^{\{B_{g'_d}\}}\right)^H \mathbf{W}_{g'_d}\Lambda_{g'_d}\right\|^2}_{\mathrm{IGI}_{g_{d,k}}} + 1\right)^{-1}\right) \tag{36}$$

6.2 Uplink achievable rate

6.2.1 Uplink achievable rate with perfect CSI

With perfect CSI, the uplink achievable rate of sensor $g_{u,k}^e$ can be written as

$$\ddot{R}_{g_{u,k}^e}(\alpha) = (1-\alpha)\log_2\left(1 + \ddot{\gamma}_{g_{u,k}^e}\right) \tag{37}$$

where $\ddot{\gamma}_{g_{u,k}^e}$ denotes the SINR of the sensor $g_{u,k}^e$. Since perfect CSI is available, $\Xi_{g_{u,k}^e}$ in (32) can be rewritten as

$$\begin{aligned}\Xi_{g_{u,k}^e} &= \ddot{\Xi}_{g_{u,k}^e} + \sigma^2\mathbf{I}_{b_u^e} \\ &= \sum_{g'^e_u \in G_u/\{g_u^e\}} \tilde{\mathbf{H}}_{g'^e_u}^{\{B_{g_u^e}\}}\Lambda_{g'^e_u}^{1/2}\left(\tilde{\mathbf{H}}_{g'^e_u}^{\{B_{g_u^e}\}}\Lambda_{g'^e_u}^{1/2}\right)^H \\ &+ \sum_{g'_d \in G_d} \tilde{\mathbf{H}}_{SI}^{\{B_{g_u^e},B_{g'_d}\}}\mathbf{W}_{g'_d}\Lambda_{g'_d}^{1/2}\left(\tilde{\mathbf{H}}_{SI}^{\{B_{g_u^e},B_{g'_d}\}}\mathbf{W}_{g'_d}\Lambda_{g'_d}^{1/2}\right)^H \\ &+\sigma^2\mathbf{I}_{b_u^e}\end{aligned} \tag{38}$$

and (32) is in generalized Rayleigh quotient form. The maximum value of (32) is obtained when

$$\ddot{\mathbf{W}}_{g_{u,k}^e} = \frac{\left(\ddot{\Xi}_{g_{u,k}^e} + \sigma^2\mathbf{I}_{b_u^e}\right)^{-1}\tilde{\mathbf{h}}_{g_{u,k}^e}^{\{B_{g_u^e}\}}}{\left\|\left(\ddot{\Xi}_{g_{u,k}^e} + \sigma^2\mathbf{I}_{b_u^e}\right)^{-1}\tilde{\mathbf{h}}_{g_{u,k}^e}^{\{B_{g_u^e}\}}\right\|} \tag{39}$$

The maximum SINR of sensor $g_{u,k}^e$ can be expressed as

$$\begin{aligned}\ddot{\gamma}_{g_{u,k}^e} &= p_{g_{u,k}^e}\left(\tilde{\mathbf{h}}_{g_{u,k}^e}^{\{B_{g_u^e}\}}\right)^H\left(\ddot{\Xi}_{g_{u,k}^e} + \sigma^2\mathbf{I}_{b_u^e}\right)^{-1}\tilde{\mathbf{h}}_{g_{u,k}^e}^{\{B_{g_u^e}\}} \\ &= \frac{\alpha}{1-\alpha}\hat{p}_{g_{d,k}^e}^H\left(\tilde{\mathbf{h}}_{g_{u,k}^e}^{\{B_{g_u^e}\}}\right)^H\left(\ddot{\Xi}_{g_{u,k}^e} + \sigma^2\mathbf{I}_{b_u^e}\right)^{-1}\tilde{\mathbf{h}}_{g_{u,k}^e}^{\{B_{g_u^e}\}}\end{aligned} \tag{40}$$

With the help of Sherman-Morrison formula [59], the achievable rate of sensor $g_{u,k}^e$ can be rewritten as

$$\ddot{R}_{g_{u,k}^e}(\alpha) = (1-\alpha)\log_2\left(1 + \frac{\alpha\hat{p}_{g_{d,k}^e}^H}{(1-\alpha)\sigma^2}\left(\left\|\tilde{\mathbf{h}}_{g_{u,k}^e}^{\{B_{g_u^e}\}}\right\|^2 - \vartheta_{g_{u,k}^e}\right)\right) \tag{41}$$

where $\vartheta_{g_{u,k}^e}$ can be expressed as

$$\vartheta_{g_{u,k}^e} = \frac{\left(\tilde{\mathbf{h}}_{g_{u,k}^e}^{\{B_{g_u^e}\}}\right)^H\ddot{\Xi}_{g_{u,k}^e}\tilde{\mathbf{h}}_{g_{u,k}^e}^{\{B_{g_u^e}\}}}{\sigma^2 + tr\left(\ddot{\Xi}_{g_{u,k}^e}\right)} \tag{42}$$

In this way, the achievable rate of sensor $g_{u,k}^e$ can be further rewritten as

$$\ddot{R}_{g_{u,k}^e}(\alpha) = (1-\alpha)\log_2\left(1 + \frac{\alpha\hat{p}_{g_{d,k}^e}^H}{(1-\alpha)\sigma^2}\left\|\tilde{\mathbf{h}}_{g_{u,k}^e}^{\{B_{g_u^e}\}}\right\|^2\right) \tag{43}$$

when $\ddot{\Xi}_{g_{u,k}^e}$ is completely eliminated, and hence, we have $\vartheta_{g_{u,k}^e} = 0$.

6.2.2 Uplink achievable rate with imperfect CSI

According to (29) and using the bounding technique in [58], the average achievable rates at the downlink user $g_{u,k}$ can be expressed as (44) at the top of the next page. The transmit power of sensor $g_{d,k}^e$ is set to $p_{g_{u,k}^e} = p_{g_{d,k}^e}^{\mathrm{H}}$. The powers of CEE, IUI, IGI, and SI are explained in the equation.

$$
\begin{aligned}
R_{g_{u,k}^e}(\alpha) = (1-\alpha)\log_2\Bigg(1 + p_{g_{u,k}^e}\|\tilde{\mathbf{h}}_{g_{u,k}^e}^{\{B_{g_u^e}\}}\|^2 \Bigg(& \underbrace{p_{g_{u,k}^e}\left|\mathbf{w}_{g_{u,k}^e}^H \Delta\tilde{\mathbf{h}}_{g_{u,k}^e}^{\{B_{g_u^e}\}}\right|^2}_{\mathrm{CEE}_{g_{u,k}^e}} \\
& + \underbrace{\sum_{k'=1,k'\neq k}^{K_{g_u^e}} p_{g_{u,k'}^e}\left|\mathbf{w}_{g_{u,k}^e}^H \Delta\tilde{\mathbf{h}}_{g_{u,k'}^e,LM}^{\{B_{g_u^e}\}}\right|^2}_{\mathrm{IUI}_{g_{u,k}^e}} + \underbrace{\sum_{g'^e_u \in G_u/\{g_u^e\}}\left\|\mathbf{w}_{g_{u,k}^e}^H \tilde{\mathbf{H}}_{g'^e_u}^{\{B_{g_u^e}\}}\Lambda_{g'^e_u}^{1/2}\right\|^2}_{\mathrm{IGI}_{g_{u,k}^e}} \\
& + \underbrace{\sum_{g'_d \in G_d}\left\|\mathbf{w}_{g_{u,k}^e}^H \tilde{\mathbf{H}}_{SI}^{\{B_{g_u^e},B_{g'_d}\}}\mathbf{W}_{g'_d}\Lambda_{g'_d}^{1/2}\right\|^2}_{\mathrm{SI}_{g_{u,k}^e}} + \|\mathbf{w}_{g_{u,k}^e}\|^2\Bigg)^{-1}\Bigg)
\end{aligned}
\tag{44}
$$

6.3 Optimization of the achievable sum rate

In particular system, the assumption of having perfect instantaneous CSI is idealistic due to the fact that the CSI at the BS, users, and sensors are obtained by estimation or feedback. Hence, the CSI is subject to estimation, feedback, delay, and quantization errors. In this subsection, we consider the problem of optimizing the system achievable sum rate with imperfect CSI.

The achievable sum rate of the system with imperfect CSI can be written as

$$
R_{\mathrm{SUM}}(\alpha) = \sum_{g_{d,k}\in \mathbf{K}_D} R_{g_{d,k}}(\alpha) + \sum_{g_{u,k}^e \in \mathbf{K}_U} R_{g_{u,k}^e}(\alpha) \tag{45}
$$

where $\mathbf{K}_D$ denotes the set of all the users and $\mathbf{K}_U$ denotes the set of all the sensors. The downlink rate $R_{g_{d,k}}$ of user $g_{d,k}$ and the uplink rate $R_{g_{u,k}^e}$ of sensor $g_{u,k}^e$ are given in (36) and (44), respectively. The system achievable sum rate maximization problem can be written as

$$
\begin{aligned}
\max_{\mathcal{P},\alpha} \quad & \sum_{g_{d,k}\in \mathbf{K}_D} R_{g_{d,k}}(\alpha) + \sum_{g_{u,k}^e \in \mathbf{K}_U} R_{g_{u,k}^e}(\alpha) \\
\text{s.t.} \quad & \begin{cases} p1: 1 > \alpha \geq \tau_u/T \\ p2: \sum_{d=1}^{K_g^E} p_{g_d^e} \leq P_d^E \\ p3: \sum_{k=1}^{K_d} p_{g_{d,k}}\left\|\mathbf{w}_{g_{d,k}}\right\|^2 \leq P_d \\ p4: p_{g_{d,k}}^e, p_{g_{d,k'}} \geq 0, k\in \mathbf{K}_U, k'\in \mathbf{K}_D \end{cases}
\end{aligned}
\tag{46}
$$

where K_g^E denotes the number of sensor groups. P_d^{E} and P_d denote the maximum transmit power constraint at the BS in *phase I* and *phase II*, respectively. $\mathcal{P} = \left\{p_{g_{d,1}},\cdots,p_{g_{d,K_d}},p_{g_1^e},\cdots,p_{g_{K_g^E}^e}\right\}$.

6.3.1 Optimize $\mathcal{P}$ for fixed α

We assume that the length of the pilot signals and energy signals are fixed, that is, τ_u and τ_d are fixed. The downlink rate $R_{g_{d,k}}$ can be written as $R_{g_{d,k}} = (1-\alpha)\log_2\left(1+\gamma_{g_{d,k}}\right)$, and the SINR at user $g_{d,k}$ can be written as $\gamma_{g_{d,k}} = \frac{p_{g_{d,k}}\mathbb{E}\left[\left[\Upsilon_{g_d}^{-1}\right]_{k,k}\right]}{\mathrm{CEE}_{g_{d,k}}+\mathrm{IUI}_{g_{d,k}}+\mathrm{IGI}_{g_{d,k}}+1}$. The uplink rate $R_{g_{u,k}^e}$ can be written as $R_{g_{u,k}^e} = (1-\alpha)\log_2\left(1+\gamma_{g_{u,k}^e}\right)$, and the SINR of user $g_{u,k}^e$ signal can be written as $\gamma_{g_{u,k}^e} = p_{g_{u,k}^e} b_{g_{u,k}^e}\left(c_{g_{u,k}^e}+d_{g_{u,k}^e}\right)^{-1}$, wherein $p_{g_{u,k}^e} = \frac{\alpha}{(1-\alpha)} p_{g_d^e} q_{g_{d,k}^e}$ and $b_{g_{u,k}^e}, c_{g_{u,k}^e}, d_{g_{u,k}^e}$ are positive variables and are unrelated to α. Specifically, $b_{g_{u,k}^e} = \left\|\tilde{\mathbf{h}}_{g_{u,k}^e}^{\{B_{g_u}\}}\right\|^2$, $c_{g_{u,k}^e} = \mathrm{CEE}_{g_{u,k}^e} + \mathrm{IUI}_{g_{u,k}^e} + \mathrm{IGI}_{g_{u,k}^e}$, and $d_{g_{u,k}^e} = \mathrm{SI}_{g_{u,k}^e} + \|\mathbf{w}_{g_{u,k}^e}^H\|^2$. We omit the user group index, and the sum rate optimization problem can be rewritten as

$$
\begin{aligned}
& \max_{\mathcal{P}} \prod_{k=1}^{K_u}\left(1+\gamma_{u,k}^e\right)\prod_{k=1}^{K_d}\left(1+\gamma_{d,k}\right) \\
& = \min_{\mathcal{P}}\left[\prod_{k=1}^{K_u}\left(1+\gamma_{u,k}^e\right)\prod_{k=1}^{K_d}\left(1+\gamma_{d,k}\right)\right]^{-1} \\
& \text{s.t.} \begin{cases} c1: \gamma_{g_{u,k}^e} \leq \frac{\alpha}{(1-\alpha)} q_{g_{d,k}^e} p_{g_d^e} b_{g_{u,k}^e}\left(c_{g_{u,k}^e}+d_{g_{u,k}^e}\right)^{-1}, k=1,\cdots,K_u \\ c2: \gamma_{g_{d,k}} \leq \frac{p_{g_{d,k}}\mathbb{E}\left[\left[\Upsilon_{g_d}^{-1}\right]_{k,k}\right]}{\mathrm{CEE}_{g_{d,k}}+\mathrm{IUI}_{g_{d,k}}+\mathrm{IGI}_{g_{d,k}}+1}, k=1,\cdots,K_u \\ p1,p2,p3,p4 \end{cases}
\end{aligned}
\tag{47}
$$

where $p1$, $p2$, $p3$, and $p4$ are power constraints given by (46). The form of SE optimization problem in (47) is also close to a geometric programming (GP) except that the target function is not in the posynomial form [60]. In this paper, we apply the technique in [61] to approximate the target and solve it with the help of convex optimization tools. Specifically, $1+\gamma_{a,k}$ can be approximated by $\lambda_{a,k}\gamma_{a,k}^{\mu_{a,k}}$ close to a point $\tilde{\gamma}_{a,k}$, where $\mu_{a,k} = \tilde{\gamma}_{a,k}\left(1+\tilde{\gamma}_{a,k}\right)^{-1}$, $\lambda_{a,k} = \tilde{\gamma}_{a,k}^{-\mu_{a,k}}\left(1+\tilde{\gamma}_{a,k}\right)$, and $a\in\{u,d\}$. As a result, (47) can be rewritten as

$$
\begin{aligned}
& \min_{P} \prod_{k=1}^{K_u}\left(\lambda_{u,k}^e\right)^{-1}(\gamma_{u,k}^e)^{-\mu_{u,k}}\prod_{k=1}^{K_d}\left(\lambda_{d,k}\right)^{-1}\gamma_{d,k}^{-\mu_{d,k}} \\
& \text{s.t.} \begin{cases} c1: \gamma_{g_{u,k}^e}\left(c_{g_{u,k}^e}+d_{g_{u,k}^e}\right)\left(\frac{\alpha}{(1-\alpha)} q_{g_{d,k}^e} p_{g_d^e} b_{g_{u,k}^e}\right)^{-1} \leq 1, k=1,\cdots,K_u \\ c2: \frac{\gamma_{g_{d,k}}\left(\mathrm{CEE}_{g_{d,k}}+\mathrm{IUI}_{g_{d,k}}+\mathrm{IGI}_{g_{d,k}}+1\right)}{p_{g_{d,k}}\left[\left[\Upsilon_{g_d}^{-1}\right]_{k,k}\right]} \leq 1, k=1,\cdots,K_d \\ p1,p2,p3,p4 \end{cases}
\end{aligned}
\tag{48}
$$

Constraints $c1$ and $c2$ can be rewritten as (49) and (50) at the top of this page.

$$
\begin{aligned}
c1: &\frac{1}{\left\|\tilde{\mathbf{h}}_{g_{u,k}^e}^{\left\{B_{g_u^e}\right\}}\right\|^2}\left(\gamma_{g_{u,k}^e}\left|\mathbf{w}_{g_{u,k}^e}^H \Delta\tilde{\mathbf{h}}_{g_{u,k}^e,LM}^{\left\{B_{g_u^e}\right\}}\right|^2\right.\\
&+\gamma_{g_{u,k}^e} p_{g_{u,k}^e}^{-1}\sum_{k'=1,k'\neq k}^{K_{g_u^e}} p_{g_{u,k'}^e}\left|\mathbf{w}_{g_{u,k}^e}^H \Delta\tilde{\mathbf{h}}_{g_{u,k'}^e,LM}^{\left\{B_{g_u^e}\right\}}\right|^2\\
&+\gamma_{g_{u,k}^e} p_{g_{u,k}^e}^{-1}\sum_{g_u^{\prime e}\in G_u/\left\{g_u^e\right\}}\sum_{k'=1}^{K_{g_u^{\prime e}}} p_{g_{u,k'}^{\prime e}}\left|\mathbf{w}_{g_{u,k}^e}^H \tilde{\mathbf{h}}_{g_{u,k'}^{\prime e}}^{\left\{B_{g_u^e}\right\}}\right|^2\\
&+\gamma_{g_{u,k}^e} p_{g_{u,k}^e}^{-1}\sum_{g_d'\in G_d}\sum_{k'=1}^{K_{g_d'}} p_{g_{d,k'}'}\left|\mathbf{w}_{g_{u,k}^e}^H \tilde{\mathbf{H}}_{SI}^{\left\{B_{g_u^e},B_{g_d'}\right\}}\mathbf{w}_{g_{d,k'}'}\right|^2\\
&\left.+\gamma_{g_{u,k}^e} p_{g_{u,k}^e}^{-1}\|\mathbf{w}_{g_{u,k}^e}^H\|^2\right)\leq 1\\
&k=1,\cdots,K_u
\end{aligned}
\tag{49}
$$

$$
\begin{aligned}
c2: &\frac{1}{\mathbb{E}\left[\left[\Upsilon_{g_d}^{-1}\right]_{k,k}\right]}\left(\gamma_{g_{d,k}}\left|\left(\Delta\tilde{\mathbf{h}}_{g_{d,k},LM}^{\left\{B_{g_d}\right\}}\right)^H \mathbf{w}_{g_{d,k}}\right|^2\right.\\
&+\gamma_{g_{d,k}} p_{g_{d,k}}^{-1}\sum_{k'=1,k'\neq k}^{K_{g_d}} p_{g_{d,k'}}\left|\left(\Delta\tilde{\mathbf{h}}_{g_{d,k},LM}^{\left\{B_{g_d}\right\}}\right)^H \mathbf{w}_{g_{d,k'}}\right|^2\\
&\left.+\gamma_{g_{d,k}} p_{g_{d,k}}^{-1}\sum_{g_d'\in G_d/\left\{g_d\right\}}\sum_{k'=1}^{K_{g_d'}} p_{g_{d,k'}'}\left|\left(\tilde{\mathbf{h}}_{g_{d,k}}^{\left\{B_{g_d'}\right\}}\right)^H \mathbf{w}_{g_{d,k'}'}^H\right|^2+\gamma_{g_{d,k}} p_{g_{d,k}}^{-1}\right)\leq 1\\
&k=1,\cdots,K_d
\end{aligned}
\tag{50}
$$

The SE optimization problem in (48) becomes a standard GP and can be solved by the **Algorithm 1**.

6.3.2 Optimize α for fixed $\mathcal{P}$

Since the downlink rate $R_{g_{d,k}}(\alpha)$ can be rewritten as $R_{g_{d,k}}(\alpha) = (1-\alpha)\zeta_{g_{d,k}}$, where $\zeta_{g_{d,k}} = \log_2\left(1+\frac{p_{d,k}\mathbb{E}\left[\left[\Upsilon_{g_d}^{-1}\right]_{k,k}\right]}{\mathrm{CEE}_{g_{d,k}}+\mathrm{IUI}_{g_{d,k}}+1}\right)$ is a variable unrelated to α. The ratio used for information transmission can be adjusted by changing the during of *phase II*, that is $(1-\alpha)T$. The uplink rate $R_{g_{u,k}^e}(\alpha)$ can be rewritten as

$$
R_{g_{u,k}^e}(\alpha) = (1-\alpha)\log_2\left(1+\gamma_{g_{u,k}^e}(\alpha)\right) \tag{51}
$$

where $\gamma_{g_{u,k}^e}(\alpha) = \frac{\alpha\tilde{b}_{g_{u,k}^e}}{(1-\alpha)}\left(\frac{\alpha\tilde{c}_{g_{u,k}^e}}{(1-\alpha)}+d_{g_{u,k}^e}\right)^{-1}$ and $\tilde{b}_{g_{u,k}^e}, \tilde{c}_{g_{u,k}^e}, d_{g_{u,k}^e}$ are positive variables and are unrelated to α. Specifically, $\tilde{b}_{g_{u,k}^e} = q_{g_{d,k}^e} p_{g_d^e}\left\|\tilde{\mathbf{h}}_{g_{u,k}^e}^{\left\{B_{g_u^e}\right\}}\right\|^2$, $\tilde{c}_{g_{u,k}^e} = \left(\mathrm{CEE}_{g_{u,k}^e}+\mathrm{IUI}_{g_{u,k}^e}+\mathrm{IGI}_{g_{u,k}^e}\right)(1-\alpha)/\alpha$, and

---

Algorithm 1 Solving problem (48) by GP

1. Initialization: Compute the initial value $\gamma_{u,k}^{e,(0)}$ and $\gamma_{d,k}^{(0)}$, $i=1$, set the value of tolerance ε and maximum iteration number L.

2. For the nth iteration:

- Compute $\mu_{u,k} = \gamma_{u,k}^{(n)}\left(1+\gamma_{u,k}^{(n)}\right)^{-1}$, $\mu_{d,k} = \gamma_{d,k}^{(n)}\left(1+\gamma_{d,k}^{(n)}\right)^{-1}$, $\lambda_{u,k}^e = \left(\gamma_{u,k}^{e,(n)}\right)^{-\mu_{u,k}}\left(1+\gamma_{u,k}^{e,(n)}\right)$ and $\lambda_{d,k} = \left(\gamma_{d,k}^{(n)}\right)^{-\mu_{d,k}}\left(1+\gamma_{d,k}^{(n)}\right)$.
- Solve problem (48) by GP:
  $\min_{P}\prod_{k=1}^{K_u}\left(\lambda_{u,k}^e\right)^{-1}(\gamma_{u,k}^e)^{-\mu_{u,k}}\prod_{k=1}^{K_d}\left(\lambda_{d,k}\right)^{-1}\gamma_{d,k}^{-\mu_{d,k}}$
  s.t. $c1, c2, p1, p2, p3, p4$

3. If $\max_{k=1,\cdots,K_u}\left|\gamma_{u,k}^{e,(i)}-\gamma_{u,k}^{e,(i-1)}\right| < \varepsilon$ and $\max_{k=1,\cdots,K_d}\left|\gamma_{d,k}^{(i)}-\gamma_{d,k}^{(i-1)}\right| < \varepsilon$, **stop. Else**, set $n=n+1$ and go back to step **2**.

$d_{g_{u,k}^e} = \mathrm{SI}_{g_{u,k}^e}+1$ are variables unrelated to α. The achievable sum rate maximization problem of α can be rewritten as

$$
\begin{aligned}
\max_{\alpha} R_{\mathrm{SUM}}(\alpha) = &(1-\alpha)\sum_{g_{d,k}\in\mathbf{K}_D}\zeta_{g_{d,k}}\\
&+\sum_{g_{u,k}^e\in\mathbf{K}_U}(1-\alpha)\log_2\left(1+\gamma_{g_{u,k}^e}(\alpha)\right)\\
\text{s.t.}\, 1 > \alpha \geq &\tau_u/T
\end{aligned}
\tag{52}
$$

Let $\zeta = \sum_{g_{d,k}\in\mathbf{K}_D}\zeta_{g_{d,k}}$ and (52) can be rewritten as

$$
\begin{aligned}
\max_{\alpha}\ R_{\mathrm{SUM}}(\alpha) = &(1-\alpha)\zeta\\
&+\sum_{g_{u,k}^e\in\mathbf{K}_U}(1-\alpha)\log_2\left(1+\gamma_{g_{u,k}^e}(\alpha)\right)\\
\text{s.t.}\, 1 > \alpha \geq &\tau_u/T
\end{aligned}
\tag{53}
$$

The partial derivative of function $R_{\mathrm{SUM}}(\alpha)$ can be expressed as

$$
\begin{aligned}
\frac{\partial R_{\mathrm{SUM}}(\alpha)}{\partial\alpha} = &-\zeta-\sum_{g_{u,k}^e\in\mathbf{K}_U}\log_2\left(1+\gamma_{g_{u,k}^e}(\alpha)\right)\\
&+(1-\alpha)\sum_{g_{u,k}^e\in\mathbf{K}_U}\frac{\tilde{b}_{g_{u,k}^e}d_{g_{u,k}^e}}{\ln(2)\left(1+\gamma_{g_{u,k}^e}(\alpha)\right)\left(\alpha\tilde{c}_{g_{u,k}^e}+(1-\alpha)\,d_{g_{u,k}^e}\right)^2}
\end{aligned}
\tag{54}
$$

The two order derivative of function $R_{\text{SUM}}(\alpha)$ can be expressed as (55) at the top of this page.

$$\frac{\partial^2 R_{\text{SUM}}(\alpha)}{\partial \alpha^2} = \sum_{g_{u,k}^e \in \mathbf{K}_U} \frac{2\tilde{b}_{g_{u,k}^e} d_{g_{u,k}^e}}{\ln(2)\left(1+\gamma_{g_{u,k}^e}(\alpha)\right)\left(\alpha\tilde{c}_{g_{u,k}^e}+(1-\alpha)d_{g_{u,k}^e}\right)^2} - (1-\alpha)\sum_{g_{u,k}^e \in \mathbf{K}_U} \frac{\tilde{b}_{g_{u,k}^e} d_{g_{u,k}^e}\left(\tilde{b}_{g_{u,k}^e} d_{g_{u,k}^e}+2\left(1+\gamma_{g_{u,k}^e}(\alpha)\right)\left(\alpha\tilde{c}_{g_{u,k}^e}+(1-\alpha)d_{g_{u,k}^e}\right)\left(\tilde{c}_{g_{u,k}^e}-d_{g_{u,k}^e}\right)\right)}{\ln(2)\left(1+\gamma_{g_{u,k}^e}(\alpha)\right)^2\left(\alpha\tilde{c}_{g_{u,k}^e}+(1-\alpha)d_{g_{u,k}^e}\right)^4} \tag{55}$$

Consistent with the assumption in [52], in this paper, we assume that the power of CEE, IUI, and IGI is much larger than SI and AWGN. Hence, we have $\frac{\partial^2 R_{\text{SUM}}(\alpha)}{\partial^2 \alpha} < 0$ for $0 < \alpha < 1$ and we can obtain the optimum α of $R_{\text{SUM}}(\alpha)$ when $\frac{\partial R_{\text{SUM}}(\alpha)}{\partial \alpha} = 0$, that is

$$\underbrace{(1-\alpha)^2 \sum_{g_{u,k}^e \in \mathbf{K}_U} \frac{\tilde{b}_{g_{u,k}^e} d_{g_{u,k}^e}}{\ln(2)\left(1+\gamma_{g_{u,k}^e}(\alpha)\right)\left(\alpha\tilde{c}_{g_{u,k}^e}+(1-\alpha)\, d_{g_{u,k}^e}\right)^2}}_{A(\alpha)} = (1-\alpha)\,\zeta + (1-\alpha)\sum_{g_{u,k}^e \in \mathbf{K}_U} \log_2\left(1+\gamma_{g_{u,k}^e}(\alpha)\right) = R_{\text{SUM}}(\alpha)|_{\mathcal{P}} \tag{56}$$

Equation (56) is a transcendental equation, and we also can solve it numerically or resort the Newton iterative method to solve the problem, that is, $\hat{\alpha}_{k+1} = \hat{\alpha}_k - \frac{\partial R_{\text{SUM}}(\alpha)}{\partial \alpha} / \frac{\partial^2 R_{\text{SUM}}(\alpha)}{\partial^2 \alpha}|_{\alpha=\hat{\alpha}_k}$. Let $\hat{\alpha}^{\text{Opt}}$ denote the solution of (56) and if $1 > \alpha^{\text{Opt}} > \tau_u/T$, then the solution of the problem in (53) is α^{Opt}. Moreover, it is interesting to see that the maximum sum rate of the system is $R_{\text{SUM}}^{\text{OPT}} = A\left(\alpha^{\text{Opt}}\right)$. Otherwise, the solution of the problem in (53) is τ_u/T, and the maximum sum rate of the system is

$$R_{\text{SUM}}^{\text{Opt}} = \sum_{g_{d,k} \in \mathbf{K}_D} R_{g_{d,k}}(\tau_u/T) + \sum_{g_{u,k}^e \in \mathbf{K}_U} R_{g_{u,k}^e}(\tau_u/T) \tag{57}$$

6.3.3 *Joint optimization of α and $\mathcal{P}$*

The joint optimal α and $\mathcal{P}$ can be obtained by finding the optimum $\mathcal{P}$ for each α and then selecting the found $\mathcal{P}$ and α that maximize the problem in (46). Hence, a one-dimensional search over α is needed. We can conclude that the required one-dimensional search can be limited to a small region of α by exploiting the structure of the problem (46) and the properties obtained in (54)–(56). Hence, the computational complexity for solving the joint optimization problem can be greatly reduced. The proposed solution to jointly optimize α and $\mathcal{P}$ is summarized in Algorithm 2.

Algorithm 2 Solving problem (46)

1. Set the value of tolerance ε.

2. Define a one-dimensional search grid of α starting from $\alpha_0 = \tau_u/T$ and with step $\Delta\alpha$.

3. Solve (48) by **Algorithm 1**.

- If $A(\alpha_k) > R_{\text{SUM}}(\alpha_k)$ and $\alpha_k < 1$, then $\alpha_{k+1} = \alpha_k + \Delta\alpha$.
- If $A(\alpha_k) < R_{\text{SUM}}(\alpha_k)$ and $\alpha_k > \tau_u/T$, then $\alpha_{k+1} = \alpha_k - \Delta\alpha$.

4. If $|A(\alpha_k) - R_{\text{SUM}}(\alpha_k)| < \varepsilon$, stop, else back to step 3.

7 Simulation

In this section, the performance of BD hybrid TS and PS SWIPT protocol in FD massive MIMO system is evaluated using the 3GPP LTE simulation model for macro-cell environment [62]. The center frequency is 2.4 GHz and system bandwidth is set to 20 MHz. Thermal noise density is set to − 174 dBm/Hz. The channel coherent time is 200 symbol times. The path loss between BS and users/sensors are modeled as $2.7 + 42.8\log_{10}(R)$ (dB), and the path loss between users/sensors are modeled as $55.78 + 40\log_{10}(R)$ (dB), where R denotes the distance. The number of scattering clusters is set to $M_u = M_d = 1$. It is assumed that the passive SIC scheme for infrastructure nodes proposed in [63] has been employed at the BS. In such scheme, the suppression is from two parts, namely (1) the path loss introduced by the 20 m separation between transmit and receive antenna arrays and (2) an additional cancelation of 45 dB provided by techniques, such as radio frequency absorber material and cross-polarization. No other active SIC scheme is used in this paper.

In this paper, we assume that the BS can obtain the perfect DOA and DOD information when users access the system. In the following simulations, we consider the fixed sensors, hence it is reasonable to assume that the BS can obtain the perfect DOA and DOD information of sensors. We first consider a scenario where the sensors and users gather perfectly in three groups and the DOA/DOD regions of users in each group are identical. We assume that each group contains 5 uplink/downlink users. The DOA/DOD regions of three user groups are $[-45°, -35°]$, $[12°, 22°]$ and $[42°, 52°]$, respectively. Since we assume $M_u = 1$, the DOA region of uplink group is $[a, b]$, that is $\left[\theta_{g_{u,k},1}^{\min}, \theta_{g_{u,k},1}^{\max}\right] = [a, b]$. Similarly, the DOA/DOD regions of three sensor groups are $[-42°, -32°]$, $[16°, 26°]$ and $[39°, 49°]$, respectively. The DOA and DOD regions of SI channel between the transmit antenna array and receive antenna array are set to $[-12°, -22°]$, $[56°, 66°]$ and $[-56°, -46°]$, $[27°, 37°]$, respectively. The resulting ABSs for all the groups satisfy the user and sensor grouping criteria.

Figure 4 illustrates the SE of the proposed BD hybrid TS and PS SWIPT protocol for FD massive MIMO system with perfect user and sensor grouping. The transmit powers of BS is 20.2 dBm. The distance between users/sensors and BS is set to 500 m. This setup ensures the average downlink receive SNR is 3 dB. The energy conversion efficiency η is set to 0.8, and the power ratio used for energy harvesting β is set to 0.8. For comparison, existing SWIPT protocols designed for massive MIMO (MM) system are also demonstrated, such as TDD SWIPT with MF precoder in [49], TDD SWIPT with ZF precoder in [50], and TDD SWIPT with MRT precoder in [51]. Meanwhile, traditional FDD MM with JSDM [7], FD MM with linear transceiver [64], FD MM with spatial SI suppression [17], and traditional BD FD MM transmission scheme [52] are also extended to the SWIPT scenario and demonstrated for comparison. Note that the FD MM with spatial SI suppression in [17], the BS requires instantaneous CSI of SI channel in order to perform SI cancelation in spatial domain. In the simulations, the uplink transmit power in [7, 17, 52, 64] are also harvested from the BS downlink transmit signal. With estimated effective BD CSI, the SE of the proposed BD hybrid TS and PS SWIPT protocol outperforms the existing transmission schemes. The reason is that, the proposed BD hybrid TS and PS SWIPT transmission scheme can reduce the required length of pilot sequence significantly, hence significant SE gain can be achieved by the proposed scheme over the TDD and FDD MM systems and traditional FD MM. The SE of the proposed BD hybrid TS and PS SWIPT protocol with perfect CSI and the capacity are also depicted in Fig. 4.

In Fig. 5, we test the robustness of the proposed BD hybrid TS and PS SWIPT protocol for FD MM system under the presence of DOA/DOD offset. We can see from Fig. 5 that the proposed BD hybrid TS and PS SWIPT protocol obtains the best performance. Moreover, the SE performance decreases with the increasing of DOA/DOD offset. This is because the proposed BD SWIPT protocol requires accurate DOA/DOD information when users and sensors are partitioned into groups. For the traditional SWIPT protocols in [49–51], linear precoders are adopted so that the user/sensor DOA/DOD information is not required.

In Fig. 6, the SE performance of the proposed BD hybrid TS and PS SWIPT protocol for FD MM system with different PS ratio β is demonstrated. In Fig. 7, the SE performance of the proposed BD hybrid TS and PS SWIPT protocol for FD MM system with different energy conversion efficiency η. We can see from Figs. 6 and 7 that the proposed BD hybrid TS and PS SWIPT protocol obtains the best performance.

Figure 8 depicts the SE of the proposed BD hybrid TS and PS SWIPT protocol for FD MM system with imperfect user and sensor grouping. In Fig. 8, there are 50 users and 50 sensors within the coverage of BS which is equipped with $N = 128$ transmit/receive antennas. The signal of each user/sensor is within a 10° DOA/DOD region which is randomly distributed in the sectors. The

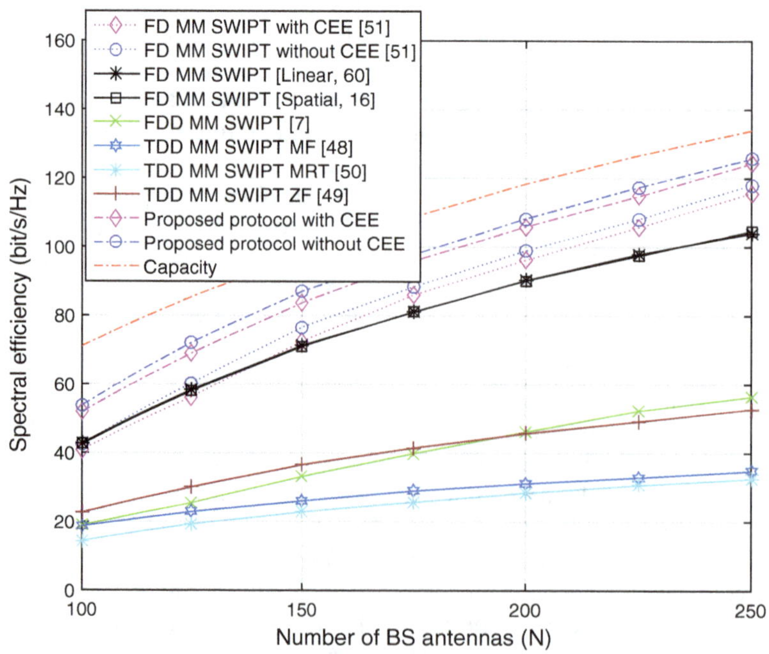

Fig. 4 SE of the proposed BD hybrid TS and PS SWIPT protocol for FD MM system with perfect user and sensor grouping, $\beta = 0.8, \eta = 0.8$

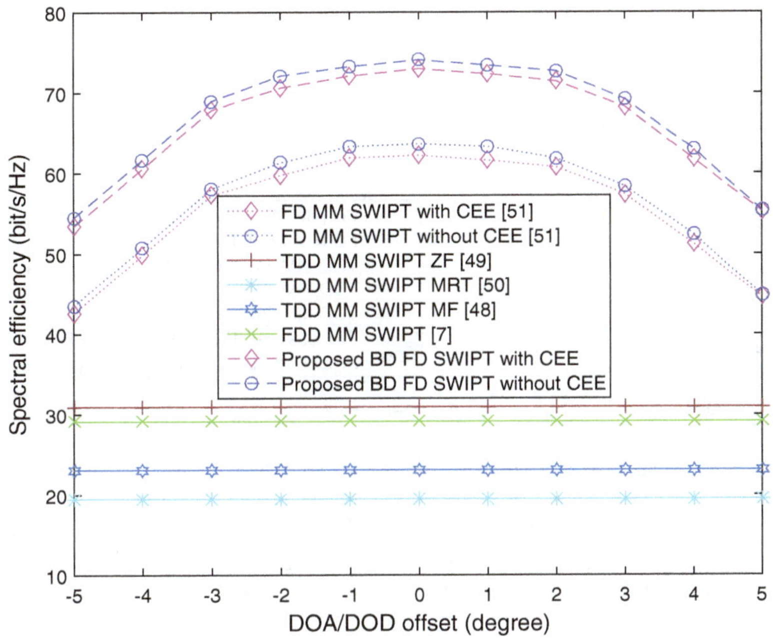

Fig. 5 SE of the proposed BD hybrid TS and PS SWIPT protocol for FD massive MIMO system with different DOA/DOD offset, $N = 128, \beta = 0.8, \eta = 0.8$

distance between uplink/downlink UE and BS is randomly distributed in the interval [200, 1000] meters. After user and sensor grouping, three user groups and three sensor groups are formed, and the users and sensors in different groups are allocated orthogonal time-frequency resources. Since users and sensors are grouped into three groups, respectively, the SE is defined as the average SEs for three groups. The number of scattering cluster for SI channel is set to $M_{SI} = 2$. We can see from Fig. 8 that the proposed BD hybrid TS and PS SWIPT protocol for FD MM system obtains the best performance. For the FD MM with linear transceiver [64], we assume that there is no

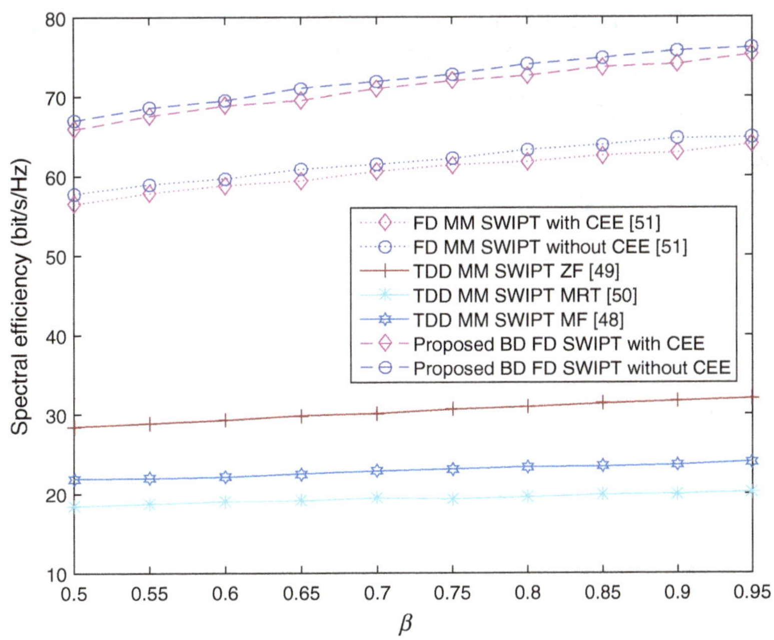

Fig. 6 SE of the proposed BD hybrid TS and PS SWIPT protocol for FD massive MIMO system with different PS ratio $\beta, N = 128, \eta = 0.8$

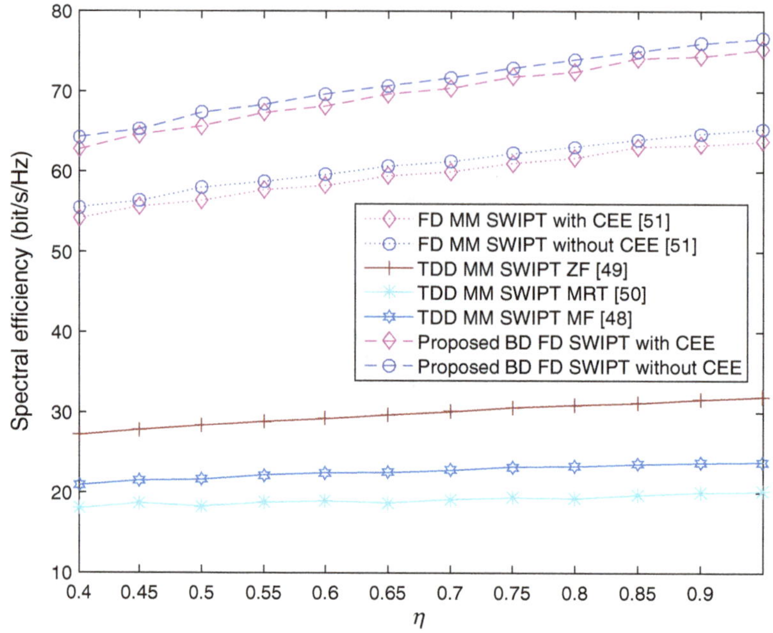

Fig. 7 Spectral efficiency of the proposed beam-domain hybrid TS and PS SWIPT protocol for FD massive MIMO system with different energy conversion efficiency η, $N = 128, \beta = 0.8$

downlink reciprocity, hence the SE performance is greatly reduced as a result of the heavy cost of pilot resources.

Figure 9 depicts the SE of the proposed BD hybrid TS and PS SWIPT protocol for FD MM system under the condition that the number of scattering clusters MSI is increase from 2 to 10. The SI signal from each scattering cluster is within a 10° DOA/DOD region which is randomly distributed $[-90°, 90°]$. Interestingly, with the increase of the number of scattering clusters M_{SI}, the SE of the proposed BD hybrid TS and PS SWIPT

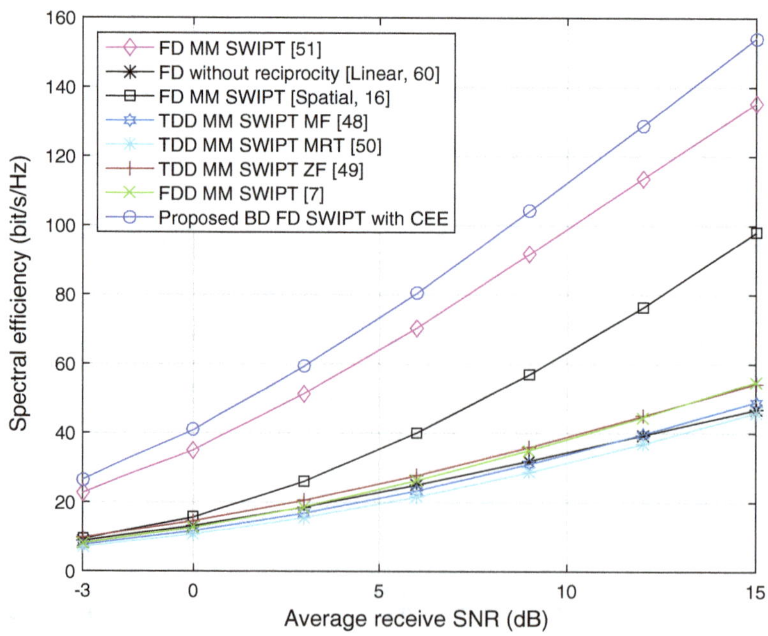

Fig. 8 SE of the proposed BD hybrid TS and PS SWIPT protocol for FD massive MIMO system with imperfect user and sensor grouping, $N = 128, \beta = 0.8, \eta = 0.8$

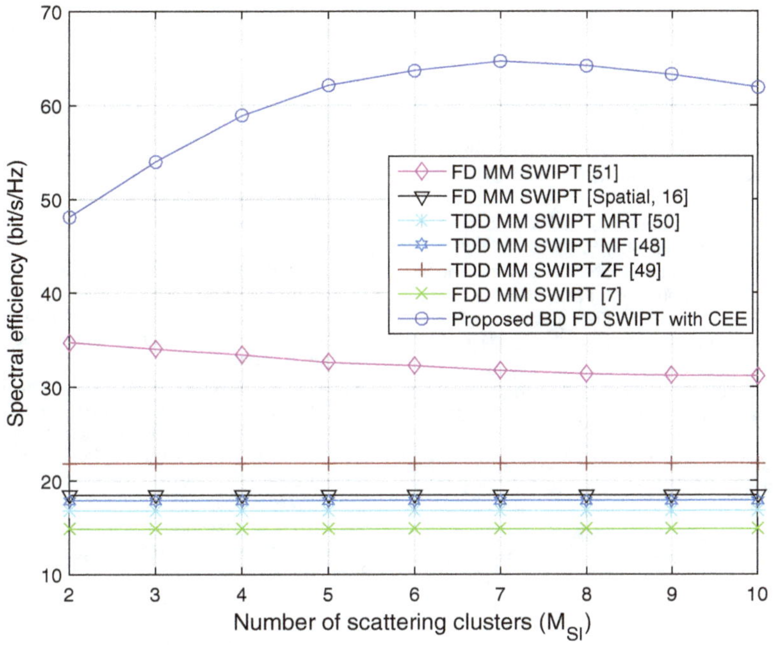

Fig. 9 SE of the proposed BD hybrid TS and PS SWIPT protocol for FD massive MIMO system with different numbers of scattering clusters M_{SI} over SI channel, $N = 128, \beta = 0.8, \eta = 0.8$

protocol for FD MM transmission scheme increases first and then decreases when $M_{SI} > 6$. Meanwhile, the proposed BD hybrid TS and PS SWIPT protocol for FD MM transmission scheme obtains the best SE performance. The SE of FD MM scheme in [52] approaches to that of the TDD/FDD MM as M_{SI} increases. This is because the numbers of uplink groups and the active beams for each uplink group decrease as M_{SI} increases.

8 Conclusions

In this paper, we propose a BD hybrid TS and PS SWIPT protocol for FD massive MIMO system. In order to reduce the pilot resource cost used for channel estimation, we resort BEM to represent massive MIMO channel in the BD. Due to the spatial sparsity of massive MIMO channel, we can compress the massive MIMO channel in the BD under certain selected basis spaces and the channel dimension required to be estimated can be greatly reduced. The hybrid TS and PS SWIPT protocol which lies in intelligently scheduling the users and sensors based on the distributions of their associated BD ABSs to mitigate SI and enhance transmission efficiency is designed. The whole BD hybrid TS and PS SWIPT protocol is divided into two phases based on the ideal of TS. The first phase is used for cellular users uplink training and sensor nodes energy harvesting as well as downlink training, wherein the cellular users transmit uplink pilots for uplink channel estimation at the BS, while the BS simultaneously transmits energy signals to the sensor nodes. Based on the idea of PS, the sensor nodes utilize the received energy signals for energy harvesting and downlink channel estimation. In the second phase, the BS forms the transmit beamformers for information transmission to the users as well as the receive beamformers for the sensors transmit their data to the BS simultaneously. By optimizing the TS ratio and transmit powers at the BS in two phases, the system achievable sum rate performance is maximized. Simulation results shown the superiority of the proposed protocol on SE compared with the existing massive MIMO SWIPT protocol.

Acknowledgements

The authors would like to thank the reviewers for their careful readings and valuable comments.

Funding

This work is supported by the Jiangsu Province Natural Science Foundation under grant BK20160079, National Natural Science Foundation of China (no. 61671472).

Authors' contributions

KX proposed the main idea and is the main writer of this paper. ZS, YW, and XX assisted in the simulations and analysis. All authors read and approved the final manuscript.

Competing interests

The authors declare that they have no competing interests.

References

1. TL Marzetta, Noncooperative cellular wireless with unlimited numbers of BS antennas. IEEE Trans. Wirel. Commun. **9**(11), 3590–3600 (2010)
2. F Rusek, D Persson, BK Lau, EG Larsson, TL Marzetta, O Edfors, F Tufvesson, Scaling up MIMO: opportunities and challenges with very large arrays. IEEE Signal Process. Mag. **30**(1), 40–60 (2013)
3. J Hoydis, S Brink, M Debbah, Massive MIMO in UL/DL of cellular networks: how many antennas do we need? IEEE J. Sel. Areas Commun. **31**, 160–171 (2013)
4. EG Larsson, F Tufvesson, O Edfors, TL Marzetta, Massive MIMO for next generation wireless systems. IEEE Commun. Mag. **52**(2), 186–195 (2014)
5. F Boccardi, RW Heath Jr, A Lozano, TL Marzetta, P Popovski, Five disruptive technology directions for 5G. IEEE Commun. Mag. **52**(2), 74–80 (2014)
6. PWC Chan, ES Lo, RR Wang, EKS Au, VKN Lau, RS Cheng, WH Mow, RD Murch, KB Letaief, The evolution path of 4G networks: FDD or TDD? IEEE Commun. Mag. **44**(12), 42–50 (2006)
7. A Adhikary, J Nam, J-Y Ahn, G Caire, Joint spatial division and multiplexing: the large-scale array regime. IEEE Trans. Info. Theory. **59**(10), 6441–6463 (2013)
8. C Sun, X Gao, S Jin, M Matthaiou, Z Ding, C Xiao, Beam division multiple access transmission for massive MIMO communications. IEEE Trans. Commun. **63**(6), 2170–2184 (2015)
9. A Liu, V Lau, Phase only RF precoding for massive MIMO systems with limited RF chains. IEEE Trans. Signal Process. **62**(17), 4505–4515 (2014)
10. D Kim, G Lee, Y Sung, Two-stage beamformer design for massive MIMO downlink by trace quotient formulation. IEEE Trans. Commun. **63**(6), 2200–2211 (2015)
11. A Sabharwal, P Schniter, D Guo, DW Bliss, S Rangarajan, R Wichman, In-band full-duplex wireless: challenges and opportunities. IEEE J.Sel. Areas Commun. **32**(9), 1637–1652 (2014)
12. D Kim, H Lee, D Hong, A survey of in-band full-duplex transmission: from the perspective of PHY and MAC layers. IEEE Commun. Surv. Tutor. **17**(4), 2017–2046 (2015). Fourthquarter
13. Z Zhang, X Chai, K Long, AV Vasilakos, L Hanzo, Full duplex techniques for 5G networks: self-interference cancellation, protocol design, and relay selection. IEEE Commun. Mag. **53**(5), 128–137 (2015)
14. Y Liao, K Bian, L Song, Z Han, Full-duplex MAC protocol design and analysis. IEEE Commun. Lett. **19**(7), 1185–1188 (2015)
15. U Ugurlu, T Riihonen, R Wichman, Optimized in-band full-duplex MIMO relay under single-stream transmission. IEEE Trans. Veh. Technol. **65**(1), 155–168 (2016)
16. Y Jang, K Min, S Park, S Choi, in *2015 IEEE International Conference on Communications (ICC)*. Spatial resource utilization to maximize uplink spectral efficiency in full-duplex massive MIMO, (London, 2015), pp. 1583–1588
17. HA Suraweera, I Krikidis, G Zheng, C Yuen, PJ Smith, Low-complexity end-to-end performance optimization in MIMO full-duplex relay systems. IEEE Trans. Wirel. Commun. **13**(2), 913–927 (2014)
18. Z Liu, W Du, D Sun, Energy and spectral efficiency tradeoff for massive MIMO systems with transmit antenna selection. IEEE Trans. Veh. Technol. **66**(5), 4453–4457 (2017)
19. Y Li, P Fan, A Leukhin, Liu L, On the spectral and energy efficiency of full-duplex small-cell wireless systems with massive MIMO. IEEE Trans. Veh. Technol. **66**(3), 2339–2353 (2017)
20. S Sudevalayam, P Kulkarni, Energy harvesting sensor nodes: survey and implications. IEEE Commun. Surv. Tutor. **13**(3), 443–46 (2011)
21. L Chen, FR Yu, H Ji, B Rong, X Li, VCM Leung, Green full-duplex self-backhaul and energy harvesting small cell networks with massive MIMO. IEEE J. Sel. Areas Commun. **34**(12), 3709–3724 (2016)
22. DK Nguyen, DNK Jayakody, S Chatzinotas, JS Thompson, J Li, in *IEEE Access, vol. 5*. Wireless energy harvesting assisted two-way cognitive relay networks: protocol design and performance analysis, (2017), pp. 21447–21460
23. W Wang, R Wang, H Mehrpouyan, N Zhao, G Zhang, in *IEEE Access, vol. 5*. Beamforming for simultaneous wireless information and power transfer in two-way relay channels, (2017), pp. 9235–9250
24. MA Marsan, G Bucalo, AD Caro, M Meo, Y Zhang, in *2013 IEEE International Conf. on Communications Workshops (ICC)*. Towards zero grid electricity networking: powering BSs with renewable energy sources, (Budapest, 2013), pp. 596–601
25. H Liu, KJ Kim, KS Kwak, HV Poor, QoS-constrained relay control for full-duplex relaying with SWIPT. IEEE Trans. Wireless Commun. **16**(5), 2936–2949 (2017)
26. G Pan, H Lei, Y Yuan, Z Ding, Performance analysis and optimization for SWIPT wireless sensor networks. IEEE Trans. Commun. **65**(5), 2291–2302 (2017)
27. F Zhou, Z Li, J Cheng, Q Li, J Si, Robust AN-aided beamforming and power splitting design for secure MISO cognitive radio with SWIPT. IEEE Trans. Wirel. Commun. **16**(4), 2450–2464 (2017)
28. AA Lu, X Gao, YR Zheng, C Xiao, in *IEEE Transactions on Communications, vol. 65, no. 7*. Linear precoder design for SWIPT in MIMO broadcasting systems with discrete input signals: manifold optimization approach, (2017), pp. 2877–2888
29. YH Suh, K Chang, A high-efficiency dual-frequency rectenna for 2.45-and 5.8-GHz wireless power transmission. IEEE Trans. Microw. Theory Tech. **50**(7), 1784–1789 (2002)
30. JR Smith, *Wirelessly powered sensor networks and computational RFID*. (Springer-Verlag New York, New York, 2013)
31. K Huang, V Lau, Enabling wireless power transfer in cellular networks: architecture, modeling and deployment. IEEE Trans. Wirel. Commun. **13**(2), 902–912 (2014)
32. LR Varshney, in *Proc. IEEE International Symposium on Information Theory (ISIT)*. 2008 Transporting information and energy simultaneously, (Toronto, 2008), pp. 1612–1616
33. P Grover, A Sahai, in *2010 IEEE International Symposium on Information Theory*. Shannon meets Tesla: wireless information and power transfer, (Austin, 2010), pp. 2363–2367
34. X Zhou, R Zhang, CK Ho, Wireless information and power transfer: architecture design and rate-energy tradeoff. IEEE Trans. Commun. **61**(11), 4754–4767 (2013)
35. X Lu, P Wang, D Niyato, DI Kim, Z Han, Wireless networks with RF energy harvesting: a contemporary survey. IEEE Commun. Surv. Tutor. **17**(2), 757–789 (2015)
36. K Huang, X Zhou, Cutting the last wires for mobile communications by microwave power transfer. IEEE Commun. Mag. **53**(6), 86–93 (2015)
37. MM Zhao, Y Cai, Q Shi, B Champagne, MJ Zhao, Robust transceiver design for MISO interference channel with energy harvesting. IEEE Trans. Signal Process. **64**(17), 4618–4633 (2016)
38. I Krikidis, S Timotheou, S Sasaki, RF energy transfer for cooperative networks: data relaying or energy harvesting? IEEE Commun. Lett. **16**(11), 1772–1775 (2012)
39. AA Nasir, X Zhou, S Durrani, RA Kennedy, Relaying protocols for wireless energy harvesting and information processing. IEEE Trans. Wirel. Commun. **12**(7), 3622–3636 (2013)
40. Y Cai, MM Zhao, Q Shi, B Champagne, MJ Zhao, Joint transceiver design algorithms for multiuser MISO relay systems with energy harvesting. IEEE Trans. Commun. **64**(10), 4147–4164 (2016)
41. Z Chen, B Xia, H Liu, in *2014 IEEE Global Conference on Signal and Information Processing (GlobalSIP)*. Wireless information and power transfer in two-way amplify-and-forward relaying channels, (Atlanta, 2014), pp. 168–172
42. K Tutuncuoglu, B Varan, A Yener, Throughput maximization for two-way relay channels with energy harvesting nodes: the impact of relaying strategies. IEEE Trans. Commun. **63**(6), 2081–2093 (2015)
43. D Li, C Shen, Z Qiu, in *2013 IEEE International Conference on Communications (ICC)*. Two-way relay beamforming for sum-rate maximization and energy harvesting, (Budapest, 2013), pp. 3155–3120
44. AA Okandeji, MRA Khandaker, KK Wong, Z Zheng, in *2016 IEEE Globecom Workshops (GC Wkshps)*. Joint transmit power and relay two-way beamforming optimization for energy-harvesting full-duplex communications, (Washington, 2016), pp. 1–6
45. L Zhang, Y Cai, M Zhao, B Champagne, L Hanzo. Nonlinear MIMO transceivers improve wireless-powered and self-interference-aided relaying, vol. 16, (2017), pp. 6953–6966
46. MM Zhao, Q Shi, Y Cai, MJ Zhao, Joint transceiver design for full-duplex cloud radio access networks with SWIPT. IEEE Trans. Wirel. Commun. **16**(9), 5644–5658 (2017)
47. MM Zhao, Y Cai, Q Shi, M Hong, B Champagne, Joint transceiver designs for full-duplex K -pair MIMO interference channel with SWIPT. IEEE Trans. Commun. **65**(2), 890–905 (2017)

48. W Wu, B Wang, Y Zeng, H Zhang, Z Yang, Z Deng, Robust secure beamforming for wireless powered full-duplex systems with self-energy recycling. IEEE Trans.Veh. Technol. **66**(11) (10055–10069)
49. L Fan, H Zhang, Y Huang, L Yang, Exploiting BS sntenna tilt for SWIPT in 3-D massive MIMO systems. IEEE Wirel. Commun. Lett. **6**(5), 666–669 (2017)
50. X Wang, J Liu, C Zhai, in *IEEE Transactions on Wireless Communications, vol. 16, no. 11*. Wireless power transfer based multi-pair two-way relaying with massive antennas, (2017), pp. 7672–7684
51. S Lee, Y Zeng, R Zhang, in *IEEE Wireless Communications Letters, vol. PP, no. 99*. Retrodirective multi-user wireless power transfer with massive MIMO, (2017), pp. 1–1
52. X Xia, K Xu, D Zhang, Y Xu, Y Wang, Beam-domain full-duplex massive MIMO: realizing co-time co-frequency uplink and downlink transmission in the cellular system. IEEE Trans. Veh. Technol. **66**(10), 8845–8862 (2017)
53. 3GPP TR 25.996, Universal mobile telecommunications system (UMTS): spatial channel model for multiple input multiple output (MIMO) simulations, v.12.0.0 (2012). www.3gpp.org. Accessed 07 Feb 2018
54. KI Pedersen, PE Mogensen, BH Fleury, A stochastic model of the temporal and azimuthal dispersion seen at the base station in outdoor propagation environments. IEEE Trans.Veh. Technol. **49**(2), 437–447 (2000)
55. J Singh, S Ramakrishna, On the feasibility of codebook-based beamforming in millimeter wave systems with multiple antenna arrays. IEEE Trans.Wirel. Commun. **14**(5), 2670–2683 (2015)
56. GB Giannakis, C Tepedelenlioglu, Basis expansion models and diversity techniques for blind identification and equalization of timevarying channels. Proc. IEEE. **86**(10), 1969–1986 (1998)
57. D Nguyen, LN Tran, et al., Precoding for full duplex multiuser MIMO systems: spectral and energy efficiency maximization. IEEE Trans. Signal Process. **61**(16), 4038–4050 (2013)
58. B Hassibi, BM Hochwald, How much training is needed in multiple-antenna wireless links? IEEE Trans. Inf. Theory. **49**(4), 951–963 (2003)
59. Hager WW, Updating the inverse of a matrix. SIAM Rev. **31**, 221–239 (1989)
60. S Boyd, L Vandenberghe, *Convex Optimization*. (Cambridge University Press, University Printing House, Cambridge, 2004)
61. PC Weeraddana, M Codreanu, M Latva-aho, A Ephremides, Resource allocation for cross-layer utility maximization in wireless networks. IEEE Trans. Veh. Technol. **60**(6), 2790–2809 (2011)
62. 3GPP TR 36.828, Further enhancements to LTE time division duplex (TDD) for downlink-uplink (DL-UL) interference management and traffic adaptation, v.11.0.0 (2012). www.3gpp.org. Accessed 07 Feb 2018
63. E Everett, A Sahai, A Sabharwal, Passive self-interference suppression for full-duplex infrastructure nodes. IEEE Trans. Wirel. Commun. **24**(2), 680–694 (2014)
64. HQ Ngo, HA Suraweera, M Matthaiou, EG Larsson, Multipair full-duplex relaying with massive arrays and linear processing. IEEE J. Sel. Areas Commun. **32**(9), 1721–1737 (2014)

2

User grouping and resource allocation in multiuser MIMO systems under SWIPT

Javier Rubio and Antonio Pascual-Iserte*

Abstract

This paper considers a broadcast multiple-input multiple-output (MIMO) network with multiple users and simultaneous wireless information and power transfer (SWIPT). In this scenario, it is assumed that some users are able to harvest power from radio frequency (RF) signals to recharge batteries through wireless power transfer from the transmitter, while others are served simultaneously with data transmission. The criterion driving the optimization and design of the system is based on the weighted sum rate for the users being served with data. At the same time, constraints stating minimum per-user harvested powers are included in the optimization problem. This paper derives the structure of the optimal transmit covariance matrices in the case where both types of users are present simultaneously in the network, particularizing the results to the cases where either only harvesting nodes or only information users are to be served. The trade-off between the achieved weighted sum rate and the powers harvested by the user terminals is analyzed and evaluated using the rate-power (R-P) region. Finally, we propose a two-stage user grouping mechanism that decides which users should be scheduled to receive information and which users should be configured to harvest energy from the RF signals in each particular scheduling period, this being one of the main contributions of this paper.

Keywords: User grouping, Energy harvesting, Simultaneous wireless information and power transfer, Multiantenna communications, Multiuser communications

1 Introduction

Currently, one of the main limiting factors of user terminals is the very limited lifetime of their batteries. One of the solutions to enhance this lifetime is based on energy harvesting technology, by means of which terminals can collect ambient energy without being physically plugged in [2] and [3]. This is especially important in scenarios where the nodes are located in places where the replacement or recharge of batteries is very difficult, costly, or even impossible (e.g., wireless sensor networks). However, this is not the only scenario that can benefit from energy harvesting technology. For example, in cellular communications, the number of users has increased exponentially, together with the rates of the communications, but the battery lifetimes are very short. In this case, energy harvesting could play a beneficial role.

*Correspondence: antonio.pascual@upc.edu

Partial results of this paper have been presented previously by the same authors at conference IEEE GLOBECOM 2013 (@2013 IEEE: some materials are reprinted, with permission, from [1]).

Department of Signal Theory and Communications, Universitat Politècnica de Catalunya (UPC), Jordi Girona 1-3, Building D5, 08034 Barcelona, Spain

Wind and solar energy compose the classical and best-known examples of sources of energy harvesting, although other technologies could also be considered, such as those applied to moving sensors (this may be the case for cellular phones) based on piezoelectric technologies. In recent years, there have also been significant advances in the use of radio frequency (RF) signals as a source of energy scavenging. Although initial experimental measurements showed that the actual strengths of the received electric fields were significant only when the distances between the transmitters and the receivers are rather short [2], current technological developments (both in terms of harvesting hardware and system features) allow for effectively taking advantage of RF energy harvesting in new scenarios [4]. In fact, this is a trend that is being adopted in the design of current and future networks based on short distances (e.g., femtocells [5]). Due to this, users will be able to be served with the higher bit rates that newer applications require. These low distances will allow for mobile terminals to be able to harvest power from the received radio signals when they are not detecting information data. This is commonly termed *wireless power transfer* (see [6] for an extensive review of this technique) and is one of the main topics of this paper.

1.1 Related work

The first work that introduced the concept of simultaneous wireless information and power transfer (SWIPT) was [7]. In that work, it was proven, for the single-antenna additive white Gaussian noise (AWGN) channel, that the data rate and power transfer are related in a nontrivial way. The extension of the previous conclusion to the frequency-selective single-antenna AWGN channels was addressed later in [8]. Much effort has been put forward lately to come up with beamforming design strategies for the SWIPT framework. In [9], the authors considered a multiple-input multiple-output (MIMO) system. In that paper, it was assumed that the transmitter was able to simultaneously transmit data and power to a single receiver. Two receiver architectures were considered able to combine both information and power sources simultaneously. In [10] and [11], the authors considered an MIMO network consisting of multiple transmitter-receiver pairs with co-channel interference. The study in [10] focused on the case with two transmitter-receiver pairs, whereas in [11], the authors generalized [10] by considering that k transmitter-receiver pairs were present. In [12], the authors considered an MIMO system with single-stream transmission. In contrast to previous works, where the system rate was optimized, the objective of the above authors was to minimize the overall power consumption with minimum signal-to-interference-plus-noise ratio (SINR) constraints and per-user harvesting constraints. Multiuser broadcast networks can also be found under the framework of multiple-input single-output (MISO) beamforming, as in [13] and [14]. The main difference between our work and previous works is that we assume a broadcast multiuser multistream MIMO network, which has not been considered before.

Although, in this paper, we assume that the channel state information (CSI) is known at the transmitter, there are some works that can be referenced in which techniques for optimizing the training under the SWIPT framework are presented [15, 16]. In particular, [15] studies the design of an efficient channel acquisition method for a point-to-point MIMO SWIPT system by exploiting the channel reciprocity. Additionally, a worst-case robust beamforming design was proposed in [17], in which imperfect CSI at the transmitter was assumed. Another strategy is to overcome this CSI feedback, as was done with implicit beamforming in [18].

In this paper, we propose some user grouping techniques in which, from frame to frame, it is decided which users will receive information data and which users will harvest energy from RF signals. There are some works in the literature that deal with user scheduling in the SWIPT framework, but they consider a single-input single-output (SISO) system. Therefore, the scheduling presented in those papers is purely the temporal scheduling of users. Among those works, [19] introduced time scheduling between information and energy transfer and derived the optimal switching policy considering time-varying co-channel interference. The receiver therefore replenished the battery opportunistically via wireless power transfer from the unintended interference and/or the intended signal sent by the transmitter. Then, in [20], the authors studied downlink multiuser scheduling for a time-slotted system with SWIPT. In particular, in each time slot, a single user is scheduled to receive information, whereas the remaining users opportunistically harvest energy from ambient signals. Finally, in [21], the authors considered a multiuser cooperative network, where M source-destination SISO pairs communicate with each other via a relay with energy harvesting capabilities. The key idea is to select a subset of those M pairs to communicate through the relay. In contrast to those works, in this paper, we present a spatial user grouping strategy since a multiuser MIMO system is considered, and multiple users therefore can be served simultaneously at each scheduling period. We also implement temporal scheduling, as those spatial user groups change over time due to the dynamics of the batteries and the historic user performance.

Finally, we want to mention that there are also several works in the literature dealing with user grouping strategies in the multiple-antenna scenario, although none of them has considered the general case addressed in this paper, that is, the problem of grouping and scheduling users in a limited-energy system with SWIPT, a multi-antenna transmitter and multiple multi-antenna receivers, and taking into account the temporal evolution of the states of the batteries. For example, in [22], a grouping strategy is developed for the case of a multiuser system, with one multi-antenna transmitter and single-antenna receivers (instead of multi-antenna receivers, as we consider in our paper) based on zero-forcing (ZF) precoding but without considering power transfer and without including the effect of the batteries. To the best of the authors' knowledge, the most recent paper related to our work is [23]. That paper addresses the same setup as [22], that is, one multi-antenna transmitter and single-antenna receivers, where the transmitter is enabled with hybrid precoding and the digital beamformers are designed according to the ZF criterion. The paper designs the transmitter by simultaneously considering the transmission of data and power through harvesting power splitting. Due to the complexity of the problem, [23] decouples the design of the user grouping (that is based on the correlation of the equivalent channels), the beamformers, and the power/harvesting parameters. The design of the power allocation and the power splitting parameters is addressed through an optimization problem, aiming at maximizing the sum rate while requiring minimum rates and harvested powers. Our paper generalizes the work of [23] by considering multiple antennas at the receivers and by not decoupling the design into several substages, which is a suboptimum approach. In this sense, we include the design of the beamformers into the optimization problem, improve the user grouping by considering the result of the optimization problem beyond the channel correlation, and explicitly take into account the states of the batteries and their time evolution in the grouping strategy. For these reasons, the techniques presented in the previous papers cannot be compared with ours due to the fact that they only consider single-antenna receivers and do not include the states of the batteries. There are more papers in the literature, but they consider even more simplified system assumptions than the previous two [22] and [23] and, therefore, are not cited here for the sake of brevity.

1.2 Contributions

In this paper, we extend the previous works by addressing a multiuser multistream MIMO system, where multiple information and energy harvesting receivers are present and where we explicitly consider other power consumption sources in the system design. The receivers are considered constrained by the system's battery dynamics, and in this sense, the batteries need to be recharged to increase their lifetimes. In the multiuser MIMO SWIPT framework, there are two groups of users to be served: one for power reception to recharge the batteries, and the other for information reception. Thus far, in the literature of MIMO beamforming techniques, the authors have considered that these two sets of users were predefined and fixed. In this paper, we propose some user grouping techniques that may change frame to frame to maximize the system throughput and/or fairness among users. Additionally, only single-stream communications have been considered for the broadcast scenario so far. The problem of maximizing the multistream sum rate for the multiuser MIMO scenario is very difficult and nonconvex [24]. For this reason, we propose the use of a conventional block diagonalization (BD) [25] simplification used extensively in the literature [26] and generalize most of the works found in the literature by considering multistream communications.

The alternative, that is, not forcing BD and allowing for the presence of interference, results in a nonconvex highly complex problem that we have addressed in our recent journal paper [27]. The complexity of that problem is such that the whole paper is dedicated exclusively to the proposal of numerical algorithms to find a local optimum of the nonconvex problem. In that paper, we assume that the user grouping is fixed and known, and we do not consider the design of those user groups, the performance evaluation of the temporal behavior of the system, the presence of any scheduler, or the presence of user batteries.

Compared to the works presented in the previous section, the main contributions of our work can be highlighted as follows:

- We consider a multiuser multistream MIMO broadcast transmission strategy in which both the transmitter and receivers are provided with multiple antennas. The system weighted sum rate with individual per-user harvesting constraints is considered in the proposed transmission strategy design. We also take into account the state of the batteries of the terminals in the proposed strategy. We study particular cases in which only information users and only harvesting users are present in the system.
- We develop an efficient algorithm that computes the optimal precoding matrices for the multiuser MIMO broadcast network setup mentioned previously.
- The fundamental (multidimensional) trade-off between system performance and (per-user) harvested energy is studied and characterized, placing emphasis on and giving specific closed-form expressions for some particular cases of interest.
- We incorporate power consumption models at the transmitter and receivers. In particular, we consider the decoding power consumption at the receivers and its impact on system performance.
- Finally, we develop harvesting-constrained user grouping schemes that employ a two-stage user scheduling mechanism that runs at different time scales. In the first stage, a subset of users are grouped to be candidates for information reception, and a subset of users are grouped to be candidates for harvesting users. Out of these selected users, in the second stage, we perform the final user information and harvesting grouping, with the aim of enhancing the system throughput and/or fairness among users.

The work developed in this paper extends our previous work presented in a conference paper [1]. The main differences and new contributions with respect to that conference version are summarized as follows. First, in this journal version, we have assumed that the system evolves over time, and we therefore have considered a generalized formulation and the inclusion of some power consumption sinks that affect the battery dynamics. Second, we have assumed that user groups are not fixed and known by the transmitter; hence, user grouping strategies have been derived, resulting from the consideration of an optimization of the system performance over time. Finally, we have included a full simulation section that evaluates the system performance over time.

1.3 Organization of the paper

The remainder of this paper is organized as follows. In Section 2, we present the system model. In Section 3, we present the formulation of the most general user grouping and resource allocation strategy. We formulate and justify the simplifications that we consider in this paper to solve such a complex problem. Section 4 covers the precoder design for simultaneous data and power transfer. We also address the characterization of the fundamental trade-off between data and power transfer. In Section 5, we present a scheduling mechanism to decide which users should be scheduled in each particular user set. The overall algorithm including all the stages, that is, the user grouping and the resource allocation, is described in detail in Section 6. Section 7 presents some numerical results of the proposed techniques. Finally, conclusions are drawn in Section 8.

1.4 Notation used in the paper

The notation that will be used in this paper is detailed in Table 1.

Table 1 Notation used in the paper

$\mathcal{A}$	Set
$\mathcal{A} = \{a_1, a_2, \dots\}$	Set $\mathcal{A}$ containing the elements $\{a_1, a_2, \dots\}$
$\lvert\mathcal{A}\rvert$	Number of elements in set $\mathcal{A}$
$a \in \mathcal{A}$	a belongs to set $\mathcal{A}$
$\mathcal{A} \setminus a$	Set resulting from subtracting a from set $\mathcal{A}$
$\emptyset$	Empty set
$\mathcal{A} \subseteq \mathcal{B}$	Set $\mathcal{A}$ is included in or equal to set $\mathcal{B}$
$\mathcal{A} \cap \mathcal{B}, \mathcal{A} \cup \mathcal{B}$	Intersection of sets $\mathcal{A}$ and $\mathcal{B}$, union of sets $\mathcal{A}$ and $\mathcal{B}$
$\mathbf{a}, \mathbf{A}$	Vector $\mathbf{a}$, matrix $\mathbf{A}$
$\mathbf{a}^T, \mathbf{A}^T$	Transpose of vector $\mathbf{a}$, matrix $\mathbf{A}$
$\mathbf{a}^H, \mathbf{A}^H$	Hermitian (transpose conjugated) of vector $\mathbf{a}$, matrix $\mathbf{A}$
$\mathrm{Tr}(\mathbf{A}), \det(\mathbf{A})$	Trace of matrix $\mathbf{A}$, determinant of matrix $\mathbf{A}$
$\mathbf{A} \succeq 0$	Matrix $\mathbf{A}$ is positive semidefinite
$\lvert\lvert\mathbf{a}\rvert\rvert$	Norm-2 of vector $\mathbf{a}$
$\mathbb{C}^{m\times n}$	Set of complex matrices of size $m \times n$
$\mathbf{I}_n$	Identity matrix of size $n \times n$
$\mathbb{E}[\cdot]$	Expectation
$=, \triangleq, \neq$	Equal, equal by definition, different
$>, \geq, <, \leq$	Higher, higher or equal, lower, lower or equal
$\log(\cdot), \exp(\cdot) = e^{(\cdot)}$	Logarithm, exponential
$n!$	Factorial of n
$\sum$	Summation
min, max	Minimum, maximum
$(x)_a^b$	$(x)_a^b = \min\{\max\{a, x\}, b\}$
a^b	a to b
$\forall$	For all
$\mathrm{maximize}_{x_1,x_2,\dots}$	Maximization with respect to variables $x_1, x_2, \dots$
$\mathrm{minimize}_{x_1,x_2,\dots}$	Minimization with respect to variables $x_1, x_2, \dots$
$x^\star$	Optimum value of x
$f^{-1}(\cdot)$	Inverse function
$x \leftarrow y$	x is updated with y

2 System model

2.1 Signal model

We consider a wireless broadcast system consisting of one base station (BS) transmitter equipped with n_T antennas and a set of K receivers, denoted as $\mathcal{U}_T = \{1, 2, \dots, K\}$, where the kth receiver is equipped with n_{R_k} antennas, as depicted in Fig. 1.

We index frames by $t \in \mathcal{T} \triangleq \{1, \dots, T\}$ with a duration of T_f seconds each. We assume block fading channels, that is, the channels remain constant within a frame but change from frame to frame. The equivalent baseband channel from the BS to the kth receiver is denoted by $\mathbf{H}_k(t) \in \mathbb{C}^{n_{R_k}\times n_T}$. It is also assumed that the set of matrices $\{\mathbf{H}_k(t)\}$ is known to the BS and to the corresponding receivers. The case of imperfect CSI is beyond the scope of the paper.

The set of users is partitioned into two subsets, as mentioned in the introduction. One of the sets contains the users that receive information, denoted as $\mathcal{U}_I(t) \subseteq \mathcal{U}_T$ and $|\mathcal{U}_I(t)| = N$, and the other set, $\mathcal{U}_E(t) \subseteq \mathcal{U}_T$, $|\mathcal{U}_E(t)| = M$, contains the users that harvest energy from the power radiated by the BS, which is used to transmit signals to the information receivers. Note that the previous sets depend on t, as the specific users in each of them may change from frame to frame. The numbers of users in each set, N and M, may change from frame to frame as well, as will be explained later in the paper. We assume that a given user is not able to simultaneously decode information and harvest energy. This forces a user to either receive information or harvest energy during the whole frame, i.e., during the scheduling period, which is a reasonable choice if the scheduling periods are short. That translates into disjoint subsets, i.e., $\mathcal{U}_I(t) \cap \mathcal{U}_E(t) = \emptyset$, $|\mathcal{U}_I(t)| + |\mathcal{U}_E(t)| \leq K$.[1] To simplify the notation when needed, we will assume that the indexing of users is such that $\mathcal{U}_I(t) = \{1, 2, \dots, N\}$ and $\mathcal{U}_E(t) = \{N+1, N+2, \dots, N+M\}$.[2] We will assume that $n_T > n_R - \min_k\{n_{R_k}\}$ is fulfilled, being $n_R = \sum_{k\in\mathcal{U}_I} n_{R_k}$.[3]

As far as the signal model is concerned, the received signal for the ith information receiver at the nth time instant within the tth frame can be modeled as:

$$\mathbf{y}_i(n,t) = \mathbf{H}_i(t)\mathbf{B}_i(t)\mathbf{x}_i(n,t) + \mathbf{H}_i(t) \sum_{\substack{k \in \mathcal{U}_I(t) \\ k \neq i}} \mathbf{B}_k(t)\mathbf{x}_k(n,t) + \mathbf{w}_i(n,t) \in \mathbb{C}^{n_{R_i}\times 1}, \forall i \in \mathcal{U}_I(t). \quad (1)$$

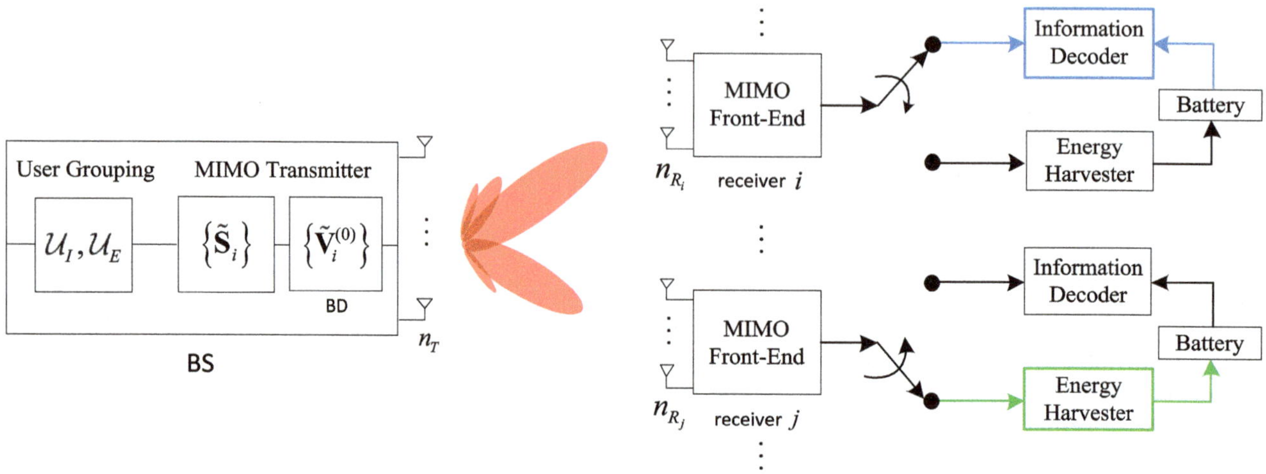

Fig. 1 Broadcast multiuser system. Schematic representation of the downlink broadcast multiuser communication system. Note that each user can switch from an information decoder receiver to an energy harvester receiver. This switching can be implemented technologically, as mentioned in papers such as [51]

In the previous notation, $\mathbf{B}_i(t)\mathbf{x}_i(n,t)$ represents the transmitted signal for user $i \in \mathcal{U}_I(t)$, where $\mathbf{B}_i(t) \in \mathbb{C}^{n_T \times n_{S_i}}$ is the precoder matrix, and $\mathbf{x}_i(t) \in \mathbb{C}^{n_{S_i} \times 1}$ represents the information symbol vector. n_{S_i} denotes the number of streams assigned to user $i \in \mathcal{U}_I(t)$, and we assume that $n_{S_i} = \min\{n_{R_i}, n_T - -(n_R - -n_{R_i})\} \forall i \in \mathcal{U}_I(t)$ is fulfilled[4]. The transmit covariance matrix is $\mathbf{S}_i(t) = \mathbf{B}_i(t)\mathbf{B}_i^H(t)$ if we assume, without loss of generality (w.l.o.g.), that $\mathbb{E}\left[\mathbf{x}_i(n,t)\mathbf{x}_i^H(n,t)\right] = \mathbf{I}_{n_{S_i}}$. $\mathbf{w}_i(n,t) \in \mathbb{C}^{n_{R_i} \times 1}$ denotes the receiver noise vector, which is considered white and Gaussian with $\mathbb{E}\left[\mathbf{w}_i(n,t)\mathbf{w}_i^H(n,t)\right] = \mathbf{I}_{n_{R_i}}$[5]. Note that the middle term of (1) is an interference term usually known as multiuser interference (MUI).

Let $\tilde{\mathbf{x}}(n,t) = \mathbf{B}(t)\mathbf{x}(n,t)$ denote the signal vector transmitted by the BS, where the joint precoding matrix is defined as $\mathbf{B}(t) = [\mathbf{B}_1(t), \ldots, \mathbf{B}_N(t)] \in \mathbb{C}^{n_T \times n_S}$, where $n_S = \sum_{i=1}^{N} n_{S_i}$ is the total number of streams of all information users, and the data vector is $\mathbf{x}(n,t) = \left[\mathbf{x}_1^T(n,t), \ldots, \mathbf{x}_N^T(n,t)\right]^T \in \mathbb{C}^{n_S \times 1}$. $\tilde{\mathbf{x}}(n,t)$ must satisfy the power constraint formulated as $\mathbb{E}\left[\|\tilde{\mathbf{x}}(n,t)\|^2\right] = \sum_{i=1}^{N} \mathrm{Tr}(\mathbf{S}_i(t)) \leq P_T$, where P_T represents the total radiated power at the BS, assuming that the information symbols of different users are independent and zero-mean.

Let us model the total power harvested by the jth user during the tth frame, denoted by $\bar{Q}_j(t)$, from all receiving antennas to be proportional to that of the equivalent baseband signal, i.e.,

$$\bar{Q}_j(t) = \zeta_j \sum_{i \in \mathcal{U}_I(t)} \mathbb{E}\left[\|\mathbf{H}_j(t)\mathbf{B}_i(t)\mathbf{x}_i(n,t)\|^2\right], \quad \forall j \in \mathcal{U}_E(t), \tag{2}$$

where ζ_j is a constant that accounts for the loss in the energy transducer when converting the harvested power to electrical power to charge the battery. Note that, for simplicity, in (2), we have omitted the harvested power due to the noise term or other external RF sources since they can be assumed negligible. Based on this, (2) can be written as:

$$\bar{Q}_j(t) = \zeta_j \sum_{i \in \mathcal{U}_I(t)} \mathrm{Tr}\left(\mathbf{H}_j(t)\mathbf{S}_i(t)\mathbf{H}_j^H(t)\right), \quad \forall j \in \mathcal{U}_E(t). \tag{3}$$

For the sake of clarity, we will drop the time and frame dependence whenever possible.

2.2 Power consumption models

The energy consumed by the transceiver can be modeled as the energy consumed by the front-end plus the energy consumed by the coding/decoding stages (omitting for the moment the power radiated by the transmitter).[6] Although other works consider battery imperfections in their models [28], we do not consider them in our work for the sake of simplicity. Note, however, that the strategy and formulation presented in this paper could be extended easily to incorporate those imperfections. In the following, we will comment briefly on the generic abstract approach followed in this paper to make the proposed strategies independent of the concrete model.

1. *Front-end consumption*: as far as the transmitter is concerned, the components that consume energy are the high-power amplifier (HPA), the mixers, the filters, and other elements of the RF chain. Concerning the receiver, the front-end consumption usually depends on the condition on the channel, i.e., the signal-to-noise ratio (SNR) (in practice, the receiver should adapt the front-end according to the

received power [29], an operation that requires some additional power). In the following, however, we assume that the component of the receiver front-end consumption that depends on the SNR is negligible, as it can be concluded from the experimental measurements and is adopted in most works [29]. We denote the energy consumed by the front-end at the transmitter and the receiver by P_c^{tx} and P_c^{rx}, respectively.

2. *Coding/decoding consumption*: it is reasonable to consider the energy consumed by the coding stage at the transmitter negligible compared to the energy consumed by the front-end. This is illustrated and commented on in papers such as [30]. For this reason, we will not include coding consumption in our models. On the other hand, the decoding consumption must be included in the models since, as shown in [31] and [32], such energy consumption is not negligible and can affect importantly the lifetime of the mobile terminal. There is a consensus about the fact that the decoding consumption increases with the data rate $R_i(t)$, $P_{\text{dec},i}(R_i(t))$. In [33], the authors presented different models for $P_{\text{dec},i}(R_i(t))$, but for the sake of generality, we will consider it a general function.

Given the previous models, the total consumption at the transmitter (omitting for the moment the radiated power) only includes the front-end consumption as mentioned previously, and it therefore is denoted as:

$$P_{\text{tot}}^{tx} = P_c^{tx}. \tag{4}$$

On the other hand, the total power consumption at the ith receiver is expressed as:

$$P_{\text{tot},i}^{rx}(R_i(t)) = P_{\text{dec},i}(R_i(t)) + P_c^{rx}. \tag{5}$$

Note that the power consumption at the receiver is limited by the current battery level, which in the following will be denoted by $C_i(t)$ for user i. According to this, the data rate of a given information user (user i) during one frame must be constrained in order not to consume more energy when decoding than the current energy available at the battery $C_i(t)$. Hence,

$$T_f(P_{\text{dec},i}(R_i(t)) + P_c^{rx}) \leq C_i(t), \tag{6}$$

which can be written in terms of a maximum rate constraint as:

$$R_i(t) \leq R_{\text{max},i}(C_i(t)), \tag{7}$$

where $R_{\text{max},i}(C_i(t)) = P_{\text{dec},i}^{-1}\left(\frac{C_i(t)}{T_f} - P_c^{rx}\right)$.

2.3 Battery dynamics

We consider that each user terminal is provided with a finite battery capacity, the level of which decreases accordingly when the user receives and decodes data. The terminals are also able to recharge their batteries by means of collecting the power dynamically coming from the BS.

The battery at the beginning of the tth frame of the ith information user served with a data rate $R_i(t-1)$ during the previous frame is denoted as:

$$C_i(t) = \left(C_i(t-1) - T_f P_{\text{tot},i}^{rx}(R_i(t-1))\right)_0^{C_{\text{max}}^i}, \quad \forall i \in \mathcal{U}_I(t), \tag{8}$$

where $(x)_a^b$ is the projection of x onto the interval $[a, b]$, i.e., $(x)_a^b = \min\{\max\{a, x\}, b\}$, C_{max}^i is the maximum battery level, and the function $P_{\text{tot,i}}^{rx}(R_i(t-1))$ was defined in (5). Note that $C_i(t)$ has units of Joules.

On the other hand, the battery at the beginning of the tth frame of the jth harvesting user is denoted as:

$$C_j(t) = \left(C_j(t-1) + T_f \bar{Q}_j(t-1) - T_f P_c^{rx}\right)_0^{C_{\text{max}}^j}, \quad \forall j \in \mathcal{U}_E(t), \tag{9}$$

where $\bar{Q}_j(t-1)$ is the power harvested during the frame $t-1$.

The receivers must inform the BS about their battery level status to make decisions on whether to serve that user with information or with power. In this paper, we assume that the feedback channel is ideal and not rate-limited.

The power consumption and battery dynamics models, which are based on the state of the art and existing literature, were also used in a similar way by the same authors of this paper in their previous work [33].

3 Joint resource allocation and user grouping formulation

In this section, we formulate the joint design of the covariance matrices $\mathbf{S}_i(t)$, the data rates $R_i(t)$, and the user grouping $\mathcal{U}_I(t)$, $\mathcal{U}_E(t)$, based on the maximization of the weighted sum rate with individual power harvesting constraints for all time instants $t \in \mathcal{T}$. Given this, the problem is formulated through the following optimization problem (this formulation generalizes the problem defined in our previous conference paper [1]):

$$\begin{array}{cc} \underset{\substack{\{R_i(t),\, \mathbf{S}_i(t)\}_{\forall i \in \mathcal{U}_I(t)},\\ \mathcal{U}_I(t),\, \mathcal{U}_E(t)}}{\text{maximize}} & \sum_{t \in \mathcal{T}} \sum_{i \in \mathcal{U}_I(t)} \omega_i(t) R_i(t) \\ \text{subject to} & \end{array} \tag{10}$$

$$
\begin{aligned}
&C1: \sum_{i\in\mathcal{U}_I(t)} \mathrm{Tr}(\mathbf{H}_j(t)\mathbf{S}_i(t)\mathbf{H}_j^H(t)) \ge Q_j, && \forall j\in\mathcal{U}_E(t),\ \forall t\in\mathcal{T}\\
&C2: \sum_{i\in\mathcal{U}_I(t)} \mathrm{Tr}(\mathbf{S}_i(t)) + P_c^{tx} \le P_{\max}, && \forall t\in\mathcal{T}\\
&C3: R_i(t) \le \log\det\left(\mathbf{I}+\mathbf{H}_i(t)\mathbf{S}_i(t)\mathbf{H}_i^H(t)\right), && \forall i\in\mathcal{U}_I(t),\ \forall t\in\mathcal{T}\\
&C4: R_i(t) \le R_{\max,i}(C_i(t)), && \forall i\in\mathcal{U}_I(t),\ \forall t\in\mathcal{T}\\
&C5: \mathbf{H}_k(t)\mathbf{S}_i(t)\mathbf{H}_k^H(t) = 0, && \forall k\neq i,\ k,i\in\mathcal{U}_I(t),\ \forall t\in\mathcal{T}\\
&C6: \mathbf{S}_i(t)\succeq 0, && \forall i\in\mathcal{U}_I(t),\ \forall t\in\mathcal{T}\\
&C7: C_i(t) = \left(C_i(t-1) - T_f P_{\mathrm{tot},i}^{rx}(R_i(t-1))\right)_0^{C_{\max}^i}, && \forall i\in\mathcal{U}_I(t),\ \forall t\in\mathcal{T}\\
&C8: C_j(t) = \left(C_j(t-1) + T_f\bar{Q}_j(t-1) - T_f P_c^{rx}\right)_0^{C_{\max}^j}, && \forall j\in\mathcal{U}_E(t),\ \forall t\in\mathcal{T},
\end{aligned}
$$

where the weights $\omega_i(t) \ge 0$ can be set to assign priorities to achieve fairness among the different users[7], $R_i(t) \le \log\det\left(\mathbf{I}+\mathbf{H}_i(t)\mathbf{S}_i(t)\mathbf{H}_i^H(t)\right)$ denotes the achievable data rate of the ith user when considering linear precoding following a BD strategy [25], $Q_j = \frac{\bar{Q}_j^{\min}}{\zeta_j}$, where $\left\{\bar{Q}_j^{\min}\right\}$ is the set of minimum power harvesting constraints, and $P_{\max}$ is the available power at the BS. In fact, BD is applied through constraint $C5$, which forces the complete cancellation of the MUI, making the whole problem more tractable (as will be shown later in the paper). Notice that constraint $C1$ is associated with the minimum power to be harvested for a given user. In the case that another external energy harvesting source was available and the amount to be harvested could be estimated (or was fully known in advance), we could subtract such value from Q_j accordingly. Constraint $C4$ assures that the information users do not spent more energy decoding the message than the current energy available at the battery.

As we have already noted, we have assumed a linear precoding approach in the system formulation. Note that the optimum transmission policy in an MIMO broadcast channel is the well-known nonlinear dirty paper coding strategy [24]. Nevertheless, that strategy has high computational demands and cannot be implemented in real time. Instead, much simpler linear transceiver designs have also been shown to achieve high capacities using much lower computational resources (see [34] for more details). Thus, for simplicity in the transmitter design, in this work, we force the precoder to be linear.

Two main difficulties arise when attempting to solve (10). First, note that the solution for all time instants has to be found jointly. The reason is that resource allocation decisions at frame t have an impact not only on that frame but also on future frames. Some researchers have attempted to solve harvesting (time-coupled) problems by assuming that the whole channel and harvesting realizations are known a priori, giving rise to offline approaches that are not implementable in real scenarios [35, 36]. As we assume that only causal knowledge of the channel and the harvesting is available, we would have to resort to dynamic programming (DP) techniques [37] to find the optimal solution of the problem (10). However, these techniques usually require the implementation of extremely high-complexity algorithms that are impractical in scenarios, where the set of variables to be optimized is large, and DP techniques therefore have been applied only in cases where the optimization variables are scalars [38, 39]. The second difficulty that we find is that the user grouping must also be optimized jointly with the covariance matrices and the data rates. The user grouping variables are discrete, and the problem therefore becomes combinatorial. The optimum solution has to be found by applying some sort of combinatorial search among all possible user groups, increasing the overall complexity exponentially.

Because we are interested in low-complexity solutions, we have to make some simplifications to problem (10) to make it more tractable, with the hope of finding a good suboptimum solution that is close to the global optimum solution of problem (10).

The first assumption that we consider is to decouple the problem in time and propose a separate per-frame optimization approach. With this approach, we solve the optimization problem at the beginning of each frame t, making decisions based on the current and past information on the battery levels. The optimization to solve is (we omit the time dependence for the sake of simplicity in the notation even though all these variables, including the information and harvesting users sets $\mathcal{U}_I$ and $\mathcal{U}_E$, change at each frame) as follows:

$$
\begin{aligned}
\underset{\substack{\{R_i,\mathbf{S}_i\}\forall i\in\mathcal{U}_I,\\ \mathcal{U}_I,\mathcal{U}_E}}{\text{maximize}}\quad & \sum_{i\in\mathcal{U}_I}\omega_i R_i\\
\text{subject to}\quad & C1: \sum_{i\in\mathcal{U}_I}\mathrm{Tr}\left(\mathbf{H}_j\mathbf{S}_i\mathbf{H}_j^H\right) \ge Q_j, && \forall j\in\mathcal{U}_E\\
& C2: \sum_{i\in\mathcal{U}_I}\mathrm{Tr}(\mathbf{S}_i) + P_c^{tx} \le P_{\max}\\
& C3: R_i \le \log\det\left(\mathbf{I}+\mathbf{H}_i\mathbf{S}_i\mathbf{H}_i^H\right), && \forall i\in\mathcal{U}_I\\
& C4: R_i \le R_{\max,i}(C_i), && \forall i\in\mathcal{U}_I\\
& C5: \mathbf{H}_k\mathbf{S}_i\mathbf{H}_k^H = 0, && \forall k\neq i,\ k,i\in\mathcal{U}_I\\
& C6: \mathbf{S}_i \succeq 0, && \forall i\in\mathcal{U}_I.
\end{aligned}
\tag{11}
$$

Problem (11), which generalizes the one addressed in our previous paper [1], as weights are included to take into account the time evolution of the achieved rates, is still very difficult to solve, as it involves continuous and integer variables. Note that for a fixed set of groups, $\mathcal{U}_I$ and $\mathcal{U}_E$, problem (11) is convex with respect to $\{R_i,\mathbf{S}_i\}$ and can be solved using standard optimization techniques. The optimum solution can be found by solving problem (11) for all possible combinations of user groups, that is, an exhaustive search should be implemented. Consider for example that $|\mathcal{U}_I| = 4$ and $|\mathcal{U}_E| = 4$ and that $K = 10$. Then, problem (11) (for a fixed $\mathcal{U}_I$ and $\mathcal{U}_E$) should be solved

$\frac{K!}{|\mathcal{U}_I|!|\mathcal{U}_E|!(K-|\mathcal{U}_I|-|\mathcal{U}_E|)!} = 3.150$ times. Clearly, the optimum solution is impractical, even for a system with a small number of users. In that sense, any technique aside from the exhaustive search may be suboptimal.

This fact motivates our second simplification: we decouple the decision of resource allocation and user grouping and propose a two-stage design strategy in which the user grouping is found based on suboptimal but less complex techniques. In other works, at the beginning of each frame, we first find the user groups $\mathcal{U}_I$ and $\mathcal{U}_E$, and then, for those fixed user groups, we solve the following convex optimization problem:

$$\begin{aligned} \underset{\{R_i, \mathbf{S}_i\}\forall i\in\mathcal{U}_I}{\text{maximize}} \quad & \sum_{i\in\mathcal{U}_I} \omega_i R_i \\ \text{subject to} \quad & C1 \ldots C6 \text{ of problem (11).} \end{aligned} \tag{12}$$

Note that due to $C5$, problem (12) is convex; otherwise, the objective function, i.e., the weighted sum rate, would not be convex due to the MUI.

In the next section, we are going to present a method to solve problem (12) for different settings. Later, in Section 5, we will present the user grouping techniques.

4 Weighted sum rate maximization with harvesting constraints

The problem presented in (12) is convex and can be solved using numerical interior point methods [40]. However, those methods usually have high computational complexity, and since we aim at finding a low-complexity solution, a customized algorithm should be developed. In some cases, it is possible to obtain the structure of the transmit covariance matrices in closed form and then develop an efficient algorithm based on that structure. Unfortunately, it is not possible to find the closed-form expression of the optimal transmit covariances for the previous problem due to the constraint $C4$. However, as we will show later, it is possible to find the transmit covariance structure of problem (12) if $C4$ is not active.

To guarantee that constraint $C4$ is not active, we will assume that the set of information users is selected by the scheduler in the first stage in a way that they have enough battery such that $R_i^\star(t) < R_{\max,i}(C_i(t))$, $\forall i \in \mathcal{U}_I$ can be guaranteed in that particular scheduling period (later, we will comment on what to do in the unlikely event of violating the previous requirement). This is a reasonable assumption since users who have very low batteries should not be selected to receive information but to harvest energy. Due to the previous simplifying assumption, constraint $C4$ will not be active, and we therefore do not consider it in the optimization problem. This assumption considerably simplifies the resolution of the problem.

Note that constraint $C5$ from the original problem (12) forces the precoder matrix $\mathbf{B}_i$ to lie in the right null space of $\tilde{\mathbf{H}}_i = \left[\mathbf{H}_1^T \;\ldots\; \mathbf{H}_{i-1}^T \; \mathbf{H}_{i+1}^T \;\ldots\; \mathbf{H}_N^T\right]^T \in \mathbb{C}^{(n_R-n_{R_i})\times n_T}$ [25]. Computing the SVD of $\tilde{\mathbf{H}}_i$ yields $\tilde{\mathbf{H}}_i = \tilde{\mathbf{U}}_i\tilde{\Lambda}_i[\,\tilde{\mathbf{V}}_i^{(1)} \;\; \tilde{\mathbf{V}}_i^{(0)}]^H$, where $\tilde{\Lambda}_i$ is a diagonal matrix containing the singular values, and $\tilde{\mathbf{V}}_i^{(0)} \in \mathbb{C}^{n_T\times(n_T-n_R+n_{R_i})}$ contains the right-singular vectors in the null space of $\tilde{\mathbf{H}}_i$. Thus, $\mathbf{B}_i$ can be written as $\mathbf{B}_i = \tilde{\mathbf{V}}_i^{(0)}\tilde{\mathbf{B}}_i$ (with $\tilde{\mathbf{B}}_i \in \mathbb{C}^{(n_T-n_R+n_{R_i})\times n_{S_i}}$), and then, $\mathbf{S}_i = \tilde{\mathbf{V}}_i^{(0)}\tilde{\mathbf{S}}_i\tilde{\mathbf{V}}_i^{(0)H}$, where $\tilde{\mathbf{S}}_i = \tilde{\mathbf{B}}_i\tilde{\mathbf{B}}_i^H$. Now, the optimization problem can be rewritten in terms of the new optimization variables $\{\tilde{\mathbf{S}}_i\}$. Let $\hat{\mathbf{H}}_i = \mathbf{H}_i\tilde{\mathbf{V}}_i^{(0)}$ and $\hat{\mathbf{H}}_{ji} = \mathbf{H}_j\tilde{\mathbf{V}}_i^{(0)}$. Note that if constraint $C4$ is not present in (12), constraint $C3$ is tight at the optimum, i.e., $R_i^\star = \log\det\left(\mathbf{I} + \hat{\mathbf{H}}_i\tilde{\mathbf{S}}_i^\star\hat{\mathbf{H}}_i^H\right)$, and thus, the objective function is directly expressed as $\sum_{i\in\mathcal{U}_I}\omega_i\log\det\left(\mathbf{I} + \hat{\mathbf{H}}_i\tilde{\mathbf{S}}_i\hat{\mathbf{H}}_i^H\right)$. Then, problem (12) (without considering $C4$) is reformulated as

$$\begin{aligned} \underset{\{\tilde{\mathbf{S}}_i\}\forall i\in\mathcal{U}_I}{\text{maximize}} \quad & \sum_{i\in\mathcal{U}_I} \omega_i \log\det\left(\mathbf{I} + \hat{\mathbf{H}}_i\tilde{\mathbf{S}}_i\hat{\mathbf{H}}_i^H\right) \\ \text{subject to} \quad & C1: \sum_{i\in\mathcal{U}_I} \mathrm{Tr}\left(\hat{\mathbf{H}}_{ji}\tilde{\mathbf{S}}_i\hat{\mathbf{H}}_{ji}^H\right) \geq Q_j, \qquad \forall j \in \mathcal{U}_E \\ & C2: \sum_{i\in\mathcal{U}_I} \mathrm{Tr}(\tilde{\mathbf{S}}_i) + P_c^{tx} \leq P_{\max} \\ & C3: \tilde{\mathbf{S}}_i \succeq 0, \qquad \forall i \in \mathcal{U}_I. \end{aligned} \tag{13}$$

The problem above can be checked to be convex since the objective function is concave and the constraints define a convex set. As a consequence, there exists a global optimal solution that can be obtained numerically by means of, for example, interior point methods [40]. However, due to the fact that (13) is convex and satisfies Slater's conditions [40], the duality gap is zero, and the problem, therefore, can be solved using tools derived from the Lagrange duality theory, and the optimal structure of the transmit covariance matrices $\{\tilde{\mathbf{S}}_i\}$ can be revealed. Let $\boldsymbol{\lambda} = \{\lambda_j\}_{j\in\mathcal{U}_E}$ be the vector of dual variables associated with constraint $C1$ and μ be the dual variable associated with constraint $C2$. The optimal solution of problem (13) is given by the following theorem in terms of $\boldsymbol{\lambda}^\star$ and $\mu^\star$.

Theorem 1 *The optimal solution of problem (13) has the following structure:*

$$\tilde{\mathbf{S}}_i^\star(\boldsymbol{\lambda}^\star, \mu^\star) = \mathbf{A}_i^{-1/2}\hat{\mathbf{V}}_i\hat{\mathbf{D}}_i\hat{\mathbf{V}}_i^H\mathbf{A}_i^{-1/2}, \tag{14}$$

where matrix $\mathbf{A}_i = \mu^\star\mathbf{I} - \sum_{j\in\mathcal{U}_E}\lambda_j^\star\hat{\mathbf{H}}_{ji}^H\hat{\mathbf{H}}_{ji}$, $\hat{\mathbf{V}}_i \in \mathbb{C}^{(n_T-n_R+n_{R_i})\times n_{S_i}}$ *is obtained from the reduced SVD of matrix* $\hat{\mathbf{H}}_i^H\mathbf{A}_i^{-1/2} = \hat{\mathbf{U}}_i\hat{\Sigma}_i^{1/2}\hat{\mathbf{V}}_i^H$, *with* $\hat{\Sigma}_i = \mathrm{diag}\left(\hat{\sigma}_{1,i}, \ldots, \hat{\sigma}_{n_{S_i},i}\right)$, $\hat{\sigma}_{1,i} \geq \hat{\sigma}_{2,i} \geq \cdots \geq \hat{\sigma}_{n_{S_i},i} > 0$, *and* $\hat{\mathbf{D}}_i = diag\left(\hat{d}_{1,i}, \ldots, \hat{d}_{n_{S_i},i}\right)$, *with* $\hat{d}_{k,i} = (\omega_i/\log(2) - 1/\hat{\sigma}_{k,i})_0^{\infty}$, $\forall i \in \mathcal{U}_I$ *and* $k = 1, \ldots, n_{S_i}$.

Proof See Appendix 1. □

Note the similarities in the precoder structure between the result presented in (14) for the multiuser case and the result found in [9] for the single-user case. In the multiuser case, we have to find a set of multipliers associated with the per-user harvesting constraints, which makes the problem more complex to solve. Finally, the optimum data rate achieved by user i is thus:

$$R_i^\star = \log\det\left(\mathbf{I}+\hat{\mathbf{H}}_i\tilde{\mathbf{S}}_i^\star\hat{\mathbf{H}}_i^H\right) = \sum_{j=1}^{n_{S_i}} \log\left(1+\hat{\sigma}_{j,i}\hat{d}_{j,i}\right), \quad \forall i \in \mathcal{U}_I. \tag{15}$$

However, the above process is still pending the computation of the optimal dual variables since we assumed in the previous development that the dual variables were given (in Theorem 1, matrix $\mathbf{A}_i$ depends on the optimal values of the Lagrange multipliers). As long as we have a closed-formed expression of the covariance matrices $\tilde{\mathbf{S}}_i(\boldsymbol{\lambda}, \mu)$ as a function of the dual variables, we can solve the dual problem of (13) by maximizing the dual function $g(\boldsymbol{\lambda}, \mu)$ subject to $\boldsymbol{\lambda} \succeq 0$, $\mu \geq 0$, and $\mathbf{A}_i \succ 0\ \forall i$. This can be addressed by applying any subgradient-type method, such as, for example, the ellipsoid method [41]. It can be shown that the subgradient of $g(\boldsymbol{\lambda}, \mu)$, denoted as $\mathbf{t}$, is given by $[\mathbf{t}]_m = Q_{N+m} - \sum_{i\in\mathcal{U}_I} \mathrm{Tr}\left(\hat{\mathbf{H}}_{(N+m)i}\tilde{\mathbf{S}}_i\hat{\mathbf{H}}_{(N+m)i}^H\right)$ for $1 \leq m \leq M$ and $[\mathbf{t}]_{M+1} = \mathrm{Tr}(\tilde{\mathbf{S}}_i) - \left(P_{\max} - P_c^{tx}\right)$ [42], which represents the subgradient of $g(\boldsymbol{\lambda}, \mu)$ with respect to λ_m and μ, respectively ($[\mathbf{t}]_k$ denotes the kth entry of vector $\mathbf{t}$), and $\tilde{\mathbf{S}}_i$ is computed as in (14) for a given $\boldsymbol{\lambda}$ and μ (for each step of the algorithm, we compute $\tilde{\mathbf{S}}_i$ just by replacing, in expression (14), the optimal values of the Lagrange multipliers by their current values). Since the duality gap is zero, when we obtain the optimal dual variables ($\boldsymbol{\lambda}^\star$ and $\mu^\star$) with the ellipsoid method, the optimal solution $\tilde{\mathbf{S}}_i^\star(\boldsymbol{\lambda}^\star, \mu^\star)$ converges to the primal optimal solution of problem (13). As a summary, the algorithm that solves problem (13) is described in Table 2 (this table was already presented in [1] but is included here for the sake of completeness).

4.1 Particular cases: scenario with only one type of user

There exists a couple of particular cases of the problem presented before in which only one type of user is present in the system. Such simplified scenarios are found in real systems and will yield simpler optimization problems with lower computational complexity in the resolution of the resource allocation algorithm. For the sake of ease of readability of the paper, the mathematical developments of both particular cases have been moved to Appendix C.

Table 2 Algorithm for solving problem (13)

1: Initialize $\boldsymbol{\lambda} \succeq 0$, $\mu \geq 0$ such that $\mu\mathbf{I} - \sum_{j\in\mathcal{U}_E} \lambda_j\hat{\mathbf{H}}_{ji}^H\hat{\mathbf{H}}_{ji} \succ 0, \forall i$
2: **Repeat**
3: Compute $\tilde{\mathbf{S}}_i(\boldsymbol{\lambda}, \mu)\ \forall i$ using (14)
4: Compute subgradient of $g(\boldsymbol{\lambda}, \mu)$:
5: $[\mathbf{t}]_m = Q_{N+m} - \sum_{i\in\mathcal{U}_I} \mathrm{Tr}(\hat{\mathbf{H}}_{(N+m)i}\tilde{\mathbf{S}}_i\hat{\mathbf{H}}_{(N+m)i}^H)$ for $1 \leq m \leq M$
6: $[\mathbf{t}]_{M+1} = \mathrm{Tr}(\tilde{\mathbf{S}}_i) - (P_{\max} - P_c^{tx})$
7: Update $\boldsymbol{\lambda}$, μ using the ellipsoid method [41] subject to the following: $\boldsymbol{\lambda} \succeq 0$, $\mu \geq 0$ and $\mu\mathbf{I} - \sum_{j\in\mathcal{U}_E} \lambda_j\hat{\mathbf{H}}_{ji}^H\hat{\mathbf{H}}_{ji} \succ 0, \forall i$
8: **Until** dual variables converge

4.2 Trade-off analysis between weighted sum rate and power constraints

In this section, we analyze the multidimensional trade-off between the objective function, that is, the weighted sum rate, and the set of power harvesting constraints. For simplicity, let us consider that $C_i(t)\ \forall i \in \mathcal{U}_I$ is high enough so that it could be assumed that $R_i^\star < R_{\max,i}$ and $R_i^\star = \log\det\left(\mathbf{I} + \hat{\mathbf{H}}_i\tilde{\mathbf{S}}_i^\star\hat{\mathbf{H}}_i^H\right)$. We would like to emphasize that, as the noise and channels are normalized, we will refer to the powers harvested by the receivers in terms of power units instead of Watts. Given this approach, we propose to use the *rate-power* (R-P) region to characterize all the achievable sum rates (in bit/s/Hz) and power harvesting (in power units) $M+1$-tuples under a given power constraint as in [9]. The R-P region of problem (13) is defined as:

$$\mathcal{C}_{\text{R-P}}\left((P_{\max} - P_c^{tx}), \{\omega_i\}\right) \triangleq \left\{ (\mathrm{SR}; \{Q_j\}) \mid \exists \{\tilde{\mathbf{S}}_i\} \text{ with } \mathrm{SR} \leq \sum_{i\in\mathcal{U}_I} \omega_i \log\det\left(\mathbf{I} + \hat{\mathbf{H}}_i\tilde{\mathbf{S}}_i\hat{\mathbf{H}}_i^H\right), \right.$$
$$\left. \sum_{i\in\mathcal{U}_I} \mathrm{Tr}\left(\hat{\mathbf{H}}_{ji}\tilde{\mathbf{S}}_i\hat{\mathbf{H}}_{ji}^H\right) \geq Q_j,\ \sum_{i\in\mathcal{U}_I} \mathrm{Tr}(\tilde{\mathbf{S}}_i)+P_c^{tx} \leq P_{\max},\ \tilde{\mathbf{S}}_i \succeq 0 \quad \forall j \in \mathcal{U}_E, \forall i \in \mathcal{U}_I \right\}. \tag{16}$$

To be able to graphically show an example of the trade-off, we restrict the cardinality of the set of harvesting users and information users to be two, i.e., $|\mathcal{U}_E| = 2$ and $|\mathcal{U}_I| = 2$, and for simplicity, we consider that $\omega_i = 1$, $\forall i \in \mathcal{U}_I$. In such a case, the trade-off region between the sum rate and the two power constraints is a three-dimensional surface. The setup taken as an example for this section is a BS with four transmit antennas and where all users have two antennas. The maximum transmission power at the BS is $P_{\max} - P_c^{tx} = 10$ W. The entries of the matrix channels are generated independently from a complex circularly symmetric Gaussian distribution with zero mean and variance equal to one.[8]

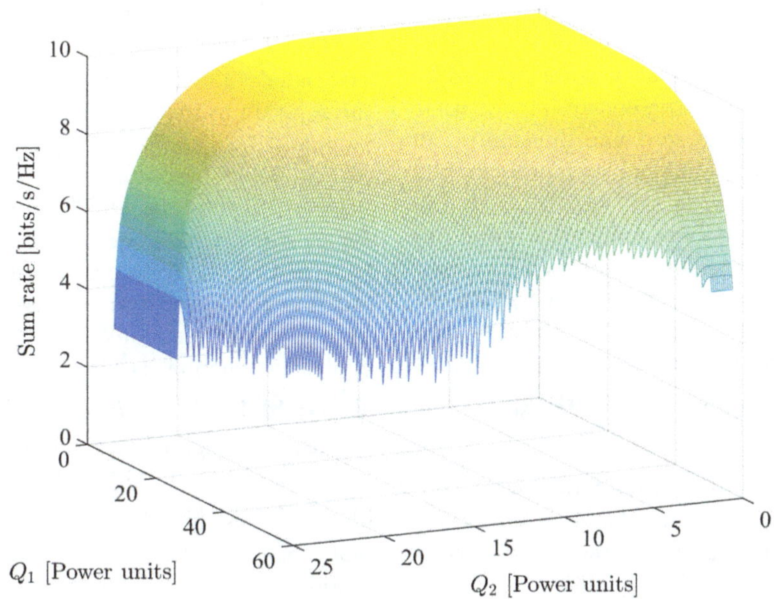

Fig. 2 Three-dimensional R-P region. Representation of the three-dimensional R-P region of problem (13). The figure represents the existing trade-off between the optimal solution of the problem, i.e., the weighted sum rate, and the two power harvesting constraints

Figure 2 depicts the three-dimensional R-P region for the previous setup. As can be appreciated, the optimal sum rate solution is jointly concave on Q_1 and Q_2, as expected [40]. The values of Q_1 and Q_2 for which the region is not defined correspond to situations where problem (13) is infeasible. To characterize the surface accurately, let us introduce the contour lines of the R-P region in Fig. 3. In the plot, when the lines are close together, the magnitude of the gradient is large. There are also some important boundary points marked in the 3D plot of the surface. Those points can be computed in a simple way and provide

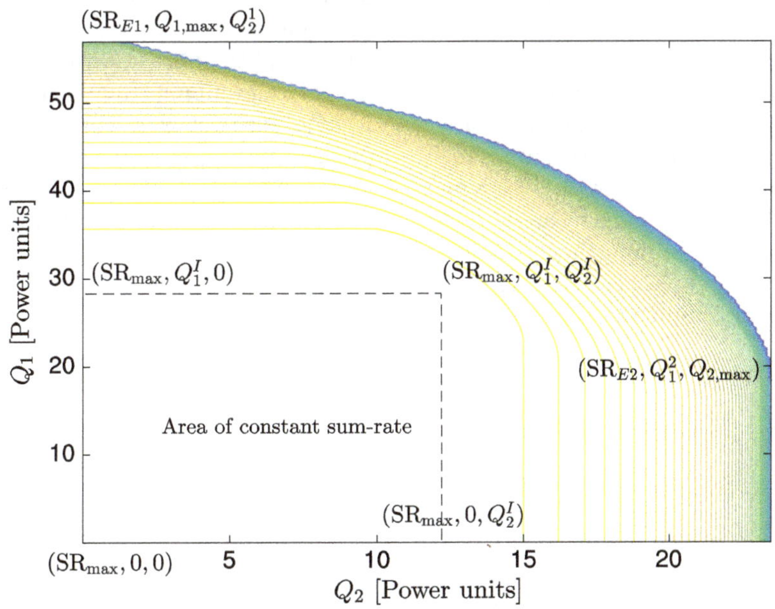

Fig. 3 Contour lines of the three-dimensional R-P region. Contour lines of the three-dimensional R-P region of problem (13). Note that the density of lines increases with the gradient of the surface, and the color indicates the value of such a surface. Note also that some important boundary characteristic points have been marked

us with useful cases that will be commented on in what follows.

Let us first start with the boundary point defined by $(\mathrm{SR}_{\max}, 0, 0)$. The power harvesting constraints for users 1 and 2 at this point are set to zero, and the solution of the problem therefore can be obtained from problem (22) (or from problem (13) with $Q_1 = Q_2 = 0$). $\mathrm{SR}_{\max}$ represents the maximum sum rate that can be achieved in this situation when no energy harvesting is imposed. The optimum covariance matrices were obtained in Section 4.1 and are denoted here as $\tilde{\mathbf{S}}^{\star}_{\mathrm{SR}_i}$ for the ith user. Following that notation, the maximum sum rate can also be expressed as $\mathrm{SR}_{\max} = \log\det\left(\mathbf{I} + \hat{\mathbf{H}}_1\tilde{\mathbf{S}}^{\star}_{\mathrm{SR}_1}\hat{\mathbf{H}}_1^H\right) + \log\det\left(\mathbf{I} + \hat{\mathbf{H}}_2\tilde{\mathbf{S}}^{\star}_{\mathrm{SR}_2}\hat{\mathbf{H}}_2^H\right)$.

Note that although when computing $\mathrm{SR}_{\max}$ we do not apply power harvesting constraints, this does not necessarily mean that the actual harvested powers are zero. In this context, we have the boundary point $(\mathrm{SR}_{\max}, Q_1^{\mathrm{I}}, 0)$, where Q_1^{I} represents the power harvested by user 1 when the precoder matrices are the ones that maximize the weighted sum rate, i.e., $Q_1^{\mathrm{I}} = \mathrm{Tr}\left(\hat{\mathbf{H}}_{11}\tilde{\mathbf{S}}^{\star}_{\mathrm{SR}_1}\hat{\mathbf{H}}_{11}^H\right) + \mathrm{Tr}\left(\hat{\mathbf{H}}_{12}\tilde{\mathbf{S}}^{\star}_{\mathrm{SR}_2}\hat{\mathbf{H}}_{12}^H\right)$. The same can be said for the boundary point $(\mathrm{SR}_{\max}, 0, Q_2^{\mathrm{I}})$, where $Q_2^{\mathrm{I}} = \mathrm{Tr}\left(\hat{\mathbf{H}}_{21}\tilde{\mathbf{S}}^{\star}_{\mathrm{SR}_1}\hat{\mathbf{H}}_{21}^H\right) + \mathrm{Tr}\left(\hat{\mathbf{H}}_{22}\tilde{\mathbf{S}}^{\star}_{\mathrm{SR}_2}\hat{\mathbf{H}}_{22}^H\right)$. Then, there is a fourth point that defines a flat surface (or tableland) of constant sum rate $\mathrm{SR}_{\max}$, which is the combination of the two previous points, $\left(\mathrm{SR}_{\max}, Q_1^{\mathrm{I}}, Q_2^{\mathrm{I}}\right)$. In other words, the tableland of constant maximum weighted sum rate $\mathrm{SR}_{\max}$ defines all possible values of harvested power constraints for which constraints $C1$ are not active and thus do not affect the optimum value of the weighted sum rate.

Now, let us consider the boundary points in terms of maximum harvested power. On top of the figure, there is the point $(\mathrm{SR}_{\mathrm{E}1}, Q_{1,\max}, Q_2^1)$. This point corresponds to the situation in which the power harvested by user 1 is a maximum or, in other words, the maximum value of Q_1 for which problem (13) is feasible, assuming no constraint on the power to be harvested by user 2. To calculate $Q_{1,\max}$, we solve the following optimization problem:

$$\begin{aligned} \underset{\tilde{\mathbf{S}}_{\mathrm{E}1}}{\text{maximize}} \quad & \mathrm{Tr}\left(\hat{\mathbf{H}}_{11}\tilde{\mathbf{S}}_{\mathrm{E}1}\hat{\mathbf{H}}_{11}^H\right) \\ \text{subject to} \quad & C1: \mathrm{Tr}(\tilde{\mathbf{S}}_{\mathrm{E}1}) + P_c^{tx} \leq P_{\max} \\ & C2: \tilde{\mathbf{S}}_{\mathrm{E}1} \succeq 0, \end{aligned} \tag{17}$$

where $\tilde{\mathbf{S}}_{\mathrm{E}1}$ represents the sum of the two covariance matrices for the information users (note that in this problem, the objective function and the constraint depend on such matrices through their sum), and the objective function is the power harvested by user 1. Now, by applying the result from Proposition 2, we obtain the solution of problem (17) as follows. Let the reduced eigen-decomposition of $\hat{\mathbf{H}}_{11}^H\hat{\mathbf{H}}_{11}$ be $\hat{\mathbf{U}}_{11}\hat{\Lambda}_{11}\hat{\mathbf{U}}_{11}^H$ such that $\hat{\mathbf{u}}_{11,\max}$ is the eigenvector associated with the maximum eigenvalue $\hat{\lambda}_{11,\max}$. Then, the solution to the previous problem is based on the following inequality: $\mathrm{Tr}\left(\tilde{\mathbf{S}}_{\mathrm{E}1}\hat{\mathbf{H}}_{11}^H\hat{\mathbf{H}}_{11}\right) \leq \hat{\lambda}_{11,\max}\mathrm{Tr}\left(\tilde{\mathbf{S}}_{\mathrm{E}1}\right) = \hat{\lambda}_{11,\max} \times (P_{\max} - P_c^{tx})$ (since at the optimum $\mathrm{Tr}\left(\tilde{\mathbf{S}}^{\star}_{\mathrm{E}1}\right) = P_{\max} - P_c^{tx}$), where such inequality becomes equality if $\tilde{\mathbf{S}}^{\star}_{\mathrm{E}1} = (P_{\max} - P_c^{tx}) \times \hat{\mathbf{u}}_{11,\max}\hat{\mathbf{u}}_{11,\max}^H$. In this case, the maximum harvested energy is accomplished by *energy beamforming*[9] (i.e., rank 1) to the best eigenmode of the equivalent channel $\hat{\mathbf{H}}_{11}^H\hat{\mathbf{H}}_{11}$. Then, we obtain $Q_{1,\max} = \mathrm{Tr}\left(\hat{\mathbf{H}}_{11}\tilde{\mathbf{S}}^{\star}_{\mathrm{E}1}\hat{\mathbf{H}}_{11}^H\right) = (P_{\max} - P_c^{tx}) \times \hat{\lambda}_{11,\max}$. According to this, the weighted sum rate obtained by solving problem (13) and $Q_1 = Q_{1,\max}, Q_2 = 0$ (denoted as $\mathrm{SR}_{\mathrm{E}1}$) is $\mathrm{SR}_{\mathrm{E}1} = \log\det\left(\mathbf{I} + \hat{\mathbf{H}}_1\tilde{\mathbf{S}}^{\star}_{\mathrm{E}1}\hat{\mathbf{H}}_1^H\right) + \log\det\left(\mathbf{I} + \hat{\mathbf{H}}_2\tilde{\mathbf{S}}^{\star}_{\mathrm{E}1}\hat{\mathbf{H}}_2^H\right)$. Note that, even though we do not apply the power harvesting constraint of user 2 when computing $\tilde{\mathbf{S}}_{\mathrm{E}1}$, it does not mean that the actual power harvested by user 2 is zero. In this context, we define the last coordinate of the point, denoted as Q_2^1, which represents the power harvested by user 2 when the covariance matrix is $\tilde{\mathbf{S}}^{\star}_{\mathrm{E}2}$, i.e., $Q_2^1 = \mathrm{Tr}\left(\hat{\mathbf{H}}_{21}\tilde{\mathbf{S}}^{\star}_{\mathrm{E}2}\hat{\mathbf{H}}_{21}^H\right)$. The same reasoning can be applied to obtain the last boundary point $(\mathrm{SR}_{\mathrm{E}2}, Q_1^2, Q_{2,\max})$ by interchanging the roles of users 1 and 2.

The remaining boundary points in the curve can be obtained by properly varying the values of Q_1 and Q_2 ($0 \leq Q_1 \leq Q_{1,\max}, 0 \leq Q_2 \leq Q_{2,\max}$) in problem (13).

5 User selection policies

Thus far, we have assumed that the two groups of users, i.e., $\mathcal{U}_I$ and $\mathcal{U}_E$, were known. The goal of this section is to propose a grouping strategy to select which users should go into each set in a way that the aggregated throughput over time is maximized. As the channels and batteries fluctuate throughout time, the users in each group may also change from frame to frame. In this section, we will assume that the values of $\{Q_j\}$ are known and fixed. The management of these values is beyond the scope of the paper (see the work in [43], where the authors propose some procedures to adjust the values of $\{Q_j\}$, considering the impact on the system performance).

As previously noted, the optimal information and harvesting grouping should be obtained by joint exhaustive search (see Section 3). This search is prohibitively complex, and suboptimum techniques therefore should be derived. The case of having only information users has been studied in the literature, and suboptimal techniques that perform close to the optimum one have been proposed [44, 45]. In this paper, to keep the overall complexity as low as possible without compromising the performance of the system, we present suboptimal techniques for the user grouping for both kinds of users, i.e., information and harvesting users. This is one of the major contributions of our paper, that is, work with users that have different objectives. Additionally, as we will show, the proposed greedy algorithms take into account that the selection of the harvesting users impacts directly the performance of the information users, that is, there is a coupling behavior between both aspects.

The overall user grouping strategy will be divided into two stages. In the first stage (that will be known as *supergrouping*), we will provide a preselection of user candidates to be in each set. This will depend primarily on the current energies available at the batteries, and it will be run at a longer time scale, every few scheduling periods or frames. For the second stage, known as *grouping*, we are going to present two different user grouping strategies that will be run at every frame. The strategy with the highest complexity provides a better performance than the simpler strategy.

In the first (simpler) approach, we will split the user grouping further into two stages. The first stage selects the information users, $\mathcal{U}_I$, from the supergrouping set $\mathcal{U}_I^S$ based on a greedy approach, whereas the second stage selects the harvesting users, $\mathcal{U}_E$, based on the already selected information users. In the second approach, we will develop a joint information-harvesting grouping strategy, which constitutes an intermediate approach between the first simple approach and the optimum approach based on exhaustive search.

5.1 User supergrouping strategy

Recall that when we derived the optimal precoder matrix in Section 4, we assumed that the optimal rates would fulfill $R_i^\star(t) < R_{\max,i}(t), \forall i \in \mathcal{U}_I$ for any particular frame, and therefore, constraints C4 in problem (12) were not active. This is achieved by preselecting the users that are to be scheduled for data transmission or battery charging. In our proposed approach, we first implement a selection of candidates to be in $\mathcal{U}_I$ and $\mathcal{U}_E$, known as $\mathcal{U}_I^S$ and $\mathcal{U}_E^S$, such that $\mathcal{U}_I \subseteq \mathcal{U}_I^S$, $\mathcal{U}_E \subseteq \mathcal{U}_E^S$, and $|\mathcal{U}_I^S| + |\mathcal{U}_E^S| = K$, and we then select the users that finally go into the sets $\mathcal{U}_I$ and $\mathcal{U}_E$. The proposed supergrouping algorithm is presented in Table 3 and works as follows: we set a threshold α such that $0 \leq \alpha \leq 1$. Then, we compute the ratio of the current battery level and the battery capacity for all users, and we then order these ratios increasingly. If the middle ratio of the previous list is greater than the value of the threshold α, we then split the overall group by half and put half of the users in $\mathcal{U}_I^S$ and the other half in $\mathcal{U}_E^S$. On the other hand, if the middle ratio of the previous list is lower than the value of α, we find the user with battery ratio closest to the value of α and put all users with lower ratios than the one closest to α in the harvesting set and the remaining users in the information set. The larger the value of α, the greater

Table 3 Algorithm to obtain the superframe sets $\mathcal{U}_I^S$ and $\mathcal{U}_E^S$

```
Set a threshold 0 ≤ α ≤ 1
Order the users increasingly with the following rule:
    C_1(t)/C^1_max ≤ C_2(t)/C^2_max ≤ ... ≤ C_{K/2}(t)/C^{K/2}_max ≤ C_{K/2+1}(t)/C^{K/2+1}_max ≤ ... ≤ C_K(t)/C^K_max
if α < C_{K/2}(t)/C^{K/2}_max
    Users {1, 2, ..., K/2} go to U_E^S
    Users {K/2 + 1, K/2 + 2, ..., K} go to U_I^S
else
    Find the user m such that m = arg min_i | C_i(t)/C^i_max − α |
    Users {1, 2, ..., m} go to U_E^S
    Users {m + 1, m + 2, ..., K} go to U_I^S
end if
```

the number of users that will be included in the harvesting set $\mathcal{U}_E^S$. Note that the BS has to know the battery levels of all users, which implies that receivers must send the battery levels through a feedback channel and, hence, the battery levels must be quantized (in [33], we addressed the problem of quantizing the battery levels and evaluated the effect on the overall system performance, and we conclude that a few bits for quantization is enough to obtain good performance).

5.2 Disjoint information and harvesting user grouping

This first approach is based on two stages. In the first stage, the selection of the information users follows a greedy approach, in which each user is added at a time and the maximization of the weighted sum rate without harvesting constraints is evaluated for all possible candidate information users with the already selected users. No harvesting users are considered at this stage.

Let us assume, for simplicity, that every information user has the same number of antennas, i.e., $n_{R_i} = N_R$, $\forall i \in \mathcal{U}_T$. The maximum number of simultaneous users to be served following the BD strategy is then $U = \lceil \frac{n_T}{N_R} \rceil$ [25]. The algorithm for selecting the information users is shown in Table 4; first, we select the user that can achieve the greatest weighted rate[10]. Then, we incorporate one user at a time into the set only if the accumulated weighted sum rate increases due to incorporating such a user (weighted sum rate evaluated with the already selected users). The algorithm ends when there is no improvement in the weighted sum rate or when the maximum number of users to be scheduled (U) is reached.

Note that the distances from the BS to the users are taken into account implicitly in the algorithm since in step 2 and step 8 of Table 4, we select users according to the rates. These rates depend on the channel matrices $\{\mathbf{H}_i\}$, and the components of these matrices, of course, will be small if the distances are large. Therefore, the distances will have a direct impact on the selection of users.

Table 4 Algorithm to obtain the set of information users $\mathcal{U}_I$

1: Set $\mathcal{U}_I = \emptyset$, $Q_i \geq 0$, and $\omega_i > 0$, $\forall i \in \mathcal{U}_T$
2: Find $i_1 = \arg\max_{\forall i \in \mathcal{U}_I^S} \max_{\mathbf{S}_i} \omega_i \log\det\left(\mathbf{I} + \mathbf{H}_i\mathbf{S}_i\mathbf{H}_i^H\right)$
   Subject to $\mathrm{Tr}(\mathbf{S}_i) \leq P_T$, $\mathbf{S}_i \succeq 0$
3: Set $f_{\text{temp}} = \omega_{i_1} \log\det\left(\mathbf{I} + \mathbf{H}_{i_1}\mathbf{S}_{i_1}\mathbf{H}_{i_1}^H\right)$
4: Set $\mathcal{U}_I \leftarrow \mathcal{U}_I \cup \{i_1\}$, $\mathcal{U}_I^S \leftarrow \mathcal{U}_I^S \setminus \{i_1\}$
5: **for** $j = 2$ to U
6:   **for** every $i \in \mathcal{U}_I^S$
7:     Let $\mathcal{U}_I^{(i)} = \mathcal{U}_I \cup \{i\}$
8:     Solve (13) without C1, and obtain $R_m^\star$, $\forall m \in \mathcal{U}_I^{(i)}$
9:     Compute $f_i = \sum_{m \in \mathcal{U}_I^{(i)}} \omega_m R_m^\star$
10:   **end for**
11:   Let $i_j = \arg\max_{i \in \mathcal{U}_I^S} f_i$
12:   **if** $f_{i_j} < f_{\text{temp}} \longrightarrow$, go to 17 (break for)
13:   **else**
14:     $\mathcal{U}_\mathcal{I} \leftarrow \mathcal{U}_I \cup \{i_j\}$, $\mathcal{U}_I^S \leftarrow \mathcal{U}_I^S \setminus \{i_j\}$
15:     Let $f_{\text{temp}} = f_{i_j}$
16:   **end if**
17: **end for**

Once we have selected the information users, we continue with the selection of the harvesting users in the second stage of this grouping strategy. The idea is to select the harvesting users so that when the resource allocation strategy is executed, they affect (reduce) the system performance as little as possible (see Section 4.2). Let $\mathbf{S}^\star = \sum_{i \in \mathcal{U}_I} \mathbf{S}_i^\star$, where $\mathcal{U}_I$ and $\{\mathbf{S}_i^\star\}_{i \in \mathcal{U}_I}$ are the information user set and the optimum covariance matrices obtained from the algorithm detailed in Table 4, respectively. The algorithm works as follows. For each harvesting user j, we evaluate and decreasingly order $\mathrm{Tr}(\mathbf{H}_j\mathbf{S}^\star\mathbf{H}_j) - Q_j$ and select the first M harvesting users according to this order. Note that in the previous expression, we are evaluating how the optimum covariance matrices of the selected information users transmit power in the geometrical direction of the channels of the harvesting users. We also take into account the minimum required power to be harvested Q_j to ensure feasibility of the solution of the resource allocation problem. The algorithm is presented in Table 5.

5.3 Joint information and harvesting user grouping

In this second approach, the selection of the information and harvesting users is coupled. Due to this joint approach, the system performance will be degraded less by the effect of having harvesting users in the system compared with the previous decoupled approach. However, the computational complexity increases as more combinations need to be evaluated.

The algorithm for selecting the information users is based on the same greedy approach that we presented before. The difference is that, now, instead of selecting the information users and then the harvesting users, we select both types of users simultaneously. For simplicity in the formulation, let us consider that M is an integer multiple of U and define $k = \frac{M}{U}$ (we will comment later on how we could apply the algorithm if that was not the case). The idea behind the algorithm is as follows. We select one information user q and obtain its optimum covariance matrix $\mathbf{S}_q^\star$. Then, we find the best k harvesting users based on the principle developed in Table 5. After that, we select another information user and repeat the same process until there is no improvement in the objective function. Due to the fact that the grouping is coupled, we consider the impact of having selected harvesting users on the future selection of information users. The specific details of the joint algorithm are presented in Table 6.

Table 5 Algorithm to obtain the set of harvesting users $\mathcal{U}_E$

1:Input: $\mathcal{U}_I$ taken from algorithm in Table 4, $\mathbf{S}^\star = \sum_{i \in \mathcal{U}_I} \mathbf{S}_i^\star$,
2:Evaluate $m_j = \mathrm{Tr}(\mathbf{H}_j\mathbf{S}^\star\mathbf{H}_j) - Q_j$, $\forall j \in \mathcal{U}_E^S$
3:Decreasingly order m_j
4:Construct $\mathcal{U}_E$ with the users corresponding to the first M ordered terms of m_j

Table 6 Algorithm to jointly obtain the set of information and harvesting users $\mathcal{U}_I, \mathcal{U}_E$

1: Set $\mathcal{U}_I = \emptyset$, $Q_i \geq 0$, and $\omega_i > 0$, $\forall i \in \mathcal{U}_T$
2: Find $i_1 = \arg\max_{\forall i \in \mathcal{U}_I^S} \max_{\mathbf{S}_i} \omega_i \log\det\left(\mathbf{I} + \mathbf{H}_i\mathbf{S}_i\mathbf{H}_i^H\right)$
   subject to $\mathrm{Tr}(\mathbf{S}_i) \leq P_T$, $\mathbf{S}_i \succeq 0$
3: Set $f_{\text{temp}} = \omega_{i_1} \log\det(\mathbf{I} + \mathbf{H}_{i_1}\mathbf{S}_{i_1}\mathbf{H}_{i_1}^H)$
4: Set $\mathcal{U}_I \leftarrow \mathcal{U}_I \cup \{i_1\}$, $\mathcal{U}_I^S \leftarrow \mathcal{U}_I^S \setminus \{i_1\}$
5: Evaluate $m_j = \mathrm{Tr}(\mathbf{H}_j\mathbf{S}_{i_1}^\star\mathbf{H}_j) - Q_j$, $\forall j \in \mathcal{U}_E^S$
6: Find the k users with highest value of m_j. Put them in set $\mathcal{H}$
7: Set $\mathcal{U}_E \leftarrow \mathcal{U}_E \cup \mathcal{H}$, $\mathcal{U}_E^S \leftarrow \mathcal{U}_E^S \setminus \mathcal{H}$, $\mathcal{H} = \emptyset$
8: **for** $j = 2$ to U
9:   **for** every $i \in \mathcal{U}_I^S$
10:     Let $\mathcal{U}_I^{(i)} = \mathcal{U}_I \cup \{i\}$
11:     Solve (13), and obtain $R_m^\star$, $\mathbf{S}_m^\star$, $\forall m \in \mathcal{U}_I^{(i)}$
12:     Compute $f_i = \sum_{m \in \mathcal{U}_I^{(i)}} \omega_m R_m^\star$
13:   **end for**
14:   Let $i_j = \arg\max_{i \in \mathcal{U}_I^S} f_i$
15:   **if** $f_{i_j} < f_{\text{temp}} \longrightarrow$, go to 23 (break for)
16:   **else**
17:     $\mathcal{U}_\mathcal{I} \leftarrow \mathcal{U}_I \cup \{i_j\}$, $\mathcal{U}_I^S \leftarrow \mathcal{U}_I^S \setminus \{i_j\}$
18:     Let $f_{\text{temp}} = f_{i_j}$
19:   **end if**
20:   Evaluate $m_j = \mathrm{Tr}(\mathbf{H}_j \sum_{i \in \mathcal{U}_I} \mathbf{S}_i^\star\mathbf{H}_j) - Q_j$, $\forall j \in \mathcal{U}_E^S$
21:   Find the k users with highest value of m_j. Put them in set $\mathcal{H}$
22:   Set $\mathcal{U}_E \leftarrow \mathcal{U}_E \cup \mathcal{H}$, $\mathcal{U}_E^S \leftarrow \mathcal{U}_E^S \setminus \mathcal{H}$, $\mathcal{H} = \emptyset$
23: **end for**

The main difference with the algorithm in Table 5 is that, now, we solve problem (13) with constraints $C1$, that is, with harvesting users, which increases the complexity of the overall grouping procedure. As addressed before, if M is not an integer multiple of U, we can introduce more harvesting users in step 21 in Table 6 in some iterations, e.g., if $M = 7$ and $U = 3$, we first select three harvesting users and then two harvesting users in the other two iterations.

6 Overall user grouping and resource allocation algorithm

In the following, we present a summary of the overall algorithm that consists of the user supergrouping, the user grouping, and the resource allocation stages presented in the previous two sections. Note that the user supergrouping is carried out every few frames, whereas the user grouping is executed at each frame. If, for some reason, the supergrouping algorithm fails in fulfilling $R_i^\star(t) < R_{\max,i}(t)$, $\forall i \in \mathcal{U}_I$ (an event that would be unlikely to happen), then for those users for which $R_i^\star(t) \geq R_{\max,i}(t)$, we just transmit information in some channel accesses of the frame until their battery is over. The overall algorithm is detailed in Table 7.

7 Results and discussion

In this section, we perform some numerical analysis of the proposed grouping and resource allocation strategies. The system comprises one transmitter with 8 antennas and 30 users ($|\mathcal{U}_T| = 30$) with 2 antennas each. The maximum radiated power is $P_{\max} = 11$ W, and the transmitter front-end consumption is $P_c^{tx} = 1$ W. Front-end power consumption at the receiver is $P_c^{rx} = 100$ mW, and the model used for decoding is exponential, i.e., $P_{\text{dec}}(R) = c_1 e^{c_2 R}$, where $c_1 = 30$, and $c_2 = 0.75$ [33]. The frame duration is equal to $T_f = 100$ ms, and the superframe duration is equal to 3 s. The channel matrices are generated randomly with i.i.d. entries distributed according to $\mathcal{CN}(0, 1)$. The noise power is normalized to 1. The effective window length for the PF scheme is $T_c = 5$. The percentage used for supergrouping is $\alpha = 0.1$. The battery capacities are generated randomly from 3000 to 10,000 energy units. As we mentioned previously, we assume that all the harvesting constraints are the same for all users and fixed for all periods to $Q_j = 50$ power units, unless stated otherwise. A strategy on how to manage and dynamically adjust the values of the $\{Q_j\}$ was proposed in [43] and is beyond the scope of this paper.

Table 7 Overall user grouping and resource allocation algorithm

Beginning of a superframe:

1: Run **user supergrouping** algorithm in Table 3: obtain sets $\mathcal{U}_I^S$ and $\mathcal{U}_E^S$

Beginning of each frame (two options):

Option 1:

2a: Run **information user grouping** algorithm in Table 4: obtain set $\mathcal{U}_I$

2b: Run **harvesting user grouping** algorithms in Table 5: obtain set $\mathcal{U}_E$

2c: Run **resource allocation** algorithm in Table 2

Option 2:

3a: Run **joint information and harvesting grouping** algorithm in Table 6: Obtain sets $\mathcal{U}_I$ and $\mathcal{U}_E$

3b: Run **resource allocation** algorithm in Table 2

End of each frame:

4: Update batteries:

$$C_i(t) = \left(C_i(t-1) - T_f P_{\text{tot},i}^{rx}(R_i^\star(t-1))\right)_0^{C_{\max}^i}, \quad \forall i \in \mathcal{U}_I$$

$$C_j(t) = \left(C_j(t-1) + T_f \bar{Q}_j(t-1) - T_f P_c^{rx}\right)_0^{C_{\max}^j}, \quad \forall j \in \mathcal{U}_E$$

5: Update weights (e.g., using a PF approach):

$$w_i(t) = \frac{1}{T_i(t)}, \quad T_i(t) = \left(1 - \frac{1}{T_c}\right) T_i(t-2) + \frac{1}{T_c} R_i^\star(t-1)$$

In the simulations, we compare our proposed two methods with two other schemes. As there are no proposals in the literature for user scheduling in the SWIPT framework, we compare our approaches with traditional schemes. In one of the schemes, we assume that the supergrouping and grouping are implemented with a round robin strategy. We will denote this strategy RR-SF/RR-F. In the other scheme, we consider that random selection of users is implemented at both levels as well. This strategy will be denoted by Ra-SF/Ra-F. On the other hand, the proposed supergrouping strategy (Table 3) will be denoted by LB, and the grouping will be denoted according to the algorithm: DHS for the decoupled approach presented in Section 5.2 (Tables 4 and 5) and CHS for the approach presented in Section 5.3 (Table 6).

7.1 Time evolution simulations

Figure 4 depicts the evolution of the battery levels of all users in the system. We can observe that for the round robin scheme, users reach their maximum battery capacity. This is because the data rates achieved are low, and thus, users use little energy for decoding. Then, in the top-right figure, we have the case where random scheduling is considered. In this case, we see how the battery evolutions of all users evolve randomly because new users are scheduled in a random fashion in each frame. Due to the battery overflows that some users experience and the randomness in the selection, this approach, as will happen with the round robin scheme, will not be very efficient in terms of aggregate throughput. The last two figures depict the battery evolution of the two proposed schemes. We observe that in both cases, the battery levels of the users are substantially lower than the ones observed in previous schemes. This reduction in battery levels is related with the large throughput achieved by the users, as will be apparent later. It is difficult to assess from these figures which proposed scheme provides better performance.

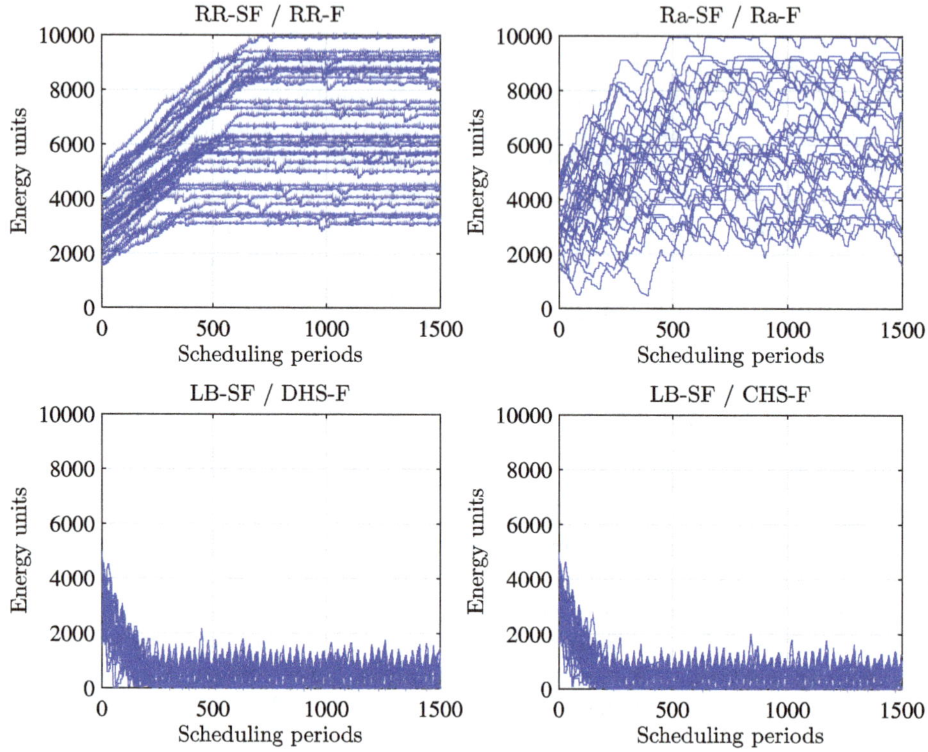

Fig. 4 Time evolution of the battery levels. Time evolution of the battery levels of all users in the system for the different approaches

Figure 5 presents the average sum rate of the system (computed as $SR(\tau) = \frac{1}{\tau}\sum_{t=1}^{\tau}\sum_{i\in\mathcal{U}_I} R_i(t)$). This metric is an estimation of the expected throughput of the system. From the figure, we see that the sum rate of the round robin and random schemes provides a stable average throughput over time but the magnitude of the throughput is not so high. Then, we see how the proposed schemes notably outperform the previous benchmarking strategies. The simpler approach, DHS, performs similar to the more complex strategy, CHS. We also plot, as benchmarks, two cases. The first one, called "no harvesting management," refers to the case in which the

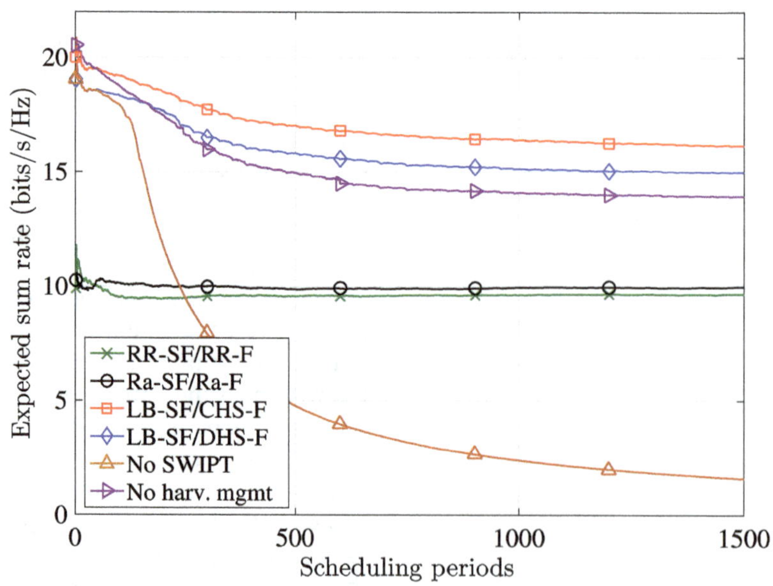

Fig. 5 Time evolution of the average sum rate. Time evolution of the average sum rate of the system for the different approaches

harvesting users are selected jointly with data users following the CHS approach, but their harvesting constraints are set to zero, $Q_j = 0, \forall j$, that is, harvesting users collect energy without imposing a constraint. In this case, the rate achieved is higher at the beginning, but the energy collected by the users is lower, having an impact on the performance as time goes on. The second case considers that no power transfer (no SWIPT) is available, and users therefore cannot recharge their batteries. In this case, the users run out of battery, and the expected sum rate therefore tends to zero.

Figure 6 shows the cumulative distribution function (CDF) of the individual data rates of the users in the system. The CDF of the no SWIPT case has a particular shape due to the fact that many users obtain zero data rate as they run out of battery. In this figure, we clearly see the benefits of the proposed user selection schemes compared to the other approaches, such as low data rate percentiles and high data rate percentiles being much better for the proposed strategies.

Finally, in Fig. 7, we depict the average evolution of the harvested power. It is interesting to note how all users tend to converge to a certain point (or the vicinity of a point). This is due to the fact that if a user is receiving much power, then its battery will increase, which will make the user more eligible to receive data, making the harvesting decrease, whereas if a user has low energy in its battery, then it is directly selected to be included in set $\mathcal{U}_E^S$. We observe that the more complex approach, CHS, is able to provide the users with larger harvested power compared to the less complex approach, DHS.

7.2 System performance simulations

In the next figures, we will show the performance of the system obtained once the algorithms have converged (i.e., after 1500 frames). The first two figures, Figs. 8 and 9, show the system performance, considering that half of the users are at a relative distance to the BS greater than for the other half of the users. In particular, Fig. 8 presents the sum of the expected sum rate for the four schemes for four different relative distances. As expected, the sum rate decreases as the distance to the BS increases. On the other hand, Fig. 9 shows the sum of the expected harvested power as a function of the relative distance. We see that if half of the users are four times farther away from the BS, the loss in harvested power is from 25 to 50%, and the relative loss is lower for the proposed schemes.

The last two figures, Figs. 10 and 11, show the performance of the system when the size of the harvesting group increases in relative terms when compared to the size of the information group, i.e., when $\frac{M}{U}$ increases. This phenomenon is interesting to evaluate since the harvesting users appear in the constraints and they negatively affect the aggregated sum rate (see trade-off in Section 4.2). However, if many users are introduced in the harvesting set, then their batteries will recharge faster, and they will be able to receive higher data rates. This is the compromise that is analyzed in the figures. First, in Fig. 10, we see the expected aggregated sum rate. As we see, for the two benchmarking approaches, Ra and RR, the sum rate decreases as $\frac{M}{U}$ increases. This is because the harvesting users are selected without considering the impact

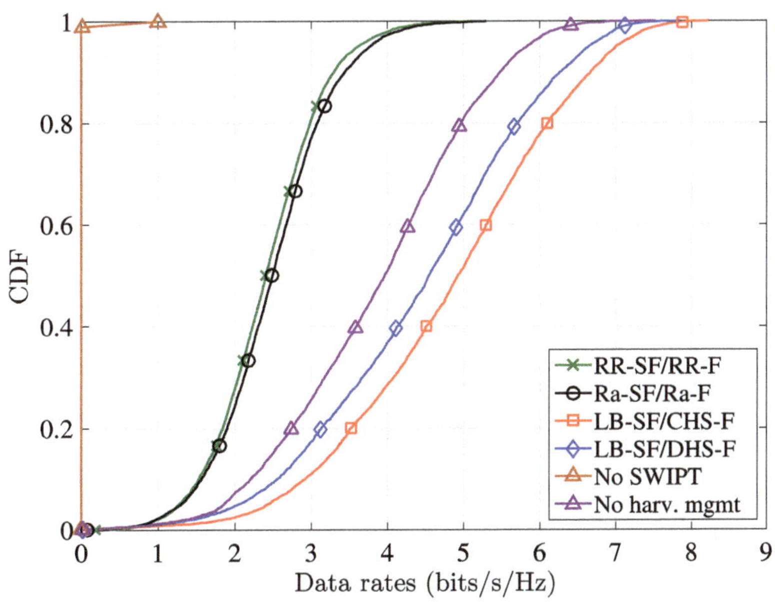

Fig. 6 CDF of the individual data rates. CDF of the individual data rates of all the users in the system for the different approaches

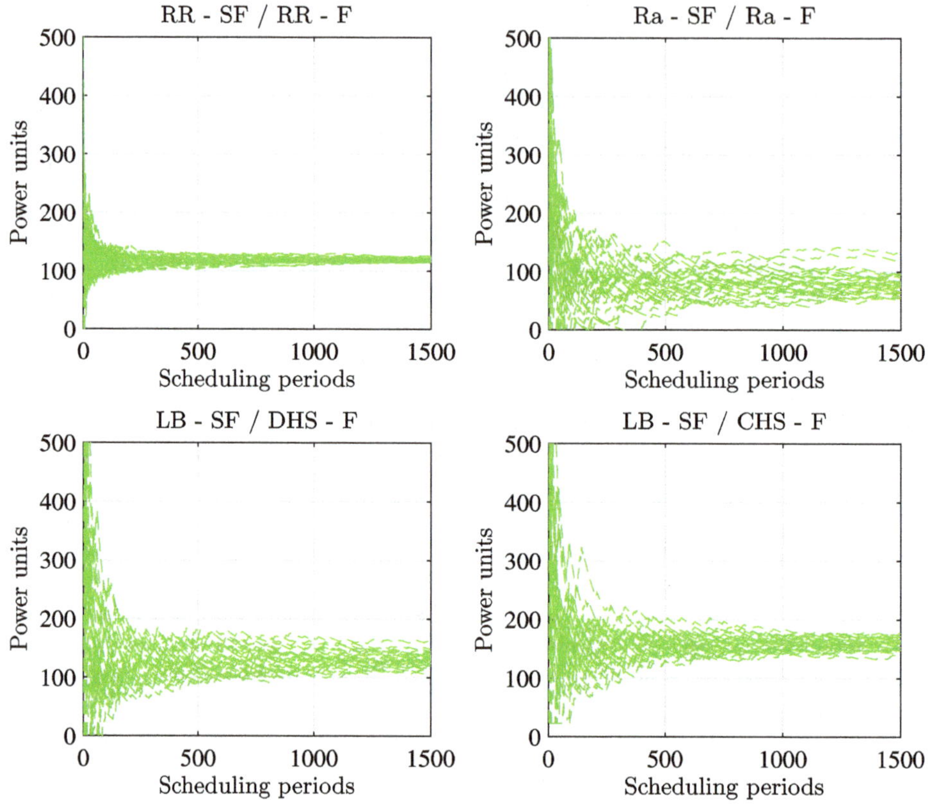

Fig. 7 Time evolution of the average harvested power. Time evolution of the average harvested power of all the users in the system for the different approaches (in power units)

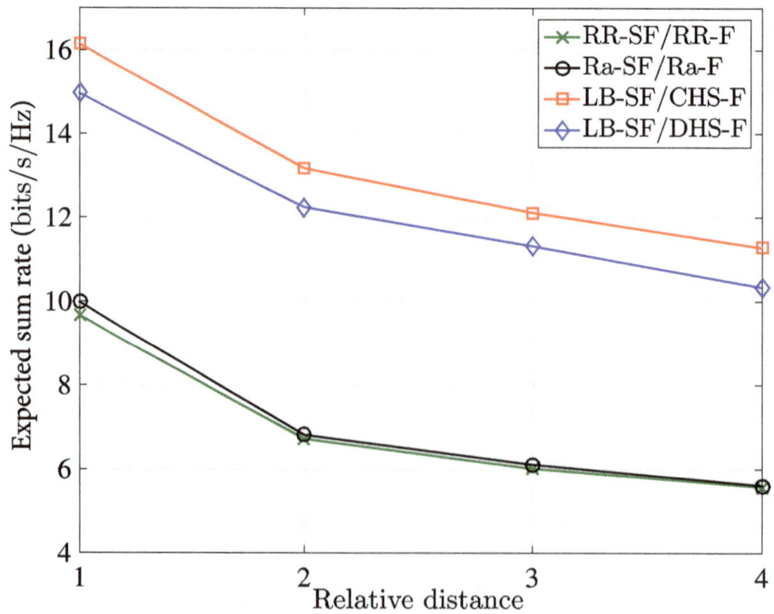

Fig. 8 Expected system sum rate vs distance to the BS. Expected system sum rate as a function of the distance to the BS

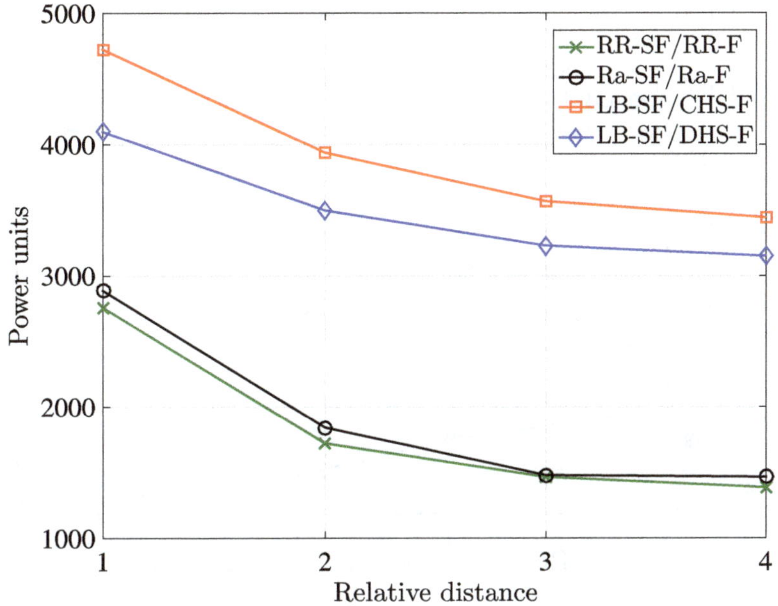

Fig. 9 Sum of expected harvested powers by all users vs distance to the BS. Sum of the expected harvested powers by all users (in power units) as a function of the distance to the BS

that they have on the objective function, and therefore, if more harvesting users are considered in the optimization problem, a lower sum rate will be achieved. In those cases, the optimization problem turns out to be infeasible many times, and therefore, the energy collected by all users also decreases, see Fig. 11. On the other hand, the aggregated sum rate increases a bit for $\frac{M}{U} = 2$ for the proposed strategies. This is due to the fact that harvesting users are selected very efficiently, and thus, the constraints associated to them are not active, i.e., they do not affect

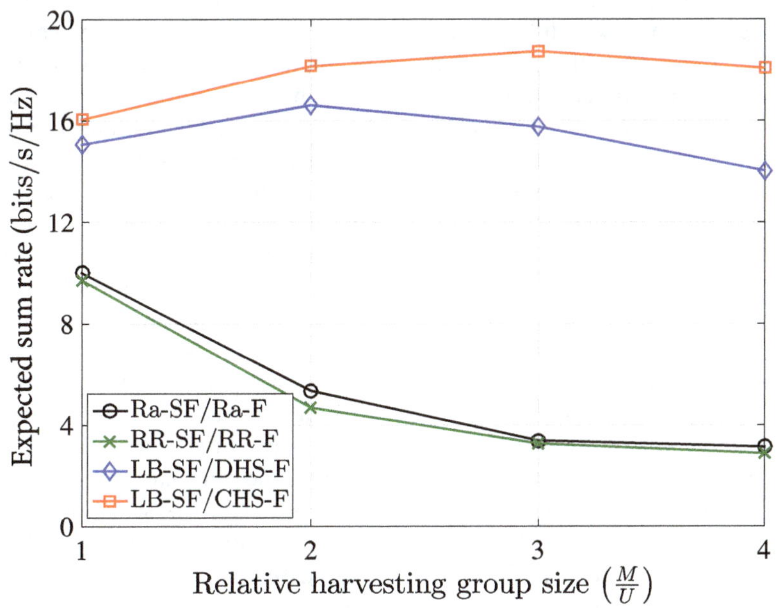

Fig. 10 Expected system sum rate vs relative size of the harvesting user group. Expected system sum rate as a function of the relative size of the harvesting user group

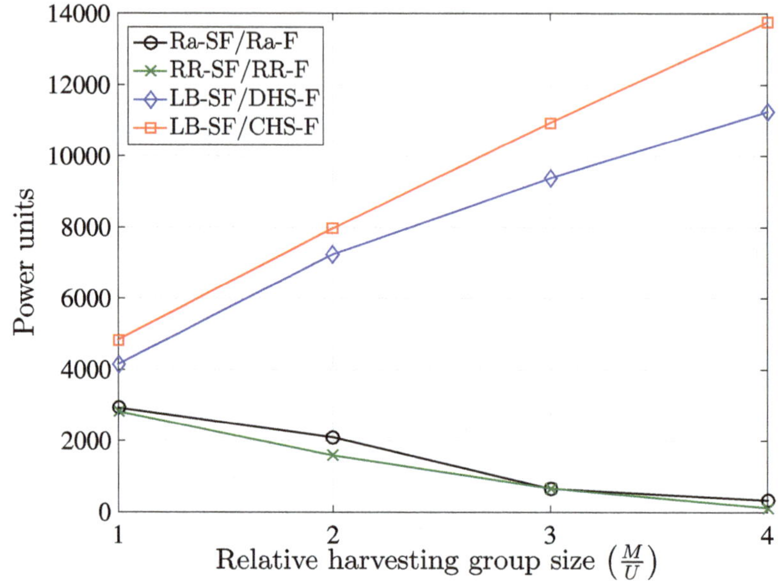

Fig. 11 Sum of expected harvested powers by all users vs relative size of the harvesting user group. Sum of the expected harvested powers by all users (in power units) as a function of the relative size of the harvesting user group

the optimum value of the objective function. Additionally, as more users are able to recharge their batteries (see Fig. 11), they can decode higher rates in future frames. Nonetheless, from a given size $\frac{M}{U}$ on, the system sum rate starts to decrease as the harvesting constraints become active, although the problem is always feasible, and users recharge their batteries, as is indirectly depicted in Fig. 11.

7.3 Computational complexity

An analytic evaluation of the computational complexity of the proposed techniques for each scheduling period is extremely difficult since these algorithms are iterative, each iteration involves the numerical solution of an optimization problem, there are discrete variables related to the grouping of users, and the solution and convergence times depend on the concrete channels associated to the users in the scenario. Because of this, we have performed a numerical evaluation of the computational complexity of the different algorithms by performing many simulations over random channels and averaging the convergence times at each scheduling period obtained in the simulator. Figure 12 shows a set of bars comparing the complexities needed for convergence of the different algorithms that require grouping, that is, RR-SF/RR-F, Ra-SF/Ra-F, LB-SF/CHS-F, and LB-SF/DHS-F. The highest bar corresponds to the algorithm requiring the highest computational complexity, which is LB-SF/CHS-F and has been labeled as the 100% reference. The other bars show the complexities associated to the other algorithms, taking as relative reference, the complexity of LB-SF/CHS-F.

8 Conclusions

This paper has studied the performance of a proposed scheduling algorithm in a multiuser MIMO broadcasting system, where wireless power transfer from BS has been considered a potential technique for energy harvesting taken from radio signals. We derived the particular structure of the optimal transmit covariance matrices and particularized the scenario where only information or harvesting users were present in the system and where both types of users coexist in the system. If only harvesting users were considered, the problem was reformulated as a feasibility problem, and we provided some proposals to be applied in case that the original problem was infeasible. Then, we addressed the multidimensional trade-off between the sum rate and the harvesting constraints in the general case. We showed that energy beamforming was optimal in the case that the power harvested by one particular user was to be maximized. Finally, we presented some user grouping techniques that allow for the BS to select the users better suited for information and those for battery replenishment in each particular frame for the case, where both types of users are present in the system. We proposed two different scheduling techniques based on a different level of computational complexity. In the first approach, we selected the information and the harvesting users separately. In the second approach, the selection of both types of users was performed jointly. The simulation results show that the aggregated throughput can be considerably improved if the proposed grouping strategy is implemented when the results are compared with those of traditional scheduling approaches.

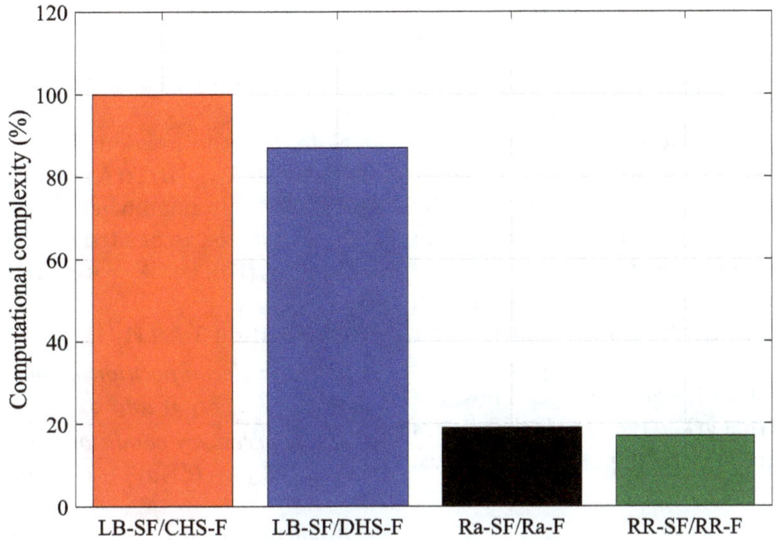

Fig. 12 Relative computational complexities. Average computational complexities of each algorithm needed for convergence at each scheduling period relative to algorithm LB-SF/CHS-F

Appendix 1

The Lagrangian of problem (13) is:

$$\mathcal{L}(\{\tilde{\mathbf{S}}_i\};\boldsymbol{\lambda},\mu) = -\sum_{i\in\mathcal{U}_I}\omega_i \log\det\left(\mathbf{I}+\hat{\mathbf{H}}_i\tilde{\mathbf{S}}_i\hat{\mathbf{H}}_i^H\right) + \sum_{j\in\mathcal{U}_E}\lambda_j\left(Q_j-\sum_{i\in\mathcal{U}_I}\mathrm{Tr}\left(\hat{\mathbf{H}}_{ji}\tilde{\mathbf{S}}_i\hat{\mathbf{H}}_{ji}^H\right)\right)+\mu\left(\sum_{i\in\mathcal{U}_I}\mathrm{Tr}(\tilde{\mathbf{S}}_i)-P_T\right), \tag{18}$$

where we have omitted constraint $C3$. The previous Lagrangian can be manipulated and transformed into:

$$\mathcal{L}(\{\tilde{\mathbf{S}}_i\};\boldsymbol{\lambda},\mu) = -\sum_{i\in\mathcal{U}_I}\omega_i \log\det\left(\mathbf{I}+\hat{\mathbf{H}}_i\tilde{\mathbf{S}}_i\hat{\mathbf{H}}_i^H\right)+\sum_{i\in\mathcal{U}_I}\mathrm{Tr}\left(\mathbf{A}_i\tilde{\mathbf{S}}_i\right)+G, \tag{19}$$

where $G = \sum_{j\in\mathcal{U}_E}\lambda_j Q_j - \mu P_T$, and $\mathbf{A}_i = \mu\mathbf{I} - \sum_{j\in\mathcal{U}_E}\lambda_j\hat{\mathbf{H}}_{ji}^H\hat{\mathbf{H}}_{ji}$. The dual function of problem (13) is defined as $g(\boldsymbol{\lambda},\mu)=\min_{\tilde{\mathbf{S}}_i\succeq 0}\mathcal{L}(\{\tilde{\mathbf{S}}_i\};\boldsymbol{\lambda},\mu)$.

Proposition 1 *To have a bounded solution of the dual function $g(\boldsymbol{\lambda},\mu)$, matrix $\mathbf{A}_i$ must be $\mathbf{A}_i \succ 0\ \forall i$; otherwise, $g(\boldsymbol{\lambda},\mu)$ is unbounded below, i.e., $g(\boldsymbol{\lambda},\mu)=-\infty$.*

Proof See Appendix 2. □

Due to the fact that matrices $\{\mathbf{A}_i\}$ are positive definite, we can assure that they can be decomposed as $\mathbf{A}_i = \mathbf{A}_i^{1/2}\mathbf{A}_i^{1/2}$ and that they always have inverses. Thus, by calling $\hat{\mathbf{S}}_i = \mathbf{A}_i^{1/2}\tilde{\mathbf{S}}_i\mathbf{A}_i^{1/2}$, the dual function can be expressed as:

$$g(\boldsymbol{\lambda},\mu)=\min_{\hat{\mathbf{S}}_i\succeq 0}\left\{-\sum_{i\in\mathcal{U}_I}\omega_i\log\det\left(\mathbf{I}+\hat{\mathbf{H}}_i\mathbf{A}_i^{-1/2}\hat{\mathbf{S}}_i\mathbf{A}_i^{-1/2}\hat{\mathbf{H}}_i^H\right)+\sum_{i\in\mathcal{U}_I}\mathrm{Tr}\left(\hat{\mathbf{S}}_i\right)+G\right\}. \tag{20}$$

The dual function in (20) can be recognized to be equivalent to the dual function of the classical maximization of the sum rate with a power constraint, where the optimum covariance matrix $\hat{\mathbf{S}}_i$ diagonalizes the equivalent channel $\hat{\mathbf{H}}_i\mathbf{A}_i^{-1/2}$ [26], i.e., $\hat{\mathbf{S}}_i = \hat{\mathbf{V}}_i\hat{\mathbf{D}}_i\hat{\mathbf{V}}_i^H$, where $\hat{\mathbf{D}}_i$ is the power allocation matrix, and its components are computed following the water-filling policy [49]. Finally, it is straightforward to show that the precoder $\mathbf{B}_i$ matrix with dimensions $n_T \times n_{S_i}$ corresponding to such covariance matrix is:

$$\mathbf{B}_i^\star = \tilde{\mathbf{V}}_i^{(0)}\mathbf{A}_i^{-1/2}\hat{\mathbf{V}}_i\hat{\mathbf{D}}_i^{1/2}. \tag{21}$$

Appendix 2

Let the eigen-decomposition of $\mathbf{A}_i$ be $\bar{\mathbf{U}}_i\bar{\Gamma}_i\bar{\mathbf{U}}_i^H$, where $\bar{\Gamma}_i$ contains the eigenvalues in decreasing order w.l.o.g. Then, the second term of the Lagrangian in (19) is $\sum_{i\in\mathcal{U}_I}\mathrm{Tr}\left(\bar{\Gamma}_i\bar{\mathbf{U}}_i^H\tilde{\mathbf{S}}_i\bar{\mathbf{U}}_i\right)$. Now, calling $\bar{\mathbf{S}}_i=\bar{\mathbf{U}}_i^H\tilde{\mathbf{S}}_i\bar{\mathbf{U}}_i$, ($\bar{\mathbf{S}}_i \succeq 0 \iff \tilde{\mathbf{S}}_i \succeq 0$), and $\hat{\hat{\mathbf{H}}}_i = \hat{\mathbf{H}}_i\bar{\mathbf{U}}_i$, we have $g(\boldsymbol{\lambda},\mu) = \min_{\bar{\mathbf{S}}_i\succeq 0} -\sum_{i\in\mathcal{U}_I}\omega_i\log\det\left(\mathbf{I}+\hat{\hat{\mathbf{H}}}_i\bar{\mathbf{S}}_i\hat{\hat{\mathbf{H}}}_i^H\right)+\sum_{i\in\mathcal{U}_I}\mathrm{Tr}\left(\bar{\Gamma}_i\bar{\mathbf{S}}_i\right)+G$. Let us take the particular structure for the covariance matrix $\bar{\mathbf{S}}_i$ as being diagonal, with all the elements equal to 0, except the last one, which is equal to P, i.e., $\bar{\mathbf{S}}_i = \mathrm{diag}(0,\ldots,P)$. Then, denoting $L_i = n_T - n_R + n_{R_i}$, the first term of the dual function becomes $-\sum_{i\in\mathcal{U}_I}\omega_i\log\left(1+P\|[\hat{\hat{\mathbf{H}}}_i]_{:,L_i}\|^2\right)$, where $[\hat{\hat{\mathbf{H}}}_i]_{:,L_i}$ denotes the L_ith column of $\hat{\hat{\mathbf{H}}}_i$. Since matrix $\hat{\hat{\mathbf{H}}}_i$ is formed by unitary rotations of a random matrix with i.i.d. entries, we can assure with probability equal to 1 that $\|[\hat{\hat{\mathbf{H}}}_i]_{:,L_i}\|^2 \neq 0$. As a conclusion, the first term of the Lagrangian is negative and decreases without bound as P increases. Let us have a look at the second term. If matrix $\mathbf{A}_i$ is not positive definite, i.e., if the lowest element (and, thus, the last component) in the diagonal of $\bar{\Gamma}_i$ is not positive, then the second term of the Lagrangian either is negative and decreases without bound as $P \to \infty$ or is zero. In both cases, and taking into account the behavior of the first term of the Lagrangian as P tends to infinity, it is concluded that the dual function is equal to $-\infty$. Thus, the only possible solution so that $g(\boldsymbol{\lambda}, \mu) \neq -\infty$ is that $\bar{\Gamma}_i$ has diagonal elements that are all strictly positive and, thus, $\mathbf{A}_i \succ 0$.

Appendix 3

System with only information users

Let us consider first the broadcast scenario with only users to be served with information and no energy harvesting users, i.e., $\mathcal{U}_E = \emptyset$. In this case, problem (12) can be expressed as:

$$\begin{aligned} \underset{\{R_i, \mathbf{S}_i\}_{\forall i \in \mathcal{U}_I}}{\text{maximize}} \quad & \sum_{i \in \mathcal{U}_I} \omega_i R_i \\ \text{subject to} \quad & C2 \ldots C6 \text{ of problem (11).} \end{aligned} \tag{22}$$

Without going into too much detail, let us say that the optimal solution to the above problem was presented in [33] and is omitted due to space limitations.

System with only harvesting users

Let us now consider the case where there are only users who want to harvest energy, i.e., $\mathcal{U}_I = \emptyset$. In this case, since there is no objective function, the optimization problem becomes a feasibility problem [40] that can be expressed as:[11]

$$\begin{aligned} \text{find} \quad & \mathbf{S} \\ \text{subject to} \quad & C1, C2, C6 \text{ of problem (11).} \end{aligned} \tag{23}$$

Note that constraints $C3$, $C4$, and $C5$ from problem (12) have no effect since the set $\mathcal{U}_I$ is empty. Notice also that, without loss of optimality, we have changed the optimization variable from a set of precoding matrices $\{\mathbf{S}_i\}$ to a single precoder matrix $\mathbf{S}$. In the following, we will present a necessary condition for feasibility of (23).

Proposition 2 ([50]) *Let* $\lambda_{\max}(\mathbf{X}) \geq \lambda_2(\mathbf{X}) \geq \cdots \geq \lambda_{\min}(\mathbf{X})$ *be the eigenvalues of the positive semidefinite matrix* $\mathbf{X}$*. Then, for any two semidefinite positive matrices,* $\mathbf{A}$ *and* $\mathbf{B}$*, we have:*

$$\lambda_j(\mathbf{AB}) \leq \lambda_{\max}(\mathbf{B})\lambda_j(\mathbf{A}) \; and \lambda_j(\mathbf{BA}) \leq \lambda_{\max}(\mathbf{B})\lambda_j(\mathbf{A}), \quad \forall j, \tag{24}$$

$$\lambda_j(\mathbf{AB}) \geq \lambda_{\min}(\mathbf{B})\lambda_j(\mathbf{A}) \; and \lambda_j(\mathbf{BA}) \geq \lambda_{\min}(\mathbf{B})\lambda_j(\mathbf{A}), \quad \forall j. \tag{25}$$

Note that the previous lemma can be generalized as $\mathrm{Tr}(\mathbf{AB}) \leq \lambda_{\max}(\mathbf{B})\mathrm{Tr}(\mathbf{A})$ since $\mathrm{Tr}(\mathbf{A}) = \sum_j \lambda_j(\mathbf{A})$. The inequality is attained when $\mathbf{A}$ has rank 1 and is built with the eigenvector associated with the maximum eigenvalue of $\mathbf{B}$, $(\mathbf{e}_{\max}(\mathbf{B}))$, i.e., $\mathbf{A} = k\, \mathbf{e}_{\max}(\mathbf{B})\mathbf{e}_{\max}(\mathbf{B})^H$.

Proposition 3 *Let* $\mathbf{H}_j^H\mathbf{H}_j = \mathbf{V}_{H,j}\Sigma_{H,j}\mathbf{V}_{H,j}^H$ *be the reduced eigenvalue decomposition of matrix* $\mathbf{H}_j^H\mathbf{H}_j$ *with* $\Sigma_{H,j} = diag(\sigma_{1,j}, \ldots, \sigma_{n_{R_j},j})$ *and* $\sigma_{1,j} \geq \sigma_{2,j} \geq \cdots \geq \sigma_{n_{R_j},j} > 0$. *Then, a necessary condition for the feasibility of problem (23) is* $(P_{\max} - P_c^{tx})\sigma_{1,j} - Q_j \geq 0, \forall j$.

Proof Just as we considered before, if the problem is feasible, at least one solution fulfills $\mathrm{Tr}(\mathbf{S}) = P_{\max} - P_c^{tx}$, and the maximum value that $\mathrm{Tr}\left(\mathbf{H}_j\mathbf{S}\mathbf{H}_j^H\right)$ can take, based on Proposition 2, is $(P_{\max} - P_c^{tx})\sigma_{1,j}$ ($\mathrm{Tr}(\mathbf{AB}) \leq \lambda_{\max}(\mathbf{B})\mathrm{Tr}(\mathbf{A})$) with $\mathbf{S} = \left(P_{\max} - P_c^{tx}\right)\mathbf{v}_{n_{R_j},j}\mathbf{v}_{n_{R_j},j}^H$, where $\mathbf{v}_{n_{R_j},j}$ is the eigenvector associated with the maximum eigenvalue $\sigma_{1,j}$ of $\mathbf{H}_j^H\mathbf{H}_j$. □

Generally, as we are not able to provide a necessary and sufficient condition, we need to solve the following convex optimization problem to test the feasibility of problem (23):

$$\begin{aligned} \underset{\mathbf{S}, \bar{P}_{\max}}{\text{minimize}} \quad & \bar{P}_{\max} \\ \text{subject to} \quad & C2: \mathrm{Tr}(\mathbf{S}) + P_c^{tx} \leq \bar{P}_{\max} \\ & C1, C6 \text{ of problem (11).} \end{aligned} \tag{26}$$

The above problem is categorized as a semidefinite optimization problem. There is no closed-form solution for the above problem, but the optimum solution can be obtained efficiently with the application of interior point methods [40]. Let us denote the optimum solution of the problem above as $\bar{P}_{\max}^\star$. Now, it only remains to check whether $\bar{P}_{\max}^\star \leq P_{\max}$ (which means that the problem is feasible) or $\bar{P}_{\max}^\star > P_{\max}$ (which implies infeasibility). If the problem is feasible, the optimum covariance matrix obtained in (26) is the matrix that fulfills all the harvesting power constraints with the minimum transmitted power. If the problem is infeasible, one possible solution would be to reduce all the power harvesting constraints $\{Q_j\}$ such that constraints $C1$ become looser until the problem becomes feasible.

Abbreviations

AWGN: Additive white Gaussian noise; BD: Block diagonalization; BS: Base station; CDF: Cumulative density function; CSI: Channel state information; DP: Dynamic programming; HPA: High-power amplifier; MIMO: Multiple-input multiple-output; MISO: Multiple-input single-output; MUI: Multiuser interference; PF: Proportional fair; R-P: Rate-power; RF: Radio frequency; SINR: Signal-to-interference-plus-noise ratio; SISO: Single-input single-output; SNR: Signal-to-noise ratio; SR: Sum rate; SVD: Singular value decomposition; SWIPT: Simultaneous wireless information and power transfer; w.l.o.g.: Without loss of generality

Funding
The research leading to these results has received funding through the grant 2017 SGR 578 (funded by the Catalan Government—Secretaria d'Universitats i Recerca, Departament d'Empresa i Coneixement, Generalitat de Catalunya, AGAUR) and project TEC2016-77148-C2-1-R (AEI/FEDER, UE): 5G&B-RUNNER-UPC (funded by the Agencia Estatal de Investigación, AEI, and Fondo Europeo de Desarrollo Regional, FEDER).

Methods/Experimental
The aim of the work presented in this paper is to develop a dynamic grouping mechanism that decides which users should be scheduled to receive information and which users should be configured to harvest energy. The design also includes the derivation of the optimal transmission covariance matrices.
The design is based on a theoretical modeling of the scenario and the signal, the definition of a mathematical optimization problem, and the proposal of an algorithm to find a suboptimal solution to that problem that can be implemented.
Finally, numerical computer simulations have been carried out to evaluate the performance of the proposed strategy based on the mathematical modeling of the setup.

Authors' contributions
JR and API put forward the idea and wrote the manuscript. JR carried out the experiments and simulations. JR and API contributed to the interpretation of the results and read and approved the final manuscript.

Competing interests
The authors declare that they have no competing interests.

References
1. J. Rubio, A. Pascual-Iserte, in *IEEE Global Communications Conference (GLOBECOM)*. Simultaneous wireless information and power transfer in multiuser MIMO systems, (2013), pp. 2755–2760. https://doi.org/10.1109/GLOCOM.2013.6831491
2. J. A. Paradiso, T. Starner, Energy scavenging for mobile wireless electronics. IEEE Comput. Pervasive. **4**(1), 18–27 (2005)
3. S. Sudevalayam, P. Kulkarni, Energy harvesting sensor nodes: survey and implications. IEEE Commun. Surv. Tutor. **13**(3), 443–461 (2011)
4. Z. Zhang, H. Pang, A. Georgiadis, C. Cecati, Wireless power transfer-an overview. IEEE Trans. Ind. Electron. **66**(2), 1044–1058 (2019)
5. V. Chandrasekhar, J. Andrews, A. Gatherer, Femtocell networks: a survey. IEEE Comm. Mag. **46**, 59–67 (2008)
6. X. Lu, et al., Wireless networks with RF energy harvesting: a contemporary survey. IEEE Commun. Surv. Tutor. **17**(2), 757–789 (2015)
7. L. R. Varshney, in *International Symposium on Information Theory*. Transporting information and energy simultaneously (IEEE, Toronto, 2008)
8. P. Grover, A. Sahai, in *International Symposium on Information Theory*. Shannon meets tesla: wireless information and power transfer (IEEE, Austin, 2010)
9. R. Zhang, C. K. Ho, MIMO broadcasting for simultaneous wireless information and power transfer. IEEE Trans. Wirel. Commun. **12**(5), 1989–2001 (2013)
10. J. Park, B. Clerckx, Joint wireless information and energy transfer in a two-user MIMO interference channel. IEEE Trans. Wirel. Comm. **12**(8), 4210–4221 (2013)
11. J. Park, B. Clerckx, Joint wireless information and energy transfer in a K-user MIMO interference channel. IEEE Trans. Wirel. Comm. **13**(10), 5781–5796 (2014)
12. Z. Zong, et al., Optimal transceiver design for SWIPT in K-user MIMO interference channels. IEEE Trans. Wirel. Comm. **15**(1), 430–445 (2016)
13. J. Xu, L. Liu, R. Zhang, Multiuser MISO beamforming for simultaneous wireless information and power transfer. IEEE Trans. Signal Proc. **62**(18), 4798–4810 (2015)
14. Q. Shi, L. Liu, W. Xu, R. Zhang, Joint transmit beamforming and receive power splitting for MISO SWIPT systems. IEEE Trans. Wirel. Comm. **13**(6), 3269–3280 (2014)
15. Y. Zeng, R. Zhang, Optimized training design for wireless energy transfer. IEEE Trans. Commun. **63**(2), 536–550 (2015)
16. J. Xu, R. Zhang, A general design framework for MIMO wireless energy transfer with limited feedback. IEEE Trans. Signal Proc. **64**(10), 2475–2488 (2016)
17. Z. Xiang, M. Tao, Robust beamforming for wireless information and power transmission. IEEE Wirel. Commun. Lett. **1**(4), 372–375 (2012)
18. C. Shepard, et al., in *ACM International Conference on Mobile Computing and Networking*. Argos: practical many-antenna base stations (ACM, Istanbul, 2012)
19. L. Liu, R. Zhang, K.-C. Chua, Wireless information transfer with opportunistic energy harvesting. IEEE Trans. Wirel. Commun. **12**(1), 288–300 (2013)
20. R. Morsi, D. S. Michalopoulos, R. Schober, Multiuser scheduling schemes for simultaneous wireless information and power transfer over fading channels. IEEE Trans. Wirel. Comm. **14**(4), 1967–1982 (2015)
21. Z. Ding, H. V. Poor, in *IEEE International Conference on Communication Systems (ICCS)*. User scheduling in wireless information and power transfer networks (IEEE, Macau, 2014)
22. W.-L. Shen, K. C.-J. Lin, M.-S. Chen, K. Tan, in *IEEE Conference on Computer Communications (INFOCOM)*. SIEVE: scalable user grouping for large MU-MIMO systems (IEEE, Hong Kong, 2015), pp. 1975–1983
23. L. Dai, B. Wang, M. Peng, S. Chen, Hybrid precoding-based millimeter-wave massive MIMO-NOMA with simultaneous wireless information and power transfer. IEEE J. Sel. Areas Comm. **37**(1), 131–141 (2019)
24. A. Goldsmith, et al., Capacity limits of MIMO channels. IEEE J. Sel. Areas Comm. **21**(5), 684–702 (2003)
25. Q. H. Spencer, et al., Zero-forcing methods for downlink spatial multiplexing in multiuser MIMO channels. IEEE Trans. Signal Proc. **52**(2), 461–471 (2004)
26. R. Zhang, Y.-C. Liang, S. Cui, Dynamic resource allocation in cognitive radio networks. IEEE Signal Process. Mag. **27**(3), 102–114 (2010)
27. J. Rubio, A. P. Iserte, D. P. Palomar, A. Goldsmith, Joint optimization of power and data transfer in multiuser MIMO systems. IEEE Trans. Signal Process. **65**(1), 212–227 (2017)
28. B. Devillers, D. Gündüz, A general framework for the optimization of energy harvesting communication systems with battery imperfections. J. Commun. Netw. **14**(2), 130–139 (2012)
29. A. R. Jensen, et al., in *IEEE Vehicular Technology Conference (VTC 2012 Fall)*. LTE UE power consumption model: for system level energy and performance optimization (IEEE, Quebec, 2012)
30. G. Auer, et al., How much energy is needed to run a wireless network? IEEE Trans. Wirel. Commun. **18**(5), 40–49 (2011)
31. P. Grover, K. Woyach, A. Sahai, Towards a communication-theoretic understanding of system-level power consumption. IEEE J. Sel. Areas Comm. **29**(8), 1744–1755 (2011)
32. P. Rost, G. Fettweis, in *IEEE Global Communications Conference (GLOBECOM), Workshop on Green Communications*. On the transmission-computation-energy tradeoff in wireless and fixed networks (IEEE, Miami, 2010), pp. 1934–1939
33. J. Rubio, A. Pascual-Iserte, Energy-aware broadcast multiuser-MIMO precoder design with imperfect channel and battery knowledge. IEEE Trans. Wirel. Commun. **13**(6), 3137–3152 (2014)
34. J. Lee, N. Jindal, in *Asilomar Conference on Signals, Systems and Computers*. Dirty paper coding vs linear precoding for MIMO broadcast channels (IEEE, Pacific Grove, 2006), pp. 779–783
35. C. Ho, R. Zhang, Optimal energy allocation for wireless communications with energy harvesting constraints. IEEE Trans. Signal Process. **60**(9), 4808–4818 (2012)
36. J. Yang, O. Ozel, S. Ulukus, Broadcasting with an energy harvesting rechargeable transmitter. IEEE Trans. Wirel. Commun. **11**(2), 571–583 (2012)
37. D. P. Bertsekas, *Dynamic Programming and Optimal Control. 3rd edn.* (Athena Scientific, Belmont, 2005)
38. P. Blasco, D. Gündüz, M. Dohler, A learning theoretic approach to energy harvesting communication system optimization. IEEE Trans. Wirel. Commun. **12**(4), 1872–1882 (2013)
39. L. M. Lopez-Ramos, A. G. Marques, J. Ramos, Jointly optimal sensing and resource allocation for multiuser interweave cognitive radios. IEEE Trans. Wirel. Commun. **13**(11), 5854–5967 (2014)

40. S. Boyd, L. Vandenbergue, *Convex optimization*. (Cambridge, Cambridge, 2004)
41. R. G. Bland, D. Goldfarb, M. J. Todd, The ellipsoid method: a survey. Oper. Res. **29**(6), 1039–1091 (1981)
42. D. P. Bertsekas, *Nonlinear programming. 2nd edn*. (Athena Scientific, Belmont, 1999)
43. J. Rubio, A. Pascual-Iserte, in *IEEE Vehicular Technology Conference (VTC) Spring*. Harvesting management in Multiuser MIMO systems with simultaneous wireless information and power transfer (IEEE, Gasglow, 2015)
44. G. Dimic, N. Sidiropoulos, On downlink beamforming with greedy user selection: performance analysis and a simple new algorithm. IEEE Trans. Signal Process. **53**(10), 3857–3868 (2005)
45. S. Sigdel, W. A. Krzymien, Simplified fair scheduling and antenna selection algorithms for multiuser MIMO orthogonal space-division mutiplexing downlink. IEEE Trans. Veh. Technol. **58**(3), 1329–1344 (2009)
46. A. Jalali, R. Padovani, R. Pankaj, in *IEEE Vehicular Technology Conference (VTC) Spring*. Data throughput of CDMA-HDR a high efficiency-high data rate personal communication wireless system (IEEE, Tokyo, 2000)
47. X. Wang, G. B. Giannakis, A. G. Marques, A unified approach to QoS-guaranteed scheduling for channel-adaptive wireless networks. Proc. IEEE. **95**(12), 2410–2431 (2007)
48. L. Liu, Y.-H. Nam, J. Zhang, in *44th Annual Conference on Information Sciences and Systems (CISS)*. Proportional fair scheduling for multi-cell multi-user MIMO systems (IEEE, Princeton, 2010)
49. T. M. Cover, J. A. Thomas, *Elements of information theory*. (Wiley, Hoboken, 2006)
50. R. Bhatia, *Perturbation bounds for matrix eigenvalues*. (Longman Scientific & Technical, Essex, 1987)
51. K. Huang, C. Zhong, G. Zhu, Some new trends in wirelessly powered communications. IEEE Wirel. Commun. **23**(2), 19–27 (2016)

3

Diversity gain of millimeter-wave massive MIMO systems with distributed antenna arrays

Dian-Wu Yue[1,2*], Shuai Xu[1] and Ha H. Nguyen[2]

Abstract

This paper is concerned with diversity gain analysis for millimeter-wave (mmWave) massive MIMO systems employing distributed antenna subarray architecture. First, for a single-user mmWave system in which the transmitter and receiver consist of K_t and K_r subarrays, respectively, a diversity gain theorem is established when the numbers of subarray antennas go to infinity. Specifically, assuming that all subchannels have the same number of propagation paths L, the theorem states that by employing such a distributed antenna subarray architecture, a diversity gain of $K_rK_tL - N_s + 1$ can be achieved, where N_s represents the number of data streams. This result means that compared to the co-located antenna architecture, using the distributed antenna subarray architecture can scale up the diversity gain proportionally to K_rK_t. The analysis of diversity gain is then extended to the multiuser scenario as well as the scenario with conventional partially connected radio-frequency structure in the literature. Simulation results obtained with the hybrid analog/digital processing corroborate the analysis results and show that the distributed subarray architecture indeed yields a significantly better diversity performance than the co-located antenna architectures.

Keywords: Millimeter-wave communications, Massive MIMO, Diversity gain, Multiplexing gain, Diversity-multiplexing trade-off, Distributed antenna subarrays, Hybrid precoding

1 Introduction

Millimeter-wave (mmWave) communication has recently gained considerable attention as a candidate technology for 5G mobile communication systems and beyond [1–5]. The main reason for this is the availability of vast spectrum in the mmWave band that is very attractive for high data rate communications. However, compared to communication systems operating at lower microwave frequencies (such as those currently used for 4G mobile communications), propagation loss in mmWave frequencies is much higher, in orders of magnitude. Fortunately, given the much smaller carrier wavelengths, mmWave communication systems can make use of compact massive antenna arrays to compensate for the increased propagation loss.

Nevertheless, the large-scale antenna arrays together with high cost and large power consumption of the mixed analog/digital signal components make it difficult to equip a separate radio-frequency (RF) chain for each antenna element and perform all the signal processing in the baseband. Therefore, research on *hybrid* analog-digital processing of precoder and combiner for mmWave communication systems has attracted considerably strong interests from both academia and industry [6–18]. In particular, a large body of work has been performed to address challenges in employing a limited number of RF chains for massive antenna arrays. For example, the authors in [6] investigated single-user precoding in massive MIMO mmWave systems and obtained the optimality of beam steering for both single-stream and multi-stream transmission scenarios. In [9], the authors showed that the hybrid processing can realize any fully digital processing exactly when the number of RF chains is twice the number of data streams.

Two architectures for connecting the RF chains in the hybrid processing that have been investigated in the literature are *fully connected* and *partially connected*. In the former, each RF chain is connected to all the antenna elements, while only a subset of antenna elements is connected to each RF chain in the latter. The partially connected architecture is more energy-efficient and implementation-friendly since the number of required

*Correspondence: dwyue@dlmu.edu.cn
[1] School of Information Science and Technology, Dalian Maritime University, Dalian 116026, Liaoning, China
[2] Department of Electrical and Computer Engineering, University of Saskatchewan, 57 Campus Drive, Saskatoon SK S7N 5A9, Canada

phase shifters can be reduced efficiently without a significant performance loss. In the conventional partially connected architecture [10–14], the antenna array is partitioned into a number of smaller disjoint subarrays, each of which is driven by a single transmission chain. More recently, a more general partially connected architecture, referred to as hybridly connected in [15] and overlapped subarray-based in [16], has been proposed. In such a hybridly connected structure, each subarray is connected to multiple RF chains while each RF chain is connected to all the antennas, which corresponds to the subarray in question. In particular, the authors in [15] show that the spectral efficiency of the hybridly connected structure is higher than that of the partially connected structure and that as the number of RF chains increases, its spectral efficiency can close to that of the fully connected structure.

Nevertheless, due to the facts that the antenna arrays in the abovementioned RF architectures are co-located and mmWave signal propagation has an important feature of multipath sparsity in both the spatial and temporal domains [19, 20], it is expected that the potentially available diversity and multiplexing gains are not large for the co-located antenna deployment. In order to enlarge the diversity gain and/or multiplexing gain in mmWave massive MIMO communication systems, this paper considers a more general antenna array architecture, called *distributed antenna subarray architecture*, which includes co-located array architecture as special cases. It is pointed out that deploying distributed antennas has shown a promising technique to increase spectral efficiency and expand the coverage of wireless communication networks [21–25]. As such, it is of great interest to consider distributed antenna deployment in the context of mmWave massive MIMO systems.

The diversity-multiplexing trade-off (DMT) is a compact and convenient framework to compare different MIMO systems in terms of the two main and related system indicators: data rate and error performance [26–29]. This trade-off was originally characterized in [26] for MIMO communication systems operating over independent and identically distributed (i.i.d.) Rayleigh fading channels. Then, the framework has ignited a lot of interests in analyzing various communication systems and under different channel models. For a massive MIMO mmWave system, how to quantify the diversity performance and characterize its DMT is a fundamental and open research problem. In particular, to the best of our knowledge, until now, there is no unified diversity gain analysis for massive MIMO mmWave systems that is applicable to both co-located and distributed antenna array architectures.

To fill this gap, this paper investigates the diversity performance of massive MIMO mmWave systems with the proposed distributed subarray architecture. The focus is on the asymptotical diversity gain analysis in order to find out the potential diversity advantage provided by multiple distributed antenna arrays. The obtained analysis can be used conveniently to compare various massive MIMO mmWave systems with different distributed antenna array structures. Our main contributions are summarized as follows: First, for a single-user system employing the proposed distributed subarray architecture, a diversity gain expression is obtained when the number of antennas at each subarray increases without bound. This expression clearly indicates that one can obtain a large diversity gain and/or multiplexing gain by employing the proposed distributed subarray architecture. Second, the diversity gain analysis is extended to the multiuser scenario with downlink and uplink transmission, as well as the single-user system employing the conventional partially connected RF structure based on the distributed subarrays. Simulation results are provided to corroborate the analysis results and show that the distributed subarray architecture yields significantly better diversity performance than the co-located single-array architecture.

The remainder of this paper is organized as follows: Section 2 describes the massive MIMO mmWave system model and hybrid processing with the distributed subarray architecture in mmWave fading channels. Section 3 provides the asymptotical diversity analysis for the single-user mmWave system. In Sections 4 and 5, the diversity gain analysis is extended to the multiuser scenario and the scenario with the conventional partially connected RF architecture, respectively. Numerical results are presented in Section 6. Section 7 finally concludes the paper.

Throughout this paper, the following notations are used. The superscripts $(\cdot)^T$ and $(\cdot)^H$ stand for transpose and conjugate transpose, respectively. Boldface upper and lower case letters denote matrices and column vectors, respectively. $\text{diag}\{a_1, a_2, \ldots, a_N\}$ stands for a diagonal matrix with diagonal elements $\{a_1, a_2, \ldots, a_N\}$. The expectation operator is denoted by $\mathbb{E}()$. $[\mathbf{A}]_{ij}$ gives the (i,j)th entry of matrix $\mathbf{A}$. $\mathbf{A} \bigotimes \mathbf{B}$ is the Kronecker product of $\mathbf{A}$ and $\mathbf{B}$. A function $a(x)$ of x is written as $o(x)$ if $\lim_{x\to 0} a(x)/x = 0$. Finally, $\mathcal{CN}(0, 1)$ represents a circularly symmetric complex Gaussian random variable with zero mean and unit variance.

2 System model

A single-user massive MIMO mmWave system is considered as shown in Fig. 1. The transmitter is equipped with a distributed antenna array to send N_s data streams to a receiver, which is also equipped with a distributed antenna array. Here, a distributed antenna array means an array consisting of several remote antenna units (RAUs) (i.e., antenna subarrays) that are distributively located, as depicted in Fig. 2. Specifically, the antenna array at the

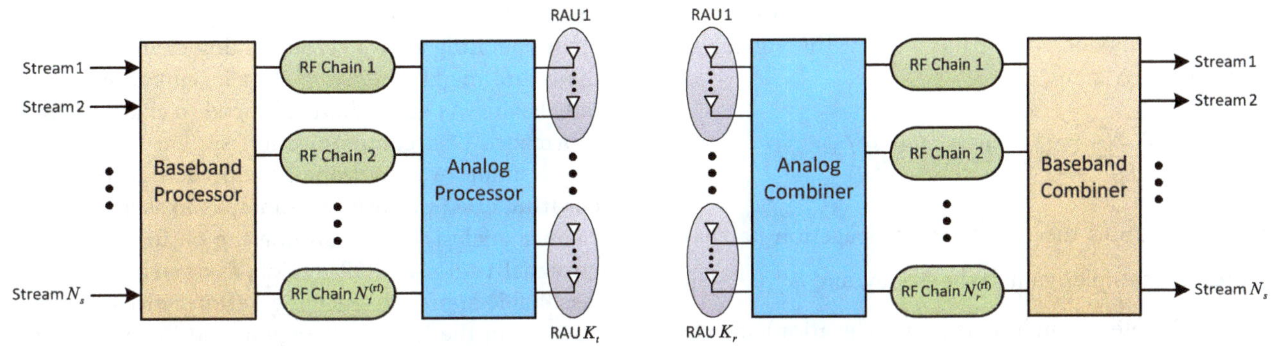

Fig. 1 Block diagram of a mmWave massive MIMO system with distributed antenna arrays

transmitter consists of K_t RAUs, each of which has N_t antennas and is connected to a baseband processing unit (BPU) via fiber. Likewise, the distributed antenna array at the receiver consists of K_r RAUs, each having N_r antennas and also being connected to a BPU by fibers. Such a MIMO system shall be referred to as a (K_t, N_t, K_r, N_r)-distributed MIMO (D-MIMO) system. When $K_t = K_r = 1$, the system reduces to a conventional co-located MIMO (C-MIMO) system.

The transmitter accepts as its input N_s data streams and is equipped with $N_t^{(\text{rf})}$ RF chains, where $N_s \leq N_t^{(\text{rf})} \leq N_t K_t$. Given $N_t^{(\text{rf})}$ transmit RF chains, the transmitter can apply a low-dimension $N_t^{(\text{rf})} \times N_s$ baseband precoder, $\mathbf{W}_t$, followed by a high-dimension $K_t N_t \times N_t^{(\text{rf})}$ RF precoder, $\mathbf{F}_t$. Note that for the baseband precoder $\mathbf{W}_t$, amplitude and phase modifications are feasible while only phase changes can be made by the RF precoder $\mathbf{F}_t$ through the use of variable phase shifters and combiners. The transmitted signal vector can be written as:

$$\mathbf{x} = \mathbf{F}_t \mathbf{W}_t \mathbf{s}, \tag{1}$$

where $\mathbf{s}$ is the $N_s \times 1$ symbol vector such that $\mathbb{E}\left[\mathbf{s}\mathbf{s}^H\right] = \frac{P}{N_s}\mathbf{I}_{N_s}$. Thus, P represents the average total input power. Considering a narrowband block fading channel, the $K_r N_r \times 1$ received signal vector is:

$$\mathbf{y} = \mathbf{H}\mathbf{F}_t \mathbf{W}_t \mathbf{s} + \mathbf{n} \tag{2}$$

where $\mathbf{H}$ is $K_r N_r \times K_t N_t$ channel matrix, and $\mathbf{n}$ is a $K_r N_r \times 1$ vector consisting of i.i.d. $\mathcal{CN}(0, 1)$ noise samples. Throughout this paper, $\mathbf{H}$ is assumed known to both the transmitter and receiver. Given that $N_r^{(\text{rf})}$ RF chains $\left(\text{where} N_s \leq N_r^{(\text{rf})} \leq N_r K_r\right)$ are used at the receiver to detect the N_s data streams, the processed signal vector can be given by:

$$\mathbf{z} = \mathbf{W}_r^H \mathbf{F}_r^H \mathbf{H}\mathbf{F}_t \mathbf{W}_t \mathbf{s} + \mathbf{W}_r^H \mathbf{F}_r^H \mathbf{n} \tag{3}$$

where $\mathbf{W}_r$ is the $N_r^{(\text{rf})} \times N_s$ baseband combining matrix, and $\mathbf{F}_r$ is the $K_r N_r \times N_r^{(\text{rf})}$ RF combining matrix.

Furthermore, according to the architecture of RAUs at the transmitting and receiving ends, $\mathbf{H}$ can be written as:

$$\mathbf{H} = \begin{bmatrix} \sqrt{g_{11}}\mathbf{H}_{11} & \cdots & \sqrt{g_{1K_t}}\mathbf{H}_{1K_t} \\ \vdots & \ddots & \vdots \\ \sqrt{g_{K_r 1}}\mathbf{H}_{K_r 1} & \cdots & \sqrt{g_{K_r K_t}}\mathbf{H}_{K_r K_t} \end{bmatrix}. \tag{4}$$

In the above expression, g_{ij} represents the large scale fading parameter between the ith RAU at the receiver and the jth RAU at the transmitter. We assume that g_{ij} is constant over many coherence time intervals. The normalized subchannel matrix $\mathbf{H}_{ij}$ represents the MIMO channel between the jth RAU at the transmitter and the ith RAU at the receiver.

Based on the extended Saleh-Valenzuela model, a clustered channel model is used often in mmWave channel modeling and standardization [6, 14, 15], and it is also adopted in this paper. As in [6], each scattering cluster is

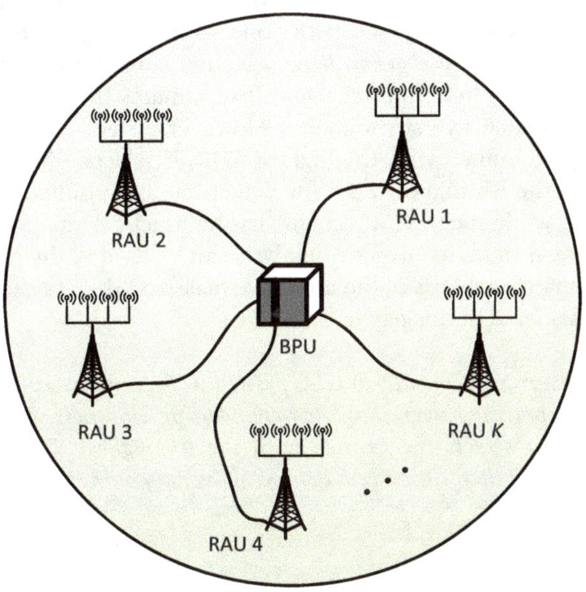

Fig. 2 Illustration of distributed antenna array deployment

assumed to contribute a single propagation path for simplicity of exposition. Using this model, the subchannel matrix $\mathbf{H}_{ij}$ is given by:

$$\mathbf{H}_{ij} = \sqrt{\frac{N_t N_r}{L_{ij}}} \sum_{l=1}^{L_{ij}} \alpha_{ij}^l \mathbf{a}_r\left(\phi_{ij}^{rl}, \theta_{ij}^{rl}\right) \mathbf{a}_t^H\left(\phi_{ij}^{tl}, \theta_{ij}^{tl}\right), \tag{5}$$

where L_{ij} denotes the number of propagation paths, α_{ij}^l denotes the complex gain of the lth ray, and $\phi_{ij}^{rl}\left(\theta_{ij}^{rl}\right)$ and $\phi_{ij}^{tl}\left(\theta_{ij}^{tl}\right)$ denote its random azimuth (elevation) angles of arrival and departure, respectively. Without loss of generality, the complex gains α_{ij}^l are assumed to be $\mathcal{CN}(0,1)$[1]. The vectors $\mathbf{a}_t\left(\phi_{ij}^{tl}, \theta_{ij}^{tl}\right)$ and $\mathbf{a}_r\left(\phi_{ij}^{rl}, \theta_{ij}^{rl}\right)$ stand for the normalized transmit/receive array response vectors at the corresponding angles of departure/arrival. The array response vector of an N-element uniform linear array (ULA) is:

$$\mathbf{a}^{\mathrm{ULA}}(\phi) = \frac{1}{\sqrt{N}}\left[1, e^{j2\pi\frac{d}{\lambda}\sin(\phi)}, \ldots, e^{j2\pi(N-1)\frac{d}{\lambda}\sin(\phi)}\right]^T \tag{6}$$

where λ is the wavelength of the carrier, and d is the inter-element spacing. It is pointed out that the angle θ is not included in the argument of $\mathbf{a}^{\mathrm{ULA}}$ since the response for an ULA is independent of the elevation angle. In contrast, for a uniform planar array (UPA), which is composed of N_v and N_h antenna elements in the vertical and horizontal directions, respectively, the array response vector is represented by:

$$\mathbf{a}^{\mathrm{UPA}}(\phi,\theta) = \mathbf{a}_h^{\mathrm{ULA}}(\phi) \otimes \mathbf{a}_v^{\mathrm{ULA}}(\theta), \tag{7}$$

where

$$\mathbf{a}_h^{\mathrm{ULA}}(\phi) = \frac{1}{\sqrt{N_h}}\left[1, e^{j2\pi\frac{d_h}{\lambda}\sin(\phi)}, \ldots, e^{j2\pi(N_h-1)\frac{d_h}{\lambda}\sin(\phi)}\right]^T \tag{8}$$

and

$$\mathbf{a}_v^{\mathrm{ULA}}(\theta) = \frac{1}{\sqrt{N_v}}\left[1, e^{j2\pi\frac{d_v}{\lambda}\sin(\theta)}, \ldots, e^{j2\pi(N^v-1)\frac{d_v}{\lambda}\sin(\theta)}\right]^T. \tag{9}$$

3 Diversity gain analysis

The most common performance metric of a digital communication system is the error probability, which can be defined either as the probability of symbol error or the probability of bit error (i.e., the bit error rate (BER)). When communicating over a fading channel, errors obviously depend on specific channel realizations. As such, the random behavior of a fading channel needs to be taken into account, which leads to the concept of average error probabilities [30]. Determining the exact expressions for the average error probabilities for a digital communication system operating over a certain fading channel is usually tedious and might not give a clear insight about the system behavior. As such, there is a need to characterize the performance of a communication system in an insightful and simple way. A popular approach is to shift the focus from exact performance analysis to asymptotic performance analysis, i.e., analyzing the performance at the high signal-to-noise (SNR) region, as done in [31]. This is a reasonable approach since the performance of practical interest is in the high SNR region, and in such a region, good approximation can be made on the exact analysis.

For most cases, the average BER function can be approximated in the high SNR region as [31]:

$$\overline{\mathrm{BER}} \approx (G_c \cdot \bar{\gamma})^{-G_d} \tag{10}$$

where $\bar{\gamma}$ stands for the average received SNR, and G_d and G_c are just the diversity and coding gains, respectively. At high SNR, the diversity gain determines the slope of the BER curve versus $\bar{\gamma}$ in a log-log scale, whereas the coding gain determines how the curve is shifted along the horizontal axis with respect to a benchmark BER curve $\bar{\gamma}^{-G_d}$. Therefore, this yields a simple parameterized average BER characterization for high SNR, which can provide meaningful insights on the system performance behavior.

In this section, the diversity gain is first examined for a *generalized selection combining*. The main result is then invoked in the diversity analysis of the distributed mmWave massive MIMO system studied in this paper.

3.1 Diversity gain of generalized selection combining

Selection combining (SC) is the most popular low-complexity combining scheme. In selection combining, the receiver estimates the SNRs of all available diversity branches and then select the one with the highest SNR for detection. For *generalized* selection combining (GSC) considered here, the receiver also estimates the SNRs of all available diversity branches. However, instead of selecting the branch with the highest SNR, it selects a branch with the lth highest SNR for detection. It is pointed out that, while such a GSC scheme has no practical interest in its own right, its diversity analysis can be used in the performance analysis of the mmWave massive MIMO system considered in this paper.

Lemma 1 *Consider a GSC system with L receive antennas operating over i.i.d. Rayleigh fading channels. If the receiver selects the branch with the lth highest SNR for detection then the system achieves diversity gain:*

$$G_d = L - l + 1. \tag{11}$$

Proof Let $F(\gamma)$ and $f(\gamma)$ be the cumulative distribution function (CDF) and probability density function (PDF) of the instantaneous SNRs in all branches, respectively. Let $\bar{\gamma}$ denote the average receive SNR of each branch. With Rayleigh fading, it follows from [31] that $F(\gamma)$ and $f(\gamma)$ can be expressed as:

$$F(\gamma) = 1 - e^{-\frac{\gamma}{\bar{\gamma}}} = \frac{\gamma}{\bar{\gamma}} + o\left(\frac{\gamma}{\bar{\gamma}}\right) \quad (12)$$

and

$$f(\gamma) = \frac{1}{\bar{\gamma}} e^{-\frac{\gamma}{\bar{\gamma}}} = \frac{1}{\bar{\gamma}} - \frac{\gamma}{\bar{\gamma}^2} + o\left(\frac{\gamma}{\bar{\gamma}}\right). \quad (13)$$

If the receiver selects the branch with the lth highest SNR for detection, then based on the theory of order statistics [32], the PDF of the instantaneous receive SNR at the receiver, denoted γ_l, is given by:

$$\begin{aligned} f_{l:L}(\gamma_l) &= \frac{L!}{(L-l)!\,(l-1)!} [F(\gamma_l)]^{L-l} [1 - F(\gamma_l)]^{l-1} f(\gamma_l) \\ &= \frac{L!}{(L-l)!\,(l-1)!} \frac{1}{\bar{\gamma}} \left(\frac{\gamma_l}{\bar{\gamma}}\right)^{L-l} + o\left(\left(\frac{\gamma_l}{\bar{\gamma}}\right)^{L-l}\right). \end{aligned} \quad (14)$$

Applying the above PDF in Proposition 1 in [31] leads to the desired result. □

Lemma 1 can be extended to the case of independent but not identically distributed (i.n.i.d.) Rayleigh fading channels and the result is stated in the next lemma.

Lemma 2 *Suppose that the GSC system with L receive antennas operates over the i.n.i.d. Rayleigh fading channels. If it selects the path with the lth highest SNR for detection, then it can achieve diversity gain:*

$$G_d = L - l + 1. \quad (15)$$

Proof Let $\bar{\gamma}_{\max}$ and $\bar{\gamma}_{\min}$ denote the maximum and minimum values of the average receive SNRs of all these L diversity paths, respectively. Furthermore, let $\mathcal{A}$ and $\mathcal{B}$ denote two GSC systems, each equipped with L receive antennas and operating over i.i.d. Rayleigh fading channels such that the average receive SNRs equal to $\bar{\gamma}_{\max}$ and $\bar{\gamma}_{\min}$, respectively. It is known from Lemma 1 that the diversity gains of these two systems are the same and equal to $L - l + 1$ if both systems select the branch with the lth highest instantaneous SNR for detection. Furthermore, since the GSC system under consideration cannot have better diversity performance than system $\mathcal{A}$ and cannot have worse diversity performance than system $\mathcal{B}$, it can then be concluded that the i.n.i.d. system must also achieve the diversity gain of $L - l + 1$. □

3.2 Diversity gain analysis of the distributed mmWave massive MIMO system

From the structure and definition of the channel matrix $\mathbf{H}$ in Section 2, there is a total of $L_s = \sum_{i=1}^{K_r} \sum_{j=1}^{K_t} L_{ij}$ propagation paths. Naturally, $\mathbf{H}$ can be decomposed into a sum of L_s rank-one matrices, each corresponding to one propagation path. Specifically, $\mathbf{H}$ can be rewritten as:

$$\mathbf{H} = \sum_{i=1}^{K_r} \sum_{j=1}^{K_t} \sum_{l=1}^{L_{ij}} \tilde{\alpha}_{ij}^l \tilde{\mathbf{a}}_r\left(\phi_{ij}^{rl}, \theta_{ij}^{rl}\right) \tilde{\mathbf{a}}_t^H\left(\phi_{ij}^{tl}, \theta_{ij}^{tl}\right), \quad (16)$$

where

$$\tilde{\alpha}_{ij}^l = \sqrt{g_{ij} \frac{N_t N_r}{L_{ij}}} \alpha_{ij}^l, \quad (17)$$

$\tilde{\mathbf{a}}_r\left(\phi_{ij}^{rl}, \theta_{ij}^{rl}\right)$ is a $K_r N_r \times 1$ vector whose bth entry is defined as:

$$\left[\tilde{\mathbf{a}}_r\left(\phi_{ij}^{rl}, \theta_{ij}^{rl}\right)\right]_b = \begin{cases} \left[\mathbf{a}_r\left(\phi_{ij}^{rl}, \theta_{ij}^{rl}\right)\right]_{b-(i-1)N_r}, & b \in Q_i^r \\ 0, & b \notin Q_i^r \end{cases} \quad (18)$$

where $Q_i^r = ((i-1)N_r, iN_r]$. And $\tilde{\mathbf{a}}_t\left(\phi_{ij}^{tl}, \theta_{ij}^{tl}\right)$ is a $K_t N_t \times 1$ vector whose bth entry is defined as:

$$\left[\tilde{\mathbf{a}}_t\left(\phi_{ij}^{tl}, \theta_{ij}^{tl}\right)\right]_b = \begin{cases} \left[\mathbf{a}_t\left(\phi_{ij}^{tl}, \theta_{ij}^{tl}\right)\right]_{b-(j-1)N_t}, & b \in Q_j^t \\ 0, & b \notin Q_j^t. \end{cases} \quad (19)$$

where $Q_j^t = ((j-1)N_t, jN_t]$.

Lemma 3 *Suppose that the antenna configurations at all RAUs are either ULA or UPA. Then, all L_s vectors $\left\{\tilde{\mathbf{a}}_r\left(\phi_{ij}^{rl}, \theta_{ij}^{rl}\right)\right\}$ are orthogonal to each other when $N_r \to \infty$. Likewise, all L_s vectors $\left\{\tilde{\mathbf{a}}_t\left(\phi_{ij}^{tl}, \theta_{ij}^{tl}\right)\right\}$ are orthogonal to each other when $N_t \to \infty$.*

Proof It follows immediately from (18) and (19) that if $u \neq v$, then vectors $\left\{\tilde{\mathbf{a}}_r\left(\phi_{up}^{rl}, \theta_{up}^{rl}\right)\right\}$ and $\left\{\tilde{\mathbf{a}}_r\left(\phi_{vq}^{rl}, \theta_{vq}^{rl}\right)\right\}$ are orthogonal. On the other hand, when $u = v$ and $p \neq q$, it is known from Lemma 1 and Corollary 2 in [6] (also see [33]) that vectors $\left\{\tilde{\mathbf{a}}_r\left(\phi_{up}^{rl}, \theta_{up}^{rl}\right)\right\}$ and $\left\{\tilde{\mathbf{a}}_r\left(\phi_{vq}^{rl}, \theta_{vq}^{rl}\right)\right\}$ are orthogonal. The proof that $\left\{\tilde{\mathbf{a}}_t\left(\phi_{ij}^{tl}, \theta_{ij}^{tl}\right)\right\}$ is a set of orthogonal vectors can be shown similarly. □

Theorem 1 *Suppose that both sets $\left\{\tilde{\mathbf{a}}_r\left(\phi_{ij}^{rl}, \theta_{ij}^{rl}\right)\right\}$ and $\left\{\tilde{\mathbf{a}}_t\left(\phi_{ij}^{tl}, \theta_{ij}^{tl}\right)\right\}$ are orthogonal vector sets when $N_r \to \infty$ and $N_t \to \infty$. Let $N_s \leq L_s$. Then, the distributed massive MIMO system with large N_r and N_t can achieve a diversity gain of:*

$$G_d = L_s - N_s + 1. \tag{20}$$

Remarks 1 *When N_t and N_r are large enough, (33) indicates that the system multiplexing gain is at most equal to L_s. This is reasonable since there exist only L_s effective singular values in the channel matrix* **H**. *Theorem 1 provides a simple diversity-multiplexing trade-off of a mmWave massive MIMO system: adding one data stream to the system decreases the diversity gain by one, whereas removing one data stream increases the diversity gain by one. Such a trade-off is useful in designing a system to meet requirements on both data rate and error performance.*

Remarks 2 *Under the case where N_t and N_r are large enough, it can be found from the proof of Theorem* 1 *that the diversity performance of the mmWave massive MIMO system only depends on the singular value set $\{\tilde{\alpha}^l\}$ and is not influenced by how sub-matrices $\{\sqrt{g_{ij}}\mathbf{H}_{ij}\}$ are placed in the channel matrix* **H** *(see further discussion of Fig.* 9 *on this point).*

Corollary 1 *Consider the scenario that the antenna configuration at each RAU is ULA. Also assume that $L_{ij} = L$ for any i and j. Let $N_s \leq K_r K_t L$. When both N_t and N_r are very large, the distributed massive MIMO system can achieve a diversity gain:*

$$G_d = K_r K_t L - N_s + 1. \tag{21}$$

In particular, when $K_r = K_t = 1$, the massive MIMO system with co-located antennas arrays can achieve a diversity gain:

$$G_d = L - N_s + 1 \tag{22}$$

Remarks 3 *Corollary* 1 *implies that for a mmWave co-located massive MIMO system, its diversity gain and multiplexing gain are limited and at most equal to the number of paths L. However, these gains can be increased by employing the distributed antenna architecture and can be scaled up proportionally to $K_r K_t$.*

4 Diversity gain analysis with the conventional partially connected structure

The previous section has analyzed the diversity gain for the massive MIMO system with the general fully connected RF architecture. This section focuses on a massive MIMO system employing the conventional partially connected RF architecture as illustrated in Fig. 3. Here, the transmitter equipped with K_t RF chains sends N_s data streams to the receiver equipped with K_r RF chains. Each of RF chains at the transmitter or receiver is connected to only one RAU. It is assumed that $N_s \leq \min\{K_t, K_r\}$. The numbers of antennas per each RAU at the transmitter and receiver are fixed as N_t and N_r, respectively. Note that $N_t \gg N_s$ and $N_r \gg N_s$. Both the transmitter and receiver employ very small digital processors and very large analog processors, represented respectively by $\mathbf{W}_t$ and $\mathbf{F}_t$ for the transmitter and $\mathbf{W}_r$ and $\mathbf{F}_r$ for the receiver.

As before, denote by **s** the transmitted symbol vector, by **H** the fading channel matrix, and by **n** the noise vector. Then, at the receiver, the processed signal vector **z** is given by (3), whereas **H** is described as in (4). Due to the partially connected RF architecture, the analog processors $\mathbf{F}_t$ and $\mathbf{F}_r$ are block diagonal matrices, expressed as:

$$\mathbf{F}_t = \operatorname{diag}\{\mathbf{f}_{t1}, \mathbf{f}_{t2}, \ldots, \mathbf{f}_{tK_t}\} \tag{23}$$

and

$$\mathbf{F}_r = \operatorname{diag}\{\mathbf{f}_{r1}, \mathbf{f}_{r2}, \ldots, \mathbf{f}_{rK_r}\} \tag{24}$$

where $\mathbf{f}_{ti}$ stands for the $N_t \times 1$ steering vector of phases for the ith RAU at the transmitter, and $\mathbf{f}_{rj}$ stands for the $N_r \times 1$ steering vector of phases for the jth RAU at the transmitter.

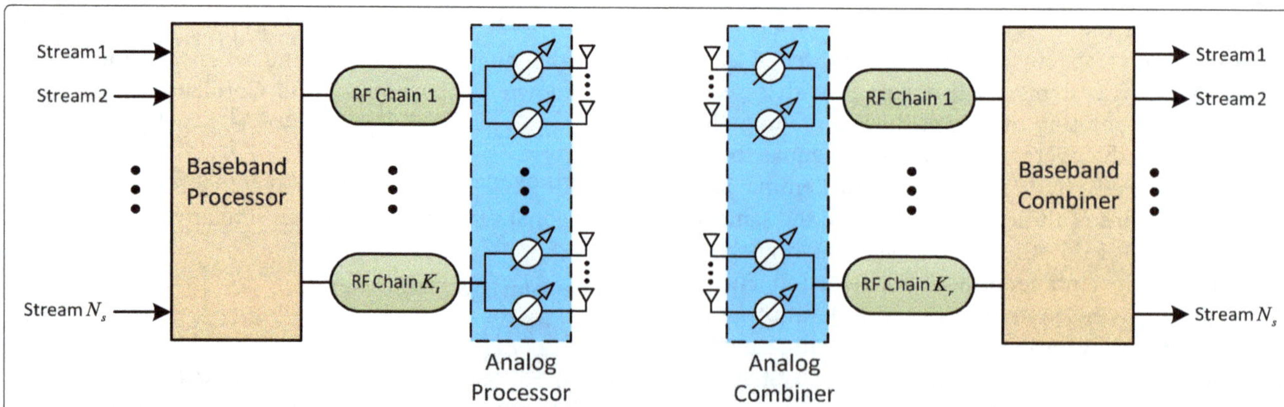

Fig. 3 Block diagram of a mmWave massive MIMO system with the conventional partially connected RF architecture

Theorem 2 *Consider the case that the antenna array configuration at each RAU is ULA and $L_{ij} = L$ for any i and j. In the limit of large N_t and N_r, the distributed massive MIMO system with partially connected RF architecture can achieve a diversity gain:*

$$G_d = (K_t - N_s + 1)(K_r - N_s + 1)L. \tag{25}$$

Remarks 4 *Comparing the diversity gain given in Corollary 1 with that given in Theorem 2 reveals that when $N_s = 1$, the diversity gains with the two systems under consideration are the same. However, when $N_s > 1$, the proposed distributed antenna system with fully connected RF architecture achieves a higher diversity gain than the system with the partially connected architecture, and the gap between the two diversity gains is $(N_s - 1)[(K_r + K_t - N_s + 1)L - 1]$.*

5 Diversity gain analysis for the multiuser scenario

This section considers the downlink transmission in a multiuser massive MIMO system as illustrated in Fig. 4. Here, the base station (BS) employs K_b RAUs with each having N_b antennas and $N_b^{(\mathrm{rf})}$ RF chains to transmit data streams to K_u mobile stations. Each mobile station (MS) is equipped with N_u antennas and $N_u^{(\mathrm{rf})}$ RF chains to support the reception of its own N_s data streams. This means that there is a total of K_uN_s data streams transmitted by the BS. The numbers of data streams are constrained as $K_uN_s \le N_b^{(\mathrm{rf})} \le K_bN_b$ for the BS and $N_s \le N_u^{(\mathrm{rf})} \le N_u$ for each MS.

At the BS, denote by $\mathbf{F}_b$ the $K_bN_b \times N_b^{(\mathrm{rf})}$ RF precoder and by $\mathbf{W}_b$ the $N_b^{(\mathrm{rf})} \times N_sK_u$ baseband precoder. With the narrowband flat fading channel model, then the received signal vector at the ith MS is given by:

$$\mathbf{y}_i = \mathbf{H}_i\mathbf{F}_b\mathbf{W}_b\mathbf{s} + \mathbf{n}_i,\ i = 1, 2, \ldots, K_u \tag{26}$$

where $\mathbf{s}$ is the signal vector for all K_u mobile stations, which satisfies $\mathbb{E}\left[\mathbf{s}\mathbf{s}^H\right] = \frac{P}{K_uN_s}\mathbf{I}_{K_uN_s}$, and P is the average transmit power. The $N_u \times 1$ vector $\mathbf{n}_i$ represents additive white Gaussian noise, whereas the $N_u \times K_bN_b$ matrix $\mathbf{H}_i$ is the channel matrix corresponding to the ith MS, whose entries $\mathbf{H}_{ij}$ are described as in Section 2. Furthermore, the signal vector after combining can be expressed as:

$$\mathbf{z}_i = \mathbf{W}_{ui}^H\mathbf{F}_{ui}^H\mathbf{H}_i\mathbf{F}_b\mathbf{W}_b\mathbf{s} + \mathbf{W}_{ui}^H\mathbf{F}_{ui}^H\mathbf{n}_i,\ i = 1, 2, \ldots, K_u \tag{27}$$

where F_{ui} is the $N_u \times N_u^{(\mathrm{rf})}$ RF combining matrix and $\mathbf{W}_{ui}$ is the $N_u^{(\mathrm{rf})} \times N_s$ baseband combining matrix for the ith MS.

Theorem 3 *Consider the case that all antenna array configurations for the downlink transmission are ULA and $L_{ij} = L$ for any i and j (i.e., all subchannels $\mathbf{H}_{ij}$ have the same number of propagation paths). In the limit of large N_b and N_u, the downlink transmission in a massive MIMO multiuser system can achieve a diversity gain:*

$$G_d = K_bL - N_s + 1. \tag{28}$$

Remarks 5 *Theorem 3 implies that when N_b and N_u are large enough, the available diversity gain G_d does not depend on the number of mobile users K_u.*

Remarks 6 *In a similar fashion, it is easy to prove that the uplink transmission in a massive MIMO multiuser system can also achieve a diversity gain $G_d = K_bL - N_s + 1$. Moreover, it can also be proved that when $L = 1$, the system diversity gain is equal to $G_d = K_b$ for the case $N_u = 1$, i.e., each MS has only one antenna.*

6 Simulation results and discussion

For all simulation results presented in this section, it is assumed that each subchannel matrix $\mathbf{H}_{ij}$ consists of $L_{ij} = L = 3$ paths, each of the large scale fading coefficients g_{ij}

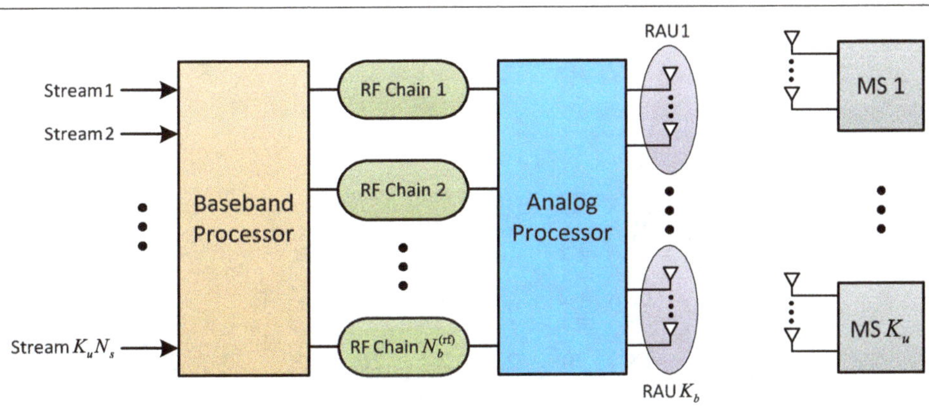

Fig. 4 Block diagram of a multiuser mmWave system with distributed antenna arrays

equals to $g = -20$ dB (except for Fig. 9), and the numbers of transmit and receive RF chains are twice the number of data streams [9] (i.e., $N_t^{(\mathrm{rf})} = N_r^{(\mathrm{rf})} = 2N_s$). It is further assumed that the variance of AWGN samples is unity, and hence, the input SNR is the same as the average input power P/N_s. For simplicity, only ULA array configuration with $d = 0.5$ is considered at RAUs and BPSK modulation is employed for each data stream. With such system configurations, the instantaneous BER is given by $Q(\sqrt{2\gamma})$ [34], where γ denotes the instantaneous receive SNR and the Q function is defined as $Q(x) = \int_x^{\infty} \exp\left(-\frac{y^2}{2}\right)\mathrm{d}y$. For ease of comparison and discussion, introduce the concept of *designed* SNR as $\overline{\mathrm{SNR}}_{\mathrm{dg}} = PN_rN_t/(N_sL)$. This means that $P = \overline{\mathrm{SNR}}_{\mathrm{dg}}N_sL/(N_rN_t)$ for a given designed SNR $\overline{\mathrm{SNR}}_{\mathrm{dg}}$. In fact, there exists a power scaling law for mmWave communications which states that the data transmit power P can be scaled down proportionally to $1/(N_rN_t)$ in order to maintain a desirable BER performance [35].

In all simulations, unless stated otherwise, there are three main steps for hybrid digital-analog processing as follows:

(a) Perform the SVD for channel matrix $\mathbf{H}$ and find the optimal overall digital precoder and combiner for N_s data streams.
(b) Form an analog precoder and an analog combiner based on the optimal overall digital precoder and combiner, respectively.
(c) Perform zero-forcing (ZF) digital detection based on the analog precoder and analog combiner and complete the data detection operation.

First studied is the diversity performance of a massive MIMO mmWave system with distributed antenna arrays. With $N_r = N_t = N = 50$ and $K_r = K_t = K = 2$, Fig. 5 plots BER curves versus the designed SNR for different numbers of data streams, $N_s = 2, 4, 6$. For comparison, the BER curve obtained in the case of co-located antenna arrays are also plotted for $N_s = 1, 2, 3$. It can be seen that even for the larger number of data streams, the BER performance with distributed antenna arrays is clearly better than that with co-located antenna arrays. Furthermore, as N_s decreases, the BER performance with either distributed or co-located antenna arrays is improved. These observations are expected and agree with Corollary 1, which states that using distributed antenna arrays yields higher diversity gains than using co-located antenna arrays. To verify exactly the diversity gain result given in Corollary 1, Fig. 6 plots the diversity gain verifying (GDV) curves produced by simulating the GSC systems. It can be seen that a BER curve with either distributed or co-located antenna arrays has the same slope in the high SNR region as the corresponding GDV curve.

Illustrated in Fig. 7 is the performance with the conventional partially connected (PC) RF architecture analyzed in Section 4. With this structure, one first carries out the SVDs for subchannel matrices $\{\mathbf{H}_{ij}\}$ rather than for the whole channel matrix $\mathbf{H}$ and then forms the analog

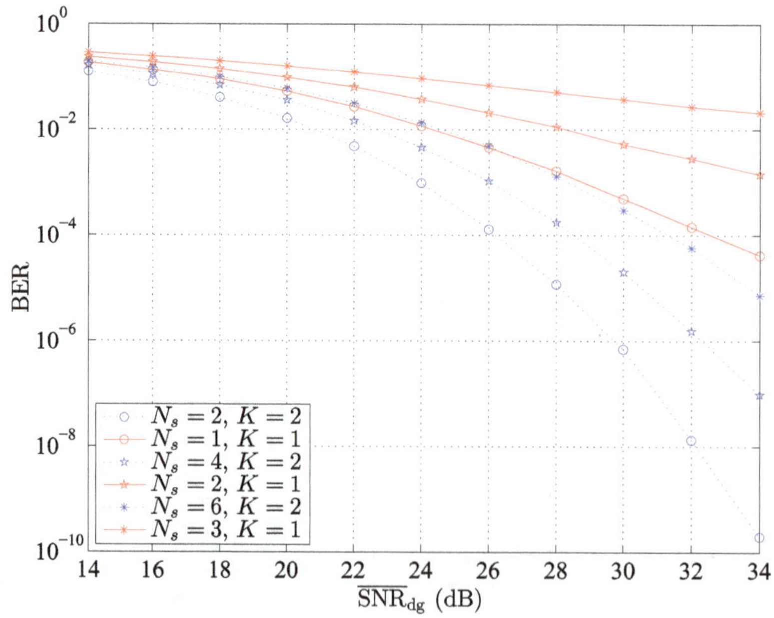

Fig. 5 BER versus designed SNR: comparison between distributed and co-located antenna array architectures

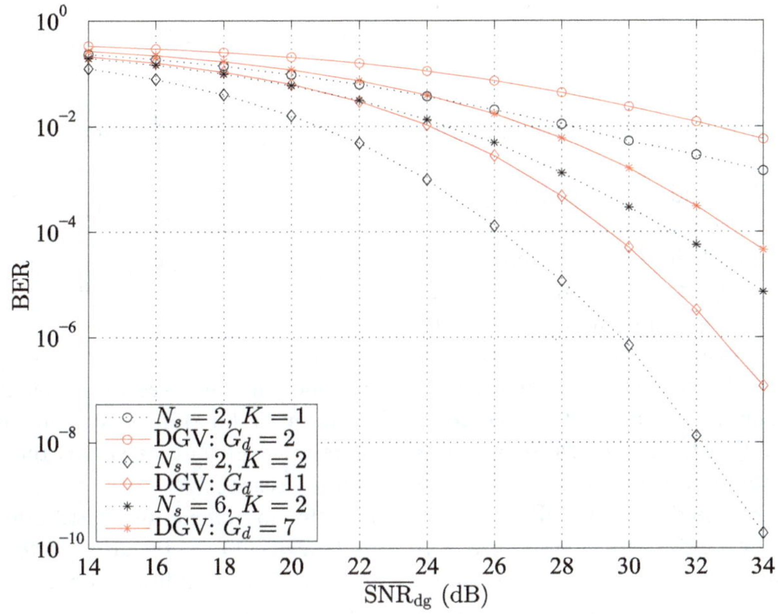

Fig. 6 BER versus designed SNR: verifying diversity gain

precoder and analog combiner. Let $K_r = K_t = K$. With $N_r = N_t = N = 50$, Fig. 7 plots the BER curves for the following four cases: $(K = 1, N_s = 1)$, $(K = 2, N_s = 2)$, $(K = 3, N_s = 3)$, and $(K = 4, N_s = 4)$. It is known from Theorem 2 that the diversity gains for the four cases are identical and equal to $G_d = L = 3$. To illustrate this, a DGV curve with diversity gain $G_d = 3$ is also plotted in this figure. It can be seen that the system with the conventional PC structure for the four cases can achieve the full diversity gain 3, while the coding gain increases when both K and N_s increase. For comparison, the BER curve obtained with the general fully connected (FC) RF

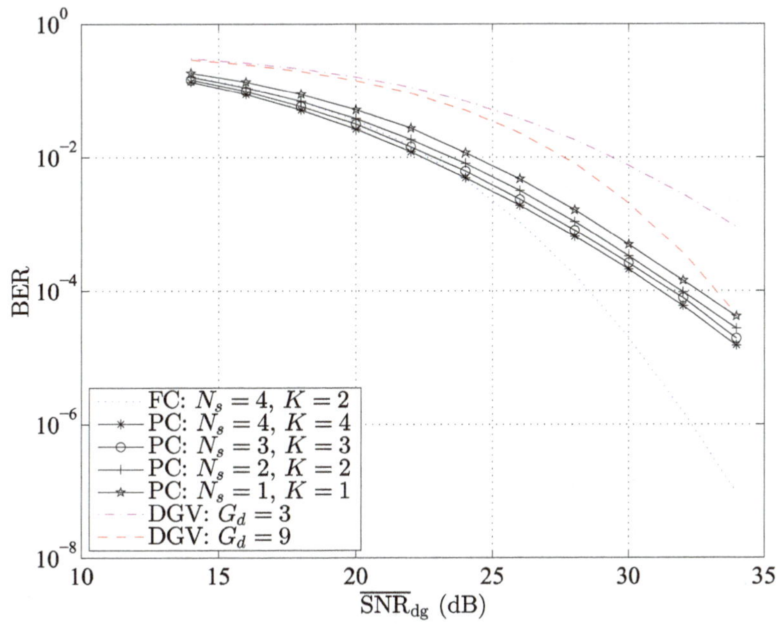

Fig. 7 BER versus designed SNR: comparion between the the proposed distributed subarray architecture and the conventional partially connected architecture

structure when $N_s = 4$ and $K = 2$ is also plotted. The theoretical limit on the diversity gain in this case is 9, which agrees well with the DGV curve having $G_d = 9$. Observe that in the high SNR region, the general FC structure yields significantly better diversity performance than the conventional PC structure.

Next, when $N_s = 1$, we consider the diversity performance with the multiuser downlink scenario where there are 5 or 10 mobile users, each having 10 antennas and each RAU at the BS is equipped with 50 antennas. Due to the fact that there is no cooperation among the users, one first carries out the SVDs for subchannel matrices $\{\mathbf{H}_i\}$ rather than for the whole channel matrix $\mathbf{H}$ and then forms the analog precoder for the BS and analog combiners for the users. Note that the BS needs to carry out ZF digital preprocessing before transmitting data. Figure 8 plots the BER curves versus the designed SNR for different numbers of subarrays at the BS, namely $K_b = 1, 3, 5$. It can be observed from this figure that as K_b increases, the diversity performance of the multiuser system improves remarkably. This is because, as established in Theorem 3, the diversity gain becomes larger with increasing K_b. Furthermore, it can be seen from Fig. 8 that the system has the same diversity gain for different numbers of users while the coding gain increases as K_u decreases. This observation agrees with Remark 5.

Finally, the diversity performance of the single-user massive MIMO mmWave system is examined under the scenario that the distributions of large scale fading coefficients, $\{g_{ij}\}$, are inhomogeneous. To this end, let $\mathbf{G} = [g_{ij}\ (\text{dB})]$ denote the large scale fading coefficient matrix. When $N_r = N_t = N = 50$ and $K_r = K_t = K = 2$, simulation is performed for the following six inhomogeneous $\mathbf{G}$:

$$\mathbf{G}_1 = \begin{bmatrix} -25 & -20 \\ -20 & -25 \end{bmatrix}, \ \mathbf{G}_2 = \begin{bmatrix} -20 & -20 \\ -25 & -25 \end{bmatrix},$$

$$\mathbf{G}_3 = \begin{bmatrix} -20 & -25 \\ -25 & -20 \end{bmatrix}, \ \mathbf{G}_4 = \begin{bmatrix} -20 & -25 \\ -20 & -25 \end{bmatrix},$$

$$\mathbf{G}_5 = \begin{bmatrix} -25 & -25 \\ -20 & -20 \end{bmatrix}, \ \mathbf{G}_6 = \begin{bmatrix} -25 & -20 \\ -25 & -20 \end{bmatrix}.$$

It can be found that the diversity performance for the six inhomogeneous cases are almost the same (see Remark 2). In order to illustrate this interesting phenomenon, Fig. 9 plots the BER curves versus the designed SNR with $\mathbf{G}_1$ and $\mathbf{G}_2$, respectively. For comparison, the two BER curves for the homogeneous distributions with $g = -20$ dB and $g = -25$ dB are also plotted. As expected, the BER curves with the inhomogeneous coefficient distributions are between the two BER curves with homogeneous coefficient distributions. It can be concluded from this figure that the case of inhomogeneous coefficient distributions has the same diversity gain as in the case of homogeneous coefficient distributions.

7 Conclusions

This paper has provided asymptotical diversity analysis for massive MIMO mmWave systems with co-located and distributed antenna architectures when the number of antennas at each subarray goes to infinity. Theoretical analysis shows that with a co-located massive antenna

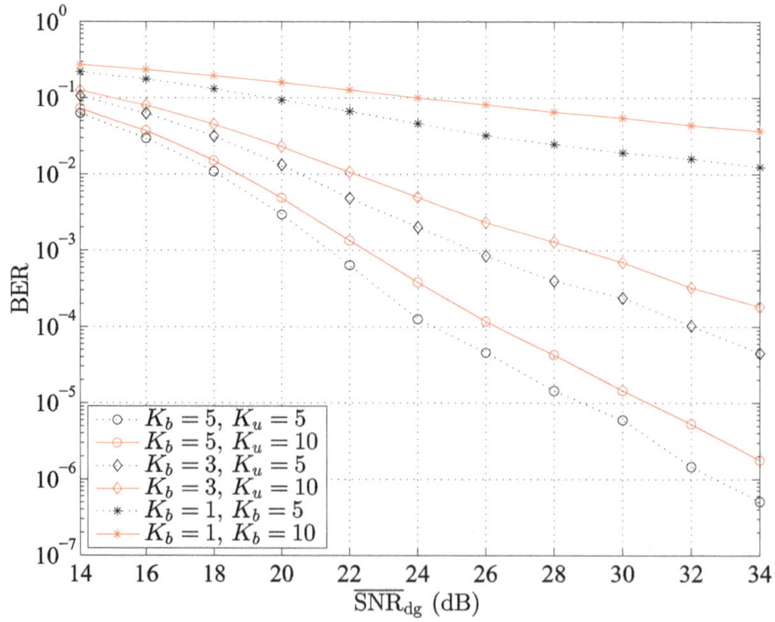

Fig. 8 BER versus designed SNR: multiuser scenario with different numbers of subarrays

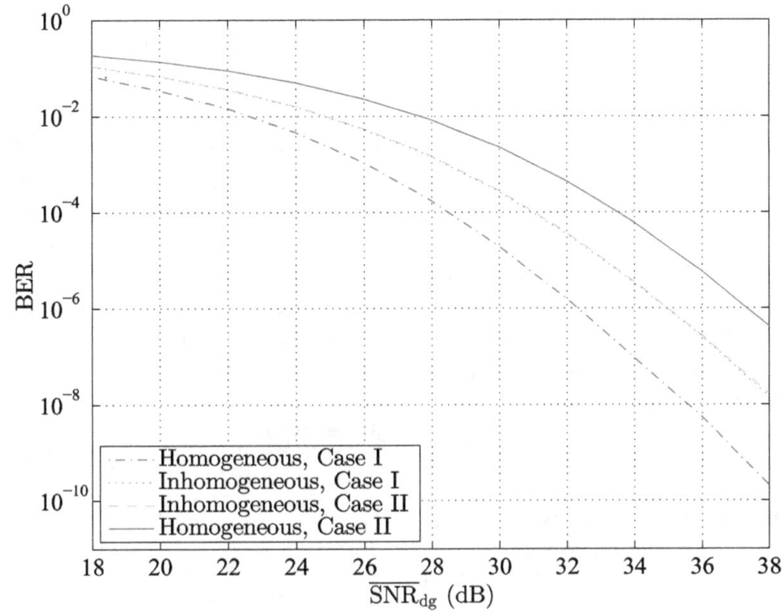

Fig. 9 BER versus designed SNR: comparison between homogeneous distributions and inhomogeneous distributions for large scale fading coefficients $\{g_{ij}\}$

array, scaling up the number of antennas of the array can increase the coding gain but not the diversity gain. However, if the array is built from distributed subarrays (RAUs), each having a very large number of antennas, then increasing the number of RAUs does increase the diversity gain and/or multiplexing gain [36, 37]. As such, the analysis leads to a novel approach to improve the diversity and multiplexing gains of massive MIMO mmWave systems. It is acknowledged that the asymptotical diversity analysis obtained in this paper is under the idealistic assumption of having perfect CSI. Performing the diversity analysis for massive MIMO mmWave systems under imperfect CSI is important and deserves further research.

Endnote

[1] The different variances of α_{ij}^l can easily accounted for by absorbing into the large scale fading coefficients g_{ij}.

Appendix 1: Proof of Theorem 1

Proof The distributed massive MIMO system can be considered as such a co-located system with L_s paths that have complex gains $\left\{\tilde{\alpha}_{ij}^l\right\}$, receive array response vectors $\left\{\tilde{\mathbf{a}}_r\left(\phi_{ij}^{rl},\theta_{ij}^{rl}\right)\right\}$, and transmit response vectors $\left\{\tilde{\mathbf{a}}_t\left(\phi_{ij}^{tl},\theta_{ij}^{tl}\right)\right\}$. Furthermore, order all paths in a decreasing order of the absolute values of the complex gains $\left\{\tilde{\alpha}_{ij}^l\right\}$. Then, the channel matrix can be written as:

$$\mathbf{H}=\sum_{l=1}^{L_s}\tilde{\alpha}^l\tilde{\mathbf{a}}_r\left(\phi^{rl},\theta^{rl}\right)\tilde{\mathbf{a}}_t\left(\phi^{tl},\theta^{tl}\right)^H, \tag{29}$$

where $\tilde{\alpha}^1\geq\tilde{\alpha}^2\geq\cdots\geq\tilde{\alpha}^{L_s}$. One can rewrite $\mathbf{H}$ in a matrix form as:

$$\mathbf{H}=\mathbf{A}_r\mathbf{D}\mathbf{A}_t^H \tag{30}$$

where $\mathbf{D}$ denotes a $L_s\times L_s$ diagonal matrix with $[\mathbf{D}]_{ll}=\tilde{\alpha}^l$, and $\mathbf{A}_r$ and $\mathbf{A}_t$ are defined as follows:

$$\mathbf{A}_r=\left[\tilde{\mathbf{a}}_r\left(\phi^{r1},\theta^{r1}\right),\tilde{\mathbf{a}}_r\left(\phi^{r2},\theta^{r2}\right),\ldots,\tilde{\mathbf{a}}_r\left(\phi^{rL_s},\theta^{rL_s}\right)\right] \tag{31}$$

and

$$\mathbf{A}_t=\left[\tilde{\mathbf{a}}_t\left(\phi^{t1},\theta^{t1}\right),\tilde{\mathbf{a}}_t\left(\phi^{t2},\theta^{t2}\right),\ldots,\tilde{\mathbf{a}}_t\left(\phi^{tL_s},\theta^{tL_s}\right)\right]. \tag{32}$$

Since both $\left\{\tilde{\mathbf{a}}_r\left(\phi^{rl},\theta^{rl}\right)\right\}$ and $\left\{\tilde{\mathbf{a}}_t\left(\phi^{tl},\theta^{tl}\right)\right\}$ are orthogonal vector sets when $N_r\rightarrow\infty$ and $N_t\rightarrow\infty$, $\mathbf{A}_r$ and $\mathbf{A}_t$ are asymptotically unitary matrices. Then, one can form a singular value decomposition (SVD) of matrix $\mathbf{H}$ as:

$$\mathbf{H}=\mathbf{U}\Sigma\mathbf{V}^H=\left[\mathbf{A}_r|\mathbf{A}_r^{\perp}\right]\Sigma\left[\tilde{\mathbf{A}}_t|\tilde{\mathbf{A}}_t^{\perp}\right]^H \tag{33}$$

where Σ is a diagonal matrix containing all singular values on its diagonal, i.e.:

$$[\Sigma]_{ll}=\begin{cases}\left|\tilde{\alpha}^l\right|, & \text{for } 1\leq l\leq L_s\\ 0, & \text{for } l>L_s\end{cases} \tag{34}$$

and the submatrix $\tilde{\mathbf{A}}_t$ is defined as:

$$\tilde{\mathbf{A}}_t=\left[\mathrm{e}^{-j\psi_1}\tilde{\mathbf{a}}_t\left(\phi^{t1},\theta^{t1}\right),\ldots,\mathrm{e}^{-j\psi_{L_s}}\tilde{\mathbf{a}}_t\left(\phi^{tL_s},\theta^{tL_s}\right)\right] \tag{35}$$

where ψ_l is the phase of complex gain $\tilde{\alpha}^l$ corresponding to the lth path. □

Based on (33), the optimal precoder and combiner are chosen respectively as:

$$[\mathbf{F}_t\mathbf{W}_t]_{\text{opt}} = \left[\mathrm{e}^{-j\psi_1}\tilde{\mathbf{a}}_t\left(\phi^{t1}, \ldots, \mathrm{e}^{-j\psi_{Ls}}\tilde{\mathbf{a}}_t\left(\phi^{tN_s}, \theta^{tN_s}\right)\right]\right. \tag{36}$$

and

$$[\mathbf{F}_r\mathbf{W}_r]_{\text{opt}} = \left[\tilde{\mathbf{a}}_r\left(\phi^{r1}, \ldots, \tilde{\mathbf{a}}_r\left(\phi^{rN_s}, \theta^{rN_s}\right)\right]\right.. \tag{37}$$

To summarize, when N_t and N_r are large enough, the massive MIMO system can employ the optimal precoder and combiner given in (36) and (37), respectively. Then, it follows from the above SVD analysis that the instantaneous SNR of the lth data stream is given by:

$$\mathrm{SNR}_l = \frac{P}{N_s}\left|\tilde{\alpha}^l\right|^2, \quad l = 1, 2, \ldots, N_s. \tag{38}$$

Now the detection of the lth data stream is equivalent to the detection in a generalized selection combining system, which selects the path with the lth highest SNR for detection. Therefore, it follows from Lemma 2 that the detection performance of the lth data stream has a diversity gain $L_s - l + 1$. Since the overall BER is the arithmetic mean of individual BERs, i.e., $\overline{\mathrm{BER}} = \frac{1}{N_s}\sum_{l=1}^{N_s}\overline{\mathrm{BER}}(l)$, the system's diversity gain equals to the diversity gain in detecting the N_sth data stream, which is the worst among all data streams. Therefore, the result in (20) is obtained. □

Appendix 2: Proof of Theorem 2

Proof When N_t and N_r are very large, the diversity gain analysis is similar to that in Theorem 1. For the first data stream that enjoys the best path, it is simple to see that its diversity gain is the largest and equal to K_rK_tL. This is because the detection of the first data stream is equivalent to a selection combining system operating with K_rK_tL paths. However, for the second data stream, due to the structure of $\mathbf{F}_t$ and $\mathbf{F}_r$, its detection is equivalent to a selection combining system operating with $(K_r - 1)(K_t - 1)L$ paths. Therefore, it can be inferred that its diversity gain is equal to $(K_r - 1)(K_t - 1)L$. Similarly, for the last data stream among the N_s data streams, its diversity gain is $(K_r - N_s + 1)(K_t - N_s + 1)L$. It then follows that the diversity gain of the whole system is just $(K_r - N_s + 1)(K_t - N_s + 1)L$. □

Appendix 3: Proof of Theorem 3

Proof For the downlink transmission in a massive MIMO multiuser system, the overall equivalent multiuser basedband channel can be written as:

$$\mathbf{H}_{\text{eq}} = \begin{bmatrix} \mathbf{F}_{u1}^H & \mathbf{0} & \cdots & \mathbf{0} \\ \mathbf{0} & \mathbf{F}_{u2}^H & \cdots & \mathbf{0} \\ \vdots & \vdots & \ddots & \vdots \\ \mathbf{0} & \mathbf{0} & \cdots & \mathbf{F}_{uK_u}^H \end{bmatrix} \begin{bmatrix} \mathbf{H}_1 \\ \mathbf{H}_2 \\ \vdots \\ \mathbf{H}_{K_u} \end{bmatrix} \mathbf{F}_b. \tag{39}$$

On the other hand, when both N_b and N_u are very large, both receive and transmit array response vector sets, $\left\{\tilde{\mathbf{a}}_r\left(\phi_{ij}^{rl}, \theta_{ij}^{rl}\right)\right\}$ and $\left\{\tilde{\mathbf{a}}_t\left(\phi_{ij}^{tl}, \theta_{ij}^{tl}\right)\right\}$, are asymptotically orthogonal. Therefore, the diversity performance for the ith user depends only on the subchannel matrix $\mathbf{H}_i$ and the choices of $\mathbf{F}_{ui}$ and $\mathbf{F}_b$. The subchannel matrix $\mathbf{H}_i$ has a total of K_bL propagation paths. Similar to the proof of Theorem 1, by employing the optimal RF precoder and combiner for the ith user, the user can achieve a maximum diversity gain $K_bL - N_s + 1$. It is then concluded that the downlink transmission can achieve a diversity gain $G_d = K_bL - N_s + 1$. □

Abbreviations

BS: Base station; BPU: Baseband processing unit; BER: Bit error rate; C-MIMO: Co-located MIMO; CDF: Cumulative distribution function; DMT: Diversity-multiplexing trade-of;f GDV: Diversity gain verifying; D-MIMO: Distributed MIMO; FC: Fully connected; GSC: Generalized selection combining; i.i.d.: Independent and identically distributed; mmWave: Millimeter-wave; MS: Mobile station; PDF: Probability density function; RF: Radio-frequency; RAU: Remote antenna unit; SNR: Signal-to-noise; SC: Selection combining; SVD: Singular value decomposition; ULA: Uniform linear array; UPA: Uniform planar array; ZF: Zero-forcing

Acknowledgements

No applicable.

Funding

This research was funded by the Fundamental Research Funds for the Central Universities grant number 3132016347 and the Natural Science Foundation of Liaoning Province grant number 201602086.

Authors' contributions

All authors read and approved the final manuscript.

Authors' information

No applicable.

Competing interests

The authors declare that they have no competing interests.

References

1. T.S. Rappaport, et al., Millimeter wave mobile communications for 5G cellular: it will work! IEEE Access. **1**, 335–349 (2013)
2. A.L. Swindlehurst, E. Ayanoglu, P. Heydari, F. Capolino, Millimeter-wave massive MIMO: the next wireless revolution? IEEE Commun. Mag. **52**(9), 56–62 (2014)
3. W. Roh, et al., Millimeter-wave beamforming as an enabling technology for 5G cellular communications: theoretical feasibility and prototype results. IEEE Commun. Mag. **52**(2), 106–113 (2014)
4. X. Liu, M. Jia, X. Zhang, W. Lu, A novel multi-channel Internet of Things based on dynamic spectrum sharing in 5G communication (2018). IEEE Internet Things Published, J., online, https://doi.org/10.1109/JIOT.2018.2847731
5. X. Liu, M. Jia, Z. Na, W. Lu, F. Li, Multi-modal cooperative spectrum sensing based on Dempster-Shafer fusion in 5G-based cognitive radio. IEEE Access. **6**, 199–208 (2018)
6. O.E. Ayach, R.W. Heath, S. Abu-Surra, S. Rajagopal, Z. Pi, in *Proc. IEEE 13th Intl. Workshop on Sig. Process. Advances in Wireless Commun. (SPAWC)*. The capacity optimality of beam steering in large millimeter wave MIMO systems, (Cesme, 2012), pp. 100–104
7. L. Liang, W. Xu, X. Dong, Low-complexity hybrid precoding in massive multiuser MIMO systems. IEEE Wirel. Commun. Lett. **3**(6), 653–656 (2014)
8. W. Ni, X. Dong, Hybrid block diagonalization for massive multiuser MIMO systems. IEEE Trans. Commun. **64**(1), 201–211 (2016)
9. F. Sohrabi, W. Yu, Hybrid digital and analog beamforming design for large-scale antenna arrays. IEEE Sel. J. Top. Sign. Process. **10**(3), 501–513 (2016)

10. O.E. Ayach, R.W. Heath, S. Rajagopal, Z. Pi, in *Proc. IEEE Glob. Commun. Conf.* Multimode precoding in millimeter wave MIMO transmitters with multiple antenna sub-arrays, (2013), pp. 3476–3480
11. J. Singh, S. Ramakrishna, in *in Proc. IEEE GLOBECOM.* On the feasibility of beamforming in millimeter wave communication systems with multiple antenna arrays, (Austin, 2014), pp. 3802–3808
12. J.A. Zhang, X. Huang, V. Dyadyuk, Y. J. Guo, Massive hybrid antenna array for millimeter-wave cellular communications. IEEE Wirel. Commun. **22**(1), 79–87 (2015)
13. S. He, C. Qi, Y. Wu, Y. Huang, Energy-efficient transceiver design for hybrid sub-array architecture MIMO systems. IEEE Access. **4**, 9895–9905 (2016)
14. N. Li, Z. Wei, H. Yang, X. Zhang, D. Yang, Hybrid precoding for mmWave massive MIMO systems with partially connected structure. IEEE Access. **5**, 2–15151 (2017)
15. D. Zhang, Y. Wang, X. Li, W. Xiang, Hybridly-connected structure for hybrid beamforming in mmWave massive MIMO systems. IEEE Trans. Commun. **66**(2), 662–674 (2017)
16. N. Song, T. Yang, H. Sun, Overlapped subarray based hybrid beamforming for millimeter wave multiuser massive MIMO. IEEE Signal Proc. Lett. **24**(5), 550–554 (2017)
17. S. Kutty, D. Sen, Beamforming for millimeter wave communications: an inclusive surey. IEEE Commun. Surv. Tutor. **18**(2), 949–973 (2016)
18. A.F. Molisch, V.V. Ratnam, S. Han, Z. Li, S.L.H. Nguyen, L. Li, K. Haneda, Hybrid beamforming for massive MIMO-a survey. IEEE Commun. Mag. **55**(9), 134–141 (2017)
19. Z. Pi, F. Khan, An introduction to millimeter-wave mobile broadband systems. IEEE Comm. Mag. **49**(6), 101–107 (2011)
20. T.S. Rappaport, F. Gutierrez, E. Ben-Dor, J.N. Murdock, Y. Qiao, J.I. Tamir, Broadband millimeter-wave propagation measurements and models using adaptive-beam antennas for outdoor urban cellular communications. IEEE Trans. Antennas Propag. **61**(4), 1850–1859 (2013)
21. M.V. Clark, T.M.W. III, L.J. Greenstein, A.J. Rustako, V. Erceg, R.S. Roman, in *in Proc. IEEE Veh. Technology Conf. (VTC'01).* Distributed versus centralized antenna arrays in broadband wireless networks, (Rhodes, 2001), pp. 33–37
22. W. Roh, A. Paulraj, in *in IEEE Veh. Technology Conf. (VTC'02).* MIMO channel capacity for the distributed antenna systems, (Vancouver, 2002), pp. 1520–1524
23. L. Dai, A comparative study on uplink sum capacity with co-located and distributed antennas. IEEE Sel. J. Areas Commun. **29**(6), 1200–1213 (2011)
24. Q. Wang, D. Debbarma, A. Lo, Z. Cao, I. Niemegeers, S.H. de Groot, Distributed antenna system for mitigating shadowing effect in 60 GHz WLAN. Wirel. Pers. Commun. **82**(2), 811–832 (2015)
25. S. Gimenez, D. Calabuig, S. Roger, J.F. Monserrat, N. Cardona, Distributed hybrid precoding for indoor deployments using millimeter wave band. Mob. Inf. Syst. **2017**(Article ID 5751809), 12 (2017)
26. L. Zheng, D.N.C. Tse, Diversity and multiplexing: a fundamental tradeoff in multiple antenna channels. IEEE Trans. Inform. Theory. **49**(5), 1073–1096 (2003)
27. D. Tse, P. Viswanath, L. Zheng, Diversity-multiplexing tradeoff in multiple access channels. IEEE Trans. Inform. Theory. **50**(9), 1859–1874 (2004)
28. M. Yuksel, E. Erkip, Multiple-antenna cooperative wireless systems: a diversity-multiplexing tradeoff perspective. IEEE Trans. Inform. Theory. **53**(10), 3371–3393 (2007)
29. D. Tse, P. Viswanath, *Fundamentals of wireless communication.* (Cambridge Univ. Press, Cambridge, 2007)
30. L.G. Ordóñz, D.P. Palomar, A. Pagès-Zamora, J.R. Fonollosa, High-SNR analytical performance of spatial multiplexing MIMO systems with CSI. IEEE Trans. Signal Process. **55**(11), 5447–5463 (2007)
31. Z. Wang, G.B. Giannakis, A simple and general parameterization quantifying performance in fading channels. IEEE Trans. Signal Process. **51**(8), 1389–1398 (2003)
32. H.A. David, *Order statistics.* (Wiley, New York, 1981)
33. J. Chen, in *Global Commun. Conf.* When does asymptotic orthogonality exist for very large arrays, (Atlanta, 2013), pp. 4251–4255
34. A. Goldsmith, *Wireless communications, Chapter 6.* (Cambrige University Press, Cambridge, 2005)
35. D.W. Yue, Y. Zhang, Y.N. Jia, Beamforming based on specular component for massive MIMO systems in Ricean fading. IEEE Wireless Commun. Lett. **4**(2), 197–200 (2015)
36. D.W. Yue, S. Xu, H.H. Nguyen, Diversity analysis of millimeter-wave massive MIMO systems (2018). arXiv preprint arXiv. 1801.00387
37. D.W. Yue, S. Xu, H.H. Nguyen, in *The 7th International conference on communications, signal processing, and systems (CSPS2018).* Diversity gain of millimeter wave massive MIMO systems with distributed antenna arrays, (Dalian, 2018)

4

SWIPT in mMIMO system with non-linear energy-harvesting terminals: protocol design and performance optimization

Kui Xu, Ming Zhang*, Jie Liu, Nan Sha, Wei Xie and Lihua Chen

Abstract

In this paper, we design the simultaneous wireless information and power transfer (SWIPT) protocol for massive multi-input multi-output (mMIMO) system with non-linear energy-harvesting (EH) terminals. In this system, the base station (BS) serves a set of uplink fixed half-duplex (HD) terminals with non-linear energy harvester. Considering the non-linearity of practical energy-harvesting circuits, we adopt the realistic non-linear EH model rather than the idealistic linear EH model. The proposed SWIPT protocol can be divided into two phases. The first phase is designed for terminals EH and downlink training. A beam domain energy beamforming method is employed for the wireless power transmission. In the second phase, the BS forms the two-layer receive beamformers for the reception of signals transmitted by terminals. In order to improve the spectral efficiency (SE) of the system, the BS transmit power- and time-switching ratios are optimized. Simulation results show the superiority of the proposed beam-domain SWIPT protocol on SE performance compared with the conventional mMIMO SWIPT protocols.

Keywords: Massive MIMO, Beam-domain channel representation, Simultaneous wireless information and power transfer, Beamforming, Achievable sum-rate

Introduction

With the rapid rise of the Internet of Things (IoT) industry [1], massive intelligent devices will be deployed in anywhere to monitor engineering structures, diagnose the physical condition of the patients, report the real-time traffic information, and so on. A major problem facing is that how to prolong the life of the energy-constrained devices, so that they could be utilized in a sustainable and low-cost way. Currently, a feasible solution is to adopt the radio frequency (RF) signal-based wireless energy transfer (WET) technique [2–4], as RF signal is controllable, stable, and efficient compared to the natural sources such as solar and wind.

As one of the key technologies of fifth-generation (5G) communication system, massive multiple-input multiple-output (mMIMO) [5] technology can obtain the diversity gain and multiplexing gain, and hence can greatly improve the spectrum efficiency (SE) of wireless networks [6–8]. Compared with traditional MIMO, the spatial resolution of mMIMO systems is significantly improved [9], which can greatly reduce various types of interference [7].

As a practical solution to improve the energy efficiency (EE) of battery-powered IoT network in 5G systems and hence extend the service lifetime of the IoT network, RF energy-harvesting (EH) technology has been widely studied [10–12]. In an actual simultaneous wireless information and power transfer (SWIPT) system [13], the base station (BS) or access point (AP) provides the receiver with concurrent information and energy supply. Currently, the practical SWIPT receiver architecture can be mainly classified as time-switching (TS) and power-splitting (PS) [14] according to the signal partition method for EH and information decoding (ID). The TS SWIPT receiver alternates between EH and ID according to the TS ratio [15], and as a result, the received RF signal is first sent to the energy harvester and then to the information decoder. For the PS SWIPT receiver, the received RF signal is split into two parts according to the PS ratio [13]; one portion is sent to the energy harvester, and the remainder is sent to the information decoder.

*Correspondence: 18805198368@139.com
College of Communication Engineering, Army Engineering University of PLA, No. 2, Biaoying Street, Nanjing, China

In [16], the TS and PS protocols were designed for the amplify-and-forward (AF) one-way single-antenna SWIPT relay networks. In [17], the SWIPT protocol was designed for one-way MIMO relay network, where the MIMO relay can simultaneously harvest energy from the signal sent by the destination and decode the signal sent by the source. Since two-way relaying (TWR) system can further improve the SE of multi-hop relay system [18–20], recently, SWIPT protocol for TWR systems were studied in many literatures. The authors in [21] studied the SWIPT protocol in the AF TWR system, wherein the RF-powered AF relay node is applied to help the two source nodes exchange information. In [22], the problem of maximizing the sum throughput in the TWR networks with wireless-powered nodes was studied. The authors in [23] considered the SWIPT protocol design in multiple relay AF TWR networks, where two source nodes harvest energy from multiple relays.

In [16, 17, 21–23], it was assumed that the amount of energy harvested by the EH circuit is linearly proportional to the received RF power, namely, the linear EH model. However, as studied in recent literature [24–29], the linear EH model is too idealistic and impractical because linear assumptions do not hold in practice. The energy conversion efficiency of an actual EH circuit is a non-linear function of the received RF signal due to various causes of non-linearity, e.g., diode non-linearity, and saturation non-linearity. In addition, the sensitivity of the energy harvester is limited [30], that is, signals with power around communications sensitivity can be decoded at a SWIPT receiver but cannot be exploited for energy-harvesting purposes. Therefore, in practice, the amount of harvested energy is clearly a non-linear function of the received power.

In [31], the SWIPT protocol for three-dimensional (3D) mMIMO system is designed, where the matched filter (MF) precoder is applied. In [32], the SWIPT protocol for multi-pair mMIMO TWR system is designed, where linear precoders, i.e., zero-forcing (ZF) and maximal ratio combining (MRC), are employed at the relay. A low-complexity SWIPT scheme with retrodirective maximum ratio transmission (MRT) beamforming is introduced in [33]. In [34], each frame is divided into three phases: the uplink channel estimation phase, the downlink wireless energy transmission phase, and the uplink wireless information transmission phase. ZF and MRC beamforming are applied at the access point. For SWIPT system with mMIMO, instantaneous full-dimensional channel state information (CSI) is required to perform transmit and receive precoding [34]. The above four works [31–34] have considered linear energy harvesters. To obtain full dimensional CSI, the uplink training overhead is linearly proportional to the number of user equipments, thus causing significant bottlenecks on the achievable SE.

Methods

In this paper, the beam-domain SWIPT protocol for mMIMO system with fixed half-duplex (HD) non-linear energy-harvesting terminals (TEs) is designed, and the system spectral efficiency performance is maximized by optimizing the TS ratio and transmit power of the BS[1]. The contributions are summarized as follows.

- We consider the SWIPT protocol design for TEs with non-linear energy harvesters in mMIMO systems. Moreover, a more practical power consumption model is employed at the TE. In particular, after the energy-harvesting phase, a portion of the harvested energy is used for necessary processing, such as the energy consumption of channel estimation, channel feedback, and energy consumption of the circuit, to achieve its primary task. The remainder energy is used for the subsequent wireless information transmission.
- Based on the idea of TS, the entire transmission process of the proposed beam-domain SWIPT protocol can be divided into two phases. The first phase is designed for TEs EH and downlink training. During this phase, the BS transmits energy signals to the TEs. The TEs utilize one portion of the received energy signals for EH and the remainder for downlink channel estimation. After obtaining the channel CSI, TEs feedback it to the BS. The beam-domain energy beamforming method is employed for wireless energy transmission. In the second phase, the BS forms the two-layer receive beamformers for the reception of signals transmitted by TEs.
- In order to improve the SE of the system, the base station transmit power and time-switching ratio are optimized. The simulation results show that compared with the traditional mMIMO SWIPT protocol, the proposed beam-domain SWIPT protocol has superior performance in SE performance.

The rest of the paper is organized as follows. The system and channel model are introduced in Section 1. Section 1 illustrates the proposed beam-domain mMIMO SWIPT transmission scheme. Section 1 analyzes the achievable sum-rate performance and proposes the optimization method. Section 1 presents the simulation results. Our conclusions are presented in Section 1.

Notations: In this paper, $\mathbb{E}(\cdot)$ denotes the expectation. $\mathbf{A}^{\{B_1,B_2\}}$ is the submatrix of $\mathbf{A}$ containing rows of set B_1 and columns of set B_2. $\mathbf{A}^{\{B,:\}}$ and $\mathbf{A}^{\{:,B\}}$ are the submatrix of $\mathbf{A}$ containing rows and columns of set B, respectively. $\delta(\cdot)$ denotes the dirac delta function. $(\cdot)^T$, $(\cdot)^*$, $(\cdot)^H$, $|\cdot|$, $\|\cdot\|$, and $tr(\cdot)$ denote transpose, conjugate, conjugate transpose, determinant, Frobenius norm, and trace of a matrix, respectively. $\lfloor\cdot\rfloor$ and $\lceil\cdot\rceil$ denote the floor and ceilings, respectively. $\mathbf{I}_V$ denotes $V \times V$ identity matrix. $\langle \cdot, \cdot \rangle$ denotes the inner product operator.

System model

We consider a mMIMO SWIPT system, where the BS serves a set of K_t fixed uplink TEs $\mathbf{K}_T = \{k_1, k_2, \cdots, k_{K_t}\}$, as shown in Fig. 1. We assume that all TEs are uniformly distributed in the coverage area of the BS. The BS is equipped with a massive antenna array with N antennas. We assume that all TEs operate in HD mode and are equipped with a single antenna. Each TE employs a non-linear energy harvester to collect energy and uses the harvested energy for subsequent data transmission.

In this paper, the frame-based SWIPT protocol is designed for mMIMO system. The length of one frame is fixed to T symbols, and we assume that T is less than the coherence interval of the fading channel. Let $\mathbf{h}_{k_i}^u \in \mathbb{C}^{N\times 1}$ and $\mathbf{h}_{k_i}^d \in \mathbb{C}^{N\times 1}$ denote the channel vector of the uplink channel between transmit antenna array of BS and TE k_i and corresponding downlink channel vector. We assume that the channel reciprocity between uplink and downlink channels can be exploited, that is, the uplink channel can be obtained by the downlink channel estimation.

Uplink/downlink channel model

In this paper, we apply the general cluster-based channel model [35], that is, the signal of TE k_i received by the BS is the sum of the contributions from M_u scatterings. Similarly, the signal of the BS received by the TE k_i is the sum of the contributions from M_d scatterings. We assume that the direction of arrival (DOA) or direction of departure (DOD) of signals from cluster m is within the region $\left[\theta_{k_i,m}^{o,\min}, \theta_{k_i,m}^{o,\max}\right], o \in \{u, d\}$, then the uplink/downlink channel vector between the BS and TE k_i can be written as [35]

$$\mathbf{h}_{k_i}^o = \sum_{m=1}^{M_o} \int_{\theta_{k_i,m}^{o,\min}}^{\theta_{k_i,m}^{o,\max}} \mathbf{a}(\theta)\, r_{k_i,m}^o(\theta)\, d\theta \tag{1}$$

where $\mathbf{a}(\theta) = \left[1, \exp\left(\frac{j2\pi d \sin(\theta)}{\lambda}\right), \cdots, \exp\left(\frac{j2\pi d(N-1)\sin(\theta)}{\lambda}\right)\right]^T$ represents the array response vector, d denotes the antenna spacing, and λ denotes the carrier wavelength. $r_{k_i,m}^o(\theta)$ denotes the complex uplink/downlink channel response gain.

The uplink/downlink channel vector $\mathbf{h}_{k_i}^o$ can be expanded by a set of uniform basis vectors $\{\mathbf{f}_1, \mathbf{f}_2, \cdots, \mathbf{f}_N\} \in \mathbb{C}^{N\times 1}$ by utilizing the basis expansion model (BEM) [36], that is,

$$\mathbf{h}_{k_i}^o = \sum_{n=1}^{N} \tilde{h}_{k_i,n}^o \mathbf{f}_n = \mathbf{F}\tilde{\mathbf{h}}_{k_i}^o \tag{2}$$

where $\mathbf{F} = [\mathbf{f}_1, \mathbf{f}_2, \cdots, \mathbf{f}_N]$, and the basis vector $\mathbf{f}_n$ is also called a beam [37]. The vector $\tilde{\mathbf{h}}_{k_i}^o = \left[\tilde{h}_{k_i,1}^o, \tilde{h}_{k_i,2}^o, \cdots, \tilde{h}_{k_i,N}^o\right]^T$ is called beam-domain (BD) channel vector [38] and can be obtained by

$$\tilde{\mathbf{h}}_{k_i}^o = \mathbf{F}^H \mathbf{h}_{k_i}^o = \sum_{m=1}^{M_o} \int_{\theta_{k_i,m}^{o,\min}}^{\theta_{k_i,m}^{o,\max}} \mathbf{F}^H \mathbf{a}(\theta)\, r_{k_i,m}^o(\theta)\, d\theta \tag{3}$$

Usually, the effective channel length of the mMIMO channel, that is, the number of elements with non-negligible channel gains in vector $\tilde{\mathbf{h}}_{k_i}^o$, is far less than the number of antennas. Therefore, applying BEM can greatly reduce the size of the channel that needs to be estimated and the number of CSIs that need feedback. Thus, for a given

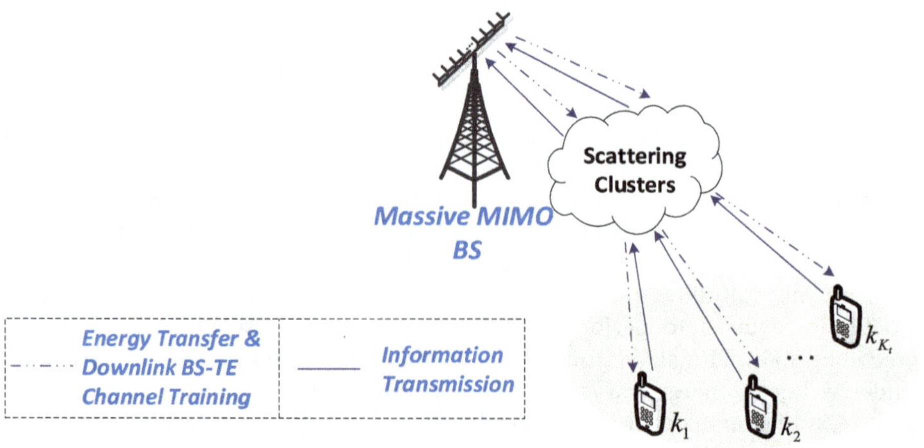

Fig. 1 Illustration of mMIMO SWIPT system

set of beams, we can approximate the channel vector $\mathbf{h}_{k_i}^o$ as [38]

$$\begin{aligned}\mathbf{h}_{k_i}^o &= \sum_{m\in B_{k_i}} \tilde{h}_{k_i,m}^o \mathbf{f}_m + \sum_{m\notin B_{k_i}} \tilde{h}_{k_i,m}^o \mathbf{f}_m \\ &\approx \sum_{m\in B_{k_i}} \tilde{h}_{k_i,m}^o \mathbf{f}_m = \mathbf{F}^{\left\{B_{k_i},:\right\}} \tilde{\mathbf{h}}_{o,k_i}^{\left\{B_{k_i}\right\}}\end{aligned} \tag{4}$$

where B_{k_i} is the index set of the selected active beams of the terminal k_i. The approximation in the middle of (4) comes from the assumption that the channel gain of the terminal k_i on the beams corresponding to the index set B_{k_i} is not negligible, and the channel gain on other beams is close to zero and can be neglected, that is, $\sum_{m\notin B_{k_i}} \tilde{h}_{k_i,m}^o \mathbf{f}_m \approx 0$. $\mathbf{F}^{\left\{B_{k_i},:\right\}}$ denotes the corresponding active beam (AB) space with columns are indexed in the set B_{k_i}. The channel vector with reduced-dimension $\tilde{\mathbf{h}}_{o,k_i}^{\left\{B_{k_i}\right\}} \in \mathbb{C}^{\left|B_{k_i}\right|\times 1}$ is named as the BD effective channel and (4) holds with equality as $N \to \infty$.

Based on (1) and (4), $\tilde{\mathbf{h}}_{o,k_i}^{\left\{B_{k_i}\right\}}$ can be written as

$$\begin{aligned}\tilde{\mathbf{h}}_{o,k_i}^{\left\{B_{k_i}\right\}} &= \left(\mathbf{F}^{\left\{B_{k_i},:\right\}}\right)^H \mathbf{h}_{k_i}^o \\ &= \sum_{m=1}^{M_o} \int_{\theta_{k_i,m}^{o,\min}}^{\theta_{k_i,m}^{o,\max}} \left(\mathbf{F}^{\left\{B_{k_i},:\right\}}\right)^H \mathbf{a}(\theta)\, r_{k_i,m}^o(\theta)\, d\theta\end{aligned} \tag{5}$$

Hence, in order to obtain $\mathbf{h}_{k_i}^o$ by channel estimation, $\left|B_{k_i}\right|$-dimension BD effective channel estimation is enough. As a result, the consumption of training resources will be significantly reduced.

As introduced in [11], all TEs can be partitioned into groups according to their AB sets. Let g_t denote the group index and B_{g_t} denote the corresponding AB set of the group g_t. Based on the orthogonality of active beam space across disjoint groups, that is, $B_{g_t} \bigcap B_{g'_t} = \emptyset$ when $g_t \neq g'_t$, the BD effective channel dimensions of TEs within group g_t is reduced to $|B_{g_t}| = b_{g_t}$.

Energy-harvesting model

At present, the total energy collected by wireless-powered terminals in the wireless charging stage can usually be modeled as two models: the ideal linear EH model, and the practical non-linear EH model.

In linear EH model [21–23], the collected energy of the terminal k_i can be expressed as

$$E_{k_i}^{\mathrm{L}} = \eta P_{k_i}^{\mathrm{EH}} T_h \tag{6}$$

where $0 < \eta < 1$ denotes the constant energy conversion efficiency for converting RF energy to electrical energy, $P_{k_i}^{\mathrm{EH}}$ denotes the total received RF power at wireless-powered user k_i, and T_h denotes the duration of energy harvesting.

In the practical non-linear EH model, the amount of the harvested energy is modeled using the logistic (or sigmoid) function, i.e., S-shaped curve, as follows [28, 29]:

$$E_{k_i}^{\mathrm{NL}} = \frac{\left[\frac{P_s}{1+\exp\left(-a\left(P_{k_i}^{\mathrm{EH}}-b\right)\right)} - P_s\Omega\right]T_h}{1-\Omega} \tag{7}$$

and

$$\Omega = \frac{1}{1+\exp(ab)} \tag{8}$$

where P_s denotes the maximum amount of harvested power when the EH circuit is saturated, and a and b are positive constants related to the circuit specification [28, 29]. Figure 2 illustrates the practical non-linear EH model given in (7) and the conventional linear EH model given in (6). We assume that all the EH TEs apply the same non-linear energy harvester.

Energy consumption model

Most of the previous works assume that all harvested energy is used for power amplifier during the wireless information transmission phase [11, 21–23]. In this work, we employ a more practical power consumption model for the TE. In particular, after the energy-harvesting phase, one portion of the harvested energy is used for necessary processing at TE to achieve its main task, and the remainder energy is used for uplink transmission during the wireless information transmission phase. Note that the similar model has also been applied in [39]. Let $E_{k_i}^{\mathrm{P}}$ denote the processing energy[2]; in the wireless information transmission phase of time period T_{I}, the transmit power of TE k_i can be expressed as

$$p_{k_i}^u = \frac{E_{k_i}^{\mathrm{NL}} - E_{k_i}^{\mathrm{P}}}{T_{\mathrm{I}}} \tag{9}$$

Note that (9) has assumed that the harvest energy is larger than the processing energy. Otherwise, $p_{k_i}^u$ should be set to zero and an outage occurs.

Proposed beam-domain SWIPT protocol

In this paper, the frame-based protocol is applied to realize beam-domain SWIPT in the cellular system with mMIMO BS. The whole transmission process is divided into N frames. Each frame can be divided into two phases based on the idea of TS, i.e., *phase I* and *phase II*, as shown in Fig. 3.

In phase I of time period αT $(0 \leq \alpha \leq 1)$, the BS transmits energy signals to all the TEs, where α denotes TS ratio. We assume that all TEs turn on the receive circuit during the interval αT. TEs utilize one portion of

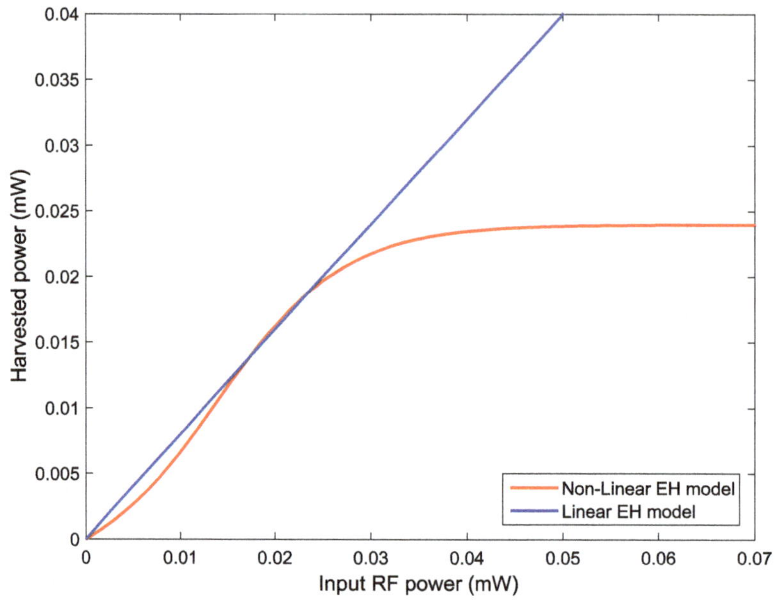

Fig. 2 A comparison between the linear EH model and the non-linear EH model. For linear EH model, $\eta = 0.8$, for non-linear EH model [28, 29], $P_s = 0.024$ mW, $a = 150$, and $b = 0.014$

the received energy signals for EH and the remainder for downlink channel estimation. In this phase, the main purpose of the protocol design is to enable TEs to collect as much energy as possible.

Specifically, the BS transmits the energy signals to all the TEs. The signal received by ith TE can be expressed as $y^S_{r,i}$. Part of the received energy signal, $\sqrt{\beta}y^S_{r,i}$, is used for EH and part of the received signal, $\sqrt{1-\beta}y^S_{r,i}$, is used for downlink channel estimation, where β denotes the PS ratio, as shown in Fig. 4. The detailed training signals design will introduced in the next subsection.

In phase II of time period $(1-\alpha)T$, the TEs feedback the uplink CSI to the BS and transmit their data during the interval $(1-\alpha)T$ using the harvested energy. Specifically, the ith TE transmits signal $\sqrt{\beta\eta}y^S_{t,i}$ to the BS. Note that the resource consumption for CSI feedback of TEs are omitted in this paper.

Energy signal design

Let $\Phi^E_{g_t} \in \mathbb{C}^{\tau_t \times K_{g_t}}$ be the orthogonal energy sequence set transmit by the BS to the TE group g_t, where K_{g_t} denotes the number of TEs in group g_t, τ_t denotes the length of

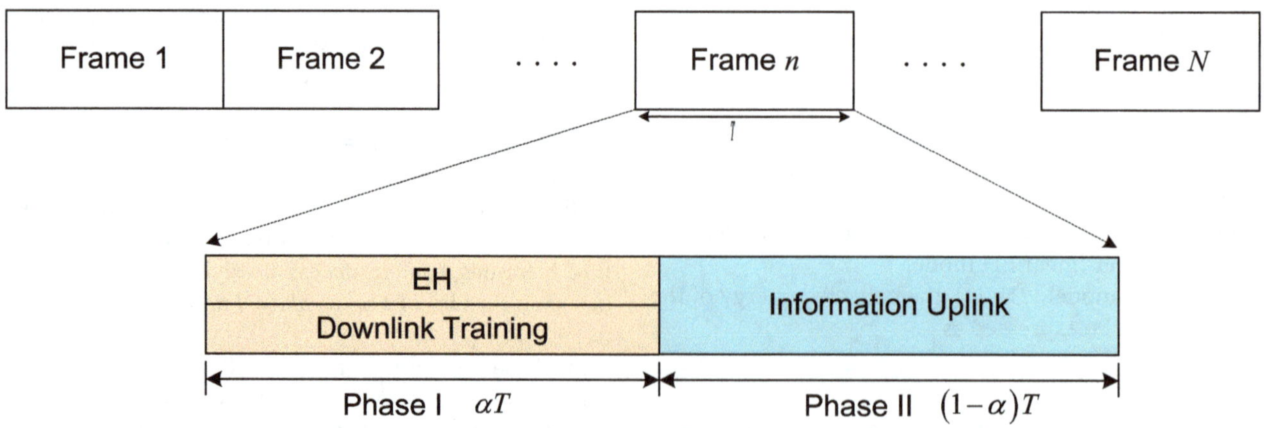

Fig. 3 Data frame structure of the proposed beam-domain SWIPT protocol

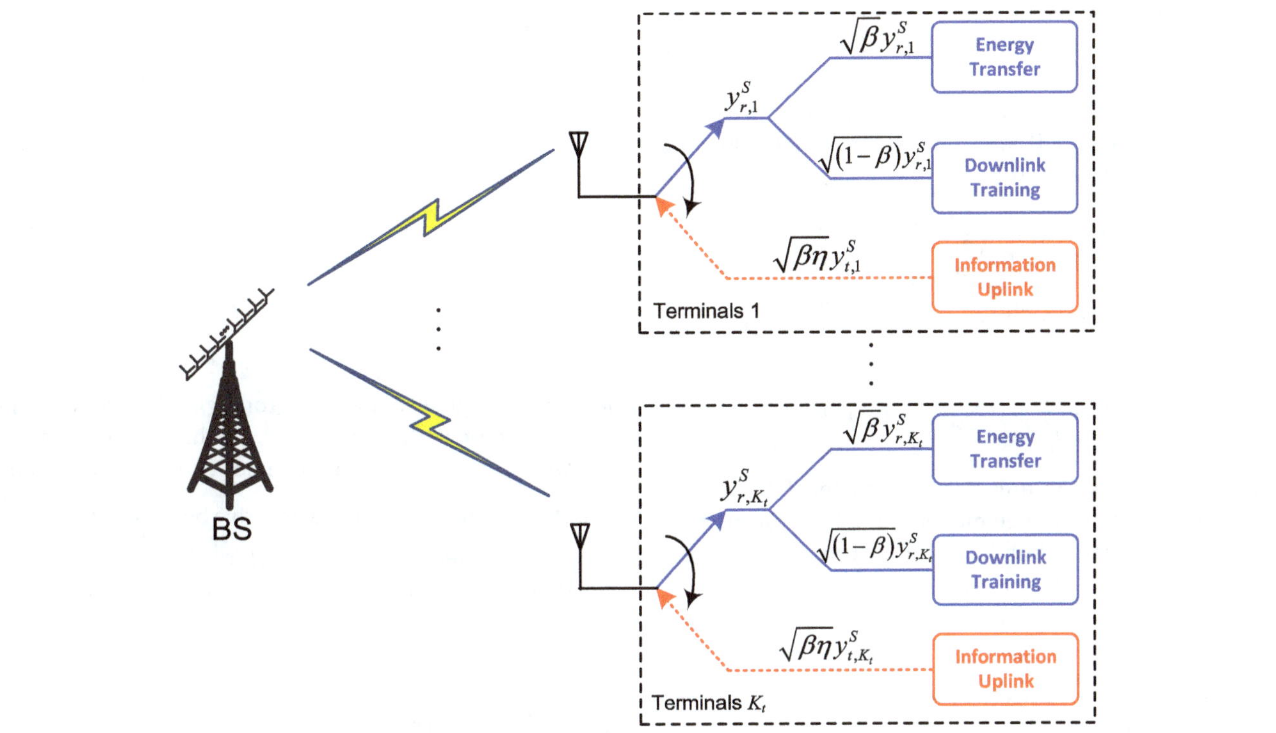

Fig. 4 Illustration of the signal flow for the proposed beam-domain SWIPT protocol

energy sequence and satisfies $\tau_t \geq b_{g_t}$, and $\left|\left[\Phi^{\mathrm{E}}_{g_t}\right]_{j,k}\right|^2 = \tau_t p^d_{g_{t,k}}$. $p^d_{g_{t,k}}$ denotes the power of each energy symbol. Let G_t denote the terminal group set, the time allocated for the TEs EH is $\alpha T \geq \max\{b_{g_t}|g_t \in G_t\}$.

Energy transfer and TE downlink channel estimation

During phase I of frame n, the received signal of TE group g_t can be expressed as

$$\mathbf{Y}^d_{g_t}[n] = \mathbf{H}^H_{d,g_t}[n] \sum_{g'_t \in G_t} \mathbf{F}^{\left\{B_{g'_t}[n],:\right\}} \left(\Phi^{\mathrm{E}}_{g'_t}\right)^T + \mathbf{N}^d_{g_t}[n] \tag{10}$$

where $\mathbf{H}_{d,g_t}[n] = \left[\mathbf{h}^d_{g_{t,1}}[n], \cdots, \mathbf{h}^d_{g_{t,K_{g_t}}}[n]\right]$ and $\mathbf{N}^d_{g_t}[n]$ denote additive white Gaussian noise (AWGN) with variance σ^2. The received signal at the $g_{t,k}$th TE can be expressed as

$$\mathbf{y}^d_{g_{t,k}}[n] = (\mathbf{h}^d_{g_{t,k}}[n])^H \sum_{g'_t \in G_t} \mathbf{F}^{\left\{B_{g'_t}[n],:\right\}} \left(\Phi^{\mathrm{E}}_{g'_t}\right)^T + \mathbf{n}^d_{g_{t,k}}[n] \tag{11}$$

where $\mathbf{n}_{g_{t,k}}[n] \sim \mathcal{CN}(0, \sigma^2)$ denotes the AWGN at TE $g_{t,k}$.

The TE harvests energy from one portion of the whole received signal, that is, $\sqrt{\beta}\mathbf{y}^d_{g_{t,k}}[n]$. For signal with large transmit power, the energy of noise is negligible in the harvested energy [16]. In the linear EH model, the harvested energy of TE $g_{t,k}$ during the frame n can be written as

$$\tilde{E}^{\mathrm{L}}_{g_{t,k}}[n] = \eta\beta \left\| (\mathbf{h}^d_{g_{t,k}}[n])^H \sum_{g'_t \in G_t} \mathbf{F}^{\left\{B_{g'_t}[n],:\right\}} \left(\Phi^{\mathrm{E}}_{g'_t}\right)^T \right\|^2 \tag{12}$$

Based on the orthogonality of active beam space across disjoint groups, that is, $\left\langle \mathbf{F}^{\{B_{g_t}[n],:\}}, \mathbf{F}^{\left\{B_{g'_t}[n],:\right\}} \right\rangle = 0$ when $B_{g_t}[n] \neq B_{g'_t}[n]$, we have the following approximation

$$\begin{aligned}\tilde{E}^{\mathrm{L}}_{g_{t,k}}[n] &\approx \eta\beta \left\| (\mathbf{h}^d_{g_{t,k}}[n])^H \mathbf{F}^{\{B_{g_t}[n],:\}} \left(\Phi^{\mathrm{E}}_{g_t}\right)^T \right\|^2 \\ &= \eta\beta\tau_t p^d_{g_{t,k}}[n] \left\| \tilde{\mathbf{h}}^{\{B_{g_t}[n]\}}_{d,g_{t,k}}[n] \right\|^2 \triangleq \eta\beta\alpha T \hat{p}^{\mathrm{H}}_{d,g_{t,k}}\end{aligned} \tag{13}$$

where $\hat{p}^{\mathrm{H}}_{d,g_{t,k}} = q_{g_{t,k}} p^d_{g_{t,k}}[n]$ and $q_{g_{t,k}} = \left\| \tilde{\mathbf{h}}^{\{B_{g_t}[n]\}}_{d,g_{t,k}}[n] \right\|^2$.

In the adopted non-linear model, the amount of the harvested energy at $g_{t,k}$ is modeled using the logistic (or sigmoid) function, i.e., S-shaped curve, as follows [28]:

$$\tilde{E}^{\mathrm{NL}}_{g_{t,k}}[n] \approx \frac{P_s \alpha T \left[\frac{1}{1+\exp\left(-a\left(\beta \hat{p}^{\mathrm{H}}_{d,g_{t,k}} - b\right)\right)} - \Omega \right]}{1 - \Omega} \tag{14}$$

Another portion of the TE $g_{t,k}$ received signal, that is, $\sqrt{1-\beta}\mathbf{y}^{d}_{g_{t,k}}[n]$, is used for the channel estimation. Since the TE knows the values of β and hence can accurately compensate it, we omit it in the following channel estimation. The least square (LS) estimator [38] of the downlink channel can be expressed as

$$\begin{aligned}\tilde{\mathbf{h}}^{\{B_{g_t}[n]\}}_{d,g_{t,k},LS}[n] &= \frac{1}{\tau_t p^{d}_{t,k}[n]}\left(\mathbf{Y}^{d}_{g_t}[n]\left(\Phi^{\mathrm{E}}_{g_t}\right)^{*}\right)^{H}\mathbf{e}^{k} \\ &= \tilde{\mathbf{h}}^{\{B_{g_t}[n]\}}_{d,g_{t,k}}[n] + \sum_{g'_t\in G_t/g_t}\tilde{\mathbf{h}}^{\{B_{g'_t}[n]\}}_{d,g_{t,k}}[n] \\ &+ \frac{1}{\tau_t p^{d}_{t,k}[n]}\left(\mathbf{N}_{g_t}[n]\left(\Phi^{\mathrm{E}}_{g_t}\right)^{*}\right)^{H}\mathbf{e}^{k}\end{aligned} \tag{15}$$

where $\mathbf{e}^{k}$ denotes the unit vector with the kth element is 1 and the rest of the elements are 0. The linear minimum mean square error (LMMSE) estimate [38] of the downlink channel can be expressed as

$$\begin{aligned}\tilde{\mathbf{h}}^{\{B_{g_t}[n]\}}_{d,g_{t,k},LM}[n] =& \sum_{g'_t\in G_t}\sum_{m=1}^{M_d}\int_{\theta^{d,\min}_{g_{t,k,m}}}^{\theta^{d,\max}_{g_{t,k,m}}}\left(\mathbf{F}^{\{B_{g_t}[n],:\}}\right)^{H}\mathbf{a}(\theta)\mathbf{a}^{H}(\theta)\mathbf{F}^{\{B_{g'_t}[n],:\}}S^{d}_{g_{t,k,m}}(\theta)\,d\theta \\ &\times\left(\sum_{g'_t,g''_t\in G_t}\sum_{m=1}^{M_d}\int_{\theta^{d,\min}_{g_{t,k,m}}}^{\theta^{d,\max}_{g_{t,k,m}}}\left(\mathbf{F}^{\{B_{g'_t}[n],:\}}\right)^{H}\mathbf{a}(\theta)\right. \\ &\left.\times\,\mathbf{a}^{H}(\theta)\,\mathbf{F}^{\{B_{g''_t}[n],:\}}S^{d}_{g_{t,k,m}}(\theta)\,d\theta + \frac{\sigma^2}{\tau_t p^{d}_{t,k}}\right)^{-1}\tilde{\mathbf{h}}^{\{B_{g_t}[n]\}}_{d,g_{t,k},LS}[n]\end{aligned} \tag{16}$$

The uplink channel CSI $\tilde{\mathbf{h}}^{\{B_{g_t}[n]\}}_{u,g_{t,k},LM}[n]$ can be obtained at the BS by using the CSI feedback[3].

TE uplink information transmission

In phase II of time period $(1-\alpha)T$, the TEs send signal using one portion of the harvested energy during phase I, and the BS uses the estimated effective channels to perform receive beamforming and receive data sent by K_T TEs. The transmit power of TE $g_{t,k}$ can be expressed as

$$p^{u}_{g_{t,k}}[n] = \frac{\breve{E}^{\mathrm{NL}}_{g_{t,k}}[n] - E^{\mathrm{P}}_{g_{t,k}}[n]}{(1-\alpha)T} \tag{17}$$

where $\breve{E}^{\mathrm{NL}}_{g_{t,k}}[n]$ denotes the currently available energy in the nth frame, which can be expressed as the difference between the cumulative harvested energy and the consumed energy during the $\{1,2,\cdots,n-1\}$th frame, that is,

$$\breve{E}^{\mathrm{NL}}_{g_{t,k}}[n] = \sum_{i=1}^{n}\tilde{E}^{\mathrm{NL}}_{g_{t,k}}[i] - \sum_{i=1}^{n-1}E^{\mathrm{P}}_{g_{t,k}}[i] - \sum_{i=1}^{n-1}(1-\alpha[i])\,Tp^{u}_{g_{t,k}}[i] \tag{18}$$

where $\tilde{E}^{\mathrm{NL}}_{g_{t,k}}[i]$, $E^{\mathrm{P}}_{k_i}[i]$, $\alpha[i]$, and $p^{u}_{g_{t,k}}[i]$ denote the harvested energy, processing energy, TS ratio, and transmit power in the ith frame, respectively.

Let $s^{u}_{g_{t,k}}[n]$ denote the transmit signal of TE $g_{t,k}$, and the transmit power of TE $g_{t,k}$ is set to $p^{u}_{g_{t,k}}[n] = \left|s^{u}_{g_{t,k}}[n]\right|^2$. The received signal at the BS can be expressed as

$$\mathbf{y}^{u}_{t}[n] = \sum_{g'_t\in G_t}\mathbf{H}_{u,g'_t}[n]\,\mathbf{s}^{u}_{g'_t}[n] + \mathbf{n}^{u}_{t}[n] \tag{19}$$

where $\mathbf{s}^{u}_{g_t}[n] = [s^{u}_{g_{t,1}}[n], s^{u}_{g_{t,2}}[n], \cdots, s^{u}_{g_{t,K_{g_t}}}[n]]^{T} \in \mathbb{C}^{K_{g_t}\times 1}$ denotes the signal sent by TE group g_t. $\mathbf{n}^{u}_{t}[n]$ denotes the AWGN vector with variance σ^2. In this paper, we apply the two-layer beamforming scheme. The first-layer beamformer is a matrix composed of active beam sets $\mathbf{F}^{\{B_{g_t}[n],:\}}$. Multiplying both sides of (19) with $\mathbf{F}^{\{B_{g_t}[n],:\}}$, the received signal of the BS from group g_t can be expressed as

$$\begin{aligned}\tilde{\mathbf{y}}^{u}_{g_t}[n] &= \mathbf{F}^{\{B_{g_t}[n],:\}}\mathbf{y}^{u}_{t}[n] \\ &= \tilde{\mathbf{H}}^{\{B_{g_t}[n],:\}}_{u,g_t}[n]\,\mathbf{s}^{u}_{g_t}[n] + \sum_{g'_t\in G_t/\{g_t\}}\tilde{\mathbf{H}}^{\{B_{g_t}[n],:\}}_{u,g'_t}[n]\,\mathbf{s}^{u}_{g'_t}[n] + \tilde{\mathbf{n}}^{u}_{g_t}[n]\end{aligned} \tag{20}$$

where $\tilde{\mathbf{n}}^{u}_{g_t}[n] = \mathbf{F}^{\{B_{g_t}[n],:\}}\mathbf{n}^{u}_{t}[n]$. To detect the signal of group g_t, the BS employs the second-layer receive beamforming, i.e., $\dot{\mathbf{y}}^{u}_{g_t}[t] = \mathbf{W}^{H}_{g_t}[n]\,\tilde{\mathbf{y}}^{u}_{g_t}[n]$ and $\mathbf{W}_{g_t}[n] = [\mathbf{w}_{g_{t,1}}[n], \mathbf{w}_{g_{t,2}}[n], \cdots, \mathbf{w}_{g_{t,K_{g_t}}}[n]] \in \mathbb{C}^{b_t[n]\times K_{g_t}[n]}$. The kth element of $\dot{\mathbf{y}}^{u}_{g_t}[n]$ can be expressed as

$$\begin{aligned}\dot{y}^{u}_{g_{t,k}}[n] =& \mathbf{w}^{H}_{g_{t,k}}[n]\,\tilde{\mathbf{h}}^{\{B_{g_t}[n]\}}_{u,g_{t,k}}[n]\,s^{u}_{g_{t,k}}[n] + \mathbf{w}^{H}_{g_t,k_{g_t}}[n]\sum_{k'=1,k'\neq k}^{K_{g_t}}\tilde{\mathbf{h}}^{\{B_{g_t}[n]\}}_{g_{t,k'}}[n]\,s^{u}_{g_{t,k'}}[n] \\ &+ \mathbf{w}^{H}_{g_{t,k}}[n]\sum_{g'_t\in G^{e}_t/\{g_t\}}\tilde{\mathbf{H}}^{\{B_{g_t}[n],:\}}_{u,g'_t}[n]\,\mathbf{s}^{u}_{g'_u}[n] + \tilde{n}^{u}_{g_t}[n]\end{aligned} \tag{21}$$

The signal-to-interference-and-noise ratio (SINR) of the BS received signal from TE $g_{t,k}$ can be expressed as

$$\mathrm{SINR}_{g_{t,k}}[n] = \frac{p^{u}_{g_{t,k}}[n]\left|\mathbf{w}^{H}_{g_{t,k}}[n]\,\tilde{\mathbf{h}}^{\{B_{g_t}[n]\}}_{u,g_{t,k}}[n]\right|^2}{\sigma^2\|\mathbf{w}_{g_{t,k}}[n]\|^2 + \Theta_{g_{t,k}}[n]} \tag{22}$$

Let $\Delta\tilde{\mathbf{h}}_{t,g_{t,k}}^{\{B_{g_t}[n]\}}$ denote the channel estimation error of $\tilde{\mathbf{h}}_{t,g_{t,k}}^{\{B_{g_t}[n]\}}$, and $\Theta_{g_{t,k}}[n]$ can be expressed as

$$\Theta_{g_{t,k}}[n] = \underbrace{p_{g_{t,k}}^{u}[n]\left|\mathbf{w}_{g_{t,k}}^{H}[n]\,\Delta\tilde{\mathbf{h}}_{t,g_{t,k}}^{\{B_{g_t}[n]\}}[n]\right|^2}_{\mathrm{CEE}_{g_{t,k}}[n]} + \underbrace{\sum_{k'=1,k'\neq k}^{K_{g_t}} p_{g_{t,k'}}^{u}[n]\left|\mathbf{w}_{g_{t,k}}^{H}[n]\,\Delta\tilde{\mathbf{h}}_{u,g_{t,k'},LM}^{\{B_{g_t}[n]\}}[n]\right|^2}_{\mathrm{IUI}_{g_{t,k}}[n]} + \underbrace{\sum_{g'_t\in G_t/\{g_t\}}\left\|\mathbf{w}_{g_{t,k}}^{H}[n]\,\tilde{\mathbf{H}}_{u,g'_t}^{\{B_{g_t}[n],:\}}[n]\,\Lambda_{g'_t}^{1/2}[n]\right\|^2}_{\mathrm{IGI}_{g_{t,k}}[n]} \tag{23}$$

where $\mathrm{CEE}_{g_{t,k}}[n]$, $\mathrm{IUI}_{g_{t,k}}[n]$, and $\mathrm{IGI}_{g_{t,k}}[n]$ denote the total power introduced by channel estimation error (CEE), inter-user interference (IUI), and inter-group interference (IGI), respectively. $\Lambda_{g_t}[n] = \mathbb{E}\left[\mathbf{s}_{g_t}^{u}[n]\,(\mathbf{s}_{g_t}^{u}[n])^H\right]$ denotes the diagonal covariance matrix of group g_t.

The optimal $\mathbf{w}_{g_{t,k}}[n]$ is the vector that maximizes the $\mathrm{SINR}_{g_{t,k}}[n]$, and, thus, the second-layer beamformer can be obtained by

$$\max_{\left\|\mathbf{w}_{g_{t,k}}[n]\right\|=1} \frac{p_{g_{t,k}}^{u}[n]\,\mathbf{w}_{g_{t,k}}^{H}[n]\,\tilde{\mathbf{h}}_{u,g_{t,k}}^{\{B_{g_t}[n]\}}[n]\left(\tilde{\mathbf{h}}_{u,g_{t,k}}^{\{B_{g_t}[n]\}}[n]\right)^H\mathbf{w}_{g_{t,k}}[n]}{\mathbf{w}_{g_{t,k}}^{H}[n]\left(\Xi_{g_{t,k}}[n]+\sigma^2\mathbf{I}_{b_t[n]}\right)\mathbf{w}_{g_{t,k}}[n]} \tag{24}$$

where $\Xi_{g_{t,k}}[n]$ can be written as

$$\Xi_{g_{t,k}}[n] = p_{g_{t,k}}^{u}[n]\,\Delta\tilde{\mathbf{h}}_{u,g_{t,k}}^{\{B_{g_t}[n]\}}[n]\left(\Delta\tilde{\mathbf{h}}_{u,g_{t,k}}^{\{B_{g_t}[n]\}}[n]\right)^H + \sum_{k'=1,k'\neq k}^{K_{g_t}} p_{g_{t,k'}}^{u}[n]\,\Delta\tilde{\mathbf{h}}_{u,g_{t,k'},LM}^{\{B_{g_t}[n]\}}[n]\left(\Delta\tilde{\mathbf{h}}_{u,g_{t,k'},LM}^{\{B_{g_t}[n]\}}[n]\right)^H + \sum_{g'_t\in G_t/\{g_t\}} \tilde{\mathbf{H}}_{u,g'_t}^{\{B_{g_t}[n],:\}}[n]\,\Lambda_{g'_t}^{1/2}[n]\left(\tilde{\mathbf{H}}_{u,g'_t}^{\{B_{g_t}[n],:\}}[n]\,\Lambda_{g'_t}^{1/2}[n]\right)^H \tag{25}$$

The problem in (24) is in generalized Rayleigh quotient form. The maximum value in (24) can be obtained when $\mathbf{w}_{g_{t,k}}[n]$ equals to the generalized eigenvector corresponding to the maximum generalized eigenvalue [11], which can be written as

$$\mathbf{w}_{g_{t,k}}[n] = \frac{\left(\Xi_{g_{t,k}}[n]+\sigma^2\mathbf{I}_{b_t[n]}\right)^{-1}\tilde{\mathbf{h}}_{u,g_{t,k},LM}^{\{B_{g_t}[n]\}}[n]}{\left\|\left(\Xi_{g_{t,k}}[n]+\sigma^2\mathbf{I}_{b_t[n]}\right)^{-1}\tilde{\mathbf{h}}_{u,g_{t,k},LM}^{\{B_{g_t}[n]\}}[n]\right\|} \tag{26}$$

Achievable sum-rate analysis and optimization

When optimizing the system achievable sum-rate, the following two constraints should be considered: Firstly, the harvested energy at the TEs should be enough for the signal transmission in the subsequent information transmission. Secondly, the minimum transmission rate of TEs should be guaranteed.

To guarantee the effective transmission of the power-limited devices, the currently available energy $\breve{E}_{g_{t,k}}^{\mathrm{NL}}[n]$ in the nth frame should be enough for the signal transmission at devices in the subsequent information transmission , that is,

$$\breve{E}_{g_{t,k}}^{\mathrm{NL}}[n] = \sum_{i=1}^{n}\tilde{E}_{g_{t,k}}^{\mathrm{NL}}[i] - \sum_{i=1}^{n-1}(1-\alpha[i])\,Tp_{g_{t,k}}^{u}[i] - \sum_{i=1}^{n-1}E_{g_{t,k}}^{\mathrm{P}}[i] \geq \eth, \forall g_{t,k}\in\mathbf{K}_U, n\in\{1,2,\cdots,n-1\} \tag{27}$$

where $\eth$ denotes the minimum transmission energy constraint of TEs.

With the help of the bounding technique [40] and according to (21), the average achievable rates at TE $g_{t,k}$ can be expressed as

$$R_{g_{t,k}}(\mathcal{P}[n],\alpha[n]) = (1-\alpha[n])\log_2\left(1+p_{g_{t,k}}^{u}[n]\,\|\tilde{\mathbf{h}}_{u,g_{t,k}}^{\{B_{g_t}[n]\}}[n]\|^2 \times\left(\underbrace{p_{g_{t,k}}^{u}[n]\left|\mathbf{w}_{g_{t,k}}^{H}[n]\,\Delta\tilde{\mathbf{h}}_{u,g_{t,k}}^{\{B_{g_t}[n]\}}[n]\right|^2}_{\mathrm{CEE}_{g_{t,k}}[n]} + \|\mathbf{w}_{g_{t,k}}[n]\|^2 + \underbrace{\sum_{k'=1,k'\neq k}^{K_{g_t}} p_{g_{t,k'}}^{u}[n]\left|\mathbf{w}_{g_{t,k}}^{H}[n]\,\Delta\tilde{\mathbf{h}}_{u,g_{t,k'},LM}^{\{B_{g_t}[n]\}}[n]\right|^2}_{\mathrm{IUI}_{g_{t,k}}[n]} + \underbrace{\sum_{g'_t\in G_t/\{g_t\}}\left\|\mathbf{w}_{g_{t,k}}^{H}[n]\,\tilde{\mathbf{H}}_{u,g'_t}^{\{B_{g_t}[n],:\}}[n]\,\Lambda_{g'_t}^{1/2}[n]\right\|^2}_{\mathrm{IGI}_{g_{t,k}}[n]}\right)^{-1}\right) \tag{28}$$

The transmit power of TE $g_{t,k}$ is set to $p_{g_{t,k}}^{u}$. The powers of CEE, IUI, and IGI are introduced in (41).

The system achievable rate can be expressed as

$$R_{\mathrm{SUM}}(\mathcal{P}[n],\alpha[n]) = \sum_{g_{t,k}[n]\in\mathbf{K}_T} R_{g_{t,k}}(\mathcal{P}[n],\alpha[n]) \tag{29}$$

where $\mathcal{P}[n] = \left\{p_{g_{t,1}}^{d}[n],\cdots,p_{g_{t,K_t}}^{d}[n]\right\}$ denotes the power set transmitted by the BS to each TE and the uplink rate $R_{g_{t,k}}$ of TE $g_{t,k}$ are given in (41). Let $R_{\min}$ denote the minimum throughput of the TEs, and the system achievable rate maximization problem can be expressed as

$$\max_{\mathcal{P}[n],\alpha[n]} \sum_{g_{t,k}\in\mathbf{K}_T} R_{g_{t,k}}(\mathcal{P}[n],\alpha[n])$$
$$\text{s.t.} \begin{cases} p1: T > \alpha[n]\,T \geq \max\{b_{g_t} | g_t \in G_t\} \\ p2: \sum_{k=1}^{K_t} p^d_{g_{t,k}}[n] \leq P^{\mathrm{E}}_d \\ p3: p^d_{g_{t,k}}[n] > 0, k \in \mathbf{K}_T \\ p4: \breve{E}^{\mathrm{NL}}_{g_{t,k}}[n] \leq P_s \alpha T, k \in \mathbf{K}_T \\ p5: \breve{E}^{\mathrm{NL}}_{g_{t,k}}[n] \geq \eth, k \in \mathbf{K}_T \\ p6: R_{g_{t,k}}(\mathcal{P},\alpha)[n] \geq R_{\min}, k \in \mathbf{K}_T \end{cases} \tag{30}$$

where P^{E}_d denotes the maximum transmit power constraint at the BS during phase I. One can check that (30) is non-convex due to the constraints in $p2 \sim p6$, and therefore it is not straightforward to obtain the globally optimal solution. To tackle this difficulty, we apply the lower bound of the available energy in the frame n, that is, $\breve{E}^{\mathrm{NL}}_{g_{t,k}}[n] \geq \tilde{E}^{\mathrm{NL}}_{g_{t,k}}[n] - E^{\mathrm{P}}_{g_{t,k}} n] \geq \eth$. Since (14) is a transcendental function, this inequality cannot be solved directly. Fortunately, we know that the collected energy $\tilde{E}^{\mathrm{NL}}_{g_{t,k}}[n]$ is bounded by $P_s\alpha T$; hence, to ensure the subsequent uplink transmission of TEs, we have $P_s\alpha T > E^{\mathrm{P}}_{g_{t,k}}[n] + \eth$, and if this condition can be met, we have

$$\alpha[n] > \frac{E^{\mathrm{P}}_{g_{t,k}}[n] + \eth}{P_s T} \triangleq \ddot{\alpha}_{g_{t,k}}[n] \tag{31}$$

where Ω is given in (8). Hence, the SE optimization problem in (30) can be rewritten as

$$\max_{\mathcal{P}[n],\alpha[n]} \sum_{g_{t,k}\in\mathbf{K}_T} R_{g_{t,k}}(\mathcal{P}[n],\alpha[n])$$
$$\text{s.t.} \begin{cases} \hat{p}1: T > \alpha[n]\,T \geq \\ \quad \max\left\{\max\{b_{g_t} | g_t \in G_t\}, \max\{\ddot{\alpha}_{g_{t,k}}[n] \,| k \in \mathbf{K}_T\}\right\} \\ \hat{p}2: \sum_{k=1}^{K_t} p^d_{g_{t,k}}[n] \leq P^{\mathrm{E}}_d \\ \hat{p}3: p^d_{g_{t,k}}[n] > 0, k \in \mathbf{K}_T \\ \hat{p}4: R_{g_{t,k}}(\mathcal{P}[n],\alpha[n]) \geq R_{\min}, k \in \mathbf{K}_T \end{cases} \tag{32}$$

In order to simplify the representation, we omit the frame index n in the following derivations.

Optimize $\mathcal{P}$ for fixed α

We fix α to $\alpha = \alpha_0$. The uplink rate $R_{g_{t,k}}$ given in (41) can be expressed as $R_{g_{t,k}} = (1-\alpha_0)\log_2\left(1+\gamma_{g_{t,k}}\right)$ and the SINR of TE $g_{t,k}$ received signal can be expressed as $\gamma_{g_{t,k}} = p^u_{g_{t,k}} b_{g_{t,k}} \left(c_{g_{t,k}} + d_{g_{t,k}}\right)^{-1}$, wherein $b_{g_{t,k}}, c_{g_{t,k}}, d_{g_{t,k}}$ are positive variables. Specifically, $b_{g_{t,k}} = \left\|\mathbf{h}^{\{\tilde{B}_{g_t}\}}_{g_{t,k}}\right\|^2$, $c_{g_{t,k}} = \mathrm{CEE}_{g_{t,k}} + \mathrm{IUI}_{g_{t,k}} + \mathrm{IGI}_{g_{t,k}}$, and $d_{g_{t,k}} = \|\mathbf{w}^H_{g_{t,k}}\|^2 = 1$. Based on (41) and omit the group index, we can rewrite the SE optimization problem as

$$\max_{\mathcal{P}} \prod_{k=1}^{K_t}\left(1+\gamma_{t,k}\right) = \min_{\mathcal{P}}\left[\prod_{k=1}^{K_t}\left(1+\gamma_{t,k}\right)\right]^{-1}$$
$$\text{s.t.} \begin{cases} c1: \gamma_{g_{t,k}} \leq p^u_{g_{t,k}} b_{g_{t,k}} \left(c_{g_{t,k}} + d_{g_{t,k}}\right)^{-1}, \\ \qquad k = 1,\cdots,K_u \\ \hat{p}2, \hat{p}3, \hat{p}4, \hat{p}5 \end{cases} \tag{33}$$

where $\hat{p}1 \sim \hat{p}4$ are constraints given by (32). Since $p^u_{g_{t,k}} = \frac{\breve{E}^{\mathrm{NL}}_{g_{t,k}}[n] - E^{\mathrm{P}}_{g_{t,k}}}{(1-\alpha_0)T} \geq \frac{\tilde{E}^{\mathrm{NL}}_{g_{t,k}} - E^{\mathrm{P}}_{g_{t,k}}}{(1-\alpha_0)T}$, $p^u_{g_{t,k}}$ is lower bound by

$$\hat{p}5: p^u_{g_{t,k}} \geq \frac{P_s\alpha_0 T\left[\frac{1}{1+\exp\left(-a\left(\beta\hat{p}^{\mathrm{H}}_{d,g_{t,k}} - b\right)\right)} - \Omega\right] - E^{\mathrm{P}}_{g_{t,k}}(1-\Omega)}{(1-\Omega)(1-\alpha_0)T} \tag{34}$$

(34) is the constraint $\hat{p}5$. We can see from Fig. 2 that $p^u_{g_{t,k}}$ is a monotonic function of $p^d_{g_{t,k}}$, the solution $\mathcal{P}$ of problem (33) is equivalent to the optimization of $\mathcal{P}^u = \left\{p^u_{g_{t,1}},\cdots,p^u_{g_{t,K_t}}\right\}$ when $\alpha = \alpha_0$ is fixed.

The form of SE optimization problem (33) is close to a geometric programming (GP) except that the target function is not in the posynomial form [41]. The convex optimization tools can be applied to solve (33) by using the technique in [42] to approximate the target. Specifically, $1+\gamma_{t,k}$ can be approximated by $\lambda_{t,k}\gamma^{\mu_{t,k}}_{t,k}$ close to a point $\hat{\gamma}_{t,k}$, where $\mu_{t,k} = \hat{\gamma}_{t,k}\left(1+\hat{\gamma}_{t,k}\right)^{-1}$ and $\lambda_{t,k} = \hat{\gamma}^{-\mu_{t,k}}_{t,k}\left(1+\hat{\gamma}_{t,k}\right)$. Thus, (33) can be rewritten as

$$\min_{\mathcal{P}^u} \prod_{k=1}^{K_t}\left(\lambda_{t,k}\right)^{-1}\left(\gamma_{t,k}\right)^{-\mu_{t,k}}$$
$$\text{s.t.} \begin{cases} c1: \gamma_{g_{t,k}}\left(c_{g_{t,k}} + d_{g_{t,k}}\right)\left(p^u_{g_{t,k}} b_{g_{t,k}}\right)^{-1} \leq 1, \\ \qquad k = 1,\cdots,K_t \\ \hat{p}2, \hat{p}3, \hat{p}4, \hat{p}5 \end{cases} \tag{35}$$

Constraints $c1$ can be re-expressed as

$$\begin{aligned} c1: &\frac{1}{\left\|\mathbf{h}^{\{\tilde{B}_{g_t}\}}_{u,g_{t,k}}\right\|^2}\left(\gamma_{g_{t,k}}\left|\mathbf{w}^H_{g_{t,k}}\Delta\tilde{\mathbf{h}}^{\{B_{g_t}\}}_{u,g_{t,k},LM}\right|^2\right. \\ &+ \gamma_{g_{t,k}}\left(p^u_{g_{t,k}}\right)^{-1}\sum_{k'=1,k'\neq k}^{K_{g_t}} p^u_{g_{t,k'}}\left|\mathbf{w}^H_{g_{t,k}}\Delta\tilde{\mathbf{h}}^{\{B_{g_t}\}}_{u,g_{t,k'},LM}\right|^2 \\ &+ \gamma_{g_{t,k}}\left(p^u_{g_{t,k}}\right)^{-1}\sum_{g'_t\in G_t/\{g_t\}}\sum_{k'=1}^{K_{g'_t}} p^u_{g'_{t,k'}}\left|\mathbf{w}^H_{g_{t,k}}\tilde{\mathbf{h}}^{\{B_{g_t}\}}_{u,g'_{t,k'}}\right|^2 \\ &\left.+\gamma_{g_{t,k}}\left(p^u_{g_{t,k}}\right)^{-1}\|\mathbf{w}^H_{g_{t,k}}\|^2\right) \leq 1, k = 1,\cdots,K_t \end{aligned} \tag{36}$$

The SE optimization problem (35) becomes a standard GP and can be solved by Algorithm 1.

Algorithm 1 Solving problem (33) by GP:

1. Initialization: Compute the initial value $\hat{\gamma}_{t,k}^{(0)}$, $i = 1$, set the value of tolerance ε and maximum iteration number L.
2. For the nth iteration:

- Compute $\mu_{t,k} = \hat{\gamma}_{t,k}^{(n)}\left(1+\hat{\gamma}_{t,k}^{(n)}\right)^{-1}$, $\lambda_{t,k} = \left(\hat{\gamma}_{t,k}^{(n)}\right)^{-\mu_{t,k}}\left(1+\hat{\gamma}_{t,k}^{(n)}\right)$.
- Solve problem (33) by GP: $\min_{\mathcal{P}^u} \prod_{k=1}^{K_t} (\lambda_{t,k})^{-1}(\gamma_{t,k})^{-\mu_{t,k}}$ s.t.$c1, \hat{p}2, \hat{p}3, \hat{p}4, \hat{p}5$

3. If $\max_{k=1,\cdots,K_t} \left|\hat{\gamma}_{t,k}^{(i)} - \hat{\gamma}_{t,k}^{(i-1)}\right| < \varepsilon$, **stop.**
Else, set $n = n+1$ and go back to step **2**.
4. Get the solution $\widetilde{\mathcal{P}}$ according to (14).

When the solution $\widetilde{\mathcal{P}}^u$ of (35) is obtained, we can get the solution $\widetilde{\mathcal{P}}$ according to (14).

Optimize α for fixed $\mathcal{P}$

We fix the transmit power of BS to $\mathcal{P} = \mathcal{P}^*$. Since the ratio used for information transmission can be adjusted by changing the duration of phase II, that is, $(1-\alpha)T$. The uplink rate $R_{g_{t,k}}(\alpha)$ can be expressed as

$$R_{g_{t,k}}(\alpha)|_{\mathcal{P}^*} = (1-\alpha)\log_2\left(1+\gamma_{g_{t,k}}(\alpha)\right) \tag{37}$$

where $\gamma_{g_{t,k}}(\alpha) = p_{g_{t,k}}^u \tilde{b}_{g_{t,k}}\left(\tilde{c}_{g_{t,k}}+1\right)^{-1}$, and $\tilde{b}_{g_{t,k}} = \|\tilde{\mathbf{h}}_{u,g_{t,k}}^{\{B_{g_t}[n]\}}\|^2$, $\tilde{c}_{g_{t,k}} = \left(\mathrm{CEE}_{g_{t,k}} + \mathrm{IUI}_{g_{t,k}} + \mathrm{IGI}_{g_{t,k}}\right)$. Due to the tremendous gain of antenna array applied in mMIMO system, it is reasonable to assume that the uplink received SINR given by (1) is much greater than 1. Hence, we can treat $\tilde{b}_{g_{t,k}}, \tilde{c}_{g_{t,k}}$ as positive variables that are irrelevant to α. From (34) we know that

$$\begin{aligned} p_{g_{t,k}}^u &\geq \frac{P_s\alpha T\left[\frac{1}{2-a\left(\beta\check{p}_{d_{g_{t,k}}}^{\mathrm{H}}-b\right)}-\Omega\right]-E_{g_{t,k}}^{\mathrm{P}}(1-\Omega)}{(1-\Omega)(1-\alpha)T} \\ &= \frac{\alpha}{(1-\alpha)}\frac{P_s\left[\frac{1}{2-a\left(\beta\check{p}_{d_{g_{t,k}}}^{\mathrm{H}}-b\right)}-\Omega\right]}{(1-\Omega)} - \frac{E_{g_{t,k}}^{\mathrm{P}}}{(1-\alpha)T} \\ &= \frac{\alpha}{(1-\alpha)}\tilde{d}_{g_{t,k}} - \frac{E_{g_{t,k}}^{\mathrm{P}}}{(1-\alpha)T} \end{aligned} \tag{38}$$

where $\check{p}_{d,g_{t,k}} = \hat{p}_{d,g_{t,k}}|_{\mathcal{P}^*}$. In order to design a computationally efficient resource allocation algorithm, we focus on a lower bound of the objective function:

$$\begin{aligned} &\max_{\alpha} \ \vec{R}_{\mathrm{SUM}}(\alpha)|_{\mathcal{P}^*} = \sum_{g_{t,k}\in\mathbf{K}_T}(1-\alpha)\log_2\left(1+\vec{\gamma}_{g_{t,k}}(\alpha)\right) \\ &\text{s.t.} \ \ 1>\alpha\geq\frac{\max\left\{\max\{b_{g_t}|g_t\in G_t\},\max\{\ddot{\alpha}_{g_{t,k}}|k\in\mathbf{K}_T\}\right\}}{T} \end{aligned} \tag{39}$$

where $\vec{\gamma}_{g_{t,k}}(\alpha) = \left(\frac{\alpha}{(1-\alpha)}\tilde{d}_{g_{t,k}} - \frac{E_{g_{t,k}}^{\mathrm{P}}}{(1-\alpha)T}\right)\tilde{b}_{g_{t,k}}\left(\tilde{c}_{g_{t,k}}+1\right)^{-1}$ denotes the lower bound of the SINR $\gamma_{g_{t,k}}(\alpha)$. For the sake of simplification, we ignore $\mathcal{P}^*$ in the following derivation. The partial derivative of $\vec{R}_{\mathrm{SUM}}(\alpha)$ can be expressed as

$$\frac{\partial\vec{R}_{\mathrm{SUM}}(\alpha)}{\partial\alpha} = -\sum_{g_{t,k}\in\mathbf{K}_T}\log_2\left(1+\vec{\gamma}_{g_{t,k}}(\alpha)\right) + \sum_{g_{t,k}\in\mathbf{K}_T}\frac{\left(\tilde{d}_{g_{t,k}}-E_{g_{t,k}}^{\mathrm{P}}/T\right)\tilde{b}_{g_{t,k}}\left(\tilde{c}_{g_{t,k}}+1\right)^{-1}}{\ln(2)\left(1+\vec{\gamma}_{g_{t,k}}(\alpha)\right)(1-\alpha)} \tag{40}$$

The two order derivative of $\vec{R}_{\mathrm{SUM}}(\alpha)$ can be expressed as

$$\frac{\partial^2\vec{R}_{\mathrm{SUM}}(\alpha)}{\partial\alpha^2} = -\sum_{g_{t,k}\in\mathbf{K}_T}\frac{\left(\tilde{d}_{g_{t,k}}-E_{g_{t,k}}^{\mathrm{P}}/T\right)^2\left(\tilde{b}_{g_{t,k}}\right)^2}{\left(\ln(2)\left(1+\vec{\gamma}_{g_{t,k}}(\alpha)\right)\left(\tilde{c}_{g_{t,k}}+1\right)\right)^2(1-\alpha)^3} \tag{41}$$

In this paper, we assume that the power of CEE, IUI, and IGI is much larger than AWGN. Hence, we have $\frac{\partial^2\vec{R}_{\mathrm{SUM}}(\alpha)}{\partial^2\alpha} < 0$ for $0 < \alpha < 1$, and we can obtain the optimal α of $R_{\mathrm{SUM}}(\alpha)$ when $\frac{\partial\vec{R}_{\mathrm{SUM}}(\alpha)}{\partial\alpha} = 0$, that is,

$$\begin{aligned} &\underbrace{\sum_{g_{t,k}\in\mathbf{K}_T}\frac{\left(\tilde{d}_{g_{t,k}}-E_{g_{t,k}}^{\mathrm{P}}/T\right)\tilde{b}_{g_{t,k}}\left(\tilde{c}_{g_{t,k}}+1\right)^{-1}}{\ln(2)\left(1+\vec{\gamma}_{g_{t,k}}(\alpha)\right)}}_{B(\alpha)} \\ &= (1-\alpha)\sum_{g_{t,k}\in\mathbf{K}_U}\log_2\left(1+\vec{\gamma}_{g_{t,k}}(\alpha)\right) \\ &= \vec{R}_{\mathrm{SUM}}(\alpha)|_{\mathcal{P}^*} \end{aligned} \tag{42}$$

(42) is a transcendental equation, we can resort to Newton iterative method to solve this problem, that is, $\hat{\alpha}_{k+1} = \hat{\alpha}_k - \frac{\partial\vec{R}_{\mathrm{SUM}}(\alpha)}{\partial\alpha}/\frac{\partial^2\vec{R}_{\mathrm{SUM}}(\alpha)}{\partial^2\alpha}|_{\alpha=\hat{\alpha}_k}$. Let $\hat{\alpha}^{\mathrm{Opt}}$ denote the solution of (42), and if $1 > \hat{\alpha}^{\mathrm{Opt}} \geq \frac{\max\left\{\max\{b_{g_t}|g_t\in G_t\},\max\{\ddot{\alpha}_{g_{t,k}}|k\in\mathbf{K}_T\}\right\}}{T}$, then the solution of (42) is $\alpha^{\mathrm{Opt}} = \hat{\alpha}^{\mathrm{Opt}}$. Moreover, we can see that the maximum sum-rate is $R_{\mathrm{SUM}}^{\mathrm{OPT}} = B(\alpha^{\mathrm{Opt}})$. Otherwise, the solution of (42) is $\alpha^{\mathrm{Opt}} = \frac{\max\left\{\max\{b_{g_t}|g_t\in G_t\},\max\{\ddot{\alpha}_{g_{t,k}}|k\in\mathbf{K}_T\}\right\}}{T}$, and the maximum sum-rate is

$$R_{\mathrm{SUM}}^{\mathrm{Opt}} = \sum_{g_{t,k}\in\mathbf{K}_T} R_{g_{t,k}}(\alpha^{\mathrm{Opt}}) \tag{43}$$

Joint optimization of α and $\mathcal{P}$

In this subsection, we optimize $\mathcal{P}$ and α alternatively to maximize $R_{\mathrm{SUM}}(\mathcal{P},\alpha)$. The detailed algorithm is summarized in Algorithm 2.

Algorithm 2 Solving problem (32):

1): Initial α and $\mathcal{P}$. Set the value of tolerance ξ.

2): For the k-th iteration:

- For fixed α, solve (33) by Algorithm 1 and get the optimal $\mathcal{P}^{(k)}$ in this iteration.
- For fixed $\mathcal{P}$, solve α using the Newton iterative method, that is,

$$\alpha^{(k)} = \alpha^{(k-1)} - \frac{\partial \vec{R}_{\mathrm{SUM}}(\alpha)}{\partial \alpha} / \frac{\partial^2 \vec{R}_{\mathrm{SUM}}(\alpha)}{\partial^2 \alpha} |_{\alpha=\alpha^{(k-1)}}.$$

If $\alpha^{(k)}$ satisfies

$$1 > \alpha^{(k)} \geq \frac{\max\left\{\max\{b_{g_t}|g_t \in G_t\}, \max\{\ddot{\alpha}_{g_{t,k}}|k \in \mathbf{K}_T\}\right\}}{T},$$

then $\alpha^{(k)} = \alpha^{(k)}$. Otherwise,

$$\alpha^{(k)} = \frac{\max\left\{\max\{b_{g_t}|g_t \in G_t\}, \max\{\ddot{\alpha}_{g_{t,k}}|k \in \mathbf{K}_T\}\right\}}{T}.$$

3): If $|B(\alpha^{(k)}) - R_{\mathrm{SUM}}(\alpha^{(k)}| < \xi$, stop, else set $k = k + 1$ and back to step 2).

In each iteration of Algorithm 2, the update of α is performed by evaluating the first-order condition numerically, whose complexity is similar with the approach in [43].

Results and discussion

In this section, the performance of the proposed beam-domain SWIPT protocol is evaluated using the 3GPP LTE simulation model in macro-cell environment [44]. The center frequency and bandwidth are set to 2.4 GHz and 20 MHz, respectively. The path loss between BS and TEs is modeled as $2.7 + 42.8\log_{10}(R)$ [dB] and the path loss between terminals are modeled as $55.78{+}40\log_{10}(R)$ [dB], where R denotes the distance. Thermal noise density is set to -174 dBm/Hz. The power ratio used for energy harvesting β is set to 0.8. The energy conversion efficiency η for linear harvester is set to 0.8. For the non-linear EH [28, 29], P_s in this paper is set to 0.024 mW, $a = 150$, and $b = 0.014$, as shown in Fig. 2. Hence, the energy conversion efficiency for non-linear EH is not a fixed number. For comparison, the performance of traditional mMIMO SWIPT protocol with MF precoder [31], with ZF precoder [32], and with MRT precoder [33] are also simulated. In the simulations, the uplink transmit power of TEs in [31–33] is also harvested from the BS downlink transmit signal. In simulations, the CSI obtained by beam-domain channel estimation is adopted in the above schemes, respectively.

Figure 5 illustrates the SE of the proposed beam-domain SWIPT protocol for mMIMO system. All the TEs are gathered into three groups, respectively. Each group contains four TEs, and the DOA regions of TEs in each group are identical. The detailed DOA regions of three groups are $[-45°, -35°]$, $[12°, 22°]$, and $[42°, 52°]$, respectively. The resulting AB sets for all the groups satisfy the TE grouping criteria in [11]. The transmit power of the BS is set to 20.2 dBm, and the distance between TEs and BS is set to 500 m. As a result, the average BS receive signal-to-noise ratio (SNR) is 10 dB. We

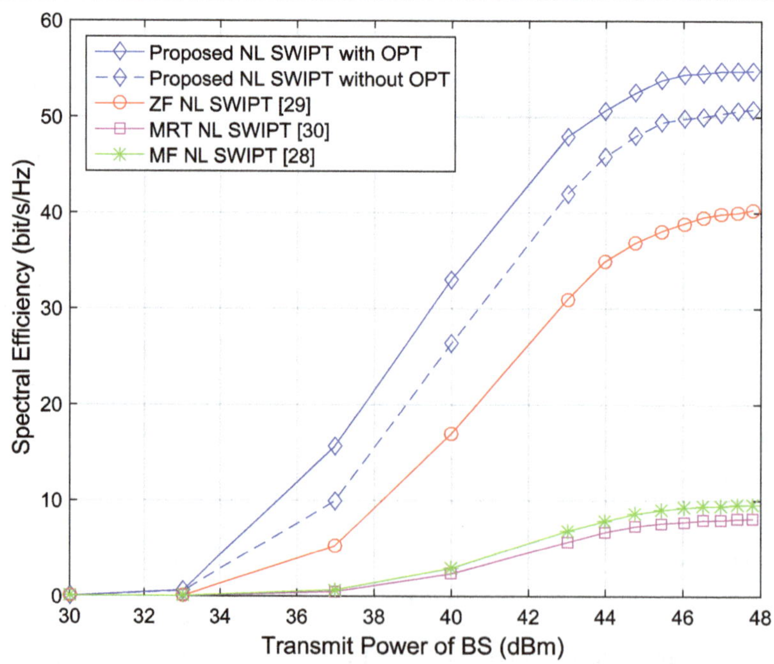

Fig. 5 Spectral efficiency of the proposed SWIPT protocol, the channel coherent time is 300 symbols, $N = 256$. The processing energy of TE is set to 10^{-8} mJ

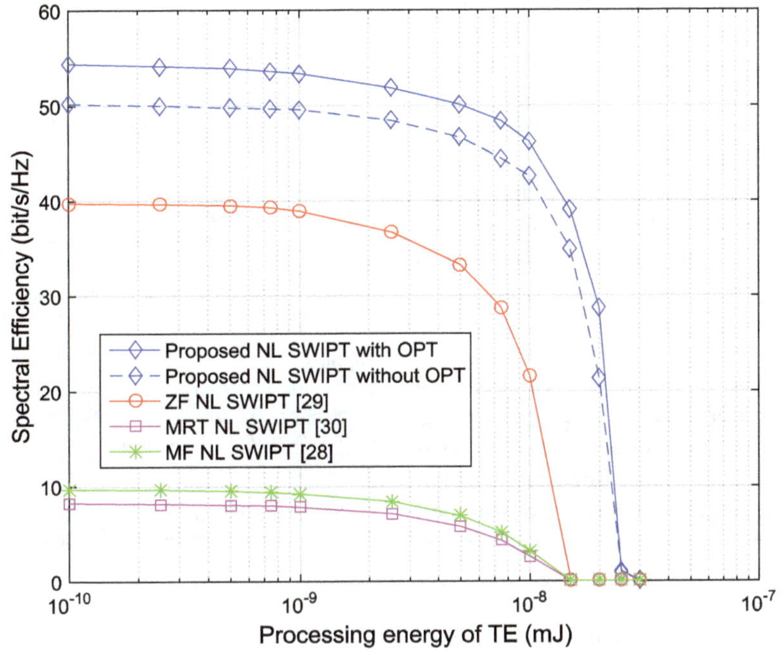

Fig. 6 Spectral efficiency of the proposed SWIPT protocol with the varying of the processing energy of TE, the channel coherent time is 300 symbols, $N = 256$. The transmit power of the BS is set to 46 dBm

can see from Fig. 5 that with estimated effective beam-domain CSI, the required length of pilot sequence can be reduced significantly; hence, significant SE gain can be achieved over the traditional mMIMO SWIPT protocols [31–33].

Figure 6 depicts the SE of the proposed beam-domain SWIPT protocol with the varying of the processing energy of TE. In Fig. 6, the number of antennas at the BS is $N = 256$ antennas. We can see from Fig. 6 that by optimizing the TS ratio and the transmit powers of the BS for

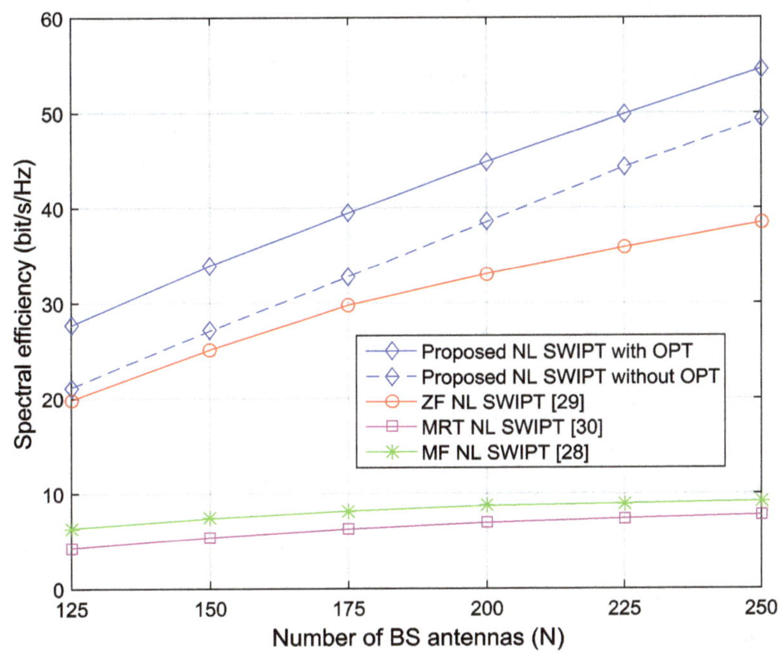

Fig. 7 Spectral efficiency of the proposed SWIPT protocol with the varying of the number of antennas, the channel coherent time is 300 symbols. The transmit power of the BS is set to 46 dBm

different terminals, the proposed beam-domain SWIPT protocol obtains the best performance. When the processing energy of TE is grater than 2.6×10^{-8} mJ, the SE performance of all the schemes is close to 0. Figure 7 depicts the SE of the proposed beam-domain SWIPT protocol with the varying of the processing energy of TE. We can conclude from Fig. 7 that the SE performance gap between the proposed scheme and traditional mMIMO SWIPT protocols [31–33] will increase as the number of antennas increases.

Conclusion

In this paper, we propose a beam-domain SWIPT protocol for mMIMO system with non-linear energy harvesting. In order to reduce the pilot and feedback resource cost used for channel estimation, we resort BEM to represent mMIMO channel. In the wireless energy transmission phase, the beam domain energy beamforming is applied. In the information transmission phase, the two-layer receive beamforming is applied for the reception of signals transmitted by TEs. In order to improve the SE of the system, the BS transmit power- and time-switching ratio are optimized. Simulation results show the superiority of the proposed protocol on spectral efficiency compared with the conventional mMIMO SWIPT protocols.

Endnotes

[1] Different from the previous works in [45, 46], in this paper, the nonlinear energy harvesting model is applied, while in [45, 46], the linear energy harvesting model is applied. A practical energy consumption model is applied in this paper. Hence, this paper will meet new problems such as more practical harvested energy and available power at TEs, the sum-rate performance optimization under new constraints.

[2] Processing energy includes all energy consumption except for TE uplink transmission, such as the energy consumption of the channel estimate, the energy consumption of the circuit, etc.

[3] This scenario is currently widely studied [43, 47]. In this paper, we use beam domain downlink channel estimation. Since the effective channel length in the beam domain is greatly compressed, the feedback overhead for downlink channel estimation will be greatly reduced.

Abbreviations

3D: Three-dimensional; 5G: 5th-generation; AB: Active beam; AF: Amplify-and-forward; AP: Access point; AWGN: Additive white Gaussian noise; BD: Beam-domain; BEM: Basis expansion model; BS: Base station; CEE: Channel estimation error; CSI: Channel state information; DOA: Direction of arrival; DOD: Direction of departure; EE: Energy efficiency; EH: Energy harvesting; GP: Geometric programming; HD: Half-duplex; ID: Information decoding; IGI: Inter-group interference; IoT: Internet of Things; IUI: Inter-user interference; LMMSE: Linear minimum mean square error; LS: Least square; MF: Matched filter; MIMO: Multi-input multi-output; MRC: Maximal ratio combining; MRT: Maximum ratio transmission; PS: Power splitting; RF: Radio frequency; SE: Spectral efficiency; SNR: Signal-to-noise ratio; SI: Self-interference; SINR: Signal-to-interference-and-noise ratio; SWIPT: Simultaneous wireless information and power transfer; TE: Terminal; TS: Time switching; TWR: Two-way relaying; WET: Wireless energy transfer; ZF: Zero-forcing

Acknowledgements

The authors would like to thank the reviewers for their careful readings and valuable comments. This work is supported by Jiangsu Province Natural Science Foundation under Grant BK20160079, National Natural Science Foundation of China (No. 61671472,61771486).

Funding

This work is supported by Jiangsu Province Natural Science Foundation under Grant (BK20160079, BK20181335), National Natural Science Foundation of China (No. 61671472,61771486).

Authors' contributions

KX is the main writer of this paper and proposed the main idea. MZ, ZS, NS, WX, and LC assisted in the simulations and analysis. All authors read and approved the final manuscript.

Authors' information

Kui Xu was born in 1982. He received the B.S. degree in wireless communications and the Ph.D. degree in software defined radio from the PLA University of Science and Technology, Nanjing, China, in 2004 and 2009, respectively. He is currently an Associate Professor with the College of Communications Engineering, Army Engineering University of PLA, Nanjing, China. Since 2013, he has been a Postdoctoral Fellow with the PLA University of Science and Technology. His research interests include broadband wireless communications, signal processing for communications, network coding, and wireless communication networks. He has authored about 50 papers in refereed journals and conference proceedings and holds five patents in China. He is currently serving on the Technical Program Committee of the IEEE WCSP 2014. He received the URSI Young Scientists Award in 2014 and the 2010 Ten Excellent Doctor Degree Dissertation Award of PLAUST. He also serves as the Reviewer of the IEEE TRANSACTIONS ON WIRELESS COMMUNICATION, the IEEE TRANSACTION VEHICLE TECHNOLOGY, the IEEE COMMUNICATIONS LETTER, the IEEE SIGNAL PROCESSING LETTERS.
Ming Zhang was born in 1992. He received the B.E. degree in communications engineering from the Xidian University in 2015, he is currently working toward the Ph.D. degree with the Institution of Communications Engineering in Army Engineering University of PLA. His research interests include MIMO techniques, heterogeneous network, full-duplex communication, and network coding.
Jie Liu was born in 1984. He received the B.E. and M.S. degrees in communications engineering from the PLA University of Science and Technology in 2006 and 2009. His research interests include MIMO techniques, heterogeneous network, full-duplex communication, and network coding.
Nan Sha was born in China, 1981. He received the B.S. degree in wireless communications and the Ph.D. degree in software defined radio from the PLA University of Science and Technology, Nanjing, China, in 2003 and 2013, respectively. He is currently an Associate Professor with the College of Communications Engineering, Army Engineering University of PLA, Nanjing, China. His research interests include broadband wireless communications, signal processing for communications, network coding, and wireless communication networks.
Wei Xie was born in 1972. He received the B.S. and Ph.D degrees in communication engineering in 1999 and 2015, respectively, from the PLA University of Science and Technology, Nanjing, China. His research interests include MIMO techniques, heterogeneous network, full-duplex communication, and network coding.
Lihua Chen was born in 1972. She received the B.S. and M.S. degrees in communication engineering in 1993 and 2005, respectively, and the Ph.D. degree in communications and information system from the PLA University of Science and Technology, Nanjing, China, in 2013. She is currently a Professor in the College of Communications Engineering, Army Engineering University of

PLA, Nanjing, China. Her research interests include new generation wireless mobile communication system, radio resource management, and network coding in wireless communication.

Competing interests

The authors declare that they have no competing interests.

References

1. F. Javed, M. K. Afzal, M. Sharif, B. Kim, Internet of Things (IoT) operating systems support, networking technologies, applications, and challenges: a comparative review. IEEE Commun. Surv. Tutor. **20**(3), 2062–2100 (2018)
2. X. Lu, P. Wang, D. Niyato, D. I. Kim, Z. Han, Wireless charging technologies: fundamentals, standards, and network applications. IEEE Commun. Surv. Tutor. **18**(2), 1413–1452 (2016)
3. I. Krikidis, S. Timotheou, S. Nikolaou, G. Zheng, D. W. K. Ng, R. Schober, Simultaneous wireless information and power transfer in modern communication systems. IEEE Commun. Mag. **52**(11), 104–110 (2014)
4. Z. Ding, et al., Application of smart antenna technologies in simultaneous wireless information and power transfer. IEEE Commun. Mag. **53**(4), 86–93 (2015)
5. F. Rusek, D. Persson, B. K. Lau, E. G. Larsson, T. L. Marzetta, O. Edfors, F. Tufvesson, Scaling up MIMO: opportunities and challenges with very large arrays. IEEE Signal Proc. Mag. **30**(1), 40–60 (2013)
6. J. Hoydis, S. Brink, M. Debbah, Massive MIMO in UL/DL of cellular networks: how many antennas do we need? IEEE J. Sel. Areas Commun. **31**, 160–171 (2013)
7. E. G. Larsson, F. Tufvesson, O. Edfors, T. L. Marzetta, Massive MIMO for next generation wireless systems. IEEE Commun. Mag. **52**(2), 186–195 (2014)
8. C. Sun, X. Gao, S. Jin, M. Matthaiou, Z. Ding, C. Xiao, Beam division multiple access transmission for massive MIMO communications. IEEE Trans. Commun. **63**(6), 2170–2184 (2015)
9. X. Xia, K. Xu, Y. Wang, Y. Xu, A 5G-enabling technology: benefits, feasibility, and limitations of in-band full-duplex mMIMO. IEEE Veh. Technol. Mag. **13**(3), 81–90 (2018)
10. D. K. Nguyen, D. N. Jayakody, S. Chatzinotas, J. Thompson, J. Li, Wireless energy harvesting assisted two-way cognitive relay networks: protocol design and performance analysis. IEEE Access. **5**, 7–21460 (2017)
11. K. Xu, Z. Shen, X. Xia, Y. Wang, D. Zhang, Hybrid time-switching and power splitting SWIPT for full-duplex massive MIMO systems: a beam-domain approach. IEEE Trans. Veh. Technol. **67**(8), 7257–7274 (2018)
12. K. Xu, Z. Shen, X. Xia, Y. Wang, D. Zhang, Beam-domain hybrid time-switching and power-splitting SWIPT in full-duplex massive MIMO system. EURASIP J. Wirel. Commun. Networking. **2018**, 1–21 (2018)
13. G. Pan, H. Lei, Y. Yuan, Z. Ding, Performance analysis and optimization for SWIPT wireless sensor networks. IEEE Trans. Commun. **65**(5), 2291–2302 (2017)
14. Z. Liu, W. Du, D. Sun, Energy and spectral efficiency tradeoff for massive MIMO systems with transmit antenna selection. IEEE Trans. Veh. Technol. **66**(5), 4453–4457 (2017)
15. K. Huang, X. Zhou, Cutting the last wires for mobile communications by microwave power transfer. IEEE Commun. Mag. **53**(6), 86–93 (2015)
16. A. A. Nasir, X. Zhou, S. Durrani, R. A. Kennedy, Relaying protocols for wireless energy harvesting and information processing. IEEE Trans. Wirel. Commun. **12**(7), 3622–3636 (2013)
17. Y. Huang, B. Clerckx, Relaying strategies for wireless-powered MIMO relay networks. IEEE Trans. Wirel. Commun. **15**(9), 6033–6047 (2016)
18. C. Li, H. J. Yang, F. Sun, J. M. Cioffi, L. Yang, Multiuser overhearing for cooperative two-way multiantenna relays. IEEE Trans. Veh. Technol. **65**(5), 3796–3802 (2016)
19. C. Li, S. Zhang, P. Liu, F. Sun, J. M. Cioffi, L. Yang, Overhearing protocol design exploiting inter-cell interference in cooperative green networks. IEEE Trans. Veh. Technol. **65**(1), 441–446 (2016)
20. C. Li, P. Liu, C. Zou, F. Sun, J. M. Cioffi, L. Yang, Spectral-efficient cellular communications with coexistent one-and two-hop transmissions. IEEE Trans. Veh. Technol. **65**(8), 6765–6772 (2016)
21. Z. Chen, B. Xia, H. Liu, in *IEEE Global Conference on Signal and Information Processing (GlobalSIP)*. Wireless information and power transfer in two-way amplify-and-forward relaying channels (IEEE Press, Piscataway, 2014), pp. 168–172
22. K. Tutuncuoglu, B. Varan, A. Yener, Throughput maximization for two-way relay channels with energy harvesting nodes: the impact of relaying strategies. IEEE Trans. Commun. **63**(6), 2081–2093 (2015)
23. D. Li, C. Shen, Z. Qiu, in *2013 IEEE International Conference on Communications (ICC)*. Two-way relay beamforming for sum-rate maximization and energy harvesting (IEEE Press, Piscataway, 2013), pp. 3155–3120
24. B. Clerckx, Wireless information and power transfer: nonlinearity, waveform design, and rate-energy tradeoff. IEEE Trans. Signal Process. **66**(4), 847–862 (2018)
25. H. Tran, G. Kaddoum, K. T. Truong, Resource allocation in SWIPT networks under a nonlinear energy harvesting model: power efficiency, user fairness, and channel nonreciprocity. IEEE Trans. Veh. Technol. **67**(9), 8466–8480 (2018)
26. T. Le, G. Kaddoum, O. Shin, Joint channel resources allocation and beamforming in energy harvesting systems. IEEE Wirel. Commun. Lett. **7**(5), 884–887 (2018)
27. H. Tran, G. Kaddoum, in 2018 *IEEE Wireless Communications and Networking Conference (WCNC)*. Green cell-less design for RF-wireless power transfer networks, (Barcelona, 2018), pp. 1–6
28. E. Boshkovska, D. W. K. Ng, N. Zlatanov, A. Koelpin, R. Schober, Robust resource allocation for MIMO wireless powered communication networks with non-linear EH model. IEEE Trans. Commun. **65**(5), 1984–1999 (2017)
29. B. Clerckx, R. Zhang, R. Schober, D.WK. Ng, D.I. Kim, H.V. Poor, Fundamentals of wireless information and power transfer: from RF energy harvester models to signal and system designs. IEEE J. Sel. Areas Commun. **37**(1), 4–33 (2019)
30. P. N. Alevizos, A. Bletsas, Sensitive and nonlinear far-field RF energy harvesting in wireless communications. IEEE Trans. Wirel. Commun. **17**(6), 3670–3685 (2018)
31. L. Fan, H. Zhang, Y. Huang, L. Yang, Exploiting BS antenna tilt for SWIPT in 3-D massive MIMO systems. IEEE Wirel. Commun. Lett. **6**(5), 666–669 (2017)
32. X. Wang, J. Liu, C. Zhai, Wireless power transfer based multi-pair two-way relaying with massive antennas. IEEE Trans. Wirel. Commun. **16**(11), 7672–7684 (2017)
33. S. Lee, Y. Zeng, R. Zhang, Retrodirective multi-user wireless power transfer with massive MIMO. IEEE Wirel. Commun. Lett. **7**(1), 54–57 (2018)
34. G. Yang, C. K. Ho, R. Zhang, Y. L. Guan, Throughput optimization for massive mimo systems powered by wireless energy transfer. IEEE J. Sel. Areas Commun. **33**(8), 1640–1650 (2015)
35. 3GPP TR 25.996, Universal mobile telecommunications system (UMTS); spatial channel model for multiple input multiple output (MIMO) simulations. v.12.0.0 (2012). www.3gpp.org. Accessed 7 Mar 2019
36. G. B. Giannakis, C. Tepedelenlioglu, Basis expansion models and diversity techniques for blind identification and equalization of timevarying channels. Proc. IEEE. **86**(10), 1969–1986 (1998)
37. A. Sayeed, Deconstructing multiantenna fading channels. IEEE Trans. Signal Process. **50**(10), 2563–2579 (2002)
38. X. Xia, K. Xu, D. Zhang, Y. Xu, Y. Wang, Beam-domain full-duplex massive MIMO: realizing co-time co-frequency uplink and downlink transmission in the cellular system. IEEE Trans. Veh. Technol. **66**(10), 8845–8862 (2017)
39. S. Kashyap, E. Björnson, E.G. Larsson, On the feasibility of wireless energy transfer using massive antenna arrays. IEEE Trans. Wirel. Commun. **15**(5), 3466–3480 (2016)
40. B. Hassibi, B. M. Hochwald, How much training is needed in multiple-antenna wireless links? IEEE Trans. Inf. Theory. **49**(4), 951–963 (2003)
41. S. Boyd, L. Vandenberghe, *Convex Optimization*. (IEEE Press, Piscataway, 2004)
42. P. C. Weeraddana, M. Codreanu, M. Latva-aho, A. Ephremides, Resource allocation for cross-layer utility maximization in wireless networks. IEEE Trans. Veh. Technol. **60**(6), 2790–2809 (2011)
43. Z. Chang, Z. Wang, X. Guo, Z. Han, T. Ristaniemi, Energy-efficient resource allocation for wireless powered massive MIMO system with imperfect CSI. IEEE Trans. Green Commun. Netw. **1**(2), 121–130 (2017)
44. 3GPP TR 36.828, Further enhancements to LTE time division duplex (TDD) for downlink-uplink (DL-UL) interference management and traffic adaptation. v.11.0.0 (2012). www.3gpp.org. Accessed 7 Mar 2019

45. K. Xu, Z. Shen, Y. Wang, X. Xia, Resource allocation for hybrid TS and PS SWIPT in massive MIMO system. Physical Communication. **28**, 201–213 (2018)
46. K. Xu, Z. Shen, Y. Wang, X. Xia, Beam-domain SWIPT in massive MIMO system with energy-constrained terminals. IET Communications. **12**(15), 1900–1909 (2018)
47. M.M. Zhao, Q. Shi, Y. Cai, M.J. Zhao, Joint transceiver design for full-duplex cloud radio access networks with SWIPT. IEEE Trans. Wirel. Commun. **16**(9), 5644–5658 (2017)

5

Performance of a dense urban massive MIMO network from a simulated ray-based channel

Mohammed Zahid Aslam[1*], Yoann Corre[1], Emil Björnson[2] and Erik G. Larsson[2]

Abstract

Massive MIMO network deployments are expected to be a key feature of the upcoming 5G communication systems. Such networks are able to achieve a high level of channel quality and can simultaneously serve multiple users with the same resources. In this paper, realistic massive MIMO channels are evaluated both in single and multi-cell environments. The favorable propagation property is evaluated in the single-cell scenario and provides perspectives on the minimal criteria required to achieve such conditions. The dense multi-cell urban scenario provides a comparison between linear, planar, circular, and cylindrical arrays to evaluate a large-scale multi-cell massive MIMO network. The system-level performance is predicted using two different kinds of channel models. First, a ray-based deterministic tool is utilized in a real North American city environment. Second, an independent and identically distributed (i.i.d.) Rayleigh fading channel model is considered, as often used in previously published massive MIMO studies. The analysis is conducted in a 16-macro-cell network with both randomly distributed outdoor and indoor users. It is shown that the physical array properties like the shape and configuration have a large impact on the throughput statistics. Although the system-level performance with i.i.d. Rayleigh fading can be close to the deterministic prediction in some situations (e.g., with large linear arrays), significant differences are noticed when considering other types of arrays. The differences in the performance of the various arrays utilizing the exact same network parameters and the same number of total antenna elements provide insights into the selection of these physical parameters for upcoming 5G networks.

Keywords: Massive MIMO, Channel modeling, Ray-based model, Performance evaluation, 5G

1 Introduction

Massive MIMO (maMIMO) technology has been shown to provide the requirements set by the 5G standard in various studies and the literature. It helps to improve the spectral efficiency and utilize the available channel resources more efficiently. Although many operators and equipment manufacturers have today conducted controlled experiments to demonstrate the possible gains of this technology, large-scale deployments of the technology and its network-level performance in real scenarios is yet to be seen.

In this paper, a realistic ray-based propagation channel has been utilized to predict the performance of a dense urban maMIMO network with different settings. Such a study makes it possible to better understand and apply the technology in realistic environments, which can significantly differ from theoretical analysis. It also permits to properly anticipate the possible applications and gains, refine the systems, and prepare the deployment strategies. Single-cell analysis considering multiple simultaneous users [1, 2] in realistic environments only provides a partial perspective of the network performance. In this paper, a multi-cell network is deployed in a real and complex environment, which is precisely represented into 3D geographical data. A combination of both outdoor and indoor users is considered to best reflect realistic conditions, with indoor users distributed in different building floors. The obtained physical channels are then utilized to perform precise system-level simulations [3] leading to the assessment of the user throughput. The simulator implements state-of-the-art signal processing schemes, including transmit precoding and power allocation techniques. It supports multi-cell multi-user scenarios and

*Correspondence: zaslam@siradel.com
[1] SIRADEL, 2 Parc de Brocéliande, St. Grégoire, France
Full list of author information is available at the end of the article

computes inter- and intra-cell interferences. It can take any kind of MIMO channel data as an input, which made the interface with the ray-based channel model straightforward.

Similar to the current 4G macro base station (BS) sites, the maMIMO base stations are considered to be located on the rooftops of dominant buildings with roughly constant inter-site distance. In maMIMO systems, the number of antenna elements can be large as compared to the existing systems, and this permits the possibility of utilizing different antenna shapes like linear, planar, circular, and cylindrical arrays. Those different shapes are considered in this study to evaluate how the performance of the network changes while all other inputs are constant. The same approach can further be applied to different complex networks with varying inter-site distances (ISD), sectorization, and even complemented with a dense small-cell network, in order to determine the best deployment strategy for given objectives into a particular environment. These complex network changes are out of scope of the presented study. The paper also explores how the chosen channel model impacts the estimated maMIMO performance. The throughputs that have derived from the ray-based channel environments are drawn against those of a basic statistical model, as often used in the evaluation studies.

The large number of BS antennas of a maMIMO system allows for large spatial degrees of freedom to serve multiple users in the same time-frequency resource. It actually contributes towards two key maMIMO physical properties known as favorable propagation and channel hardening as described in [4] and [5], respectively. Together, these properties should enable a maMIMO system to spatially separate all the simultaneous users [2] (even if closely located) and provide stable stationary channel conditions, even with small user movements. These conditions are satisfied when very large arrays are considered. But in reality, the array size of a maMIMO system is limited, although large. Especially, when considering 3D arrays like cylindrical arrays, the aperture of the antenna array is limited as compared to a linear array. With these physical limitations, it is important to assess and demonstrate the performance of practical maMIMO systems in real propagation conditions.

The performance of the maMIMO system depends on the cross-correlation between channels, either the multiple channels from all array elements to a single user or the channels from one base station to the simultaneous users (that may cause mutual interference) [2]. For accurate predictions, this requires the channel models to be spatially consistent, i.e., to be able to reproduce variations and correlations between two antenna positions. Independent and identically distributed (i.i.d.) channel models do not support such spatial correlation and can therefore not accurately predict maMIMO channels. Some other models like geometry-based stochastic channel models (GSCM) create spatially correlated channels, thus might be appropriate; however, they are not able to manage the specificities of a particular physical environment [2]. The ray-based deterministic channels also provide inherent correlation, due to the computation of the physical wave propagation and interactions, but have the additional advantage that they can be run in real environments, using geographical map data. Therefore, they are relevant to assess various kinds of network topologies in a very accurate manner.

Section 2 describes briefly the system-level maMIMO simulator and its usage to obtain the rates that are achievable by the maMIMO system. Section 3 details the single-cell and multi-cell multi-user scenarios and setup, located in a North American dense urban area. Section 4 studies the multi-cell maMIMO channel characteristics. Section 5 presents the evaluation of the single-cell study with regard to favorable propagation. Section 6 discusses the system-level maMIMO performance for different antenna array shapes and configurations along with different channel models. Finally, Section 7 provides some conclusions, perspectives, and future work to be conducted on maMIMO system-level studies, with 5G radio planning as an objective.

2 Massive MIMO system-level simulator

The first open-source system-level simulator for maMIMO systems was delivered as supplementary material to the book [3]. This MATLAB simulator supports the state-of-the-art methods for channel estimation, receive combining, transmit precoding, and power allocation, as well as quantifying the corresponding achievable rates. While some of these methods are tailored for statistical channel models, with known first- and second-order moments, in this paper, we use methods that support any channel model. More precisely, we consider least-square channel estimation (under the presence of pilot contamination) and regularized zero-forcing (RZF) precoding. The power allocation is selected to maximize the product of the users' effective SINRs, which is essentially the same as maximizing the sum rate, except that there is a "bias" towards giving higher rates to the most unfortunate users than they would get with pure sum-rate maximization (which could lead to zero rate for the weakest users). The rates presented in this paper are obtained for full-buffer transmission. We have modified the original simulator from [3] to support our setup, for example, by including BS user association.

In a maMIMO system, the number of simultaneous users is typically much smaller than the number of BS antennas, because there is a limited number of pilot sequences available for channel estimation. In this paper, we have considered a maximum of 20 active users that can be connected to any single BS simultaneously. If there are more users, then the system has the possibility to select and allocate some of them on different time slots or frequency sub-bands (maMIMO scheduling). Once a set of users is selected for simultaneous allocation, i.e., at the same time and frequency, the BS allocates 1 of the 20 pilots uniformly at random to the users. Each pilot is only used once per cell, but will be reused across cells, which leads to pilot contamination. The different propagation and interference conditions experienced by the users are partly mitigated by carefully attributing the powers.

3 Simulation methodology and scenario

To evaluate the performance of a maMIMO system, two different scenarios have been considered. A single-cell multiple simultaneous user environment is first evaluated for the minimum conditions similar to favorable propagation (FP) that can be achieved. Then, a multi-cell network of maMIMO macro BS is considered to evaluate the performance of the network considering signal processing schemes for different antenna array shapes. The first scenario provides insights into the minimum horizontal separation required for a maMIMO BS to be successfully able to differentiate between users. Whereas, the second scenario evaluates these results in a large-scale realistic maMIMO environment with 3D spatial distributions of the users.

The selected scenarios utilizes 3D geographic map data from New York City, which includes buildings of varying heights. The selected area contains various different geographic features like narrow and wide streets, open squares, and cross sections. This map data makes it possible to obtain realistic and highly variable channel predictions for the considered scenarios.

To evaluate the favorable propagation conditions, three different sites as shown in Fig. 1, are selected. At each site, two simultaneous users are considered to be located in close proximity to each other. In Fig. 2, a depiction of the gradually increasing linear distance between the users is shown. The deterministic channels are then extracted from the ray-tracing tool for these user positions in close proximity. These channels are then evaluated for the level of favorable propagation in Section 5.

Further, to analyze the network-level performance, a network of macro-cell maMIMO BSs is considered. The carrier frequency is 3.5 GHz. All the BSs are equipped with directional antenna elements with a horizontal half power beamwidth (HPBW) of 90°, and a vertical HPBW of 20°, except in the results of Fig. 14 where this element is compared with another having a vertical HPBW of 60°. All the antennas are vertically polarized with a maximum gain of 13.3 dBi. The 3D beam pattern of the antenna

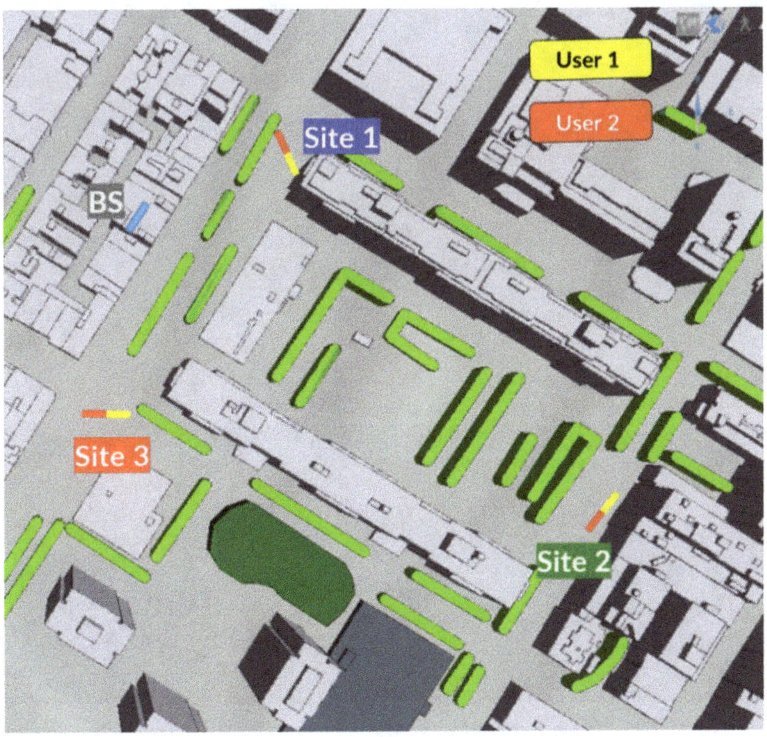

Fig. 1 Single-cell macro base station scenario

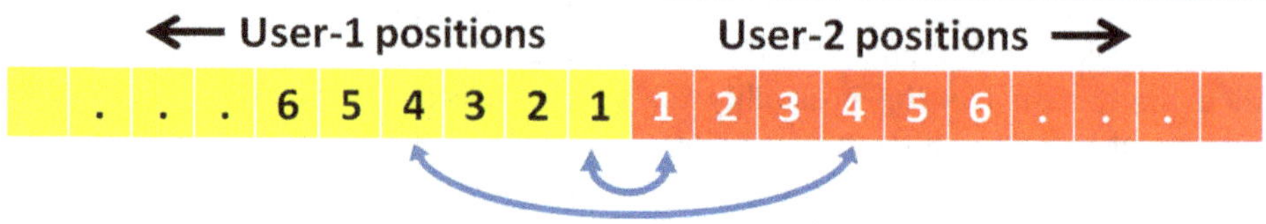

Fig. 2 Illustration of the separation of the simultaneous users in the single-cell macro base station scenario

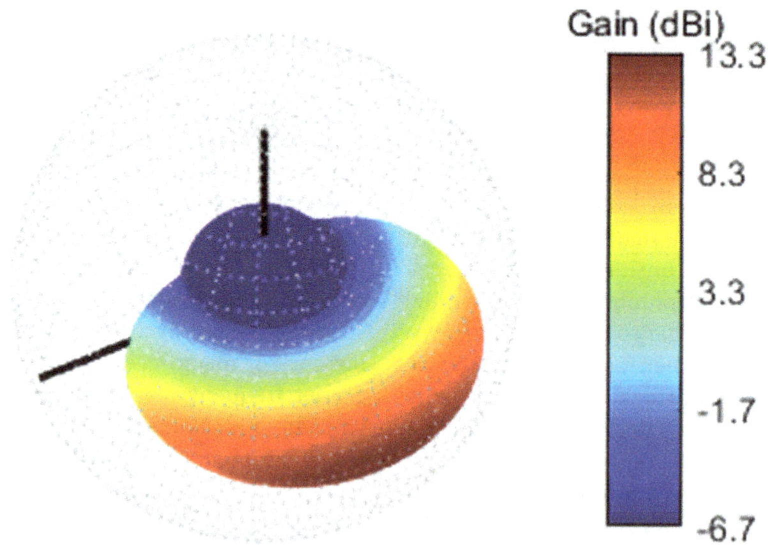

Fig. 3 Directional antenna array pattern of each antenna element in the maMIMO array

Fig. 4 Massive MIMO BS deployment area. The BSs are depicted by blue dots. The study area is 1.35 km^2. The different colors in the streets represent the color corresponding to their best serving BS for that area

used is shown in Fig. 3. The total transmit power before the application of maMIMO power allocation schemes is 43 dBm over a 20-MHz bandwidth. The network is composed of 16 maMIMO BSs located on top of dominant (or quasi dominant) buildings such that the area of interest shown in Fig. 4 is fully covered. Each BS array is formed of 192 antennas. The average inter-site distance (ISD) is 320 m over an area of 1.6 km^2. The furthest BSs (diagonally opposite) are located 1.365 km apart. The user equipment (UE) consists of a single omni-directional antenna with vertical polarization, located 1.5 m above the ground/floor level. Both indoor users located at multiple floors and outdoor users are simultaneously considered.

The objective of evaluating the network-level performance of a maMIMO system in real environments is to predict the maMIMO gains as accurately as possible. Such an approach also enables the comparison of physical parameters like different antenna shapes and their corresponding performance in the same network. Results in [1] and [6] suggest that the rate performance in many real propagation environments is comparable to the i.i.d. Rayleigh fading. Some explanations of the reason for this is given in [4]. This is due to the favorable propagation and channel hardening properties of maMIMO, which appear as the number of antennas at the BS increase. In reality, the number of BS antennas are limited (far from infinity) and deployed in a limited amount of space, depending on the antenna configuration. This limits the ability of large antenna arrays to observe various channel components. This is demonstrated in this work by comparing several antenna sizes containing the same number of elements using a ray-based deterministic model.

4 Multi-cell massive MIMO channel characterization

The complex propagation environment creates rich multipath components, which are predicted by the physical volcano urban ray-based model [7]. These multi-paths affect the maMIMO performance, either by creating spatial separability of users located close to each other or, on the contrary, by producing cross-correlation between users reached by the same canyoning or diffracting phenomenon. The use of a physical channel model does intrinsically bring all correlation factors, between cells and users, which are required for accurate maMIMO assessment.

In Fig. 5, an example of two different users connected to the same BS is shown. It can be seen that the propagation paths to user 1 are significantly different to those to user 2. Since user 1 is located in an open square, the propagation environment is well spread out and multiple strong propagation paths are available to the user; this would constitute a strong channel. In case of user 2, however, the propagation environment is limited due to the narrow street, and multiple strong propagation paths are not possible to be connected to this user. Both these users can be simultaneously served by the same BS since the channels are very different, and the BS must optimize its available resources to provide the best performance. This

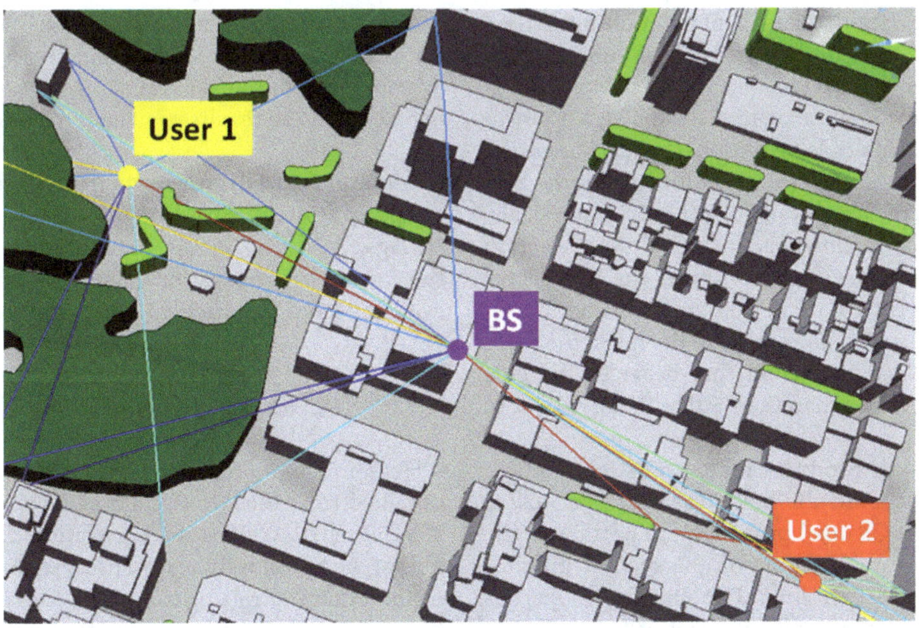

Fig. 5 Example of two users connected to the same BS having very different propagation paths

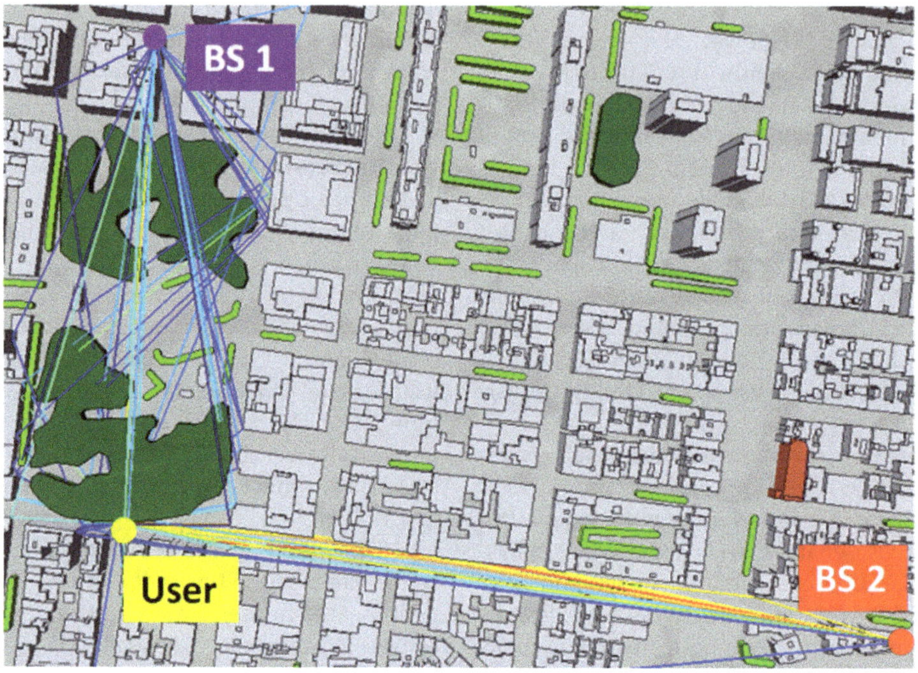

Fig. 6 Example of single user connected to multiple BSs by different propagation paths (canyoning/open square)

is achieved by jointly selecting the precoding and power allocation at the BS. In this paper, the maximum product SINR power allocation scheme is used along with the RZF transmit precoder.

In Fig. 6, an example of a single user with connections to two different BSs is shown. The user location is chosen such that the different propagation environments can be demonstrated. The propagation paths to the user from BS-1 in the large square contain multiple strong line-of-sight (LoS) and reflected paths. From BS-2, strong propagation paths are limited due to the urban canyon effect in the narrow street that reduces the signal power significantly due to multiple reflections. In such cases, the serving BS must be selected such that it provides the best received power to the user, even if another BS is physically located closer. This is known as the best server selection and is used in this paper. Figure 4 shows the best server selection by different colors in the streets, which correspond to a particular BS. To maximize the rates, the users are always connected to the BS that provides the best channels to them rather than the BS that is the closest to the user.

The channel model produces multi-paths from reflections and diffractions on the building facade, as well as rooftop diffraction. Channel properties, in particular, the departure angles, are depending on the specific physical situation of each BS-user link. The channel properties are predicted from an antenna element at the center of the BS array and then extrapolated to the whole array, by assuming the multi-paths are stationary (i.e., persistent along with constant amplitude).

In Fig. 7, an illustration of the maMIMO BS connected to a random selection of simultaneous users located at different horizontal and vertical locations in the New York City map data is shown. The indoor users are located at different floors of multiple buildings of varying heights, and user-2 is an outdoor user on the street level. Although the users are located at different heights, from the macro BS location, a larger horizontal spread can be observed as compared to the vertical spread. The impact of this is explained in the results provided in Section 6.

5 Results and discussions from favorable propagation simulations

In Figs. 8 and 9, Eq. (1) [2] has been used to evaluate the favorable propagation in the deterministic channels obtained from the three different sites from Fig. 9.

$$\text{FPmetric} = \frac{|h_i{}^H h_k|^2}{\|h_i\|^2 \|h_k\|^2} \tag{1}$$

Here, h_i and h_k represent the different channel vectors obtained from the two simultaneous closely spaced users. The FPmetric verifies the correlation between the channels of the two closely spaced users such that a lower value of the FPmetric relates to better FP-like conditions. Further details on the equation and favorable propagation can be found in [2–4].

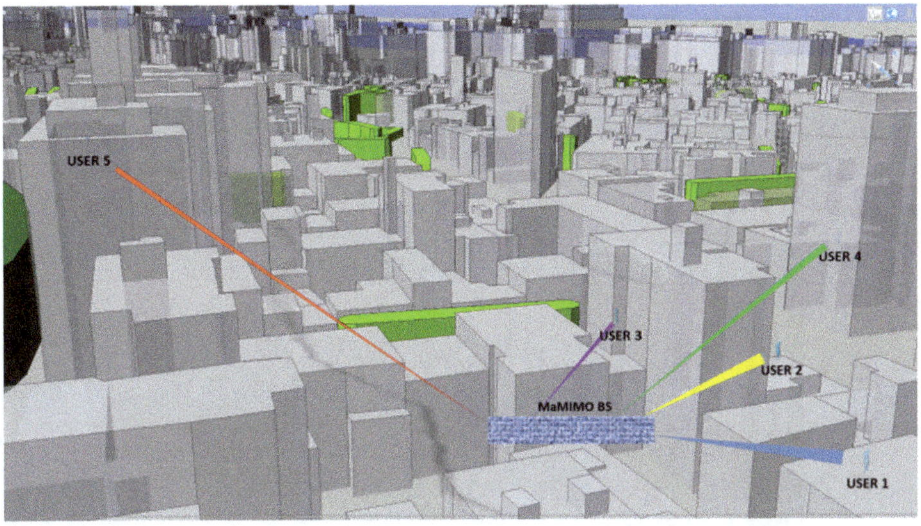

Fig. 7 Illustration of the outdoor and indoor user distribution on multiple floors of a typical North American city

When the geometric distance between the two users is considered, the propagation channels cannot be distinguished until a distance of 20 wavelengths (corresponding to 3 m for a frequency of 2 GHz). For the given scenarios, as the minimum distance between the two users increases to 20 wavelengths, FP-like conditions can be observed.

The distance between the BS and the user is also important to be able to differentiate the user channels, as observed from the BS. Two simultaneous users at different distances from the BS may experience different FP-like conditions. For this, the angle between the two users as observed from the BS provides a better perspective of the FP-like conditions. In Fig. 1, the users located at site 2 achieve FP-like conditions at the smallest angle. This is due to the distance between the BS and the users which is the largest in this location. In all three locations, FP-like conditions are achieved for an angular separation of 2.5°, between the users.

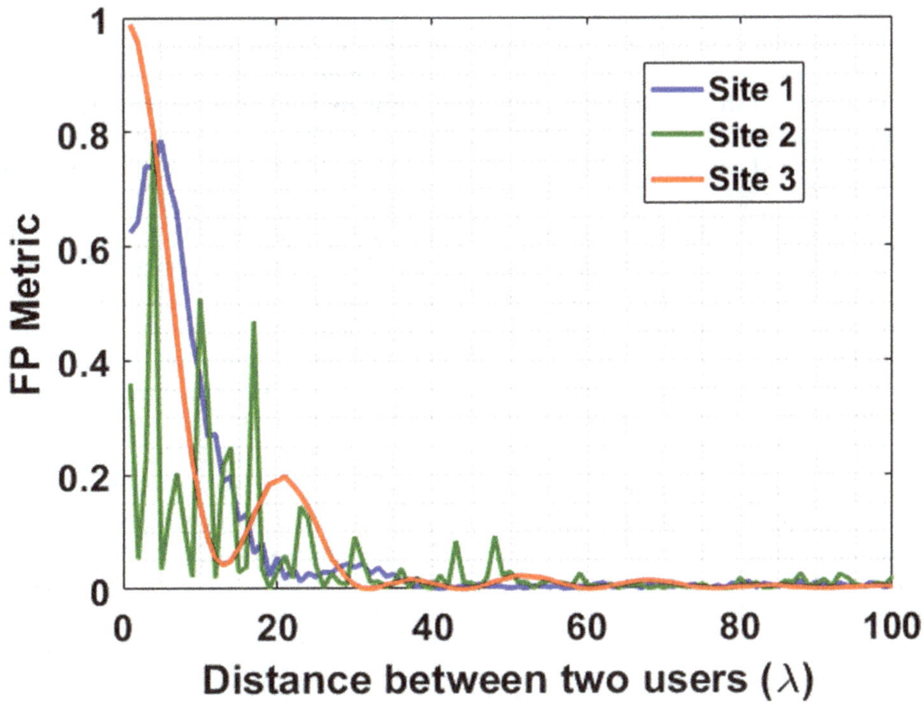

Fig. 8 Minimum distance required between the two users to obtain FP-like conditions

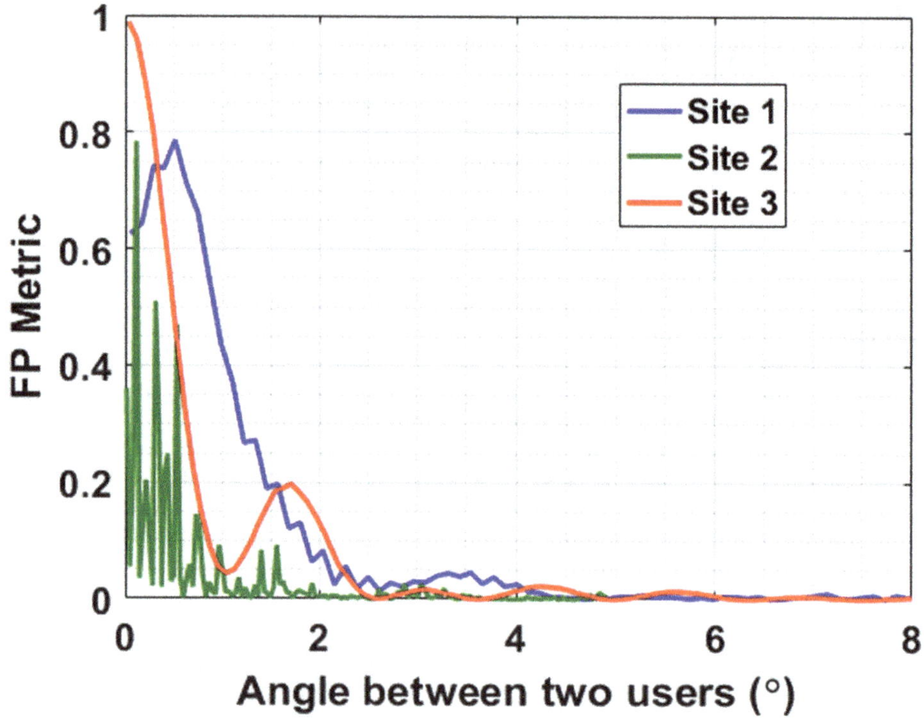

Fig. 9 Minimum angle required between the two users to obtain FP-like conditions

6 Results and discussions from massive MIMO multi-cell simulations

The performance statistics are obtained from a Monte-Carlo procedure. For the area covered by the 16 cells, 150 users are randomly dropped at each iteration. These user positions are a subset of 320 random user positions that are uniformly distributed in the entire computation area. The maMIMO propagation from all BSs to all 320 user positions are pre-computed. Then, each iteration is conducted in two successive steps; first, the maMIMO channel matrices are created for all BS-user combinations (based on pre-computed data); then, the system-level simulation is run as described in Section 2, including cell selection, precoding, power allocation optimization, and computation of achievable rates.

The system-level simulator presented in [3] is applied to the multi-cell maMIMO scenario described in Section 2. The channel data is obtained either from the statistical model (i.i.d. Rayleigh) or from importing the channel realizations created by the ray-based model. Several antenna array configurations are implemented and tested: linear, planar, circular, and cylindrical. Communication-theoretic data rate expressions given in [3] are used to evaluate the maMIMO performance.

All the considered arrays are uniform, and the adjacent antenna elements are separated by half wavelength. The linear array is a single row of 192 antenna elements that are all directed in the same direction. It is more than 8 m long at 3.5 GHz. Due to this size, the practical installation of a 3.5-GHz linear array is unlikely, but this configuration is a particular case for which it is important to assess and compare the performance; that is why it is included in our study. The circular array is a single circular row of 192 antennas that face outwards from the circle with different but uniform orientations around the circle. Its physical diameter is 2.6 m, which remains quite large even for a macro base station. The planar array consists of 24 antenna elements in each of the 8 rows for a total of 192 antenna elements. Its dimensions are obviously reasonable for a practical deployment, with a length of 1 m and a height of 0.34 m. The cylindrical array is obtained similarly by considering a 24 × 8 setup with circular shape. The cylindrical antenna has the smallest dimensions among all the considered antennas; its diameter is less than 0.3 m. It is interesting to note that although the physical number of antenna elements is the same for the different array shapes, the dimensions can vary. These changes can significantly impact the performance of the network and have to be considered for accurate evaluations.

In Fig. 10, the cumulative distributive function (CDF) of the downlink (DL) throughput/rate per UE [3] is given for different antenna array configurations. In each simulation, outdoor users are randomly dropped in the considered

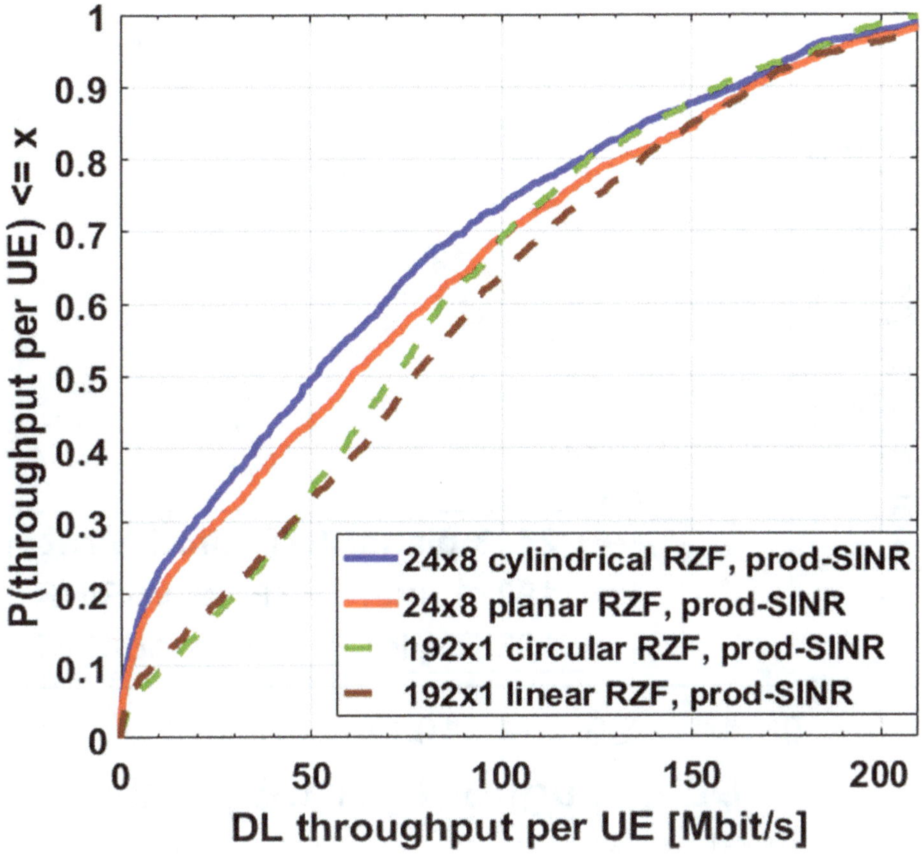

Fig. 10 DL throughput per UE for different antenna configurations considering regularized zero-forcing precoding and prod-SINR power control optimization

area. At the median percentile, the linear array performs the best with a throughput per UE of 76 Mbit/s. The cylindrical array performs the worst with a throughput per UE of 52 Mbit/s. This difference can be explained by the spatial aperture of each antenna array. The cylindrical array containing 192 antennas (24 × 8) has the smallest radius (dimension) of all the selected configurations. This is followed by the planar, circular, and linear arrays, which have similar median performance. At the highest percentiles, however, the performance of all different array configurations converges when the users all have good/favorable channel conditions in the outdoor scenario.

In Fig. 11, the i.i.d. Rayleigh fading channel is considered with the same multi-cell system-level simulator from [3]. The channel is defined only by the number of antenna elements for each user position and no correlation along the antenna coefficients or between user positions are considered. This model has the advantage of being easily implemented and analyzed, since there are closed-form rate expressions available in the literature [3], but the accuracy of the results is limited. In order to fairly compare the i.i.d. Rayleigh fading results with the deterministic setup, the path losses obtained from the deterministic setup are also used in the i.i.d. Rayleigh fading setup. This process makes sure that the path losses between the two different channel models are the same, and therefore, we are able to compare the impact of the two channel models in terms of angular diversity and cross-link correlation. It can be seen from Fig. 11 that the performance of the normalized i.i.d. Rayleigh channel is similar to that of the deterministic channel in case of the linear array. At the median percentile, both the i.i.d. Rayleigh fading channel and the deterministic channel give a throughput of 80 Mbit/s per UE. However, if we consider the planar array, a significant loss in throughput can be observed. When considering the deterministic model, a DL throughput/rate per UE of 60 Mbit/s is obtained at the median value for the planar array. This difference of 20 Mbit/s between the planar and linear arrays is not seen in the i.i.d. Rayleigh model. The difference that is not captured by the i.i.d. Rayleigh model is due to the smaller aperture in the horizontal direction of the planar array.

For the sake of a comprehensible comparison, we were able to set the path losses in both deterministic and i.i.d. Rayleigh models to be the same. However, in reality, if we only consider i.i.d. Rayleigh fading channels, we normally

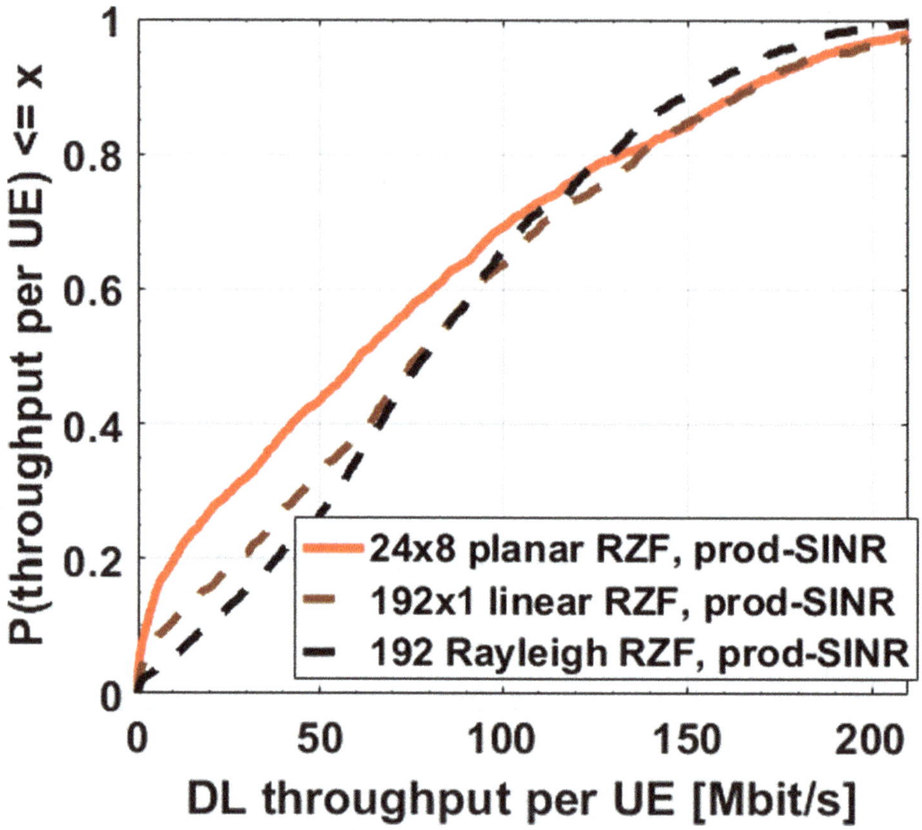

Fig. 11 DL throughput per UE for comparison of ray-based deterministic channel and i.i.d. Rayleigh based channel as applied to system-level simulations

do not have access to accurate models and the path loss information is obtained from more simple approaches (empirical models) which has to be chosen very carefully as small variations in the selection of the model can lead to very different results.

The largest throughput per UE utilizing the deterministic channels corresponds to the linear array followed by the arrays with the largest horizontal dimension. This can be explained as all the users considered are distributed in the horizontal plane (only located outdoors) and the vertical angle of the propagation paths does not vary much from one user to another. Therefore, the BS is mainly required to serve users by separating them in the horizontal direction. The addition of vertical antenna elements then does not contribute as much to the throughput.

In the last stage of our study, we considered a 3D user distribution, with 75% indoor users located at a random floor inside the buildings. The varying building heights are properly considered, so the highest buildings have more users, being uniformly distributed from the ground to the top floor. Due to the height dimension, even two users located at the same 2D location can be separated, as the minimum distance between two floors (3 m) could be sufficient to separate the users. This provides an additional degree of freedom to the BS to separate closely spaced users and provide higher throughput.

This new scenario, where only the ray-based model is involved, investigates how the insertion of the vertical dimension in the user distribution affects the estimated maMIMO performance.

Both user distributions (outdoor only, or mixed) are considered in Fig. 12. The linear antenna array performs the best in both the user distributions as compared to the planar array. Although the indoor users provide better separability to the BS, the attenuation caused due to the outdoor-to-indoor propagation loss results in lower values of the data throughput. The linear array performance is expected to be better in the outdoor scenario as discussed in relation to Fig. 10; however, similar performance is noticed in the scenario with indoor users as well. This is due to the fact that the horizontal dimension is significantly dominant as compared to the vertical dimension, even when multi-floor users are simulated. At each simultaneous time-frequency slot, each BS is connected to a maximum of ten users. These users are located at different locations of the macro-cell, and although they are located at different heights, the horizontal spread of the users is greater. Indeed, if all the simultaneous users are located

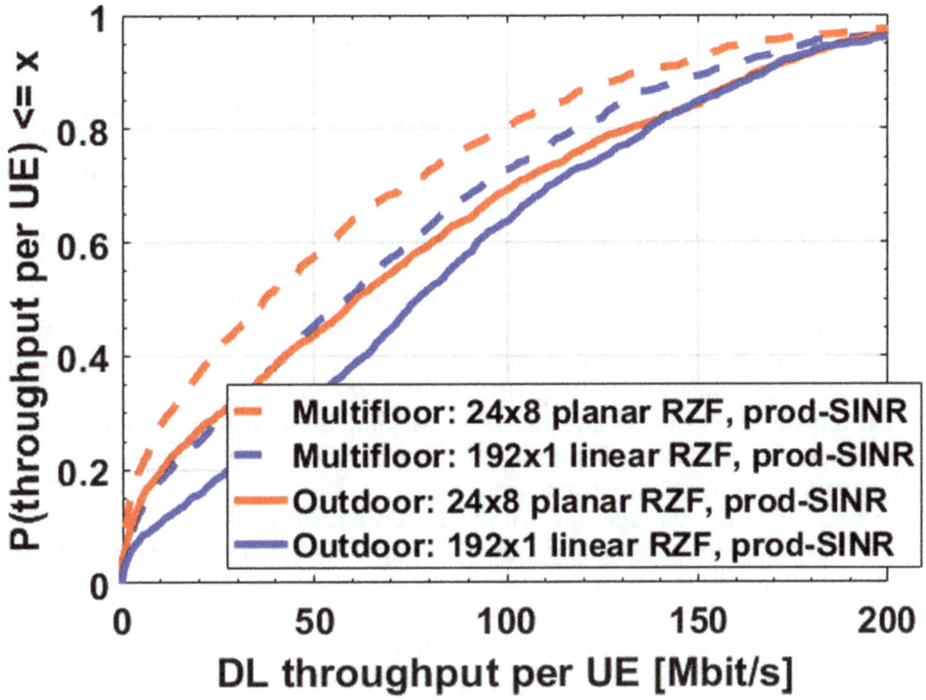

Fig. 12 DL throughput per UE between two different user distributions utilizing the identical antenna arrays at the BS

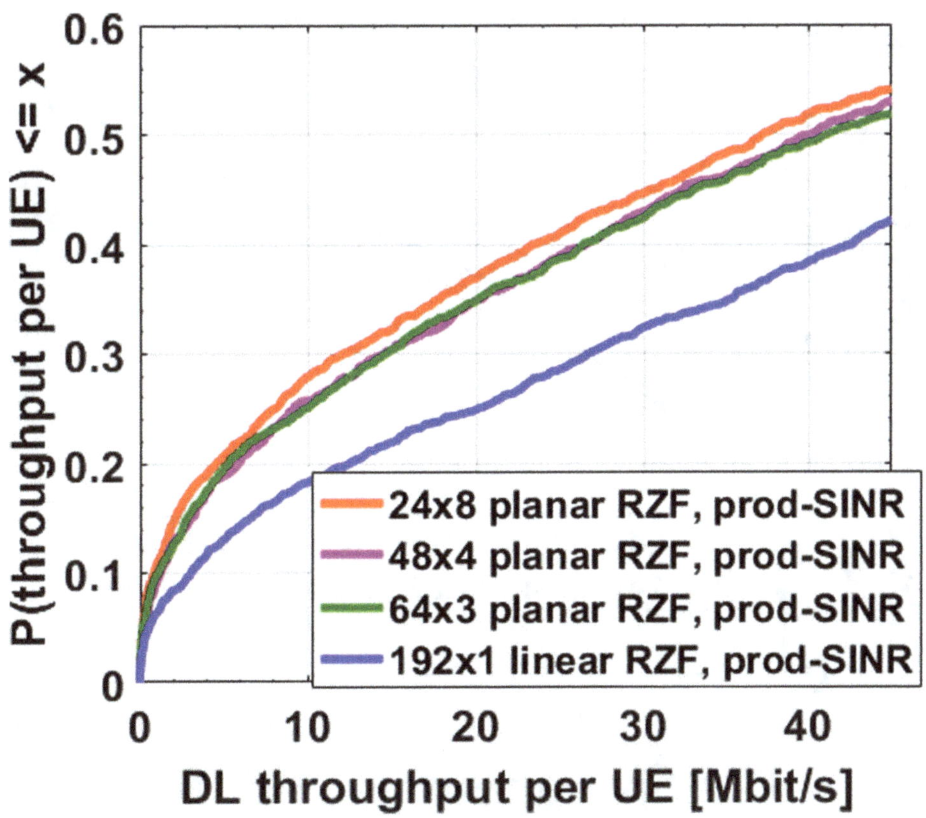

Fig. 13 DL throughput per UE between different antenna array configurations all containing 192 elements

in the same building, this could be different. As shown in Section 5, the minimum distance or angle between two closely spaced users can be significantly small to achieve FP-like conditions, and the minimum separation of 3 m between different floors of a building may be sufficient to separate users in the vertical direction. Therefore, the worst case scenario would be when all the users are located in the same building present at the cell edge. However, since the probability of such an event is very small, we can conclude that the horizontal dimension still retains the significant role in macro-cell scenarios.

In Fig. 13, the impact of decreasing the horizontal dimension of the antenna array while maintaining the same number of total antenna elements is illustrated. Clearly, the linear array performs significantly better than the planar arrays due to its large horizontal aperture. The planar array with the smallest horizontal aperture with 24 horizontal elements has the lowest throughput performance with the other intermediate arrays in between. However, this does not mean that there is no impact of the vertical dimension and it must also be considered. In Fig. 14, 2 different types of individual antenna elements used in the arrays have been compared. The first element is the same as used in the rest of the study with a beamwidth of 90° in the horizontal direction and 20° in the vertical direction. The second element has a larger vertical beamwidth of 60° and the same horizontal beamwidth. At the 40th percentile, when all the elements of the planar array are considered with the larger vertical beamwidth, a throughput of 20.5 Mbit/s is achieved whereas the smaller vertical beamwidth provides a throughput of 24 Mbit/s. It is interesting to note that the signal processing provides digital beamforming as well, and the impact of the antenna element beamwidth is limited. However, these results provide an insight into the careful selection of the physical parameters of the antenna arrays to provide the best possible performance. Such information about the physical parameters of the antenna arrays can also be useful in designing future hybrid antenna arrays that utilize fewer radio-frequency chains as well.

7 Conclusions

Practical maMIMO systems will have a large but limited number of antennas, which corresponds to limited spatial degrees of freedom and dimensions for the antenna array. The performance of the maMIMO system at the cell and system levels depends on the array structure but also on the propagation characteristics. This paper explored a mechanism to evaluate the favorable propagation conditions in a maMIMO channel utilizing

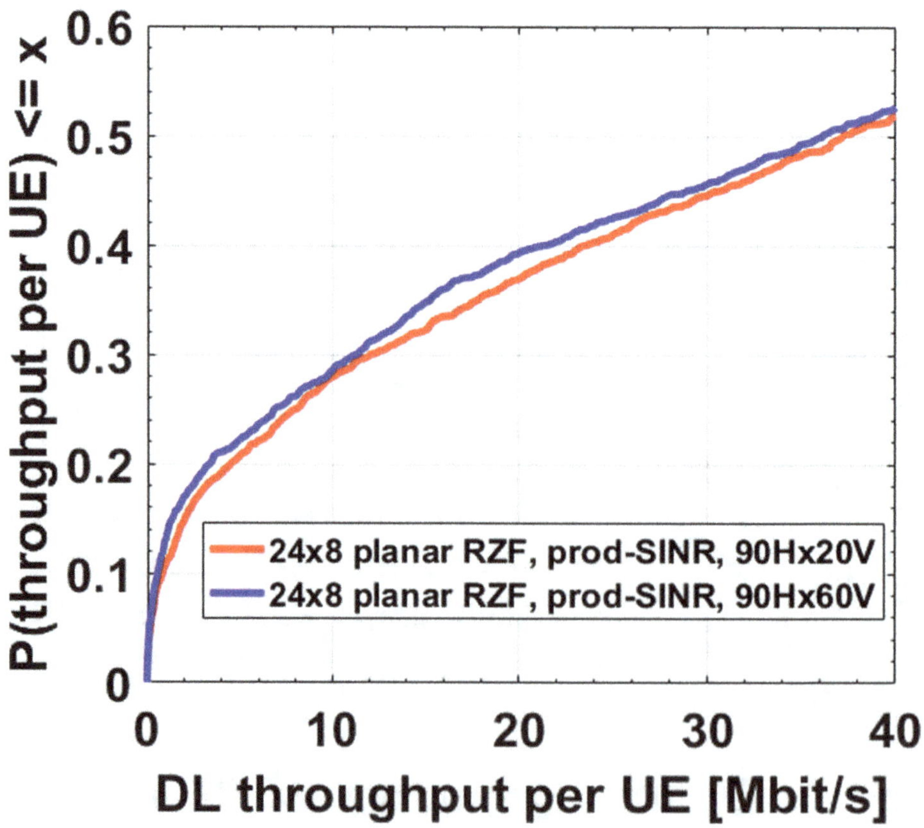

Fig. 14 DL throughput per UE between two different antenna elements with different physical vertical beamwidth

deterministic ray-based channels along with new simulation procedures to reach accurate prediction of a maMIMO network considering highly realistic user distributions in real environments.

The maMIMO channel responses from a ray-based deterministic tool were obtained and used for system-level simulations that include precoding and multi-user power allocation. A multi-cell macro deployment of 16 maMIMO BSs in a dense urban environment was evaluated. At the median percentile, the antenna array with the largest spatial aperture (linear) has the best performance as compared with the cylindrical array (smallest extent) which has the worst performance.

It is shown that at i.i.d. Rayleigh fading channels whose path losses have been set to the same values as those considered in the deterministic model provide rather similar throughput as the deterministic channel when considering the linear array. This can be explained by the fact that the large size of the linear array is able to see various different channel components that is closer to the uncorrelated channels represented by the i.i.d. channels. In the case of the planar array, the impact due to the limited horizontal dimension of the array is not captured by the i.i.d. Rayleigh model and leads to significantly different results.

Finally, the performance of a realistic maMIMO network considering 3D user distributions was provided. The performance of the maMIMO network for different configurations considering the layout of antenna array and also for different physical antenna patterns is provided. The results show that even when considering users located vertically on different floors, the linear array with large horizontal aperture performs better than the planar array. The reason for such performance is due to the physical locations of the simultaneous users in macrocells. Even when located at different heights, the number of simultaneous users is larger in the horizontal direction as compared to the vertical direction. Indeed, this could be different when considering small cells or a BS providing access to a fixed group of users like in a single large building. All the network parameters for these different results remain identical however just based on the selection of the antenna shape, layout of the array, and even the type of the beam pattern; differences can be observed in the overall performance of the network. A very large antenna array consisting of 192 elements has been considered, and these differences are expected to be larger when considering arrays with smaller dimensions. During the deployments of real maMIMO networks, such physical parameters must also be considered along with optimal signal processing schemes to obtain the best possible network performance. Another perspective of this work is also to predict deterministic maMIMO performance heatmaps, as will be required in 5G radio planning.

Funding

This work was supported in part by the European Commission through the H2020-MSCA ETN-5Gwireless project under Grant Agreement 641985.

Authors' contributions

All of the authors participated in the proposal and conception of the current study; they all read and approved the final manuscript.

Competing interests

The authors declare that they have no competing interests.

Author details

[1] SIRADEL, 2 Parc de Brocéliande, St. Grégoire, France. [2] Linköping University, 58183 Linköping, Sweden.

References

1. X. Gao, O. Edfors, F. Tufvesson, E. G. Larsson, Massive MIMO in real propagation environments: do all antennas contribute equally?. IEEE Trans. Commun. **63**, 3917–3928 (2015)
2. M. Z. Aslam, Y. Corre, E. Björnson, Y. Lostanlen, Massive MIMO Channel Performance Analysis Considering Separation of Simultaneous Users. Paper presented at the 22nd International ITG Workshop on Smart Antennas - WSA 2018, Bochum, Germany, 14-16 March 2018 (2018)
3. E. Björnson, J. Hoydis, L. Sanguinetti, Massive MIMO networks: spectral, energy, and hardware efficiency. Found. Trends Signal Process. **48**(11), 158–335 (2017)
4. H. Q. Ngo, E. G. Larsson, T. L. Marzetta, Aspects of favorable propagation in massive MIMO. Paper presented in 2014 Proceedings of the 22nd European Signal Processing Conference (EUSIPCO), Sept 2014, 76–80
5. B. M. Hochwald, T. L. Marzetta, V. Tarokh, Multiple antenna channel hardening and its implications for rate feedback and scheduling. IEEE Trans. Inf. Theory. **50**(9), 1893–1909 (2014)
6. X. Gao, O. Edfors, F. Rusek, F. Tufvesson, Massive MIMO performance evaluation based on measured propagation data. IEEE Trans. Wirel. Commun. **14**(7), 3899–3911 (2015)
7. Y. Corre, Y. Lostanlen, Three-dimensional urban EM wave propagation model for radio network planning and optimization over large areas. IEEE Trans. Veh. Technol. **58**(7), 3112–3123 (2009)

6

Multiuser 3D massive MIMO transmission in full-duplex cellular system

Dongmei Zhang[1], Ximing Wang[1*], Kui Xu[1], Yijun Yang[2] and Wei Xie[1]

Abstract

In this paper, we provide an effective multiuser transmission scheme in three-dimensional (3D) massive multiple-input multiple-output (MIMO) cellular systems, where the full-duplex (FD) base station (BS) equips two separate large-scale uniform planar antenna array (UPA). In order to reduce the computational and implementation complexity, we investigate the characteristic of the beam-domain (BD) 3D massive MIMO channels and self-interference (SI) channel. We propose the 3D multiuser beamspace transmission (MUBT) scheme that requires the spatial angular information of the users and the SI channel. We show that, due to the reduced dimension property of the effective beamspace channel, the overhead for channel estimation is reduced. Furthermore, a user scheduling algorithm is proposed to enable the 3D MUBT scheme in the FD systems. Finally, both the theoretical analysis and simulation results show the the SI can be effectively reduced and demonstrate the effectiveness and superiority of the proposed 3D MUBT scheme on spectral efficiency compared with the conventional half-duplex (HD) and FD transmission schemes.

Keywords: Multiuser cellular system, 3D massive MIMO, Full-duplex, Beamspace channel representation, Minimum mean square error (MMSE) channel estimation, Beamforming

1 Introduction

With the booming development of smart terminals and the growing demand for new mobile services, the mobile Internet traffic will continue to grow exponentially, which will increase by roughly 1000 times beyond 2020. Multiple-input multiple-output (MIMO) techniques can obtain multiplexing gain, diversity gain, and antenna gain by exploiting the spatial dimension of wireless resources, which as a result can significantly enhance the capacity and reliability in wireless communications [1]. In order to further improve the spectral efficiency (SE) of the communication systems, massive MIMO has attracted considerable attention. Massive MIMO is first advocated in [2], which can make the simple linear precoder tend to optimal and eliminate the noise along with uncorrelated interference by simply scaling up the numbers of antenna elements at the base station (BS). In this way, the SE of massive MIMO systems is improved dramatically because more users can be served in the same time-frequency resource. Furthermore, massive MIMO enables users to reduce the transmit power arbitrarily without compromising the SE [3].

However, in the practical massive MIMO systems, all these performance gains are profitted from the channel state information (CSI) that is available. Time-division duplex (TDD) seems to be more suitable for massive MIMO systems since the channel reciprocity can be exploited to obtain the instantaneous downlink (DL) CSI through uplink (UL) training [2]; thus, the overhead for channel estimation is linear with the number of user antennas. For this reason, a great deal of research work has been done for TDD [3–7]. However, the imperfect calibration between the UL/DL radio frequency chains [8] and pilot contamination [9, 10] in the practical TDD systems limits the performance gains of massive MIMO, which as a result motivates the research on frequency-division duplex (FDD) massive MIMO systems [11–13]. Since the channel reciprocity does not hold for FDD systems, the training overhead for DL estimation scales linearly with the number of BS antennas. This poses a heavy burden on user equipments and the feedback links in the system where the BS is equipped with a large number of antennas. One attempt is to adopt the closed-loop training scenarios to sequentially design the optimal beam patterns

*Correspondence: lgdxwxm@sina.com
[1]Army Engineering University of PLA, No. 88, Houbiaoying Street, Nanjing, China
Full list of author information is available at the end of the article

[11]. Another alternative method is to exploit the low-rank property of channel covariance matrix in massive MIMO systems [12, 13]. Utilizing the correlation between channels, the reduced dimension effective channels can be obtained by the eigen-decomposition of channel covariance matrix, which can reduce the overhead of training and feedback.

Furthermore, most of the prior works assume the large-scale antenna array deployed along the horizontal axis, which is not applicable for most BSs due to the limited space on the roof or mast. Three-dimensional (3D) massive MIMO, which is referred to as full-dimension MIMO, can overcome this practical challenge because large-scale antennas are arranged in both the horizontal and vertical dimensions when two-dimensional (2D) uniform planar antenna array (UPA) are deployed. In this way, the extra degrees of freedom for vertical dimension can be exploited to improve the capability of serving 3D distributed users, as illustrated in Fig. 1. In [14], the channel correlation matrix based on a generic ray-tracing 3D channel model is investigated, and the study shows that the channel correlation matrix can be well approximated by the Kronecker production of the correlations in horizontal and vertical directions. References [12, 15–17] investigate the DL transmission for 3D massive MIMO systems. Based on both statistical and instantaneous CSI, [12] proposes a joint spatial division and multiplexing with 3D precoding scheme. By exploiting the Kronecker structure of the 3D MIMO channel matrix, Wang et al. [15] propose a 2D precoding scheme to fully exploit the degrees of freedom provided by the vertical dimension, which can reduce the multiuser interference (MUI) and inter-cell interference (ICI). Li et al. [16, 17] assumes that the BS has only the statistical CSI of each user. Through eigen-decomposition of horizontal and vertical correlation matrices, the beamforming vector for each user are obtained. Moreover, the space division multiple access transmission scheme based on the 3D beamforming is proposed.

Both TDD and FDD systems are half-duplex (HD) systems that assign orthogonal time or frequency resources to the UL and DL, which theoretically leads to half of the time-frequency resources wasted. With the development of signal and hardware processing techniques, recently, the full-duplex (FD) techniques become one of the attractive options for wireless communications. The severe self-interference (SI) due to the signal leakage from the transmitter to the receiver is one of the major challenges for FD communications, which results in the rapid development of various kinds of SI cancellation (SIC) techniques [18–20]. Due to the FD transmission, the DL users may suffer from the interference transmitted by UL users in cellular systems, which can be controlled by using different beams [21–23]. In order to achieve higher SE performance, a lot of research has been done on FD massive MIMO technology [24–28]. Xia et al. [28] shows that adopting linear beamforming techniques can significantly reduce the adverse impact of SI, and FD massive MIMO systems outperform the HD counterpart in SE even without active SI cancellation.

In this paper, we assume that two separate large-scale UPAs are deployed at the BS, and investigate 3D multiuser transmission in the FD massive MIMO cellular system. The contributions of our works are as follows:

- The beam-domain (BD) characteristic of the 3D channel is investigated, including UL, DL, and SI channels. In the BD channel, most of the power is

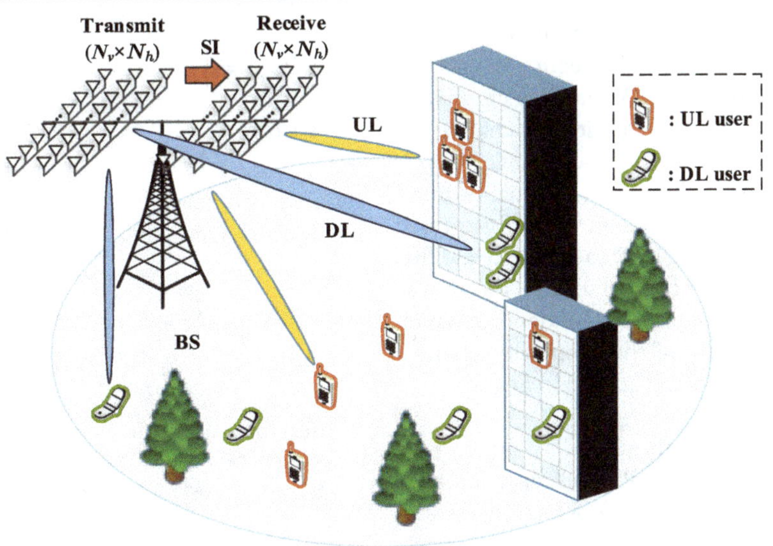

Fig. 1 Single-cell multiuser 3D massive MIMO system

concentrated on each user's own beamspace. The algorithm of calculating each user's best beamspace is provided.

- By exploiting the low-dimension property of the effective beamspace channels in the BD, we propose the 3D multiuser beamspace transmission (MUBT) scheme which processes the signals in each users's beamspaces. Utilizing the orthogonality of beamspaces between groups, we show that our the 3D MUBT can eliminate the inter-group interference (IGI), and significantly reduce the overhead of channels estimation.
- The SI channel in 3D massive MIMO is modeled, and the beamspaces of the SI channel are derived. Based on the beamspaces of SI channel and features of co-channel interference (CCI) within the cellular systems, we introduce a simple cell partitioning strategy and propose a user-scheduling algorithm. Both the theoretical analysis and simulation results show that the proposed scheme reduce the SI effectively.

Notation: We use boldface uppercase letter $\mathbf{A}$ and boldface lowercase letter $\mathbf{a}$ to represent matrix and column vector, respectively. $\mathbb{E}(\cdot)$, $\|\cdot\|$, $(\cdot)^*$, $(\cdot)^H$, $\Phi(\cdot)$ and $\mathrm{Tr}(\cdot)$ stand for the expectation, the Euclidean norm, the conjugate, the conjugate transpose, the spectral radius of a matrix and trace of a matrix, respectively. $|\mathcal{B}|$ denotes the cardinality of a set $\mathcal{B}$. $[\mathbf{A}]_{i,j}$ denotes the ith-row and jth-column entry of matrix $\mathbf{A}$. $\mathbf{I}_N$ represents the $N\times N$ identity matrix. $\mathcal{CN}(\mu,\sigma^2)$ stands for the complex-Gaussian distribution with mean of μ and variance of σ^2. $\mathbf{e}_i$ represents a vector whose ith entry is 1 and the other are 0. $\delta(\cdot)$is the Dirac delta function.

2 Methods

The outline of this paper is as follows: Section 3 describes the system and the channels model. In Section 4, the BD characteristics of 3D massive MIMO channels are investigated and the beamspace representation of channel is given. In Section 5, we detail the proposed 3D MUBT scheme, which includes user scheduling, channel estimation, and data transmission. In Section 6, we perform various simulation results using MATLAB to demonstrate the effectiveness of the proposed scheme, and furthermore, the existing approaches are compared under the same setting to show the superiority of the proposed scheme. Finally, the concluding remarks are offered in Section 7.

3 System and channel models

As shown in Fig. 1, we consider a 3D multiuser massive MIMO system, where K_u UL users and K_d DL users are distributed in 3D space. We assume that all the users work in HD mode and equipped with single antenna. The FD BS deploys two separate large-scale UPAs[1] (one for transmitting and one for receiving), and both of them has N_v antennas in the vertical dimension and N_h antennas in the horizontal dimension and each dimension is a uniform linear antenna array (ULA). Let d_v and d_h denote the distance of adjacent antennas in the vertical and horizontal directions.

3.1 UL and DL channel models

In this paper, we consider a 3D massive MIMO channel model and assume that the channels between the BS and users passes through a great number of rays [29]. Therefore, the channel of UL user k_u can be written as:

$$\mathbf{G}_{k_u} = \iint_{(\theta_{k_u},\varphi_{k_u})\in\Delta_{k_u}} r_u\left(\theta_{k_u},\varphi_{k_u}\right)\mathbf{A}\left(\theta_{k_u},\varphi_{k_u}\right)d\theta_{k_u}d\varphi_{k_u} \quad (1)$$

where θ_{k_u} and φ_{k_u} represent the direction of arrival (DoA) of user k_u in vertical and horizontal directions; $\Delta_{k_u} = \left\{\left(\theta_{k_u},\varphi_{k_u}\right)\mid\theta_{k_u}\in\left[\theta_{k_u}^{\min},\theta_{k_u}^{\max}\right],\varphi_{k_u}\in\left[\varphi_{k_u}^{\min},\varphi_{k_u}^{\max}\right]\right\}$ denote the angular spread (AS) range resulting from scatters in the vertical and horizontal directions; and $r_u\left(\theta_{k_u},\varphi_{k_u}\right)$ denotes the UL complex response gain. Here, $\mathbf{A}\left(\theta_{k_u},\varphi_{k_u}\right)\in\mathbb{C}^{N_v\times N_h}$ is the array response matrix which can be expressed as:

$$\mathbf{A}\left(\theta_{k_u},\varphi_{k_u}\right) = \begin{bmatrix} 1 & \cdots & e^{j(N_h-1)h_{k_u}} \\ e^{jv_{k_u}} & \cdots & e^{j[v_{k_u}+(N_h-1)h_{k_u}]} \\ \vdots & \ddots & \vdots \\ e^{j(N_v-1)v_{k_u}} & \cdots & e^{j[(N_v-1)v_{k_u}+(N_h-1)h_{k_u}]} \end{bmatrix}, \quad (2)$$

where $v_{k_u} = \frac{2\pi d_v}{\lambda}\cos\theta_{k_u}$ and $h_{k_u} = \frac{2\pi d_h}{\lambda}\sin\theta_{k_u}\cos\varphi_{k_u}$, and λ is the carrier wavelength. Define

$$\begin{aligned} \mathbf{a}\left(v_{k_u}\right) &= \left[1, e^{jv_{k_u}}, \cdots, e^{j(N_v-1)v_{k_u}}\right]^T, \\ \mathbf{b}\left(h_{k_u}\right) &= \left[1, e^{jh_{k_u}}, \cdots, e^{j(N_h-1)h_{k_u}}\right]^T, \end{aligned} \quad (3)$$

then we can rewrite $\mathbf{A}\left(\theta_{k_u},\varphi_{k_u}\right) = \mathbf{a}\left(v_{k_u}\right)\mathbf{b}\left(h_{k_u}\right)^T$. As a result, the UL channel matrix can be rewritten as:

$$\mathbf{G}_{k_u} = \iint_{(\theta_{k_u},\varphi_{k_u})\in\Delta_{k_u}} r_u\left(\theta_{k_u},\varphi_{k_u}\right)\mathbf{a}\left(v_{k_u}\right)\mathbf{b}\left(h_{k_u}\right)^T d\theta_{k_u}d\varphi_{k_u}. \quad (4)$$

Therefore, the equivalent vectorial form of $\mathbf{G}_{k_u}$ is:

$$\mathbf{g}_{k_u} = \mathrm{vec}\left(\mathbf{G}_{k_u}\right) = \iint_{(\theta_{k_u},\varphi_{k_u})\in\Delta_{k_u}} r_u\left(\theta_{k_u},\varphi_{k_u}\right)\mathbf{b}\left(h_{k_u}\right)\otimes\mathbf{a}\left(v_{k_u}\right)d\theta_{k_u}d\varphi_{k_u}. \quad (5)$$

Similarly, define θ_{k_d} and φ_{k_d} as the direction of departure (DoD) of DL user k_d in vertical and horizontal dimension, and let $\Delta_{k_d} = \left\{ \left(\theta_{k_d}, \varphi_{k_d}\right) | \, \theta_{k_d} \in \left[\theta_{k_d}^{\min}, \theta_{k_d}^{\max}\right], \varphi_{k_d} \in [\varphi_{k_d}^{\min}, \varphi_{k_d}^{\max}] \right\}$ and $r_d\left(\theta_{k_d}, \varphi_{k_d}\right)$ be the AS range of DoD and DL response gain; then, the channel vector of user k_d can be expressed as:

$$\mathbf{g}_{k_d} = \iint\limits_{\left(\theta_{k_d}, \varphi_{k_d}\right) \in \Delta_{k_d}} r_d\left(\theta_{k_d}, \varphi_{k_d}\right) \mathbf{b}\left(h_{k_d}\right) \otimes \mathbf{a}\left(v_{k_d}\right) d\theta_{k_d} d\varphi_{k_d}. \tag{6}$$

In this paper, the AS range Δ_{k_u} and Δ_{k_d} are assumed to be relatively narrow, which is reasonable when the BS is set at a relatively high altitude [13].

3.2 3D SI channel model

We assume that some SIC techniques have been adopted at the BS, e.g., passive SI cancellation technique, so that there exists no light-of-sight (LOS) signals from transmit antenna array to receive antenna array and the two arrays are independent. Moreover, we assume that there are several scatters (e.g., tall buildings) surrounding the arrays; then, the SI signal consists of several non-ligth-of-sight (NLOS) rays.

Let $\mathbf{G}_{SI} \in C^{N_v N_h \times N_v N_h}$ be the channel from transmit antenna array to receive antenna array, i.e., SI channel. According to the Kronecker stochastic channel model [14, 30], the SI channel can be modeled as

$$\mathbf{G}_{SI} = \mathbf{C}_R^{\frac{1}{2}} \mathbf{G}_w \left(\mathbf{C}_T^{\frac{1}{2}}\right)^T \tag{7}$$

where $\mathbf{G}_w$ is a $N_v N_h \times N_v N_h$ random matrix with independent and identical distributed (i.i.d.) $\mathcal{CN}\left(0, \beta_{SI}\right)$ elements. Here, β_{SI} can be understood as the level of the residual SI, which is related to the path loss of the SI channel and the capability of the SIC techniques. $\mathbf{C}_R$ and $\mathbf{C}_T$ denote the receive and transmit spatial correlation matrices which are also the corresponding covariance matrices. In this paper, we assume that only the antenna-domain (passive) SIC techniques are adopted and no analog or digital-domain (active) SIC techniques are used [19].

Now, we need to find out the expressions for every entry of $\mathbf{C}_R$ and $\mathbf{C}_T$, which are expressed as $\mathbf{C}_a, a \in \{R, T\}$ for simplicity of illustration. Define the correlation between (k, l)-th and (p, q)-th antenna element of antenna array as $\mathbf{C}_a^{(k,l),(p,q)}$. According to [14], $\mathbf{C}_a^{(k,l),(p,q)}$ can be approximately written as a product of vertical and horizontal correlations; then, the correlation matrix can be expressed as

$$\mathbf{C}_a = \mathbf{C}_{ah} \otimes \mathbf{C}_{av} \tag{8}$$

where $\mathbf{C}_{av}$ and $\mathbf{C}_{ah}$ are the vertical and horizontal correlation with

$$\begin{aligned} \left[\mathbf{C}_{ah}\right]_{l,q} &= \frac{1}{\sqrt{A}} e^{-\frac{B^2 \cos^2 \varphi_a}{2A}} e^{j\frac{C \cos \varphi_a}{A}} e^{-\frac{(C \sin \varphi_a \sigma_a)^2}{2A}} \\ \left[\mathbf{C}_{av}\right]_{k,p} &= e^{j\frac{2\pi d_v}{\lambda}(p-k)\cos\theta_a} e^{-\frac{1}{2}\left(\xi_a \frac{2\pi d_v}{\lambda}\right)^2 (p-k)^2 \sin^2\theta_a} \end{aligned} \tag{9}$$

where $A = \left(\xi_a \frac{2\pi d_h}{\lambda}(q-l)\cos\theta_a\right)^2 (\sin\varphi_a\sigma_a)^2 + 1$, $B = \xi_a \frac{2\pi d_h}{\lambda}(q-l)\cos\theta_a$ and $C = \frac{2\pi d_h}{\lambda}(q-l)\sin\theta_a$. Here, θ_a, φ_a, ξ_a, and σ_a denote the DoA ($a = R$) or DoD ($a = T$) in vertical and horizontal directions and the normal distribution variance of the angular perturbation in vertical and horizontal directions. Note that, from (9), we can see that $\mathbf{C}_a$ is Hermitian matrix.

4 Beam-domain characteristics of 3D massive MIMO channel

In the cellular systems, due to the fact that the antenna array at the BS are generally set at a relative high building where the scatters around are relatively sparse, therefore, most of the energy of the channels concentrate on several spatial directions [31], i.e., several beams of signals. Hence, in order to realize multiuser transmission in the 3D massive MIMO system, we first investigate the physical beam characteristic of 3D massive MIMO channels.

4.1 Beamspaces of UL and DL channels

To get started, we introduce the following lemma:

Lemma 1 *The channels of UL user k_u and i_u or DL user k_d and i_d ($k \neq i$) are asymptotically orthogonal if their vertical or horizontal AS ranges are disjoint, i.e.,*

$$\begin{aligned} &\lim_{N_v, N_h \to \infty} \frac{1}{N_v N_h} \mathrm{vec}\left(\mathbf{G}_{k_x}\right)^H \mathrm{vec}\left(\mathbf{G}_{i_x}\right) = 0 \Leftrightarrow \\ &\left(\left[\theta_{k_x}^{\min}, \theta_{k_x}^{\max}\right] \cap \left[\theta_{i_x}^{\min}, \theta_{i_x}^{\max}\right]\right) \cap \left(\left[\varphi_{k_x}^{\min}, \varphi_{k_x}^{\max}\right] \cap \left[\varphi_{i_x}^{\min}, \varphi_{i_x}^{\max}\right]\right) = \varnothing \end{aligned} \tag{10}$$

where $x \in \{u, d\}$ represents UL and DL, respectively.

Proof Using the property of $\mathrm{vec}\left(\mathbf{ABC}\right) = \left(\mathbf{C}^T \otimes \mathbf{A}\right) \mathrm{vec}\left(\mathbf{B}\right)$ and

$$\begin{aligned} \lim_{N_v, N_h \to \infty} &= \frac{1}{N_v N_h} \left[\mathbf{b}\left(h_{k_x}\right) \otimes \mathbf{a}\left(v_{k_x}\right)\right]^H \left[\mathbf{b}\left(h_{i_x}\right) \otimes \mathbf{a}\left(v_{i_x}\right)\right] \\ &= \frac{1}{N_v N_h} \left[\mathbf{b}\left(h_{k_x}\right)^H \mathbf{b}\left(h_{i_x}\right)\right] \otimes \left[\mathbf{a}\left(v_{k_x}\right)^H \mathbf{a}\left(v_{i_x}\right)\right] \\ &= \delta\left(h_{k_x} - h_{i_x}\right) \cdot \delta\left(v_{k_x} - v_{i_x}\right), \end{aligned} \tag{11}$$

Lemma 1 can be easily obtained. □

From Lemma 1, we can see that in 3D massive MIMO systems, both the vertical and horizontal resolution are high so that users have different DoD or DoA can be simultaneously scheduled, e.g., users in the same building having the same horizontal angle while different in the vertical dimension. Note that, in 2D massive MIMO system where ULA is deployed at the BS, users can only be distinguished in horizontal plane.

To investigate the physical beam property of 3D massive MIMO, we first define $N_v \times N_v$ and $N_h \times N_h$ normalized DFT matrices $\mathbf{F}_{N_v}$ and $\mathbf{F}_{N_h}$ whose (i,k)-th entry is $[\mathbf{F}_x]_{i,k} = \frac{1}{\sqrt{x}} e^{-j\frac{2\pi}{x}ik}$, $x \in \{N_v, N_h\}$, and introduce the following lemma:

Lemma 2 *Define* $\tilde{\mathbf{G}}_{k_x} = \mathbf{F}_{N_v}\mathbf{G}_{k_x}\mathbf{F}_{N_h}$ *($x \in \{u,d\}$) as the BD channel, then the entries of* $\tilde{\mathbf{G}}_{k_x}$ *have non-negligible values only at limited 2D points as* $N_v, N_h \to \infty$ *which constitutes the beamspace of channel* $\mathbf{G}_{k_x}$.

Proof We first assume that the UL or DL channels only consist of single ray, i.e., $\mathbf{G}_{k_x} = r_x\left(\theta_{k_x}, \varphi_{k_x}\right)\mathbf{A}\left(\theta_{k_x}, \varphi_{k_x}\right) = r_{k_x}\mathbf{a}\left(v_{k_x}\right)\mathbf{b}\left(h_{k_x}\right)^T$, then the (p,q)-th entry of $\tilde{\mathbf{G}}_{k_x}$ is

$$\begin{aligned}\left[\tilde{\mathbf{G}}_{k_x}\right]_{p,q} &= \frac{r_{k_x}}{\sqrt{N_v N_h}} \sum_{m=0}^{N_v-1} e^{-j\left(\frac{2\pi}{N_v}pm - \frac{2\pi d_v}{\lambda} m \cos\theta_{k_x}\right)} \\ &\times \sum_{n=0}^{N_h-1} e^{-j\left(\frac{2\pi}{N_h}qn - \frac{2\pi d_h}{\lambda} n \sin\theta_{k_x}\cos\varphi_{k_x}\right)} \\ &= \frac{r_{k_x}}{\sqrt{N_v N_h}} e^{-j\frac{N_v-1}{2}\alpha_{k_x}} \frac{\sin\left(N_v\alpha_{k_x}/2\right)}{\sin\left(\alpha_{k_x}/2\right)} \\ &\times e^{-j\frac{N_h-1}{2}\beta_{k_x}} \frac{\sin\left(N_h\beta_{k_x}/2\right)}{\sin\left(\beta_{k_x}/2\right)}\end{aligned} \tag{12}$$

where $\alpha_{k_x} = \frac{2\pi}{N_v}p - \frac{2\pi d_v}{\lambda}\cos\theta_{k_x}$ and $\beta_{k_x} = \frac{2\pi}{N_h}q - \frac{2\pi d_h}{\lambda}\sin\theta_{k_x}\cos\varphi_{k_x}$. Thus, as $N_v, N_h \to \infty$, we can obtain:

$$\begin{aligned}\lim_{N_v,N_h\to\infty} \left|\left[\tilde{\mathbf{G}}_{k_x}\right]_{p,q}\right| &= \sqrt{N_v N_h}\left|r_{k_x}\right| \\ \times\, \delta\left(\frac{p}{N_v} - \frac{d_v}{\lambda}\cos\theta_{k_x}\right) &\delta\left(\frac{q}{N_h} - \frac{d_h}{\lambda}\sin\theta_{k_x}\cos\varphi_{k_x}\right).\end{aligned} \tag{13}$$

Apparently, for any DoA or DoD $\left(\theta_{k_x}, \varphi_{k_x}\right)$, there is a (p_0, q_0) to satisfy (13) nonzero, i.e., $p_0 = N_v\frac{d_v}{\lambda}\cos\theta_{k_x}$ and $q_0 = N_h\frac{d_h}{\lambda}\sin\theta_{k_x}\cos\varphi_{k_x}$. Hence, the point (p_0, q_0) indicates the direction of the signal, i.e., $\theta_{k_x} = \arccos\frac{p_0\lambda}{d_v N_v}$ and $\varphi_{k_x} = \arccos\frac{q_0\lambda}{d_h N_h \sin\theta_{k_x}}$. Particularly, p_0 and q_0 are usually not integers, which as a result will cause the power at point (p_0, q_0) to leak to the surrounding points.

Then, we consider the case of large number of rays, i.e., $\mathbf{G}_{k_x} = \iint_{(\theta_{k_x},\varphi_{k_x})\in\Delta_{k_x}} r_x\left(\theta_{k_x}, \varphi_{k_x}\right)\mathbf{A}\left(\theta_{k_x}, \varphi_{k_x}\right)d\theta_{k_x}d\varphi_{k_x}$. According to (13), it can be concluded that each ray of $\mathbf{G}_{k_x}$ is corresponding to a (or limited) nonzero point as $N_v, N_h \to \infty$, and is restricted in a relatively small range under the narrow AS range condition, which is given by (14), shown at the top of the next page, where the $\mathcal{B}_{k_x}$ is called the beamspace of channel $\mathbf{G}_{k_x}$. This completes the proof of Lemma 2.

$$\begin{aligned}\mathcal{B}_{k_x} = \Bigg\{(p,q)\,\Bigg|&\left\lfloor N_v\frac{d_v}{\lambda}\left(\cos\theta_{k_x}\right)_{\min}\right\rfloor \\ &\leq p \leq \left\lceil N_v\frac{d_v}{\lambda}\left(\cos\theta_{k_x}\right)_{\max}\right\rceil, \left\lfloor N_h\frac{d_h}{\lambda}\left(\sin\theta_{k_x}\cos\varphi_{k_x}\right)_{\min}\right\rfloor \\ &\leq q \leq \left\lceil N_h\frac{d_h}{\lambda}\left(\sin\theta_{k_x}\cos\varphi_{k_x}\right)_{\max}\right\rceil, \forall\left(\theta_{k_x}, \varphi_{k_x}\right) \\ &\in \Delta_{k_x}\Bigg\}, \text{as } N_v, N_h \to \infty.\end{aligned} \tag{14}$$

□

Based on Lemma 2, the channel matrix can be approximately represented by

$$\mathbf{G}_{k_x} = \mathbf{F}_{N_v}^H\tilde{\mathbf{G}}_{k_x}\mathbf{F}_{N_h}^H \approx \left(\mathbf{F}_{N_v}^H\right)_{:,\Omega_p}\tilde{\mathbf{G}}_{k_x}^{\{\mathcal{B}_{k_x}\}}\left(\mathbf{F}_{N_h}^H\right)_{\Omega_q,:} \tag{15}$$

where Ω_p and Ω_q are the indexes of the columns and rows given by $\mathcal{B}_{k_x}$, and $\tilde{\mathbf{G}}_{k_x}^{\{\mathcal{B}_{k_x}\}}$ is the beamspace channel whose dimension has been efficiently reduced from $N_v \times N_h$ to $\left|\mathcal{B}_{k_x}\right|$, where $\left|\mathcal{B}_{k_x}\right|$ denotes the cardinality of the beamspace $\mathcal{B}_{k_x}$. Note that, according to Lemma 2 it can be obtained that (15) becomes equality when N_v and N_h tend to infinity. Then, the beamspace channel can be written as:

$$\tilde{\mathbf{G}}_{k_x}^{\{\mathcal{B}_{k_x}\}} = \mathbf{F}_{N_v}^{\left\{\mathcal{B}_{k_x}^{\text{row}}\right\}}\mathbf{G}_{k_x}\mathbf{F}_{N_h}^{\left\{\mathcal{B}_{k_x}^{\text{col}}\right\}} \tag{16}$$

where $\mathcal{B}_{k_x}^{\text{row}} \triangleq \{(\Omega_p, :)\}$ and $\mathcal{B}_{k_x}^{\text{col}} \triangleq \{(:, \Omega_q)\}$ are called the vertical beamspace and horizontal beamspace which consist of the vertical beams and horizontal beams in $\mathcal{B}_{k_x}$, respectively.

Corollary 1 *In the large* N_v *and* N_h *regime, if the beamspaces of user* k_x *and* i_x *($k \neq i, x \in \{u,d\}$) are non-overlapping, then the BD channels of* k_x *and* i_x *are orthogonal, which mathematically can be expressed as*

$$\lim_{N_v,N_h\to\infty}\mathbf{F}_{N_v}^{\left\{\mathcal{B}_{i_x}^{row}\right\}}\mathbf{G}_{k_x}\mathbf{F}_{N_h}^{\left\{\mathcal{B}_{i_x}^{col}\right\}} = \lim_{N_v,N_h\to\infty}\mathbf{F}_{N_v}^{\left\{\mathcal{B}_{k_x}^{row}\right\}}\mathbf{G}_{i_x}\mathbf{F}_{N_h}^{\left\{\mathcal{B}_{k_x}^{col}\right\}} = 0 \tag{17}$$

The conclusion in Corollary 1 indicates that if the beamspaces of different users are orthogonal, the users in the BD are "invisible" to each other, i.e., the signals transmitted on their own beamspace channels won't affect other users.

To illustrate the concept of beamspace, an example of DL BD channels $\tilde{\mathbf{G}}_{k_d}$ whose DoDs are $(21°, 8°)$, $(41°, 25°)$, and $(70°, 8°)$ are depicted in Fig. 2. In most cases in the reality, the AS in the vertical domain is relatively small due to the height of the BS [32]; therefore, we set the AS in the horizontal directions δ_h is 10° and in the vertical directions δ_v is 5°. Obviously, the channel power is concentrated on the beamspace, but with the finite number of antennas, the power might leak to the around points.

In order to facilitate the operation, we rewrite the BD channel into equivalent vectorial form as $\tilde{\mathbf{g}}_{k_x} = \mathrm{vec}\left(\tilde{\mathbf{G}}_{k_x}\right) = \mathrm{vec}\left(\mathbf{F}_{N_v}\mathbf{G}_{k_x}\mathbf{F}_{N_h}\right)$. Using the property $\mathrm{vec}\,(\mathbf{ABC}) = \left(\mathbf{C}^T \otimes \mathbf{A}\right)\mathrm{vec}\,(\mathbf{B})$, we can get

$$\tilde{\mathbf{g}}_{k_x} = \left(\mathbf{F}_{N_h}^T \otimes \mathbf{F}_{N_v}\right)\mathbf{g}_{k_x} = \mathbf{F}\mathbf{g}_{k_x} \tag{18}$$

where $\mathbf{F} = \left(\mathbf{F}_{N_h}^T \otimes \mathbf{F}_{N_v}\right)$ and $x \in \{u, d\}$. Similarly, the beamspace channel vector is $\tilde{\mathbf{g}}_{k_x}^{\mathrm{vec}\{\mathcal{B}_{k_x}\}} = \mathrm{vec}\left(\tilde{\mathbf{G}}_{k_x}^{\{\mathcal{B}_{k_x}\}}\right)$,

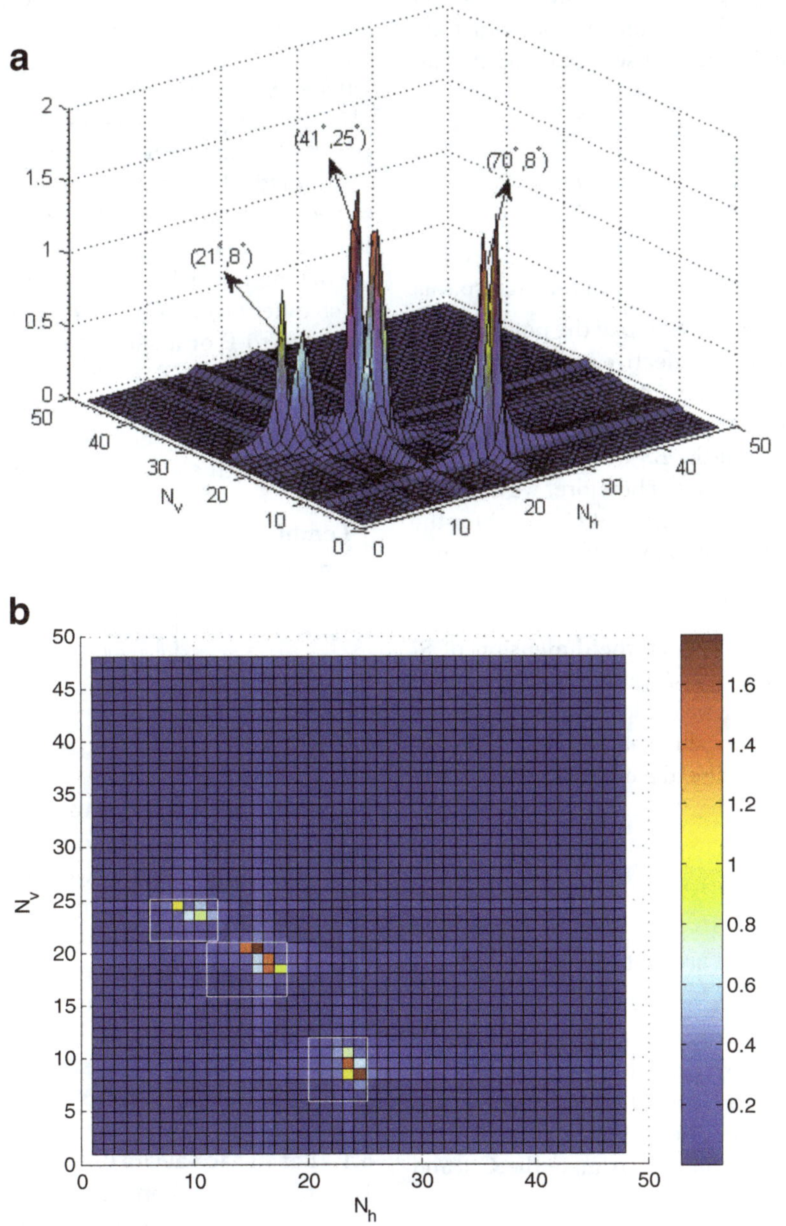

Fig. 2 Example of BD channels and their beamspaces with $N_v = N_h = 48$, the AS in the horizontal directions δ_h is 10° and in the vertical directions δ_v is 5°. **a** The BD channels $\tilde{\mathbf{G}}_{k_d}$ whose DoDs are (21°, 8°), (41°, 25°), and (70°, 8°). **b** The top view of **a** and the beamspace of each channel which is generated by (14) and is marked with rectangular frame

where vec$\{\mathcal{B}_{k_x}\}$ is the vectorial beamspace which is denoted as B_{k_x}. Then, we have

$$\tilde{\mathbf{g}}_{k_x}^{\{B_{k_x}\}} = \mathbf{F}^{\{B_{k_x}\}}\mathbf{g}_{k_x} = \left[\left(\mathbf{F}_{N_h}^{\left\{\mathcal{B}_{k_x}^{\text{col}}\right\}}\right)^T \otimes \mathbf{F}_{N_v}^{\left\{\mathcal{B}_{k_x}^{\text{row}}\right\}}\right]\mathbf{g}_{k_x}. \quad (19)$$

From Lemma 2, we can see that, with the practical finite number of antennas, most power of the beamspace channel are concentrated around the points as (14) shows, and the dimension of beamspace is also related to the AS range of the signal rays. For keeping the required channel power of the beamspace channel, the dominant beams that are selected for the transmission at the user k_x are defined as:

$$\mathcal{B}_{k_x} = \left\{1 \le i \le N_v, 1 \le j \le N_h : \left|\left[\tilde{\mathbf{G}}_{k_x}\right]_{i,j}\right|^2 \ge \xi_{k_x} \max_{i,j}\left|\left[\tilde{\mathbf{G}}_{k_x}\right]_{i,j}\right|^2\right\} \quad (20)$$

where threshold $\xi_{k_x} \in (0,1)$ which can be selected so that $\tilde{\mathbf{G}}_{k_x}^{\{\mathcal{B}_{k_x}\}}$ captures a significant fraction ρ of the power of $\tilde{\mathbf{G}}_{k_x}$ (e.g., $\rho \ge 0.9$). ρ is the required effective power ratio close to 1. Since $\mathcal{B}_{k_x}$ is only related to the direction information, i.e., $(\theta_{k_x}, \varphi_{k_x})$, which varies over a relatively long time and can be accurately obtained at the BS with little hindrance through long-term feedback [33]. Therefore, we can pre-establish an off-line data table of different $(\theta_{k_x}, \varphi_{k_x})$ for the $\mathcal{B}_{k_x}$ in order to decrease the complexity of computation.

4.2 Beamspace of SI channel

In FD 3D massive MIMO systems, the dimension of SI channel is ultra-high ($N_vN_h \times N_vN_h$); therefore, the compressibility of $\mathbf{G}_{SI}$ is urgent to be explored and exploited.

We first introduce an eigenvalue characteristic of large-scale UPA. Using the eigenvalue decomposition, $\mathbf{C}_{av}$ and $\mathbf{C}_{ah}$ can be expressed as:

$$\begin{aligned}\mathbf{C}_{av} &= \mathbf{U}_{av}\mathbf{D}_{av}\mathbf{U}_{av}^H \\ \mathbf{C}_{ah} &= \mathbf{U}_{ah}\mathbf{D}_{ah}\mathbf{U}_{ah}^H\end{aligned} \quad (21)$$

where $a \in \{R, T\}$, $\mathbf{U}_{av}$ and $\mathbf{U}_{ah}$ are unitary matrices, $\mathbf{D}_{av} = \text{diag}\left\{\lambda_{av}^{(1)}, \cdots, \lambda_{av}^{(N_v)}\right\}$ and $\mathbf{D}_{ah} = \text{diag}\left\{\lambda_{ah}^{(1)}, \cdots, \lambda_{ah}^{(N_h)}\right\}$.

According to [12], for the large dimension of ULA, the eigenvector matrix can be approximated by a unitary DFT matrix. Since UPA is composed of ULAs in each row and column, and under the assumption of large dimension of antenna array at the BS, we can approximate the $\mathbf{C}_{av}$ and $\mathbf{C}_{ah}$ as

$$\begin{aligned}\mathbf{C}_{av} &\overset{N_v\to\infty}{=} \mathbf{F}_{N_v}\mathbf{D}_{av}\mathbf{F}_{N_v}^H, \\ \mathbf{C}_{ah} &\overset{N_h\to\infty}{=} \mathbf{F}_{N_h}\mathbf{D}_{ah}\mathbf{F}_{N_h}^H,\end{aligned} \quad (22)$$

and the eigenvalues $\mathbf{D}_{av}$ and $\mathbf{D}_{ah}$ have the following characteristic:

$$\begin{aligned}\mathbf{D}_{av} &= \text{diag}\left\{0, \cdots, 0, \lambda_{av}^{(i)}, \cdots, \lambda_{av}^{(k)}, 0, \cdots, 0\right\} \\ \mathbf{D}_{ah} &= \text{diag}\left\{0, \cdots, 0, \lambda_{ah}^{(t)}, \cdots, \lambda_{ah}^{(u)}, 0, \cdots, 0\right\}\end{aligned} \quad (23)$$

where λ_{av} and λ_{ah} are non-negligible. Then, we can get the dominate domain of $\mathbf{D}_a$ as:

$$\mathbf{D}_a^{\{\Lambda_a\}} = \text{diag}\left\{\underbrace{\lambda_{av}^{(i)}\lambda_{ah}^{(t)}, \cdots, \lambda_{av}^{(i)}\lambda_{ah}^{(u)}, \cdots, \lambda_{av}^{(k)}\lambda_{ah}^{(u)}}_{(k-i)\times(u-t)}\right\} \quad (24)$$

where $\mathbf{D}_a = \mathbf{D}_{ah} \otimes \mathbf{D}_{av}$, $\Lambda_a = \left\{(\Omega_a, \Omega_a) \,|\, m \in \Omega_a, [\mathbf{D}_a]_{m,m} \neq 0\right\}$ and Ω_a is the set that contains the indexes of rows (columns) of the dominate domain of $\mathbf{D}_a$.

In practical FD systems, the SI signal is the main bottleneck of further improving the SE; thus, minimizing the harmful effect of SI is the primary task. In order to realize 3D multiuser FD transmission, we first define the BD SI channel $\tilde{\mathbf{G}}_{SI}$ as $\tilde{\mathbf{G}}_{SI} = \mathbf{F}_r\mathbf{G}_{SI}\mathbf{F}_t^H$, where $\mathbf{F}_r$ and $\mathbf{F}_t$ can be equal to $\mathbf{F}$ or a submatrix of $\mathbf{F}$. Let $\Lambda_a^{\text{row}} = \{(\Omega_a, :)\}$, $\Lambda_a^{\text{col}} = \{(:, \Omega_a)\}$, $\mathbf{P} = \mathbf{F}^2$ be a permutation matrix and $\mathbf{d} = (1, 2, \cdots, N_vN_h)^T$ denotes the indexes of rows of $\mathbf{D}_R$. Then, the SI channel gain $\mathbb{E}\left[\|\mathbf{G}_{SI}\|^2\right]$ can be effectively weakened by the following lemma:

Lemma 3 *Define* $B_T = \Lambda_T^{row}$ *and* $B_R = \left\{\left((\mathbf{Pd})^{\{\Lambda_R^{row}\}}, :\right)\right\}$ *as the beamspaces of the transmit and receive SI, then* $\mathbb{E}\left[\left\|\tilde{\mathbf{G}}_{SI}\right\|^2\right]$ *tends to zero if* $\mathbf{F}_t = \mathbf{F}^{\{B_u\}}$ *and* $\mathbf{F}_r = \mathbf{F}^{\{B_d\}}$, *where* $B_u \cap B_T = \emptyset$ *and* $B_d \cap B_R = \emptyset$.

Proof See Appendix: Proof of Lemma 3. □

5 3D multiuser beamspace transmission scenario

With 2D large-scale antenna arrays deployed at the BS, the spatial resolution of vertical and horizontal directions in 3D massive MIMO systems are high. We can exploit the slow time-varying angular information such as DoA, DoD, and channel correlation to efficiently estimate channels and transmit data. In this section, we detail our proposed 3D MUBT scheme. Based on user grouping scheme and beamspace channel estimation method, we show the scheme is capable of eliminating the SI, IGI, and MUI.

5.1 Beamspace-aware user scheduling

Due to the simultaneous transmission of UL and DL, the existence of SI and CCI restricts the overall performance of FD cellular systems. However, we can exploit the orthogonality of beamspace to avoid the SI, and by user scheduling strategy, the CCI can be ignored. To begin with, the user are partitioned into groups on the basis of the following criteria:

Criterion: Allocate the UL/DL users whose beamspaces are the same to the same group, while the beamspaces of different UL/DL groups are orthogonal, i.e.,

$$B_{g_i^u} \cap B_{g_j^u} = \emptyset, B_{g_i^d} \cap B_{g_j^d} = \emptyset \tag{25}$$

where $B_{g_i^u}$ and $B_{g_j^u}$ are the beamspaces of UL group g_i^u and g_j^u, $B_{g_i^d}$ and $B_{g_j^d}$ are the beamspaces of DL group g_i^d and g_j^d.

To simplify the illustration, the cardinalities of the beamspaces are set to be the same, i.e., $\left|B_{g_i^u}\right| = \left|B_{g_j^u}\right| = b_u$ and $\left|B_{g_i^d}\right| = \left|B_{g_j^d}\right| = b_d$. Assume that $K_{g_i^x}$ is the number of users in group g_i^x, and $g_{i,k}^x$ denotes the index of the kth user in group g_i^x, where $x \in \{u, d\}$. In the FD cellular systems, if the co-channel UL and DL users are too close to each other, the CCI may cause a serious trouble to the DL users' signal-to-interference ratio. Therefore, we provide a simple user scheduling algorithm to mitigate the SI and attenuate the CCI. The main idea of our scheduling strategy to suppress the CCI is to locate the co-channel users apart. Within the cellular systems, the sufficient distance isolation between co-channel users can significantly reduce the CCI due to the plenty of obstructions such as buildings and trees, which as a result makes the CCI negligible compared to the signals transmitted from the high-altitude BS [2]

Firstly, we partition the cell into four areas $\mathcal{A}_1$, $\mathcal{A}_2$, $\mathcal{A}_3$, and $\mathcal{A}_4$ as Fig. 3 shows, then the user scheduling algorithm is proposed as the following.

Algorithm 1 User Scheduling Based on the Distance and SI Beamspaces Isolation

1: Calculate the beamspaces of all UL and DL users through (14) and (20).

2: Partition UL and DL users into groups according to the beamspace-aware grouping criterion (25). Let $\mathcal{G}_u$ and $\mathcal{G}_d$ be the set of UL and DL groups which satisfy $\mathcal{G}_u = \left\{g_i^u \left| B_{g_i^u} \bigcap B_T = \varnothing, i = 1, \cdots, G_u \right.\right\}$ and $\mathcal{G}_d = \left\{g_i^d \left| B_{g_i^d} \bigcap B_R = \varnothing, i = 1, \cdots, G_d \right.\right\}$, respectively.

3: Schedule G_u UL groups in location area $\mathcal{A}$. **If** $\mathcal{A} = \mathcal{A}_1\mathcal{A}_2$, **then** schedule the DL users in $\mathcal{A}_4$; **else if** $\mathcal{A} = \mathcal{A}_3\mathcal{A}_4$, **then** schedule the DL users in $\mathcal{A}_1$.

4: **If** all UL groups' beamspaces are orthogonal to B_T, **then** schedule G_d DL groups; **else**, schedule G_d DL groups from $\mathcal{G}_d$ so as to eliminate the SI in the beamspace of g_i^u.

5: To include all the possibility, ***in the orthogonal time/frequency resources***, schedule G_d DL groups, and use the similar method in step 3 to determine in which area the UL groups are scheduled. **If** all DL groups' beamspaces are orthogonal to B_R, **then** schedule G_u UL groups; **else**, schedule G_u UL groups from $\mathcal{G}_u$.

6: **Repeat** 3, 4, 5.

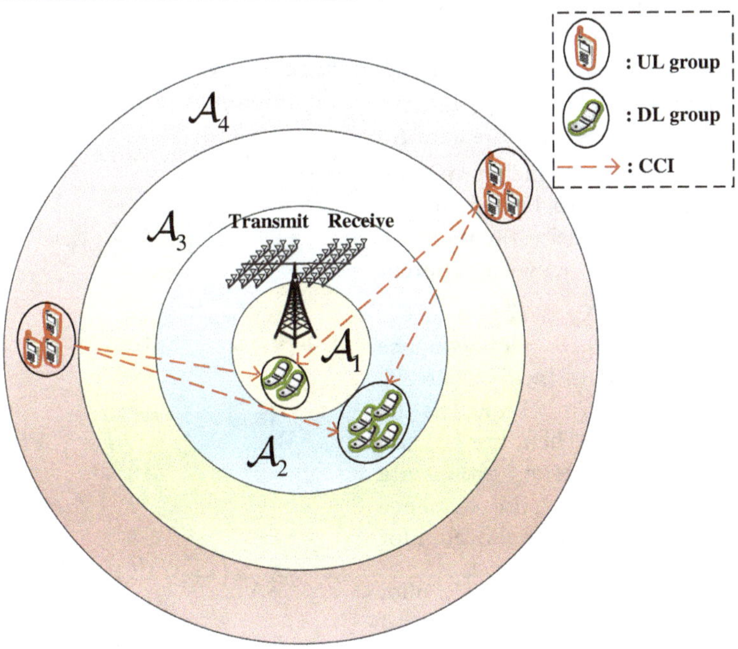

Fig. 3 Illustration of CCI between UL and DL groups under the cell partitioning scheme

5.2 Effective beamspace channel estimation

During the estimation phase of the FD system, all UL users transmit the pilot sequences to the BS, and simultaneously, the BS transmits the pilot sequences to the DL users. Then, the received pilot signals at the BS and the ith DL group are written as:

$$\mathbf{Y}_u = \sum_{i=1}^{G_u} \mathbf{G}_{g_i^u} \mathbf{X}_{g_i^u} + \mathbf{G}_{SI} \sum_{l=1}^{G_d} \tilde{\mathbf{X}}_{g_l^d} + \mathbf{N}_u \tag{26}$$

$$\mathbf{Y}_{g_i^d} = \mathbf{G}_{g_i^d}^H \sum_{i=1}^{G_d} \tilde{\mathbf{X}}_{g_i^d} + \sum_{j=1}^{G_u} \mathbf{H}_{g_i^d, g_j^u} \mathbf{X}_{g_j^u} + \mathbf{N}_{g_i^d} \tag{27}$$

where $\mathbf{G}_{g_i^u} = \left[\mathbf{g}_{g_{i,1}^u}, \cdots, \mathbf{g}_{g_{i,K_{g_i^u}}^u}\right] \in \mathbb{C}^{N_v N_h \times K_{g_i^u}}$ and $\mathbf{G}_{g_i^d} = \left[\mathbf{g}_{g_{i,1}^d}, \cdots, \mathbf{g}_{g_{i,K_{g_i^d}}^d}\right] \in \mathbb{C}^{N_v N_h \times K_{g_i^d}}$ are the channel matrices between the BS and the ith UL and DL user group, respectively. $\mathbf{H}_{g_i^d, g_j^u}$ is the co-channel matrix between jth UL group and ith DL group. The second term of the right side of (27) is the CCI resulting from the simultaneous UL and DL transmission at the same frequency. Based on the user scheduling algorithm, the CCI can be ignored in the FD scenario, which as a result is omitted in the later analysis. $\mathbf{X}_{g_i^u} \in \mathbb{C}^{K_{g_i^u} \times \tau_u}$ is pilot sequences of the ith UL group, τ_u denotes the length of pilot sequence for each UL user, $\tilde{\mathbf{X}}_{g_i^d} = \left(\mathbf{F}^{\left\{B_{g_i^d}\right\}}\right)^H \mathbf{X}_{g_i^d}$ denotes the precoded DL pilot sequence where $\mathbf{X}_{g_i^d} \in \mathbb{C}^{b_d \times \tau_d}$ is the DL beamspace pilot sequences for the ith DL group, τ_d is the length of DL pilot sequence, $\mathbf{N}_u$ and $\mathbf{N}_{g_i^d}$ are the additive white Gaussian noise (AWGN) matrices which have independent and identically distributed $\mathcal{CN}(0, 1)$ elements. In order to reduce the overhead of channel estimation, we can reuse the same pilot sequences between different groups. Let $\mathbf{X}_u \in \mathbb{C}^{\max_{1 \le i \le G_u} K_{g_i^u} \times \tau_u}$ and $\mathbf{X}_d \in \mathbb{C}^{b_d \times \tau_d}$ be the orthogonal UL pilot sequence set and DL pilot sequence set which satisfy $\mathbf{X}_a \mathbf{X}_a^H = \tau_a P_p^a \mathbf{I}, a \in \{u, d\}$, where P_p^a is the transmit power of each pilot symbol and $\mathbf{I}$ is the identity matrix with appropriate size. Then, the UL and DL pilot sequences assigned to the different groups are assumed to share the same UL and DL pilot sequence set. Mathematically, the UL pilot sequences allocated for group g_i^u is given by $\mathbf{X}_{g_i^u} = \mathbf{X}_u^{\left\{1:K_{g_i^u},:\right\}}$, and the DL pilot sequences allocated for group g_i^d is given by $\mathbf{X}_{g_i^d} = \mathbf{X}_d^{\{:,1:\tau_u\}}$ if $\tau_d \ge \tau_u$ and $\mathbf{X}_{g_i^d} = \left[\mathbf{X}_d \ \mathbf{0}_{b_d \times (\tau_u - \tau_d)}\right]$ if $\tau_d \le \tau_u$.

5.2.1 *UL beamspace channel estimation*

In order to eliminate the IGI, which is also the pilot contamination, we multiply both sides of (26) with $\mathbf{F}^{\left\{B_{g_i^u}\right\}}$ and we have:

$$\begin{aligned}\tilde{\mathbf{Y}}_{g_i^u} =& \mathbf{G}_{g_i^u}^{\left\{B_{g_i^u}\right\}} \mathbf{X}_{g_i^u} + \sum_{j=1, j\neq i}^{G_u} \mathbf{G}_{g_j^u}^{\left\{B_{g_i^u}\right\}} \mathbf{X}_{g_j^u} + \mathbf{F}^{\left\{B_{g_i^u}\right\}} \mathbf{N}_u \\ &+ \underbrace{\sum_{l=1}^{G_d} \mathbf{F}^{\left\{B_{g_i^u}\right\}} \mathbf{G}_{SI} \left(\mathbf{F}^{\left\{B_{g_l^d}\right\}}\right)^H \mathbf{X}_{g_l^d}}_{\text{SI}}.\end{aligned} \tag{28}$$

Then, the BS obtains the observation of UL beamspace channel as:

$$\begin{aligned}\mathbf{y}_{g_{i,k}^u}^{\text{ob}} =& \tilde{\mathbf{g}}_{g_{i,k}^u}^{\left\{B_{g_i^u}\right\}} + \sum_{j=1, j\neq i}^{G_u} \tilde{\mathbf{g}}_{g_{j,k}^u}^{\left\{B_{g_i^u}\right\}} + \frac{1}{\tau_u P_p^u} \left(\mathbf{F}^{\left\{B_{g_i^u}\right\}}\right)^H \mathbf{N}_u \mathbf{X}_{g_i^u}^H \mathbf{e}_k \\ &+ \underbrace{\frac{1}{\tau_u P_p^u} \sum_{l=1}^{G_d} \mathbf{F}^{\left\{B_{g_i^u}\right\}} \mathbf{G}_{SI} \left(\mathbf{F}^{\left\{B_{g_l^d}\right\}}\right)^H \mathbf{x} \tilde{\mathbf{X}}_{g_l^d} \mathbf{X}_{g_i^u}^H \mathbf{e}_k}_{\text{SI}}\end{aligned} \tag{29}$$

The second term of the right side of (29) is the pilot contamination between the kth users in all UL groups due to the reuse of the same pilot sequence. Note that, according to the grouping criterion and Corollary 1, the pilot contamination approaches to zero as $N_v, N_h \to \infty$. The last term is the SI caused by the simultaneous UL and DL training. Note that if users are scheduled by Algorithm 1, according to Lemma 3, the SI can be wiped out in the large N_v, N_h regime. Based on the observation in (29), the minimum mean square error (MMSE) estimation of the beamspace channel $\tilde{\mathbf{g}}_{g_{i,k}^u}^{\left\{B_{g_i^u}\right\}}$ can be obtained as:

$$\hat{\tilde{\mathbf{g}}}_{g_{i,k}^u}^{\left\{B_{g_i^u}\right\}} = \mathbf{R}_{g_{i,k}^u}^{\left\{B_{g_i^u}\right\}} \mathbf{C}_{k_u}^{-1} \mathbf{y}_{g_{i,k}^u}^{\text{ob}} \tag{30}$$

where $\mathbf{R}_{g_{i,k}^u}^{\left\{B_{g_i^u}\right\}}$ is given by:

$$\begin{aligned}\mathbf{R}_{g_{i,k}^u}^{\left\{B_{g_i^u}\right\}} =& \iint_{\left(\theta_{g_{i,k}^u}, \varphi_{g_{i,k}^u}\right) \in \Delta_{k_u}} \mathbf{F}^{\left\{B_{g_i^u}\right\}} \left(\mathbf{b}\left(h_{g_{i,k}^u}\right) \mathbf{b}^H\left(h_{g_{i,k}^u}\right)\right) \\ &\otimes \left(\mathbf{a}\left(v_{g_{i,k}^u}\right) \mathbf{a}^H\left(v_{g_{i,k}^u}\right)\right) \left(\mathbf{F}^{\left\{B_{g_i^u}\right\}}\right)^H \\ &\times S_u\left(\theta_{g_{i,k}^u}, \varphi_{g_{i,k}^u}\right) d\theta_{g_{i,k}^u} d\varphi_{g_{i,k}^u}\end{aligned} \tag{31}$$

where $S_u\left(\theta_{g_{i,k}^u},\varphi_{g_{i,k}^u}\right)=r_u\left(\theta_{g_{i,k}^u},\varphi_{g_{i,k}^u}\right)r_u^H\left(\theta_{g_{i,k}^u},\varphi_{g_{i,k}^u}\right)$, and

$$\mathbf{C}_{k_u}=\sum_{l=1}^{G_u}\mathbf{R}_{g_{l,k}^u}^{\left\{B_{g_i^u}\right\}}+\frac{1}{\tau_u P_p^u}\mathbf{I}+\frac{1}{\left(\tau_u P_p^u\right)^2}\sum_{j=1}^{G_d}\sum_{l=1}^{G_d}\mathbf{F}^{\left\{B_{g_i^u}\right\}}\mathbf{G}_{SI}\left(\mathbf{F}^{\left\{B_{g_l^d}\right\}}\right)^H\tilde{\mathbf{X}}_{g_l^d}\mathbf{X}_{g_i^u}^H\mathbf{e}_k\times\mathbf{e}_k^H\mathbf{X}_{g_i^u}\tilde{\mathbf{X}}_{g_j^d}^H\mathbf{F}^{\left\{B_{g_j^d}\right\}}\mathbf{G}_{SI}^H\left(\mathbf{F}^{\left\{B_{g_i^u}\right\}}\right)^H \tag{32}$$

Based on the orthogonal property of MMSE estimation [34], the channel eatimation error is $\bar{\tilde{\mathbf{g}}}_{g_{i,k}^u}^{\left\{B_{g_i^u}\right\}}=\tilde{\mathbf{g}}_{g_{i,k}^u}^{\left\{B_{g_i^u}\right\}}-\hat{\tilde{\mathbf{g}}}_{g_{i,k}^u}^{\left\{B_{g_i^u}\right\}}$, and the covariance of $\bar{\tilde{\mathbf{g}}}_{g_{i,k}^u}^{\left\{B_{g_i^u}\right\}}$ is:

$$\bar{\mathbf{R}}_{g_{i,k}^u}^{\left\{B_{g_i^u}\right\}}=\mathbf{R}_{g_{i,k}^u}^{\left\{B_{g_i^u}\right\}}-\mathbf{R}_{g_{i,k}^u}^{\left\{B_{g_i^u}\right\}}\mathbf{C}_{k_u}^{-1}\mathbf{R}_{g_{i,k}^u}^{\left\{B_{g_i^u}\right\}}. \tag{33}$$

5.2.2 DL beamspace channel estimation

(27) can be rewritten as:

$$\mathbf{Y}_{g_i^d}=\left(\mathbf{G}_{g_i^d}^{\left\{B_{g_i^d}\right\}}\right)^H\mathbf{X}_{g_i^d}+\underbrace{\sum_{j=1,j\neq i}^{G_d}\left(\mathbf{G}_{g_i^d}^{\left\{B_{g_j^d}\right\}}\right)^H\mathbf{X}_{g_j^d}}_{\text{pilot contamination}}+\mathbf{N}_{g_i^d}. \tag{34}$$

Then, the observation of the DL beamspace channel is written as

$$\mathbf{y}_{g_{i,k}^d}^{\text{ob}}=\frac{1}{\tau_d P_p^d}\left(\mathbf{Y}_{g_i^d}\mathbf{X}_{g_i^d}^H\right)^H\mathbf{e}_k=\tilde{\mathbf{g}}_{g_{i,k}^d}^{\left\{B_{g_i^d}\right\}}+\underbrace{\sum_{j=1,j\neq i}^{G_d}\tilde{\mathbf{g}}_{g_{i,k}^d}^{\left\{B_{g_j^d}\right\}}}_{\text{pilot contamination}}+\frac{1}{\tau_d P_p^d}\left(\mathbf{N}_{g_i^d}\mathbf{X}_{g_i^d}^H\right)^H\mathbf{e}_k \tag{35}$$

The second term of the right side of (35) is the pilot contamination caused by the reuse of the same DL pilot sequence set over all DL groups, and it tends to vanish as $N_v, N_h\to\infty$ according to the grouping criterion and Corollary 1.

Using the similar method as UL estimation, we have:

$$\hat{\tilde{\mathbf{g}}}_{g_{i,k}^d}^{\left\{B_{g_i^d}\right\}}=\mathbf{R}_{g_{i,k}^d}^{\left\{B_{g_i^d}\right\}}\mathbf{C}_{k_d}^{-1}\mathbf{y}_{g_{i,k}^d}^{\text{ob}} \tag{36}$$

where

$$\mathbf{R}_{g_{i,k}^d}^{\left\{B_{g_i^d}\right\}}=\sum_{l=1}^{G_d}\iint_{\left(\theta_{g_{i,k}^d},\varphi_{g_{i,k}^d}\right)\in\Delta_{k_d}}\mathbf{F}^{\left\{B_{g_i^d}\right\}}\left(\mathbf{b}\left(h_{g_{i,k}^d}\right)\mathbf{b}^H\left(h_{g_{i,k}^d}\right)\right)\otimes\left(\mathbf{a}\left(v_{g_{i,k}^d}\right)\mathbf{a}^H\left(v_{g_{i,k}^d}\right)\right)\left(\mathbf{F}^{\left\{B_{g_l^d}\right\}}\right)^H\times S_d\left(\theta_{g_{i,k}^d},\varphi_{g_{i,k}^d}\right)d\theta_{g_{i,k}^d}\,d\varphi_{g_{i,k}^d} \tag{37}$$

and

$$\mathbf{C}_{k_d}=\sum_{j=1}^{G_d}\mathbf{R}_{g_{i,k}^d}^{\left\{B_{g_j^d}\right\}}+\frac{1}{\tau_d P_p^d}\mathbf{I}=\sum_{j=1}^{G_d}\sum_{l=1}^{G_d}\iint_{\left(\theta_{g_{i,k}^d},\varphi_{g_{i,k}^d}\right)\in\Delta_{k_d}}\mathbf{F}^{\left\{B_{g_j^d}\right\}}\left(\mathbf{b}\left(h_{g_{i,k}^d}\right)\mathbf{b}^H\left(h_{g_{i,k}^d}\right)\right)\otimes\left(\mathbf{a}\left(v_{g_{i,k}^d}\right)\mathbf{a}^H\left(v_{g_{i,k}^d}\right)\right)\left(\mathbf{F}^{\left\{B_{g_l^d}\right\}}\right)^H\times S_d\left(\theta_{g_{i,k}^d},\varphi_{g_{i,k}^d}\right)d\theta_{g_{i,k}^d}\,d\varphi_{g_{i,k}^d}+\frac{1}{\tau_d P_p^d}\mathbf{I}. \tag{38}$$

Similarly, the covariance of DL channel estimation error $\bar{\tilde{\mathbf{g}}}_{g_{i,k}^d}^{\left\{B_{g_i^d}\right\}}$ is

$$\bar{\mathbf{R}}_{g_{i,k}^d}^{\left\{B_{g_i^d}\right\}}=\mathbf{R}_{g_{i,k}^d}^{\left\{B_{g_i^d}\right\}}-\mathbf{R}_{g_{i,k}^d}^{\left\{B_{g_i^d}\right\}}\mathbf{C}_{k_d}^{-1}\mathbf{R}_{g_{i,k}^d}^{\left\{B_{g_i^d}\right\}}. \tag{39}$$

5.3 Data transmission

At the data transmission stage[3], UL and DL users transmit signals in the same time-frequency resource. In our transmission scheme, the data are transmitted in each group's own beamspace. In the BD, the received signal at the BS from the ith UL group and the received signal at the ith DL group can be written as:

$$\tilde{\mathbf{y}}_{g_i^u}=\tilde{\mathbf{G}}_{g_i^u}^{\left\{B_{g_i^u}\right\}}\mathbf{s}_{g_i^u}+\underbrace{\sum_{j=1,j\neq i}^{G_u}\tilde{\mathbf{G}}_{g_j^u}^{\left\{B_{g_i^u}\right\}}\mathbf{s}_{g_j^u}}_{\text{IGI}}+\tilde{\mathbf{n}}_{g_i^u}+\underbrace{\sum_{j=1}^{G_d}\mathbf{F}^{\left\{B_{g_i^u}\right\}}\mathbf{G}_{SI}\left(\mathbf{F}^{\left\{B_{g_j^d}\right\}}\right)^H\mathbf{x}_{g_i^d}}_{\text{SI}}. \tag{40}$$

$$\mathbf{y}_{g_i^d} = \left(\tilde{\mathbf{G}}_{g_i^d}^{\left\{B_{g_i^d}\right\}}\right)^H \mathbf{x}_{g_i^d} + \underbrace{\sum_{j=1,j\neq i}^{G_d} \left(\tilde{\mathbf{G}}_{g_i^d}^{\left\{B_{g_j^d}\right\}}\right)^H \mathbf{x}_{g_j^d}}_{\text{IGI}} + \mathbf{n}_{g_i^d} \tag{41}$$

where $\mathbf{s}_{g_i^u} \in \mathbb{C}^{K_{g_i^u}\times 1}$ and $\mathbf{x}_{g_i^d} \in \mathbb{C}^{b_d\times 1}$ denote the transmit data of the ith UL group and the BD precoded data for users in the ith DL group, respectively, and $\tilde{\mathbf{n}}_{g_i^u} = \mathbf{F}^{\left\{B_{g_i^u}\right\}}\mathbf{n}_{g_i^u}$. Note that the IGIs in (40) and (41) tend to zero as $N_v, N_h \to \infty$.

For massive MIMO systems, linear signal processing techniques have already been shown to have good performance [2]. Therefore, in this paper, define $\mathbf{W}_{g_i^u} = \left[\mathbf{w}_{g_{i,1}^u}, \cdots, \mathbf{w}_{g_{i,K_{g_i^u}}^u}\right] \in \mathbb{C}^{b_u\times K_{g_i^u}}$ and $\mathbf{W}_{g_i^d} = \left[\mathbf{w}_{g_{i,1}^d}, \cdots, \mathbf{w}_{g_{i,K_{g_i^d}}^d}\right] \in \mathbb{C}^{b_d\times K_{g_i^d}}$ as the linear UL detector and DL precoder at the BS. Multiplying (40) with $\mathbf{W}_{g_i^u}^H$ we have $\mathbf{y}_{g_i^u} = \mathbf{W}_{g_i^u}^H \tilde{\mathbf{y}}_{g_i^u}$. Then, the received signal of user $g_{i,k}^u$ in beamspace $B_{g_i^u}$ can be written as

$$\begin{aligned} y_{g_{i,k}^u} =& \mathbf{w}_{g_{i,k}^u}^H \tilde{\mathbf{g}}_{g_{i,k}^u}^{\left\{B_{g_i^u}\right\}} s_{g_{i,k}^u} + \underbrace{\mathbf{w}_{g_{i,k}^u}^H \sum_{k'=1,k'\neq k}^{K_{g_i^u}} \tilde{\mathbf{g}}_{g_{i,k'}^u}^{\left\{B_{g_i^u}\right\}} s_{g_{i,k'}^u}}_{\text{MUI}} \\ &+ \underbrace{\mathbf{w}_{g_{i,k}^u}^H \sum_{j=1,j\neq i}^{G_u} \tilde{\mathbf{G}}_{g_j^u}^{\left\{B_{g_i^u}\right\}} \mathbf{s}_{g_j^u}}_{\text{IGI}} + \mathbf{w}_{g_{i,k}^u}^H \tilde{\mathbf{n}}_{g_i^u} \\ &+ \underbrace{\mathbf{w}_{g_{i,k}^u}^H \sum_{l=1}^{K_{g_d}} \mathbf{F}^{\left\{B_{g_i^u}\right\}} \mathbf{G}_{SI} \left(\mathbf{F}^{\left\{B_{g_l^d}\right\}}\right)^H \mathbf{W}_{g_l^d} \mathbf{s}_{g_l^d}}_{\text{SI}} \end{aligned} \tag{42}$$

which is capable of decoding the data of user $g_{i,k}^u$, and the second term of the right side of (42) is the MUI within the group. For DL, the BD precoded data can be written as $\mathbf{x}_{g_i^d} = \mathbf{W}_{g_i^d}\mathbf{s}_{g_i^d}$; substituting it into (41), we can express the received signal at the user $g_{i,k}^d$ in $B_{g_i^d}$ as:

$$\begin{aligned} y_{g_{i,k}^d} =& \left(\tilde{\mathbf{g}}_{g_{i,k}^d}^{\left\{B_{g_i^d}\right\}}\right)^H \mathbf{w}_{g_{i,k}^d} s_{g_{i,k}^d} + \underbrace{\sum_{j=1,j\neq i}^{G_d} \left(\tilde{\mathbf{g}}_{g_{i,k}^d}^{\left\{B_{g_j^d}\right\}}\right)^H \mathbf{W}_{g_j^d}\mathbf{s}_{g_j^d}}_{\text{IGI}} \\ &+ \underbrace{\left(\tilde{\mathbf{g}}_{g_{i,k}^d}^{\left\{B_{g_i^d}\right\}}\right)^H \sum_{k'=1,k'\neq k}^{K_{g_i^d}} \mathbf{w}_{g_{i,k'}^d} s_{g_{i,k'}^d}}_{\text{MUI}} + n_{g_{i,k}^d} \end{aligned} \tag{43}$$

We adopt the zero-forcing (ZF) beamforming scheme to eliminate the MUI. The CSI of UL and DL beamspace channels is assumed to be obtained by (30) and (36); then, the beamforming matrices can be expressed as:

$$\begin{aligned} \mathbf{W}_{g_i^u} &= \hat{\tilde{\mathbf{H}}}_{g_i^u}^{\left\{B_{g_i^u}\right\}} \left(\left(\hat{\tilde{\mathbf{H}}}_{g_i^u}^{\left\{B_{g_i^u}\right\}}\right)^H \hat{\tilde{\mathbf{H}}}_{g_i^u}^{\left\{B_{g_i^u}\right\}}\right)^{-1} \\ \mathbf{W}_{g_i^d} &= \hat{\tilde{\mathbf{H}}}_{g_i^d}^{\left\{B_{g_i^d}\right\}} \left(\left(\hat{\tilde{\mathbf{H}}}_{g_i^d}^{\left\{B_{g_i^d}\right\}}\right)^H \hat{\tilde{\mathbf{H}}}_{g_i^d}^{\left\{B_{g_i^d}\right\}}\right)^{-1} \Gamma_{g_i^d}^{-\frac{1}{2}} \end{aligned} \tag{44}$$

where $\Gamma_{g_i^d} = \operatorname{diag}\left\{\beta_{i,1}, \beta_{i,2}, \cdots, \beta_{i,K_{g_i^d}}\right\} \in \mathbb{C}^{K_{g_i^d}\times K_{g_i^d}}$ is a diagonal matrix for the normalization of the precoder.

Based on (42) and according to [35], we can obtain the signal-to-interference-plus-noise ratio (SINR) of user $g_{i,k}^u$ as (45), shown at the top of the next page, where $P_{g_{i,k}^u}$ is the transmit power of user $g_{i,k}^u$, $\mathbf{P}_{g_j^u} = \mathbb{E}\left[\mathbf{s}_{g_j^u}\mathbf{s}_{g_j^u}^H\right]$ and $\mathbf{P}_{g_l^d} = \mathbb{E}\left[\mathbf{s}_{g_l^d}\mathbf{s}_{g_l^d}^H\right]$ are the transmit power matrices, $\Psi_l = \mathbf{F}^{\left\{B_{g_i^u}\right\}}\mathbf{G}_{SI}\left(\mathbf{F}^{\left\{B_{g_l^d}\right\}}\right)^H \mathbf{W}_{g_l^d}$. According to (43) and the similar method in [35], the SINR of user $g_{i,k}^d$ can be expressed as (46), shown the top of next page.

$$\gamma_{g_{i,k}^u} = \frac{P_{g_{i,k}^u}}{\mathbf{w}_{g_{i,k}^u}^H \left[\sum_{k'=1}^{K_{g_i^u}} P_{g_{i,k'}^u} \Delta\tilde{\mathbf{g}}_{g_{i,k'}^u}^{\left\{B_{g_i^u}\right\}} \left(\Delta\tilde{\mathbf{g}}_{g_{i,k'}^u}^{\left\{B_{g_i^u}\right\}}\right)^H + \sum_{j=1,j\neq i}^{G_u} \tilde{\mathbf{G}}_{g_j^u}^{\left\{B_{g_i^u}\right\}} \mathbf{P}_{g_j^u} \left(\tilde{\mathbf{G}}_{g_j^u}^{\left\{B_{g_i^u}\right\}}\right)^H + \sum_{l=1}^{G_d} \Psi_l \mathbf{P}_{g_l^d} \Psi_l^H + \mathbf{I}_{b_u}\right] \mathbf{w}_{g_{i,k}^u}} \tag{45}$$

$$\gamma_{g_{i,k}^d} = \frac{P_{g_{i,k}^d}\left|\left(\tilde{\mathbf{g}}_{g_{i,k}^d}^{\left\{B_{g_i^d}\right\}}\right)^H \mathbf{w}_{g_{i,k}^d}\right|^2}{\left|\left(\Delta\tilde{\mathbf{g}}_{g_{i,k'}^d}^{\left\{B_{g_i^d}\right\}}\right)^H \sum_{k'=1}^{K_{g_i^d}} P_{g_{i,k'}^d}^{\frac{1}{2}} \mathbf{w}_{g_{i,k'}^d}\right|^2 + \left\|\sum_{j=1,j\neq i}^{G_d}\left(\tilde{\mathbf{g}}_{g_{i,k}^d}^{\left\{B_{g_j^d}\right\}}\right)^H \mathbf{W}_{g_j^d}\mathbf{P}_{g_l^d}^{\frac{1}{2}}\right\|^2 + 1}. \tag{46}$$

Eventually, the achievable rates of user $g_{i,k}^u$ and $g_{i,k}^d$ are expressed as:

$$R_{g_{i,k}^a} = \frac{T - \max(\tau_u, \tau_d)}{T}\mathbb{E}\left[\log_2\left(1 + \gamma_{g_{i,k}^a}\right)\right], a \in \{u, d\} \tag{47}$$

6 Simulation results

In this section, we present the performance of the proposed scheme through simulation results. The parameters of the simulated single-cell multi-user systems are listed in Table 1. For path loss, we consider the 3GPP LTE simulation model [36]. It is assumed that the transmit and receive antenna arrays are separated in distance and some passive SIC techniques, such as radio frequency absorber material, are adopted so that the level of residual SI $\beta_{SI} = 15$ dB. We consider that both the UL and DL users are scheduled by Algorithm 1 and are gathered into three groups. We assume that each group has 8 users and the range of DoA or DoD of users within each group is the same.

6.1 Channel estimation

In order to evaluate the performance of our proposed channel estimation method, we define the average individual MSE as the performance metric which is given by:

$$\kappa_a = \frac{1}{K_a}\sum_{i=1}^{G_a}\sum_{k=1}^{K_{g_i^a}} \operatorname{tr}\left(\bar{\mathbf{R}}_{g_{i,k}^a}^{\left\{B_{g_i^a}\right\}}\right), a \in \{u, d\} \tag{48}$$

Figures 4 and 5 present the comparison of the UL and DL channel estimation performance between the proposed beamspace channel estimation and the conventional full-dimension channel estimation as a function of average UL and DL receive SNR, respectively. In our proposed HD/FD beamspace channel estimation scheme, the length of pilot sequence is related to the dimension of beamspace; hence, we set $\tau_u = \tau_d = \tau = \max\{b_u, b_d\}$. Under this setup, the default length of pilot sequence is $\tau = 42$. We also compare the cases of $\tau = 54$ and $\tau = 72$. From the numerical results, we can see that the DL performance of the proposed scheme outperforms the conventional full-dimension method and the UL performance of the proposed scheme is better in the low-SNR region. What is more, comparing the proposed HD and FD scenarios, we can observe that in the low-SNR case, the performance is very close between HD and FD. As the SNR grows, the harmful impact of SI increases, which brings down the performance of FD scenario.

Note that, in the HD massive MIMO systems, the required minimum length of pilot sequence in the conventional TDD systems with DL reciprocity [5] and FDD systems [12] are $\sum_{i=1}^{G_u} K_{g_i^u} + \sum_{i=1}^{G_d} K_{g_i^d} = 48$ and $\sum_{i=1}^{G_u} K_{g_i^u} + b_d = 66$ (approximate), respectively. In the FD massive MIMO systems, [24] assumes that transmit and receive radio-frequency chains share the transmit antennas at the BS so that DL CSI can be obtained by utilizing the reciprocity which is at the cost of hardware, and the required minimum length of pilot sequence is $\sum_{i=1}^{G_u} K_{g_i^u} + \sum_{i=1}^{G_d} K_{g_i^d} = 48$. In addition, we know that the computational complexity of MMSE estimation is $\mathcal{O}\left(n^3\right)$ due to the matrix inversion operation where n is the dimension of the channel covariance matrix, which lays a heavy burden on the system if n is large, especially for massive MIMO system. For example, the computational complexity of the conventional full-dimension MMSE estimation [6] is $\mathcal{O}\left((N_v N_h)^3\right) = \mathcal{O}\left(625^3\right)$, which is very high. Nevertheless, the complexity of our low-dimension beamspace channel estimation scheme is $\mathcal{O}\left([\max(b_u, b_d)]^3\right)$ which is $\mathcal{O}\left(30^3\right)$. From this perspective, our 3D MUBT scheme is competitive in 3D massive MIMO system application.

Table 1 Simulation parameters

Parameters values	
Path loss between BS and users	2.7 + 42.8 $\log_{10}(L)$ [dB], L: distance in meter
Number of UL and DL users $K_d = K_u$	24
Antenna array size $N_h \times N_v$	25 × 25
UPA array element spacing $d_h = d_v$	$\lambda/2$
AS in the horizontal directions δ_h	10°
AS in the vertical directions δ_v	5°
Required effective channel power ratio ρ	0.9
Channel coherence time T	200 symbols

6.2 Data transmission

The SE in this subsection is defined as $SE = SE_u + SE_d$ where $SE_a = \frac{1}{2}\sum_{i=1}^{G_a}\sum_{k=1}^{K_{g_i^a}} \frac{T-\tau_a}{T}\log_2\left(1 + \gamma_{g_{i,k}^a}\right)$, $a \in \{u, d\}$, the factor $\frac{1}{2}$ is for the HD mode which does not exist in the FD case. To validate the feasibility and illustrate the effectiveness of the FD 3D MUBT scheme, Fig. 6 compares the SEs versus average receive SNR in different scenarios. It can be seen that the FD 3D MUBT achieves the best performance both in the low SNR and high SNR region. The reason is that even though assuming the DL reciprocity, the training overhead in [24] is still larger than the proposed

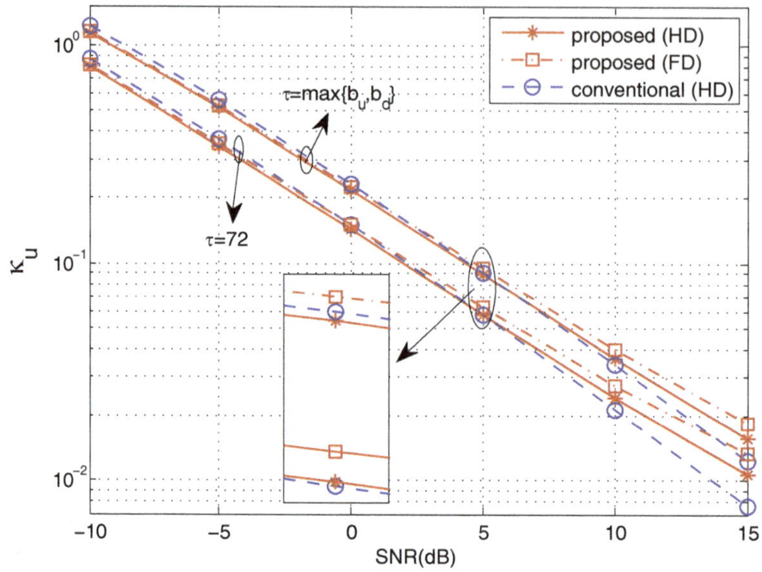

Fig. 4 Comparison of the UL channel estimation performance between the beamspace channel estimation and conventional estimation versus average receive SNRs

scheme. Besides, by using the user scheduling algorithm, the anti-SI capability of the proposed scheme outperforms the linear transceiver. Nevertheless, as the SNR grows, the performance gap between the FD 3D MUBT scheme and the FD scheme with linear transceiver decreases, and that between the HD 3D MUBT scheme and TDD case increases. The reason of them both is that as the SNR increases, the truncation error limits the performance of the proposed 3D MUBT scheme. What is more, the HD 3D MUBT outperforms the FDD [12] mainly due to the less training overhead.

Meantime, the comparisons that with less number of antennas (15×15) and no user scheduling are made. As expected, with less number of antennas, the overlap of beamspaces will increase which will increase the IGI and SI. Similarly, without the user scheduling algorithm, the

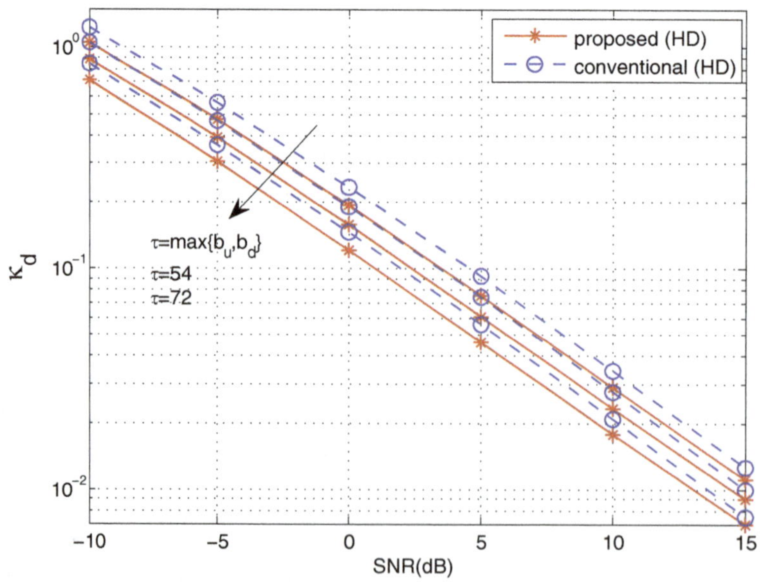

Fig. 5 Comparison of the DL channel estimation performance between the beamspace channel estimation and conventional estimation versus average receive SNRs

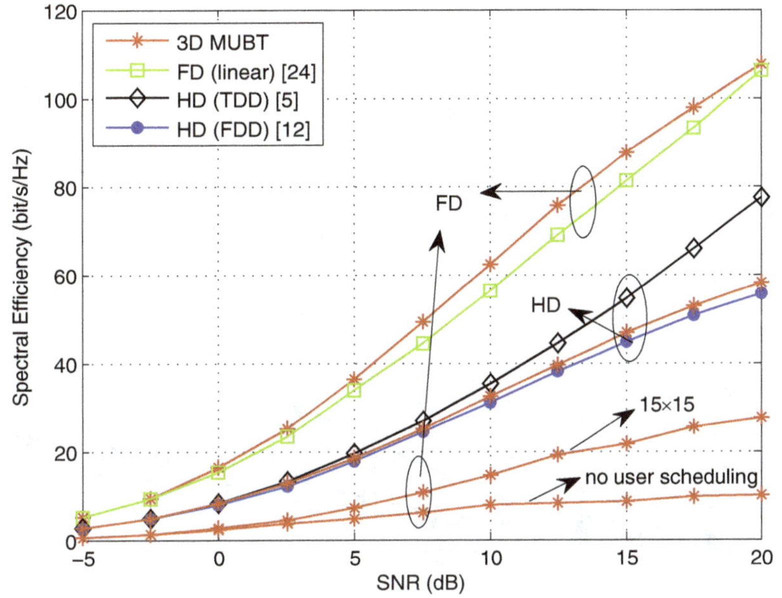

Fig. 6 SEs versus average receive SNRs in different scenarios

orthogonality of beamspaces between different groups of simultaneously scheduled users and SI cannot be guaranteed. And the CCI will also damage the system.

In the previous simulations, the β_{SI} is set to the fixed value. To investigate how the SI level influences the SE performance in the FD massive MIMO systems, we compare the performance between HD system and FD system with different SI level β_{SI} in Fig. 7. As expected, the performance of FD systems outperform that of HD systems when β_{SI} level is low and reduces as β_{SI} grows. It is observed that in the low β_{SI} regions, the linear FD scheme achieves the better SE performance, which is because when the SI is relatively small, the channel estimation error, IGI and AWGN restrict the SE performance mainly; then, as a result, the linear scheme outperforms the proposed scheme due to the truncation error. For

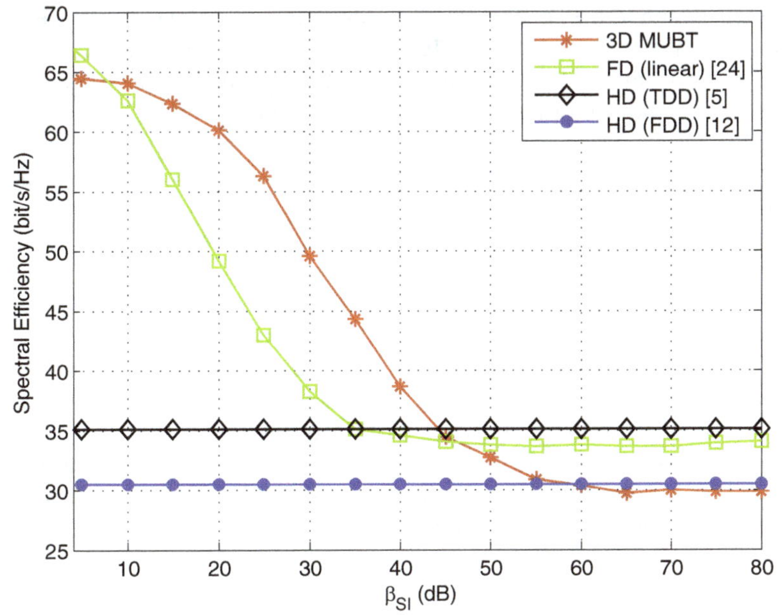

Fig. 7 SEs versus β_{SI} in different scenarios with average receive SNR= 10dB

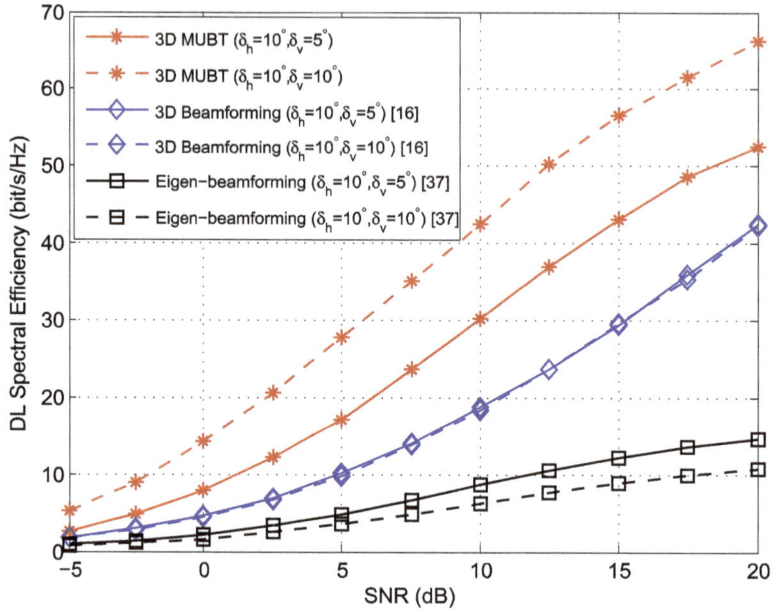

Fig. 8 SEs versus average DL receive SNRs in different 3D precoding scenarios

the similar reason, when the SI becomes the dominate, the impact of the SI on the two schemes is almost the same; therefore, the linear FD is superior when the SI level is high. However, when the β_{SI} is within (8, 45), the FD 3D MUBT scheme outperforms the linear FD, which indicates that with the user scheduling strategy, the FD 3D MUBT scheme has better anti-SI capability. In other words, applying the FD 3D MUBT scheme can lower the requirement for the SIC techniques. Take $SE = 49$ bit/s/Hz for example, the corresponding β_{SI} for the FD scheme with linear transceiver and the FD 3D MUBT scheme are 20 and 30dB, which indicates that the SIC capability in conventional FD system needs 10dB more than that in the proposed scheme and will bring the extra hardware cost and complexity.

6.3 3D precoding schemes

According to [14], the channel correlation matrix can be well approximated by the Kronecker production of the correlations in horizontal and vertical directions, i.e., $\mathbf{C} = \mathbf{C}_h \otimes \mathbf{C}_v$. Based on this, [37] designs a eigen-beamforming scheme. This scheme first chooses the principle eigenvector of $\mathbf{C}_v$, to perform the vertical dimension beamforming, then use the eigenvector matrix of $\mathbf{C}_h$ to do the horizontal dimension beamforming. In [16], Li et al. exploit the eigenvalue characteristic of large-scale UPA as discussed in Section 5.2 (23), and select the optimal DL beamforming vector for the kth user as $\mathbf{b}_k = \left(\mathbf{F}_{N_h}\right)_{:,n} \otimes \left(\mathbf{F}_{N_v}\right)_{:,m}$ to maximize the average signal-to-leakage-plus-noise ratio, where n and m are the indexes of columns where the maximum eigenvalues of $\mathbf{C}_h$ and $\mathbf{C}_v$ are, respectively. In Fig. 8, we investigate the DL transmission with different 3D precoding schemes under the different conditions of δ_v. Here, we get rid of the 1 / 2 pre-log factor of the DL SE for the sake of illustration. We can see that the proposed scheme achieves the best performance. The performance of eigen-beamforming scheme decreases when δ_v increase. This is because the power of the channel is concentrated around the dominant eigenmode, and as the vertical AS increase, the power of the principle eigenvalue decrease. Moreover, the 3D beamforming scheme is insensitive to δ_v under this setup[4]. However, we can see that the SE performance of the proposed scheme increase as δ_v grows; this can be explained by (14). The dimension of the beamspace channel increases as AS grows while keep the distance between the beamspaces of groups and the IGI at acceptable levels, which as a result improves the SE.

7 Conclusions

In this paper, we propose a 3D MUBT scheme for FD cellular systems. By adopting the property of beamspace, the 3D MUBT scheme can not only efficiently mitigate the SI due to FD transmission, but also significantly reduce the overhead of channel estimation. The simulation results show that the proposed 3D MUBT scheme outperforms the FD scheme with linear transceiver and the HD (TDD/FDD) schemes in massive MIMO cellular system.

Endnotes

[1] Note that in FD techniques, the antenna configuration can also be the shared type, which only needs one

antenna array to accomplish transmission and reception [19]. Nevertheless, due to the serious cross-talk within the antennas, it is impractical in MIMO system for now.

[2] We assume that the radius of the cell is 1000m. According to the 3GPP LTE BS-to-user and user-to-user path loss models for macrocell environment [36], the path loss between BS to users and users to users are $\mathrm{PL} = 2.7 + 42.8\log_{10}(R_{\mathrm{BS-UE}})$ and $\mathrm{PL} = 55.78 + 40\log_{10}(R_{\mathrm{UE-UE}})$, respectively, where $R_{\mathrm{BS-UE}}$ and $R_{\mathrm{UE-UE}}$ are distances in meter. When the distance between BS and users is 400m (the shortest distance between users and users is 350m according to the Algorithm 1), the interference channel between the two users is about 43dB weaker than the useful channel.

[3] The estimated DL CSI feedback process is also important and is studied in a lot of literatures [11, 38, 39]. Choi et al. [39] shows that the influence of channel feedback noise and errors can be made negligible with respect to the channel estimation errors especially when the SNR is high, hence in this paper we consider the ideal DL CSI feedback for simplicity, as assumed in [12].

[4] In fact, the simulation results in [16] have the similar conclusion with the eigen-beamforming scheme, but when the number of user in each group is small, the decrease of the SE performance is inconspicuous.

Appendix: Proof of Lemma 3

Since both $\mathbf{F}_{N_v}$ and $\mathbf{F}_{N_h}$ are unitary, then $\mathbf{F}$ is a unitary matrix, based on which we can obtain

$$\mathbb{E}\left[\left\|\tilde{\mathbf{G}}_{SI}\right\|^2\right] = \mathbb{E}\left[\left\|\mathbf{F}_r\mathbf{G}_{SI}\mathbf{F}_t^H\right\|^2\right] = \mathbb{E}\left[\|\mathbf{G}_{SI}\|^2\right]. \quad (49)$$

The vectorized BD SI channel is $\mathbf{g}_{SI} = \mathrm{vec}(\mathbf{G}_{SI})$, and the vectorized BD SI channel is $\tilde{\mathbf{g}}_{SI} = \mathrm{vec}\left(\tilde{\mathbf{G}}_{SI}\right)$. According to (7), (8), and (22), $\mathbf{g}_{SI}$ can be represented as

$$\begin{aligned}
\mathbf{g}_{SI} &= \mathrm{vec}(\mathbf{G}_{SI}) = \left(\mathbf{C}_T^{\frac{1}{2}} \otimes \mathbf{C}_R^{\frac{1}{2}}\right)\mathrm{vec}(\mathbf{G}_w) \\
&= \left(\mathbf{C}_{Th}^{\frac{1}{2}} \otimes \mathbf{C}_{Tv}^{\frac{1}{2}}\right) \otimes \left(\mathbf{C}_{Rh}^{\frac{1}{2}} \otimes \mathbf{C}_{Rv}^{\frac{1}{2}}\right)\mathbf{g}_w \\
&\approx \left(\mathbf{F}_{N_h}\mathbf{D}_{Th}^{\frac{1}{2}}\mathbf{F}_{N_h}^H\right) \otimes \left(\mathbf{F}_{N_v}\mathbf{D}_{Tv}^{\frac{1}{2}}\mathbf{F}_{N_v}^H\right) \otimes \left(\mathbf{F}_{N_h}\mathbf{D}_{Rh}^{\frac{1}{2}}\mathbf{F}_{N_h}^H\right) \\
&\otimes \left(\mathbf{F}_{N_v}\mathbf{D}_{Rv}^{\frac{1}{2}}\mathbf{F}_{N_v}^H\right)\mathbf{g}_w \\
&= \left(\mathbf{F}\mathbf{D}_T^{\frac{1}{2}}\mathbf{F}^H\right) \otimes \left(\mathbf{F}\mathbf{D}_R^{\frac{1}{2}}\mathbf{F}^H\right)\mathbf{g}_w
\end{aligned} \quad (50)$$

Based on (49) and the property $\mathbb{E}\left[\left\|\tilde{\mathbf{G}}_{SI}\right\|^2\right] = \mathbb{E}\left[\|\tilde{\mathbf{g}}_{SI}\|^2\right]$, we can get

$$\begin{aligned}
\mathbb{E}\left[\left\|\tilde{\mathbf{G}}_{SI}\right\|^2\right] &= \mathbb{E}\left[\left\|\mathrm{vec}\left(\mathbf{F}_r\mathbf{G}_{SI}\mathbf{F}_t^H\right)\right\|^2\right] = \mathbb{E}\left[\left\|\left(\mathbf{F}_t^* \otimes \mathbf{F}_r\right)\mathbf{g}_{SI}\right\|^2\right] \\
&\overset{N_v, N_h \to \infty}{=} \mathbb{E}\left[\left\|\left(\mathbf{F}_t^* \otimes \mathbf{F}_r\right)\left(\mathbf{F}\mathbf{D}_T^{\frac{1}{2}}\mathbf{F}^H\right) \otimes \left(\mathbf{F}\mathbf{D}_R^{\frac{1}{2}}\mathbf{F}^H\right)\mathbf{g}_w\right\|^2\right]
\end{aligned} \quad (51)$$

Then, we have

$$\begin{aligned}
\mathbb{E}\left[\left\|\tilde{\mathbf{G}}_{SI}\right\|^2\right] &\overset{N_v, N_h \to \infty}{=} \mathbf{g}_w^H\left(\mathbf{F}\mathbf{D}_T^{\frac{1}{2}}\mathbf{F}^H\mathbf{F}_t^T\right) \otimes \left(\mathbf{F}\mathbf{D}_R^{\frac{1}{2}}\mathbf{F}^H\mathbf{F}_r^H\right) \\
&\times \left(\mathbf{F}_t^*\mathbf{F}\mathbf{D}_T^{\frac{1}{2}}\mathbf{F}^H\right) \otimes \left(\mathbf{F}_r\mathbf{F}\mathbf{D}_R^{\frac{1}{2}}\mathbf{F}^H\right)\mathbf{g}_w
\end{aligned} \quad (52)$$

According to (29), $\mathbb{E}\left[\left\|\tilde{\mathbf{G}}_{SI}\right\|^2\right]$ can be effectively represented. Take the last two brackets of the right side of (34) for example, which is denoted as I_1 and I_2, we have:

$$I_1 = \mathbf{F}_t^*\mathbf{F}\mathbf{D}_T^{\frac{1}{2}}\mathbf{F}^H \approx \left(\mathbf{F}^{\{B_u\}}\right)^*\mathbf{F}^{\left\{\Lambda_T^{\mathrm{col}}\right\}}\left(\mathbf{D}_T^{\frac{1}{2}}\right)^{\{\Lambda_T\}}\left(\mathbf{F}^H\right)^{\left\{\Lambda_T^{\mathrm{row}}\right\}} \quad (53)$$

and

$$I_2 = \mathbf{F}_r\mathbf{F}\mathbf{D}_R^{\frac{1}{2}}\mathbf{F}^H \approx \mathbf{F}^{\{B_d\}}\mathbf{F}^{\left\{\Lambda_R^{\mathrm{col}}\right\}}\left(\mathbf{D}_R^{\frac{1}{2}}\right)^{\{\Lambda_R\}}\left(\mathbf{F}^H\right)^{\left\{\Lambda_R^{\mathrm{row}}\right\}} \quad (54)$$

where $B_u = \Lambda_T^{\mathrm{row}}$, $B_d = \left\{\left((\mathbf{Pd})^{\left\{\Lambda_R^{\mathrm{row}}\right\}}, :\right)\right\}$ are the beamspaces of transmit SI and receive SI, respectively, $\mathbf{P} = \mathbf{F}^2$ is the permutation matrix and $\mathbf{d} = (1, 2, \cdots, N_vN_h)^T$ denotes the indexes of rows of $\mathbf{D}_R$. It can be observed that (53) and (54) become equality as N_v and N_h tend to infinity. The other brackets can be similarly approximated as I_1 and I_2 do. In that way, if $B_u \cap B_T = \emptyset$ and $B_d \cap B_R = \emptyset$ are satisfied, $\mathbb{E}\left[\left\|\tilde{\mathbf{G}}_{SI}\right\|^2\right]$ will approach zero as $N_v, N_h \to \infty$. The proof is completed.

Abbreviations

2D: Two-dimensional; 3D: Three-dimensional; AS: Angular spread; BS: Base station; CCI: Co-channel interference; CSI: Channel state information; DoA: Direction of arrival; DoD: Direction of departure; DL: Downlink; FD: Full-duplex; FDD: Frequency-division duplex; HD: Half-duplex; ICI: Inter-cell interference; IGI: Inter-group interference; LOS: Light-of-sight; MIMO: Multiple-input multiple-output; MMSE: Minimum mean square error; MUBT: Multiuser beamspace transmission; MUI: Multiuser interference; NLOS: Non-ligth-of-sight; SE: Spectral efficiency; SI: Self-interference; SIC: SI cancellation; SINR: Signal-to-interference-plus-noise ratio; TDD: Time-division duplex; UL: Uplink; ULA: Uniform linear antenna array; UPA: Uniform planar antenna array; ZF: Zero-forcing

Funding

This research was supported by the National Natural Science Foundation of China (No. 61671472), Jiangsu Province Natural Science Foundation

(BK20160079), and National Natural Science Foundation of China (No. 61371123, No. 91438115).

Authors' contributions

XW is the main writer of this paper. He proposed the main idea, conducted the simulations, and analyzed it. DZ, KX, and WM assisted the review of this manuscript. All authors read and approved the final manuscript.

Competing interests

The authors declare that they have no competing interests.

Author details

[1]Army Engineering University of PLA, No. 88, Houbiaoying Street, Nanjing, China. [2]Unit 32020 of PLA, Wuhan, China.

References

1. J Mietzner, R Schober, L Lampe, WH Gerstacker, PA Hoeher, Multiple-antenna techniques for wireless communications - a comprehensive literature survey. IEEE Commun. Surv. Tutorials. **11**(2), 87–105 (2009)
2. T Marzetta, Noncooperative cellular wireless with unlimited numbers of base station antennas. IEEE Trans. Wirel. Commun. **9**(11), 3590–3600 (2010)
3. HQ Ngo, E Larsson, T Marzetta, Energy and spectral efficiency of very large multiuser MIMO systems. IEEE Trans. Commun. **61**(4), 1436–1449 (2013)
4. F Fernandes, A Ashikhmin, TL Marzetta, Inter-cell interference in noncooperative TDD large scale antenna systems. IEEE J. Sel. Areas Commun. **31**(2), 192–201 (2013)
5. J Hoydis, S Brink, M Debbah, Massive MIMO in UL/DL of cellular networks: how many antennas do we need. IEEE J. Sel. Areas Commun. **31**(2), 160–171 (2013)
6. L You, X Gao, XG Xia, N Ma, Y Peng, Pilot reuse for massive MIMO transmission over spatially correlated rayleigh fading channels. IEEE Trans. Wireless Commun. **14**(6), 3352–3366 (2015)
7. H Cui, L Song, B Jiao, Multi-pair two-way amplify-and-forward relaying with very large number of relay antennas. IEEE Trans. Wireless Commun. **13**(5), 2636–2645 (2014)
8. RCD Lamare, Massive MIMO systems: signal processing challenges and future trends. *URSI Radio Science Bulletin*. **86**(4), 8–20 (2017)
9. J Jose, A Ashikhmin, TL Marzetta, S Vishwanath, Pilot contamination and precoding in multi-cell TDD systems. IEEE Trans. Wireless Commun. **10**(8), 2640–2651 (2011)
10. N Krishnan, RD Yates, NB Mandayam, Uplink linear receivers for multi-cell multiuser mimo with pilot contamination: large system analysis. IEEE Trans. Wireless Commun. **13**(8), 4360–4373 (2014)
11. A Duly, T Kim, D Love, J Krogmeier, Closed-loop beam alignment for massive MIMO channel estimation. IEEE Commun. Lett. **18**(8), 1439–1442 (2014)
12. A Adhikary, J Nam, J Ahn, G Caire, Joint spatial division and multiplexing-the large-scale array regime. IEEE Trans. Inf. Theory. **59**(10), 6441–6463 (2013)
13. C Sun, X Gao, S Jin, M Matthaiou, Z Ding, C Xiao, Beam division multiple access transmission for massive mimo communications. IEEE Trans. Commun. **63**(6), 2170–2184 (2015)
14. D Ying, FW Vook, TA Thomas, DJ Love, A Ghosh, in *IEEE International Conference on Commun. (ICC)*. Kronecker product correlation model and limited feedback codebook design in a 3D channel model (IEEE, Sydney, 2014), pp. 5865–5870
15. Z Wang, W Liu, C Qian, S Chen, L Hanzo, Two-dimensional precoding for 3-d massive mimo. IEEE Trans. Veh. Technol. **66**(6), 5485–5490 (2017)
16. X Li, S Jin, X Gao, RW Heath, Three-dimensional beamforming for large-scale fd-mimo systems exploiting statistical channel state Information. IEEE Trans. Veh. Technol. **65**(11), 8992–9005 (2016)
17. X Li, S Jin, HA Suraweera, J Hou, X Gao, Statistical 3-d beamforming for large-scale mimo downlink systems over rician fading channels. IEEE Trans. Commun. **64**(4), 1529–1543 (2016)
18. Z Zhang, X Chai, K Long, AV Vasilakos, L Hanzo, Full duplex techniques for 5G networks: self-interference cancellation, protocol design, and relay selection. IEEE Commun. Mag. **53**(5), 128–137 (2015)
19. D Kim, H Lee, D Hong, A survey of in-band full-duplex transmission: from the perspective of PHY and MAC layers. IEEE Commun. Surv. Tutorials. **17**(4), 2017–2046 (2015)
20. A Sabharwal, P Schniter, D Guo, DW Bliss, S Rangarajan, R Wichman, In-band full-duplex wireless: challenges and opportunities. IEEE J. Sel. Areas Commun. **32**(9), 1637–1652 (2014)
21. C Psomas, M Mohammadi, I Krikidis, HA Suraweera, Impact of directionality on interference mitigation in full-duplex cellular networks. IEEE Trans. Wireless Commun. **16**(1), 487–502 (2017)
22. M Mohammadi, HA Suraweera, Y Cao, I Krikidis, C Tellambura, Full-duplex radio for uplink/downlink wireless access with spatially random nodes. IEEE Trans. Commun. **63**(12), 5250–5266 (2015)
23. J Lee, TQS Quek, Hybrid full-/half-duplex system analysis in heterogeneous wireless networks. IEEE Trans. Wirel. Commun. **14**(5), 2883–2895 (2015)
24. HQ Ngo, HA Suraweera, M Matthaiou, EG Larsson, Multipair full-duplex relaying with massive arrays and linear processing. IEEE J. Sel. Areas Commun. **32**(9), 1721–1737 (2014)
25. X Xia, D Zhang, K Xu, W Ma, Y Xu, Hardware Impairments Aware Transceiver for Full-Duplex Massive MIMO Relaying. IEEE Trans. Signal Process. **63**(24), 6565–6580 (2015)
26. M Mohammadi, BK Chalise, HA Suraweera, Z Ding, in *IEEE International Conference on Communications*. Wireless information and power transfer in full-duplex systems with massive antenna arrays (IEEE, Paris, 2017), pp. 1–6
27. Y Li, P Fan, A Leukhin, L Liu, On the Spectral and energy efficiency of full-duplex small-cell wireless systems with massive mimo. IEEE Trans. Veh. Technol. **66**(3), 2339–2353 (2017)
28. X Xia, Y Xu, K Xu, D Zhang, W Ma, Full-duplex massive mimo AF relaying with semiblind gain control. IEEE Trans. Veh. Technol. **65**(7), 5797–5804 (2016)
29. H Yin, D Gesbert, M Filippou, Y Liu, A coordinated approach to channel estimation in large-scale multiple-antenna systems. IEEE J. Sel. Areas Commun. **31**(2), 264–273 (2013)
30. JP Kermoal, L Schumacher, KI Pedersen, PE Mogensen, F Frederiksen, A stochastic MIMO radio channel model with experimental validation. IEEE J. Sel. Areas Commun. **20**(6), 1211–1226 (2002)
31. B Clerckx, C Oestges, *MIMO Wireless Networks: Channels, Techniques and Standards for Multi-Antenna, Multi-User and Multi-Cell Systems. 2nd.* (Academic Press, Oxford, 2013)
32. J Wang, R Zhang, W Duan, SX Lu, L Cai, in *Proc. ICC 2014*. Angular spread measurement and modeling for 3D MIMO in urban macrocellular radio channels (IEEE, Sydney, 2014), pp. 20–25
33. MA Maddah-Ali, DNC Tse, Completely stale transmitter channel state information is still very useful. IEEE Trans. Inform. Theory. **58**(7), 4418–4432 (2012)
34. T Kailath, AH Sayed, B Hassibi, *Linear Estimation*. (Upper Saddle River, Prentice-Hall, 2000)
35. B Hassibi, BM Hochwald, How much training is needed in multiple-antenna wireless links? IEEE Trans. Inf. Theory. **49**(4), 951–963 (2003)
36. 3GPP TR 36.828, Further enhancements to LTE time division duplex (TDD) for downlink-uplink (DL-UL) interference management and traffic adaptation. (2012). v.11.0.0. www.3gpp.org. Accessed 7 Aug 2018
37. Y Song, S Nagata, H Jiang, L Chen, in *IEEE International Conference on Communications*. CSI-RS design for 3D MIMO in future LTE-Advanced (IEEE, Sydney, 2014), pp. 5101–5106
38. G Caire, N Jindal, M Kobayashi, N Ravindran, Multiuser MIMO achievable rates with downlink training and channel state feedback. IEEE Trans. Inf. Theory. **56**(6), 2845–2866 (2010)
39. J Choi, D Love, T Kim, Trellis-extended codebooks and successive phase adjustment: A path from LTE-Advanced to FDD massive MIMO systems. IEEE Trans. Wirel. Commun. **14**(4), 2007–2016 (2015)

7

A low-complexity algorithm for the joint antenna selection and user scheduling in multi-cell multi-user downlink massive MIMO systems

Saidiwaerdi Maimaiti*, Gang Chuai, Weidong Gao, Kaisa Zhang, Xuewen Liu and Zhiwei Si

Abstract

The massive MIMO (multiple-input multiple-output) technology plays a key role in the next-generation (5G) wireless communication systems, which are equipped with a large number of antennas at the base station (BS) of a network to improve cell capacity for network communication systems. However, activating a large number of BS antennas needs a large number of radio-frequency (RF) chains that introduce the high cost of the hardware and high power consumption. Our objective is to achieve the optimal combination subset of BS antennas and users to approach the maximum cell capacity, simultaneously. However, the optimal solution to this problem can be achieved by using an exhaustive search (ES) algorithm by considering all possible combinations of BS antennas and users, which leads to the exponential growth of the combinatorial complexity with the increasing of the number of BS antennas and active users. Thus, the ES algorithm cannot be used in massive MIMO systems because of its high computational complexity. Hence, considering the trade-off between network performance and computational complexity, we proposed a low-complexity joint antenna selection and user scheduling (JASUS) method based on Adaptive Markov Chain Monte Carlo (AMCMC) algorithm for multi-cell multi-user massive MIMO downlink systems. AMCMC algorithm is helpful for selecting combination subset of antennas and users to approach the maximum cell capacity with consideration of the multi-cell interference. Performance analysis and simulation results show that AMCMC algorithm performs extremely closely to ES-based JASUS algorithm. Compared with other algorithms in our experiments, the higher cell capacity and near-optimal system performance can be obtained by using the AMCMC algorithm. At the same time, the computational complexity is reduced significantly by combining with AMCMC.

Keywords: 5G, Massive MIMO systems, Antenna selection, User scheduling, Adaptive Markov chain Monte Carlo algorithm

1 Introduction

In order to satisfy the rapidly increasing requirements for high data rate in current wireless communication systems, a new massive MIMO (multiple-input multiple-output) technology was introduced in [1–3]. Massive MIMO technique plays a key role to enhance the cell capacity without increasing system bandwidth or base station (BS) transmission power for the 5G network systems [4]. The key idea of the massive MIMO technique is to install a large amount of transmit antennas at the BS of a cellular and provide services for several users sharing the same spectrum resources. However, as the number of BS antennas and users increases, the combination complexity and hardware cost also increase dramatically. Therefore, when the numbers of BS transmit antenna and active users are extremely large, the joint antenna selection and user scheduling (JASUS) algorithm [5–8] can be adopted as an approach to decide the radio frequency (RF) chain configuration to improve the cell capacity in multi-cell massive multi-user MIMO systems.

In a practical network, one of the key challenges in multi-cell multi-user massive MIMO systems is the

* Correspondence: saidi216@bupt.edu.cn
Key Laboratory of Universal Wireless Communications, Ministry of Education, Beijing University of Posts and Telecommunications, Beijing 100876, People's Republic of China

hardware cost and power consumption because the element of each antenna needs a complete RF chain that consists of RF amplifiers and analog-to-digital converters, which are very pricey and are the main elements of the power consumption at the BS [9]. Different schemes were used in many types of research, such as hybrid precoding and spatial modulation, to reduce the cost of the hardware and the power consumption of the system [10]. One of the best schemes to solve this problem is to applying antenna selection [11–13] to decide optimal subset of BS transmit antennas for decreasing the required number of high pricey RF chains while decreasing the resulting network performance loss.

However, in multi-cell multi-user massive MIMO systems, only a limited number of transmit antennas are selected to provide services for active users scheduled. Hence, if the number of users exceeds, the number of selected transmit antennas, user scheduling must be performed because different wireless channels have different properties. High cell capacity can be obtained by scheduling users with the high channel quality. Therefore, the research of JASUS method for multi-cell multi-user massive MIMO systems is necessary.

Recently, only a few types of research have studied a low complexity JASUS for downlink massive multi-user MIMO systems. Benmimoune et al. [14] proposed a two-step JASUS scheme for downlink multi-user massive MIMO systems. It successively closed unnecessary antennas and removes undesired users which contribute little to system performance. However, due to the high computational complexity, this algorithm can only be employed to the scenarios with a smaller number of candidate antennas and user sets. Thus, using this algorithm in practical multi-cell multi-user massive MIMO system scenarios is difficult. Olyaee et al. [15] proposed a JASUS method based on zero-forcing (ZF) precoding algorithm for single-cell multi-user massive MIMO downlink systems. Though the ZF precoding method has a high system performance, it also has a very high computational complexity. For distributed downlink multi-user massive MIMO system, a JASUS method was proposed in [16] by Xu. et al. It successively obtains the majority of gain with limited backhaul capacity. Lee et al. [17] proposed a random antenna selection algorithm, the algorithm can provide significant capacity efficiency gain, but it is difficult to use for multi-cell multi-user massive MIMO systems. However, the above researches focused on single-cell multi-user massive MIMO systems. Thus, the research of JASUS method for multi-cell multi-user massive MIMO systems remains a largely open area. Therefore, the novel JASUS algorithm with considered multi-cell interference, which causes no or only a few decreases of system performance, represents a new promising research topic.

In this paper, we consider the problem of JASUS in multi-cell multi-user massive MIMO downlink system operating with TDD mode. Considering the trade-off between cell capacity and complexity, we proposed a low-complexity algorithm for JASUS method based on AMCMC. In our proposed method, only a small subset of BS transmit antennas is selected to serve predetermined active users, thus reduces the number of RF chains, avoids uneconomical hardware costs, and reduces power consumption caused by the selection of unnecessary transmit antennas to provide the required services. The main contributions of our work are as follows.

1. A low complexity JASUS method based on AMCMC algorithm is proposed for downlink multi-cell multi-user massive MIMO systems. AMCMC algorithm is helpful for selecting combination subset of antennas and users to approach the maximum cell capacity while decreasing the resulting network performance loss.
2. In this paper, we proposed updating rules for the selection probability of each base station transmit antenna and the scheduling probability for each user. In addition, we also proposed a new projection strategy to satisfy the constraints of antenna and user selection.
3. Performance analysis and simulation results show that our proposed algorithm produced promising results. Compared with ES-based JASUS algorithm, the proposed algorithm achieved comparable performance with low complexity. In addition, the AMCMC-based JASUS algorithm outperforms greedy-based JASUS and norm-based JASUS methods in terms of cell capacity and SER (symbol error rate) performance.

Notation: Symbol $\mathbb{C}$ denotes the set of complex numbers, vectors are denoted by using lower-case bold letters, matrices are denoted by using bold letters, $|.|$ denotes the absolute value of a scalar, $\|\cdot\|_F$ denotes the Frobenius norm function, and (.) represents the binomial coefficient.

The remaining content is organized as follows. In Section 2, the system model and capacity maximize problem formulation are described. In Section 3, we formulate the problem of JASUS method based on AMCMC in multi-cell multi-user massive MIMO systems. Section 4 presents the simulation setup and assumption. In Section 5, we discuss the simulation results and analyze the complexity; finally, this work is concluded in Section 6.

2 System model and problem formulation

In this part, we simply give the system model for multi-cell massive multi-user

MIMO downlink systems with the system capacity formulation model with consideration of the multi-cell interference.

2.1 System model

As shown in Fig. 1, the considered scenario is a multi-cell multi-user massive MIMO downlink system operating in TDD mode, and the cell capacity maximizing problem is studied with consideration of the inter-cell interference. The system is composed of B hexagonal cells. All B BSs, where $B = \{1, 2, \ldots, B\}$ are installed with M antennas and serves U $(M \geq U \geq 1)$ single-antenna users in each cell. The block-fading channel model is assumed. We assume that BS can select N transmit antennas among the M transmit antennas and schedule $K(K \leq N)$ users among the U users within the cell to be served simultaneously. The channel vector $\mathbf{g}_{iju} \in \mathbb{C}^M$ from the jth BS and user u in cell i can be expressed as

$$\mathbf{g}_{iju} = \sqrt{\beta_{iju}}\mathbf{h}_{iju} \tag{1}$$

where β_{iju} denotes the large scale channel fading between jth BS and user u in cell i, including shadowing and path loss. $\mathbf{h}_{iju}$ is the small-scale fading vector, and $\mathbf{h}_{iju} = [h_{iju1}, h_{iju2}, \ldots, h_{ijuM}]^T \in \mathbb{C}^M$. Then, the overall downlink transmission matrix $\mathbf{G}_{ij} \in \mathbb{C}^{M \times U}$ between the BS in cell j and all users in cell i can be expressed as

$$\mathbf{G}_{ij} = \left[\mathbf{g}_{ij1}, \mathbf{g}_{ij2}, \ldots, \mathbf{g}_{ijU}\right] = \mathbf{H}_{ij}\mathbf{D}_{ij}^{\frac{1}{2}} \in \mathbb{C}^{M \times U} \tag{2}$$

where $\mathbf{H}_{ij} = [\mathbf{h}_{ij1}, \mathbf{h}_{ij2}, \ldots, \mathbf{h}_{ijU}] \in \mathbb{C}^{M \times U}$ is the overall small-scale fading matrix and $\mathbf{D}_{ij} = \text{diag}(\beta_{ij1}, \beta_{ij2}, \ldots, \beta_{ijU})$.

Our objective is to find optimum combinations subset of BS antennas and users to approach the maximum cell capacity while decreasing the resulting network performance loss. Furthermore, we will decrease the number of expensive RF chains and avoid the uneconomic costs of the hardware and decrease power consumption caused by selecting undesired antennas to provide the requirement of service.

2.2 Problem formulation

In the downlink system, the signal received by users in cell i can be written as

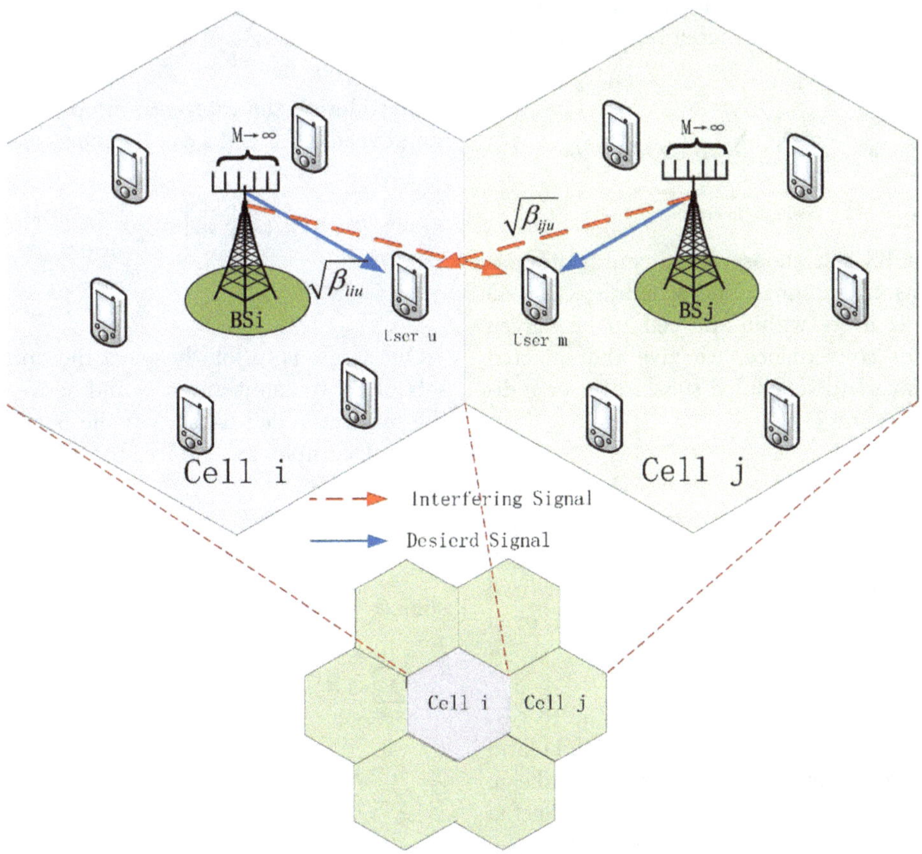

Fig. 1. Illustration of the downlink multi-cell multi-user massive MIMO network transmission model

$$\mathrm{r}_i = \sqrt{p_t}\sum_{j=1}^{B}\mathbf{G}_{ij}^{H}\mathbf{W}_j s_j + \mathbf{n}_i \tag{3}$$

where p_t denotes the transmitted power, $\mathbf{s}_j = [s_{j1}, s_{j2}, ..., s_{jU}]^T \in \mathbb{C}^U$ is the transmit signals for users in cell j, $\mathbf{W}_j \in \mathbb{C}^{M\times U}$ is the precoding matrix of BS in cell j, and $\mathbf{n}_i = [n_{i1}, n_{i2}, ..., n_{iu}]^T \in \mathbb{C}^U$ is the noise vector at the uth user in cell i. The downlink signal received by user u in cell i can be written as

$$r_{iu} = \underbrace{\sqrt{p_t}\mathbf{g}_{iiu}^{H}\mathbf{w}_{iu}s_{iu}}_{\text{desired signal}} + \underbrace{\sqrt{p_t}\sum_{b\neq u}\mathbf{g}_{iib}^{H}\mathbf{w}_{ib}s_{ib}}_{\text{intra-cell interference}} + \underbrace{\sqrt{p_t}\sum_{j\neq i}\sum_{b=1}^{U}\mathbf{g}_{iju}^{H}\mathbf{w}_{jb}s_{jb}}_{\text{inter-cell interference}} + n_{iu} \tag{4}$$

where $\mathbf{w}_{jb}$ is the bth column of precoding matrix $\mathbf{W}_j$. Formula (4) including the desired signal for the user u in cell i, intra-cell interference signal (comes from other users in the same cell), and inter-cell interference signal (comes from other cells), respectively.

Since we used the zero-forcing (ZF) methods and supposing that the channel state information (CSI) is perfectly known at the BS, the intra-cell interference signal (second term) in function (4) drops to zero according to prevenient works [14, 18–20]. Therefore, function (4) becomes

$$r_{iu} = \underbrace{\sqrt{p_t}\mathbf{g}_{iiu}^{H}\mathbf{w}_{iu}s_{iu}}_{\text{desired signal}} + \underbrace{\sqrt{p_t}\sum_{j\neq i}\sum_{b=1}^{U}\mathbf{g}_{iju}^{H}\mathbf{w}_{jb}s_{jb}}_{\text{inter-cell interference}} + n_{iu} \tag{5}$$

We assume that BS can choose N transmit antennas among the M transmit antennas, and schedule $K(K \leq N)$ users among the U users within the cell to be served simultaneously. For convenience, we give the selected subset of the antenna and scheduled subset of user indicator functions are $\boldsymbol{\omega}$ and $\overline{\boldsymbol{\varpi}}$,

$$\boldsymbol{\omega} = \{\omega_m\}_{m=1}^{M} \tag{6}$$

$$\overline{\boldsymbol{\varpi}} = \{\varpi_u\}_{u=1}^{U} \tag{7}$$

where $\boldsymbol{\omega}$ and $\overline{\boldsymbol{\varpi}}$ are binary vectors that include two values 0 and 1 to indicate if a given antenna or a given user is selected. (e.g., 1 → selected, 0 → unselected).

For making an easy description, we will define to two indicator functions, which are $I_m(\boldsymbol{\omega}) \triangleq \omega_m \in \{0, 1\}$ and $I_u(\overline{\boldsymbol{\varpi}}) \triangleq \varpi_u \in \{0, 1\}$, respectively. We use these to indicate whether the mth BS antenna and the uth user are selected or not, respectively. $\mathbf{s}_{i\{\overline{\boldsymbol{\varpi}}\}} \in \mathbb{C}^K$ denotes transmit signal vector, sub-block channel matrix of corresponding denotes by $\mathbf{G}_{ij\{\overline{\boldsymbol{\varpi}}, \boldsymbol{\omega}\}} \in \mathbb{C}^{N\times K}$ and $\mathbf{n}_{i\{\overline{\boldsymbol{\varpi}}\}} \in \mathbb{C}^K$ is noise vector, respectively. Finally, in order to denote the joint antenna and user selection, we employ the 2-tuple $\boldsymbol{\Omega} \triangleq (\overline{\boldsymbol{\varpi}}; \boldsymbol{\omega})$. In order to have an easy explanation, we will interchangeably use $\boldsymbol{\Omega}$ and $(\overline{\boldsymbol{\varpi}}; \boldsymbol{\omega})$ for the following part. After using antenna selection and user scheduling method, function (5) becomes

$$\begin{aligned} \mathrm{r}_{i\{\varpi_u\}} = {} & \sqrt{p_t}\mathbf{g}_{ii\{\overline{\boldsymbol{\varpi}},\boldsymbol{\omega}\}}^{H}\mathbf{w}_{i\{\varpi_u\}}s_{i\{\varpi_u\}} \\ & + \sqrt{p_t}\sum_{j\neq i}\sum_{b=1}^{U}\mathbf{g}_{ij\{\overline{\boldsymbol{\varpi}},\boldsymbol{\omega}\}}^{H}\mathbf{w}_{j\{\varpi_b\}}s_{j\{\varpi_b\}} \\ & + n_{i\{\varpi_u\}} \end{aligned} \tag{8}$$

where $\mathbf{w}_{j\{\varpi_b\}}$ is bth column of precoding matrix $\mathbf{W}_{j\{\overline{\boldsymbol{\varpi}}, \boldsymbol{\omega}\}} \in \mathbb{C}^{N\times K}$ and $n_{i\{\varpi_u\}}$ is the noise at the uth user in cell i.

2.3 Capacity of massive MIMO

According to the aforementioned discussion, the received signal-to-interference-plus-noise ratio (SINR) for the user $u \in U$ (which is connected to cell i) with a selected channel vector $\mathbf{g}_{ij\{\overline{\boldsymbol{\varpi}}, \boldsymbol{\omega}\}}$ can be written as

$$\mathrm{SINR}_{i\{\varpi_u\}} = \frac{p_t\left|\mathbf{g}_{ii\{\overline{\boldsymbol{\varpi}},\boldsymbol{\omega}\}}^{H}\mathbf{w}_{i\{\varpi_u\}}\right|^2}{\sum_{j\neq i}p_t\left|\mathbf{g}_{ij\{\overline{\boldsymbol{\varpi}},\boldsymbol{\omega}\}}^{H}\mathbf{w}_{j\{\varpi_u\}}\right|^2 + \left|n_{i\{\varpi_u\}}\right|^2} \tag{9}$$

Considering the inter-cell interference, the formula of sum capacity for cell i can be expressed as

$$C_{\mathrm{sum}}^{i}\left(\mathbf{G}_{\{\overline{\boldsymbol{\varpi}},\boldsymbol{\omega}\}}\right) = \log_2\left(\det\left(I + \frac{p_t\left|\mathbf{G}_{ii\{\overline{\boldsymbol{\varpi}},\boldsymbol{\omega}\}}^{H}\mathbf{W}_{i\{\overline{\boldsymbol{\varpi}},\boldsymbol{\omega}\}}\right|^2}{\sum_{j\neq i}p_t\left|\mathbf{G}_{ij\{\overline{\boldsymbol{\varpi}},\boldsymbol{\omega}\}}^{H}\mathbf{W}_{j\{\overline{\boldsymbol{\varpi}},\boldsymbol{\omega}\}}\right|^2 + \mathbf{n}_{i\{\overline{\boldsymbol{\varpi}}\}}\mathbf{n}_{i\{\overline{\boldsymbol{\varpi}}\}}^{H}}\right)\right) \tag{10}$$

Our target is to jointly select the optimal combination sets of BS transmit antenna and active user to approach the maximum cell capacity while decreasing the computational complexity. Hence, the problem of JASUS can be written as

$$\Phi_C = \underset{\overline{\boldsymbol{\varpi}},\boldsymbol{\omega}}{\text{maximize}}\ C_{\mathrm{sum}}^{i}\left(\mathbf{G}_{\{\overline{\boldsymbol{\varpi}},\boldsymbol{\omega}\}}\right) \tag{11}$$

subject to

$$\sum_{u=1}^{U} I_u(\overline{\boldsymbol{\varpi}}) = K \tag{12}$$

$$\sum_{m=1}^{M} I_m(\boldsymbol{\omega}) = N \tag{13}$$

Addressing the aforementioned problem by employing an ES method needs to evaluating the cell capacity of

$\phi \triangleq C_M^N \times C_U^K$ joint antenna and user combinations, where C_M^N and C_U^K are the binomial coefficient. This fact indicates that the ES method cannot be used in current the massive MIMO systems where U and M are very numerous, it leads to high computational complexity. Thus, a low complexity algorithm for JASUS is needed in order to obtain the best network performance with low computational complexity.

Obviously, formula (11) serves as the target function in this study. Therefore, address the problem (11) in multi-cell multi-user massive MIMO systems, we need to solve the following three main problems:

1. Inter-cell interference: the first problem is the inter-cell interference coming from other cells. In order to solve this problem, cells can adjust their precoding matrices, thus can eliminate or decrease the interference from all users. In order to make better use of the coordinated massive MIMO technology, a group of users in the cells should be scheduled so that each group of users in the cell has the biggest spatial separation with the interference channels of users in neighboring cells.
2. Computational complexity: the second problem is how to obtain the best combination subset of antenna and user in each cell with lower computational complexity so as to decrease or eliminate the inter-cell interference and maximize the sum capacity of all cells. We know that in multi-cell massive MIMO systems, the various path loss between the antennas coming from neighboring cells and users coming from target cell also bring to much computational complexity same to the antenna selection and user scheduling. Thus, the computational complexity of the ES algorithm becomes very large than of a single-cell JASUS.
3. CSI feedback cost: for massive MIMO systems, the perfect CSI feedback depends mainly on the number of active antennas and the users they support. Hence, in order to centralize processing for selecting antennas and scheduling users across cells, the BS needs to exchange the overall CSI of overall combined subset antennas and users at each scheduling period, it brings more burden for BS. In addition, when the number of BS antennas and users in each cell increases, the cost of CSI increases accordingly.

On the base of the aforementioned discussion, a low complexity scheme is needed from the practical point of view for JASUS in multi-cell multi-user massive MIMO scenarios to reduce the complexity of function (11) while decreasing the cost of the CSI feedback.

3 Joint antenna selection and user scheduling algorithm

In this part, we presented two suboptimal iterative algorithms for JASUS before a discussion of the proposed AMCMC method.

3.1 Norm-based JASUS algorithm

Firstly, we presented a norm-based JASUS method for addressing the objective function (10). The norm-based JASUS scheme maximizes $\|\mathbf{G}_{\{\overline{\boldsymbol{\omega}}, \boldsymbol{\omega}\}}\|_F$, where $\|.\|_F$ is the Frobenius norm function. Let $C_{\text{sum}}^{\text{NB}}(\mathbf{G}_{\{\overline{\boldsymbol{\omega}}, \boldsymbol{\omega}\}}) = \|\mathbf{G}_{\{\overline{\boldsymbol{\omega}}, \boldsymbol{\omega}\}}\|_F$. This scheme including initialization step and iterative updating step, respectively. This both steps use the vector norm as criteria which considerably decrease each iterative computation complexity. The norm-based JASUS problem is modeled as

$$\omega_{\text{NB}} = \underset{\overline{\boldsymbol{\omega}}, \boldsymbol{\omega}}{\arg\max} \quad C_{\text{sum}}^{\text{NB}}\left(\mathbf{G}_{\{\overline{\boldsymbol{\omega}}, \boldsymbol{\omega}\}}\right) \tag{14}$$

where ω_{NB} is the combination selection indicator. Nevertheless, it has still a problem that when performing transmit antenna selection and user scheduling simply based on the F-norm criteria, it would sacrifice some cell capacity. In sum, the norm-based JASUS method has extremely low complexity, but it cannot guarantee a high sum capacity performance.

3.2 Greedy-based JASUS algorithm

In order to enhance the cell capacity over norm-based JASUS, we presented a greedy-based JASUS algorithm. Unlike to the norm-based JASUS algorithm, the greedy-based JASUS algorithm maximizes the cell capacity in each step. This method also includes initialization step and iterative updating step. The greedy-based JASUS problem is modeled as

$$\omega_{\text{GR}} = \underset{\overline{\boldsymbol{\omega}}, \boldsymbol{\omega}}{\arg\max} \quad C_{\text{sum}}^{\text{GR}}\left(\mathbf{G}_{\{\overline{\boldsymbol{\omega}}, \boldsymbol{\omega}\}}\right) \tag{15}$$

where ω_{GR} is the combination selection indicator. Compared with the norm-based JASUS algorithm, this method has a good capacity performance, but it has high computational complexity.

3.3 AMCMC-based JASUS algorithm

Although the greedy-based JASUS method improved the cell capacity, it ignores the computational complexity of all system. For the actual network communication system, it will not have commercial value or attraction. Hence, considering the trade-off between cell capacity and complexity, we proposed a low-complexity JASUS method based on AMCMC algorithm and its description is as follows.

MCMC [21] is a method of generating random samples, which is often used to calculate statistical estimation, marginal probability, and conditional probability. MCMC algorithms depend on (Markov) sequences with limit distributions corresponding to interest distributions. In the past decades, it has been widely used in many fields such as engineering and statistics [22]. The key idea of the MCMC method is that Markov chains are simulated in state space X, and the stable distribution of the chains is the target distribution π [23].

In order to address our objection problem in (11), we must tackle tow essential issues when applying AMCMC algorithm. The first problem is how to provide a proposal distribution of candidate samples L_{MCMC}. The second problem is how to design the most suitable updating rule for the proposal distribution.

3.3.1 Derivation of the candidate sampling distribution for the MCMC method

The biggest advantage of the MCMC method is that it can search the "elite samples" instead of exhaustively searching the whole samples. At iteration t, the samples $\{\mathbf{\Omega}_{\ell,t} = \{\overline{\boldsymbol{\omega}}_{\ell,t}, \boldsymbol{\omega}_{\ell,t}\}\}_{\ell=1}^{L_{\mathrm{MCMC}}}$ from an MCMC method can be employed to estimate the maximum value of the target function $C_{\mathrm{sum}}^{i}(\mathbf{G}_{\{\overline{\boldsymbol{\omega}}, \boldsymbol{\omega}\}})$

$$\Phi_C^* = \underset{\mathbf{\Omega}_{\ell,t} \triangleq (\overline{\boldsymbol{\omega}}_{\ell,t};\boldsymbol{\omega}_{\ell,t});\quad \ell=1,2,\ldots,L_{\mathrm{MCMC}}}{\arg\max} C_{\mathrm{sum}}^{i}\left(\mathbf{G}_{\{\overline{\boldsymbol{\omega}}_{\ell,t},\boldsymbol{\omega}_{\ell,t}\}}\right) \tag{16}$$

where L_{MCMC} denotes the total number of samples, and Φ_C^* is the estimated value of formula (11).

Given that scheduled of per user ϖ_u and selection of each antenna ω_m are binary variables, we use the Boltzmann distribution of the objective function $C_{\mathrm{sum}}^{i}(\mathbf{G}_{\{\overline{\boldsymbol{\omega}}_{\ell,t},\boldsymbol{\omega}_{\ell,t}\}})$ with a suitable temperature τ

$$\underset{\mathbf{\Omega}_{\ell,t} \triangleq (\overline{\boldsymbol{\omega}}_{\ell,t},\boldsymbol{\omega}_{\ell,t})}{\pi(\mathbf{\Omega}_{\ell,t})} = \frac{\exp\{(C_{\mathrm{sum}}^{i}(\mathbf{\Omega}_{\ell,t})/\tau)\}}{\Gamma} \tag{17}$$

where $\Gamma = \sum_{\mathbf{\Omega}_{\ell,t} \triangleq (\overline{\boldsymbol{\omega}}_{\ell,t},\boldsymbol{\omega}_{\ell,t})} \exp\{(C_{\mathrm{sum}}^{i}(\mathbf{\Omega}_{\ell,t})/\tau)\}$ is a normalization constant in the MCMC method that can be neglected. Thus, maximizing $C_{\mathrm{sum}}^{i}(\mathbf{\Omega}_{\ell,t})$ is equivalent to maximizing $\pi(\mathbf{\Omega}_{\ell,t})$, and $\pi(\mathbf{\Omega}_{\ell,t})$ is the target distribution.

In order to prove the MCMC method for searching the distribution $\pi(\mathbf{\Omega}_{\ell,t})$, we use a MIS (metropolized independence sampler) [23], which is a generic MCMC method. The step is as follow. An initial value $\mathbf{\Omega}_{[0],t}$ is chosen randomly. Given the current sample $\mathbf{\Omega}_{[\ell],t}$,

- A candidate sample $\mathbf{\Omega}_{[\mathrm{new}],t}$ is drawn from proposal distribution $\Psi(\mathbf{\Omega}_{\ell,t}; \mathbf{R}_{t-1})$.
- Simulate u uniform[0, 1], and according to the accepting probability $\alpha(\mathbf{\Omega}_{[\ell],t}, \mathbf{\Omega}_{[\mathrm{new}],t})$, let

$$\mathbf{\Omega}_{[\ell+1],t} = \begin{cases} \mathbf{\Omega}_{[\mathrm{new}],t} & \text{if} \quad u \le \alpha(\mathbf{\Omega}_{[\ell],t}, \mathbf{\Omega}_{[\mathrm{new}],t}) \\ \mathbf{\Omega}_{[\ell],t} & \text{otherwise} \end{cases} \tag{18}$$

Where

$$\alpha(\mathbf{\Omega}_{[\ell],t}, \mathbf{\Omega}_{[\mathrm{new}],t}) = \min\left\{1, \frac{\pi(\mathbf{\Omega}_{[\mathrm{new}],t})/\Psi(\mathbf{\Omega}_{[\mathrm{new}],t})}{\pi(\mathbf{\Omega}_{[\ell],t})/\Psi(\mathbf{\Omega}_{[\ell],t})}\right\} \tag{19}$$

After L_{MCMC} iterations, we can achieve a set of samples $\{\mathbf{\Omega}_{[0],t}, \mathbf{\Omega}_{[1],t}, \ldots, \mathbf{\Omega}_{[L_{\mathrm{MCMC}}],t}\}$, which is subjected to distribution $\pi(\mathbf{\Omega}_{\ell,t})$.

3.3.2 Derivation of updating rule for the AMCMC algorithm

In this part, we provide updating rule for the proposal distribution. For the AMCMC method, the joint proposal distribution is proportional to the product of Bernoulli distributions, namely

$$\Psi(\mathbf{\Omega}_{\ell,t}; \mathbf{R}_{t-1}) \triangleq \underbrace{\frac{\prod_{u=1}^{U} p_{u,t-1}^{I_u(\overline{\boldsymbol{\omega}}_{\ell,t})}\left(1-p_{u,t-1}\right)^{1-I_u(\overline{\boldsymbol{\omega}}_{\ell,t})}}{\Gamma'}}_{\triangleq \Psi(\overline{\boldsymbol{\omega}}_{\ell,t};\mathbf{P}_{t-1})} \times \underbrace{\frac{\prod_{m=1}^{M} g_{m,t-1}^{I_m(\boldsymbol{\omega}_{\ell,t})}\left(1-g_{m,t-1}\right)^{1-I_m(\boldsymbol{\omega}_{\ell,t})}}{\Gamma'}}_{\triangleq \Psi(\boldsymbol{\omega}_{\ell,t};\mathbf{g}_{t-1})} \tag{20}$$

where p_u denotes the probability of the uth user being selected for communicating with the BS. That is, $\varpi_u \sim \mathrm{Ber}(p_u)$ for $u = 1, 2, \ldots, U$, and g_m is the probability of the mth BS antenna being selected. That is, $\omega_m \sim \mathrm{Ber}(g_m)$ for $m = 1, 2, \ldots, M$. We use the indicator functions $I_m(\boldsymbol{\omega}_{\ell,t})$ and $I_u(\overline{\boldsymbol{\omega}}_{\ell,t})$ to indicate whether the mth BS antenna and the uth user are selected or not, respectively. $\mathbf{R}_{t-1} \triangleq \{\mathbf{P}_{t-1}, \mathbf{g}_{t-1}\}$, where $\mathbf{P}_t = \{p_{u,t}\}_{u=1}^{U}$, $\mathbf{g}_t = \{g_{m,t}\}_{m=1}^{M}$, and Γ' is a normalization constant that can be ignored in the AMCMC. The adaptation scheme is employed to adjust the parameterized proposal distribution $\Psi(\mathbf{\Omega}_{\ell,t}; \mathbf{R}_{t-1})$ and minimize the Kullback-Leibler divergence [24, 25] between the target distribution $\pi(\mathbf{\Omega}_{\ell,t})$ and the proposal distribution $\Psi(\mathbf{\Omega}_{\ell,t}; \mathbf{R}_{t-1})$, namely

$$D[\pi(\mathbf{\Omega}_{\ell,t}) \| \Psi(\mathbf{\Omega}_{\ell,t}; \mathbf{R}_{t-1})] = \sum_{\ell=1}^{L_{\mathrm{MCMC}}} \pi(\mathbf{\Omega}_{\ell,t}) \times \log\left(\frac{\pi(\mathbf{\Omega}_{\ell,t})}{\Psi(\mathbf{\Omega}_{\ell,t}; \mathbf{R}_{t-1})}\right) \tag{21}$$

It is observed that $\tilde{D} = \pi(\mathbf{\Omega}_{\ell,t}) \times \log\pi(\mathbf{\Omega}_{\ell,t}) - D[\pi(\mathbf{\Omega}_{\ell,t}) \| \Psi(\mathbf{\Omega}_{\ell,t};\mathbf{R}_{t-1})]$ is a convex function [26]. Hence, the minimization of the Kullback-Leibler divergence $D[\pi(\mathbf{\Omega}_{\ell,t}) \| \Psi(\mathbf{\Omega}_{\ell,t};\mathbf{R}_{t-1})]$ w.r.t. $\mathbf{R}$ can be achieved when $\partial\tilde{D}/\partial\mathbf{R} = 0$. Thus, $\tilde{D}$ can be written as

$$\tilde{D} = \sum_{\ell=1}^{L_{\mathrm{MCMC}}} \pi(\mathbf{\Omega}_{\ell,t}) \times \log\Psi(\mathbf{\Omega}_{\ell,t};\mathbf{R}_{t-1}) \tag{22}$$

We set the partial derivative of (22) to zero with respect to $\mathbf{R}$. Then, the Eq. (22) can be written as

$$\frac{\partial}{\partial\mathbf{R}} \sum_{\ell=1}^{L_{\mathrm{MCMC}}} \pi(\mathbf{\Omega}_{\ell,t}) \times \log\Psi(\mathbf{\Omega}_{\ell,t};\mathbf{R}_{t-1}) = 0 \tag{23}$$

where the partial derivatives of $\log\Psi(\mathbf{\Omega}_{\ell,t};\mathbf{R}_{t-1})$ with respect to p_u and g_m are respectively given by

$$\frac{\partial}{\partial p_u} \log\Psi(\mathbf{\Omega}_{\ell,t};\mathbf{R}_{t-1}) = \frac{I_u(\overline{\boldsymbol{\omega}}_{\ell,t}) - p_{u,t-1}}{\left(1-p_{u,t-1}\right)p_{u,t-1}} \tag{24}$$

$$\frac{\partial}{\partial g_m} \log\Psi(\mathbf{\Omega}_{\ell,t};\mathbf{R}_{t-1}) = \frac{I_m(\boldsymbol{\omega}_{\ell,t}) - g_{\mathrm{m},t-1}}{\left(1-g_{\mathrm{m},t-1}\right)g_{\mathrm{m},t-1}} \tag{25}$$

By substituting (24) and (25) into (23), we obtain

$$\sum_{\ell=1}^{L_{\mathrm{MCMC}}} \pi(\mathbf{\Omega}_{\ell,t}) \times \frac{I_u(\overline{\boldsymbol{\omega}}_{\ell,t}) - p_{u,t-1}}{\left(1-p_{u,t-1}\right)p_{u,t-1}} = 0 \quad \text{for} \quad u = 1, 2, \ldots, U \tag{26}$$

$$\sum_{\ell=1}^{L_{\mathrm{MCMC}}} \pi(\mathbf{\Omega}_{\ell,t}) \times \frac{I_m(\boldsymbol{\omega}_{\ell,t}) - g_{\mathrm{m},t-1}}{\left(1-g_{\mathrm{m},t-1}\right)g_{\mathrm{m},t-1}} = 0 \quad \textit{for} \quad m = 1, 2, \ldots, M \tag{27}$$

Given a number of samples $\{\mathbf{\Omega}_{\ell,t} = \{\overline{\boldsymbol{\omega}}_{\ell,t}, \boldsymbol{\omega}_{\ell,t}\}\}_{\ell=1}^{L_{\mathrm{MCMC}}}$ drawn from target distribution $\pi(\mathbf{\Omega}_{\ell,t})$, the Monte Carlo estimate of $\partial\tilde{D}/\partial p_u$ and $\partial\tilde{D}/\partial g_m$ are

$$\frac{1}{L_{\mathrm{MCMC}}}\left[\frac{1}{\left(1-p_{u,t-1}\right)p_{u,t-1}} \sum_{\ell=1}^{L_{\mathrm{MCMC}}} I_u(\overline{\boldsymbol{\omega}}_{\ell,t}) - p_{u,t-1}\right] \tag{28}$$

$$\frac{1}{L_{\mathrm{MCMC}}}\left[\frac{1}{\left(1-g_{\mathrm{m},t-1}\right)g_{\mathrm{m},t-1}} \sum_{\ell=1}^{L_{\mathrm{MCMC}}} I_m(\boldsymbol{\omega}_{\ell,t}) - g_{\mathrm{m},t-1}\right] \tag{29}$$

Applying the Robbins-Monro stochastic approximation scheme [26], we can achieve the recursive update function to close to the root of $\partial\tilde{D}/\partial p_u = 0$ and $\partial\tilde{D}/\partial g_m = 0$, namely

$$p_{u,t} = p_{u,t-1} + \frac{r_t}{\left(1-p_{u,t-1}\right)p_{u,t-1}} \times \left(\frac{1}{L_{\mathrm{MCMC}}} \sum_{\ell=1}^{L_{\mathrm{MCMC}}} I_u(\overline{\boldsymbol{\omega}}_{\ell,t}) - p_{u,t-1}\right) \tag{30}$$

$$g_{\mathrm{m},t} = g_{\mathrm{m},t-1} + \frac{r_t}{\left(1-g_{\mathrm{m},t-1}\right)g_{\mathrm{m},t-1}} \times \left(\frac{1}{L_{\mathrm{MCMC}}} \sum_{\ell=1}^{L_{\mathrm{MCMC}}} I_m(\boldsymbol{\omega}_{\ell,t}) - g_{\mathrm{m},t-1}\right) \tag{31}$$

where r_t denotes the sequence of decreasing step sizes [27]. In addition, we can simplify formulas (30) and (31), because $(1 - p_{u,\ t-1})p_{u,\ t-1}$ and $(1 - g_{\mathrm{m},\ t-1})g_{\mathrm{m},\ t-1}$ has no significant impact on the convergence of (30) and (31). Hence, Eqs. (30) and (31) becomes

$$p_{u,t} = p_{u,t-1} + r_t \times \left. \frac{1}{L_{\mathrm{MCMC}}} \sum_{\ell=1}^{L_{\mathrm{MCMC}}} I_u(\overline{\boldsymbol{\omega}}_{\ell,t}) - p_{u,t-1}\right) \tag{32}$$

$$g_{\mathrm{m},t} = g_{\mathrm{m},t-1} + r_t \times \left. \frac{1}{L_{\mathrm{MCMC}}} \sum_{\ell=1}^{L_{\mathrm{MCMC}}} I_m(\boldsymbol{\omega}_{\ell,t}) - g_{\mathrm{m},t-1}\right) \tag{33}$$

The updated proposal distribution Eqs. (32) and (33) are iteratively used with the objective to close to the target distribution.

3.4 Constraints for the AMCMC-based JASUS problem

For the JASUS problem, the vectors $\overline{\boldsymbol{\omega}}$ and $\boldsymbol{\omega}$ are subject to constraint functions (12) and (13), respectively. However, employing functions (32) and (33) with MIS to generate samples, but we cannot ensure that the samples meet the constraints (12) and (13). In order to ensure that the samples drawn from (32) and (33) meet the constraints (12) and (13), we propose a new projection strategy. For convenience, we only introduce the proposed projection strategy for $\boldsymbol{\omega}$ because the similar projection strategy can be used to $\overline{\boldsymbol{\omega}}$.

Assume the sample $\boldsymbol{\omega}_{\ell,\ t}$ drawn from $\Psi(\cdot;\mathbf{g}_{t-1})$ at the tth iteration. We define the two sets, which are $\phi_0 = \{\mathrm{m} : I_m(\boldsymbol{\omega}_{\ell,\ t}) = 0\}$ and $\phi_1 = \{\mathrm{m} : I_m(\boldsymbol{\omega}_{\ell,\ t}) = 1\}$, respectively. We use these to collect the indices for the unselected and selected BS antennas, respectively. The following projection strategy is applied if $\sum_{m=1}^{M} I_m(\boldsymbol{\omega}_{\ell,t}) \neq N$:

- If $\sum_{m=1}^{M} I_m(\boldsymbol{\omega}_{\ell,t}) < N$, then the proposed projection strategy sequentially selects the BS antenna with the biggest probability from the set ϕ_0 to the set ϕ_1 until

$|\phi_1| = N$, where ϕ_1 is the number of elements of the set ϕ_1.

- If $\sum_{m=1}^{M} I_m(\boldsymbol{\omega}_{\ell,t}) > N$, then the BS antennas with the smallest probability in the set ϕ_1 are closed sequentially according to the proposed projection strategy until $|\phi_1| = N$.

3.5 Convergence analysis for the AMCMC-based JASUS algorithm

In order to obtain a higher convergence rate, the probability parameters $\mathbf{R}_{t-1}$ of the proposal distribution $\Psi(\boldsymbol{\Omega}_{\ell,t}; \mathbf{R}_{t-1})$ are adjusted. In this paper, we use literature [28] to explain the convergence problem, because our proposed method gives a similar description of the convergence problem and proves the effectiveness of the method through analysis. Besides, the complexity of the MCMC algorithm has been proved to be only related to sample size L_{MCMC} in [22]. The proposed adaptive strategy requires less sample size and iteration times, which can significantly improve the convergence speed of the MCMC algorithm.

3.6 Constrained AMCMC-based JASUS algorithm

On the base of the aforementioned discussion, we can be written the proposed AMCMC-based JASUS algorithm by the following steps. At iteration t, L_{MCMC} samples $\{\boldsymbol{\Omega}_{\ell,t} = \{\bar{\boldsymbol{\omega}}_{\ell,t}, \boldsymbol{\omega}_{\ell,t}\}\}_{\ell=1}^{L_{\text{MCMC}}}$ from the MCMC method are can be generated by employing MIS according to proposal distribution $\Psi(\boldsymbol{\Omega}_{\ell,t}; \mathbf{R}_{t-1})$. Then, the new proposal distribution $\Psi(\boldsymbol{\Omega}_{\ell,t}; \mathbf{R}_t)$ will be updated by the Kullback-Leibler divergence until it approach the target distribution $\pi(\boldsymbol{\Omega}_{\ell,t})$. The detailed AMCMC-based JASUS algorithm is described as follows.

Algorithm 1 : AMCMC-based JASUS algorithm

Step1: initialize $\mathbf{R}_0 \triangleq \{\mathbf{P}_0, \mathbf{g}_0\}$. Then, let $\boldsymbol{\Omega}_{[0],t} = \max_{\boldsymbol{\Omega}_{\ell,t} \triangleq (\bar{\boldsymbol{\omega}}_{\ell,t}; \boldsymbol{\omega}_{\ell,t})} C_{sum}^{i}(\boldsymbol{\Omega}_{\ell,t})$, after that, let $\Phi_C^* = \boldsymbol{\Omega}_{[0],t}$ and set $\mathbf{P}_0 = \{\mathrm{p}_{u,0}\}_{u=1}^{U}$ with $\mathrm{p}_{u,0} = 1/2$ and set $\mathbf{g}_0 = \{\mathrm{g}_{m,0}\}_{m=1}^{M}$, where $g_{\mathrm{m},0} = 1/2$ for density function $\Psi(\boldsymbol{\Omega}_{\ell}; \mathbf{R}_{[0]})$, respectively. Set t :=1.

Step2: we use proposal distribution $\Psi(\boldsymbol{\Omega}_{\ell,t}; \mathbf{R}_{t-1})$ to draw a small set of samples $\{\boldsymbol{\Omega}_{\ell,t} = \{\bar{\boldsymbol{\omega}}_{\ell,t}, \boldsymbol{\omega}_{\ell,t}\}\}_{\ell=1}^{L_{MCMC}}$ from the target function $\pi(\boldsymbol{\Omega}_{\ell,t})$ and guarantee that each sample $\boldsymbol{\Omega}_{\ell,t} = \{\bar{\boldsymbol{\omega}}_{\ell,t}, \boldsymbol{\omega}_{\ell,t}\}$ is feasible by using the our proposed projection strategy introduced in Section 3.4.

Step3: update $\mathbf{R}_t = \{\mathbf{P}_t, \mathbf{g}_t\}$, where $\mathbf{P}_t = \{p_{u,t}\}_{u=1}^{U}$ and $\mathbf{g}_t = \{\mathrm{g}_{\mathrm{m},t}\}_{m=1}^{M}$ via formula (32) and (33).

Step4: if $\pi(\boldsymbol{\Omega}_{\ell,t}) > \pi(\Phi_C^*)$ for $\ell = 1, 2, ..., \mathrm{L}_{MCMC}$, then $\Phi_C^* = \Phi_C^{(\ell)}$.

Step5: stop this simulation when the predefined maximum number of iterations is satisfied; otherwise, set $t := t + 1$, and go back to Step 2.

4 Simulation configuration

In this section, the simulation configuration and simulation parameters are described. The considered scenario is a multi-cell multi-user massive MIMO downlink system operating in TDD mode with $B = 7$, as shown in Fig. 2. The simulation is done with a static network simulator. The key simulation parameters are summarized in Table 1. In this simulation, we assume that the CSI is perfectly known at the transmitter, the total power is uniformly allocated among the transmit antennas. The system composed of B hexagonal cells. All B BSs, where $B = \{1, 2, \ldots, B\}$ are installed with M antennas and serve U single-antenna users in each cell. Each BS is located at the cell center while U single-antenna users are randomly located in the cell. There is no user movement and handover during the simulation process.

5 Simulation results and analysis

In this section, we provide numerical results and computational complexity analysis of the proposed algorithm by simulative evaluation.

5.1 Performance evaluation

As can be seen the result from Fig. 3, the multi-cell multi-user massive MIMO downlink system using different JASUS method at various SINR. We find that the sum capacity of the AMCMC-based JASUS algorithm is very close to the maximum capacity result obtained by the ES-based JASUS algorithm with a wide range of SINRs. For example, when the SINR is 20 dB, the achieved values of cell capacity by using ES and AMCMC algorithms are

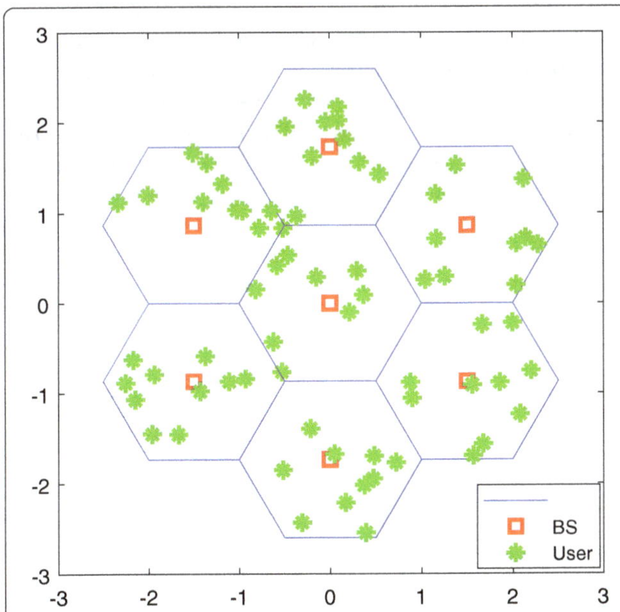

Fig. 2 Multi-cell multi-user massive MIMO system with $B = 7$ and $K = 10$

Table 1 Simulation parameters setting

Parameters	Values
No. of cells in the simulation B	7
Average no. of UEs in one cell	$2 \leq U \leq 50$
Number of BS antennas M	$10 \leq M \leq 64$
Inter-site distance	500 m
Cell radius	295 m
Path loss	$128.1 + 37.6 \times \log_{10}$(distance(km))[dB]
Each BS transmission power	10 dB
Shadowing standard derivation	7 dB
Noise spectral density	− 174 dBm/Hz
Users' speed	0
System bandwidth	20 MHz

50.7 and 49.9 b/s/Hz, respectively. Ninety-eight percent of the optimal capacity is obtained by our proposed method. The result shows that the AMCMC-based JASUS method has a good performance compared to the greedy-based JASUS method with the wide range of SINRs, both AMCMC-based JASUS and greedy-based JASUS methods have better capacity performances compared to the norm-based JASUS algorithm. This simulation result shows that, when the SINR is 30 dB, ES, AMCMC, and greedy algorithms enhanced cell capacity of approximately 9.1, 8.2, and 5.8 b/s/Hz, respectively.

Figure 4 shows the increase of sum cell capacity during each iteration for AMCMC-based JASUS algorithm with SINR = 20 dB. As can be seen from Fig. 4, we find that the AMCMC-based JASUS converges after about $t = 30$ iterations. As expected, the sum cell capacity obtained by AMCMC-based JASUS strictly monotonically increased with a number of iteration.

Now, the assumed scenario is a multi-cell multi-user massive MIMO downlink system with 50 active users ($U = 50$) at SINR = 20 dB, and we assume a different number of transmit antenna, M, from 16 to 60. Sixteen antennas ($N = 16$) were selected to be used by the transmitter and ten users ($K = 10$) were served. As can be seen in the result from Fig. 5, the cell capacity difference between the ES-based JASUS and AMCMC-based JASUS scheme is relatively small, and the cell capacity achieved by the aforementioned algorithms slightly grows with M. In summary, when numbers of the selected BS antennas (N) and scheduled users (K) are confirmed, the increase of the number of transmitting antennas (M) has little effect on the system capacity performance. Thus, it can be proven from the result that activation of more transmit antennas at the BS side is unnecessary.

A cell capacity performance comparison of each JASUS algorithms with various numbers of the selected antennas at SINR = 20 dB is shown in Fig. 6. The various

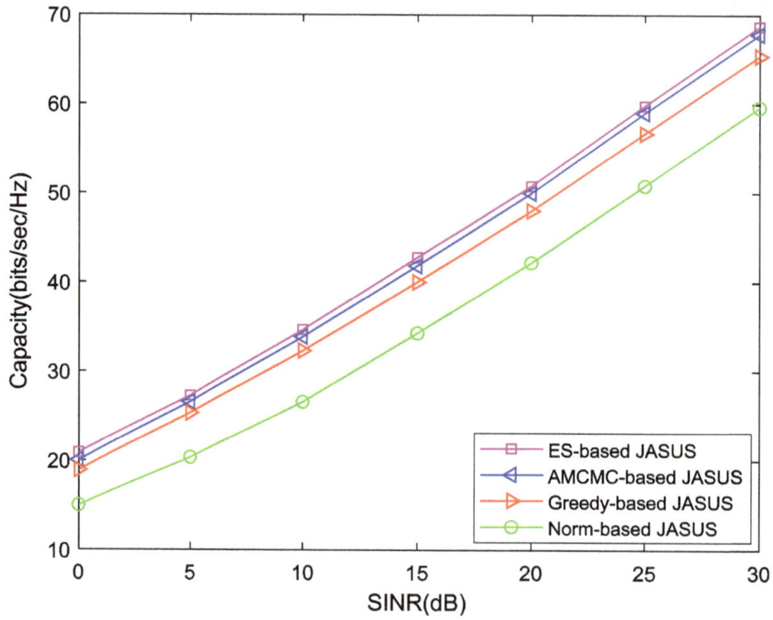

Fig. 3 Ergodic capacity versus SINR with $B = 7$, $M = 64$, $U = 50$, $N = 16$, and $K = 10$

numbers of the selected antennas ($12 \leq N \leq 24$) correspond to the different maximum cell capacity of the networks. Compared with the norm-based JASUS, the JASUS algorithms, which are ES, AMCMC, and greedy, had more significant enhancement for the system performance. For example, the maximum cell capacity enhancements, which are approximately 2.5, 2.2, and 1.7 b/s/Hz at SINR = 20 dB, are achieved when 18 ($N = 18$) BS antennas are selected. From the figure, it can be observed, when $N > 18$, the cell capacity achieved by the JASUS algorithms is slightly growing when the number of selected antenna goes large. Thus, it can be proven from the result that when numbers of the scheduled users K are confirmed, the system capacity sequentially increases until the numbers of selected BS antenna close to the $N = 18$, when $N > 18$, the increasing number of selected antennas has no significant effect on the system capacity performance. Therefore, the results show that more antenna selection is unnecessary at the base station. Thus, the proposed algorithm is demonstrated to be effective. In addition, we considerably decreased system cost and power consumption while approach the maximum cell capacity by selected suitable transmit antennas at BS side.

Figure 7 shows that the cell capacity different K at SINR = 20 dB, for user scheduling and with transmit antenna selection. It can be observed that the cell sum capacity increases with increasing of user K. The different numbers of the scheduled users ($2 \leq K \leq 16$) correspond to the different maximum cell capacity of the networks. Compared with the norm-based result, ES, AMCMC, and greedy algorithms enhanced the system performance. For example, the maximum cell capacity enhancement, which are approximately 6.2,

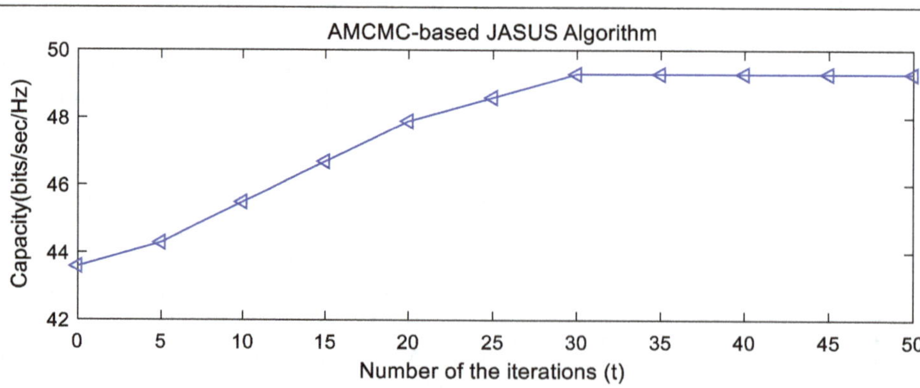

Fig. 4 Ergodic capacity achieved in iterative by AMCMC-based JASUS with $B = 7$, $M = 64$, $U = 50$, $N = 16$, $K = 10$, and SINR = 20 dB

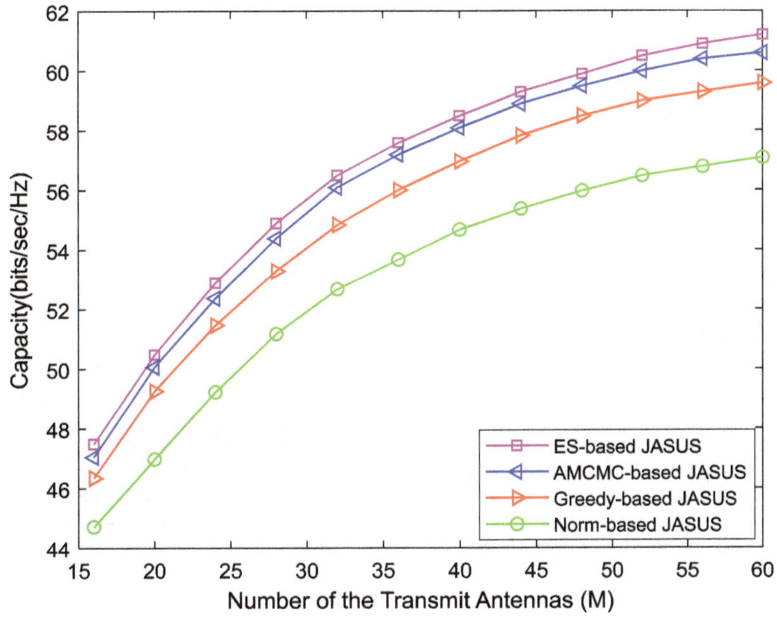

Fig. 5 Ergodic capacity of different algorithms with versus transmit antenna numbers with $B = 7$, $U = 50$, $N = 16$, $K = 10$, and SINR = 20 dB

5.5, and 4.0 b/s/Hz at SINR = 20 dB, are achieved when ten ($K = 10$) users are scheduled. From the figure, it can be observed that when $K > 10$, the cell capacity achieved by the aforementioned algorithms slightly grows when the number of the scheduled user goes large. This result proves that the behavior of JASUS algorithm does not change drastically when the scheduled user number becomes large.

Finally, we discuss on SER of different JASUS algorithms. A 16-QAM scheme is used with a ZF receiver. The SER performance of the linear ZF receiver system is shown in Fig. 8. Compared with the results of norm-based JASUS algorithm, ES, AMMC, and greedy algorithms improved the SER of the system. Same to the case of the cell capacity performance, the system SER of the AMCMC-based JASUS algorithm is close to that of

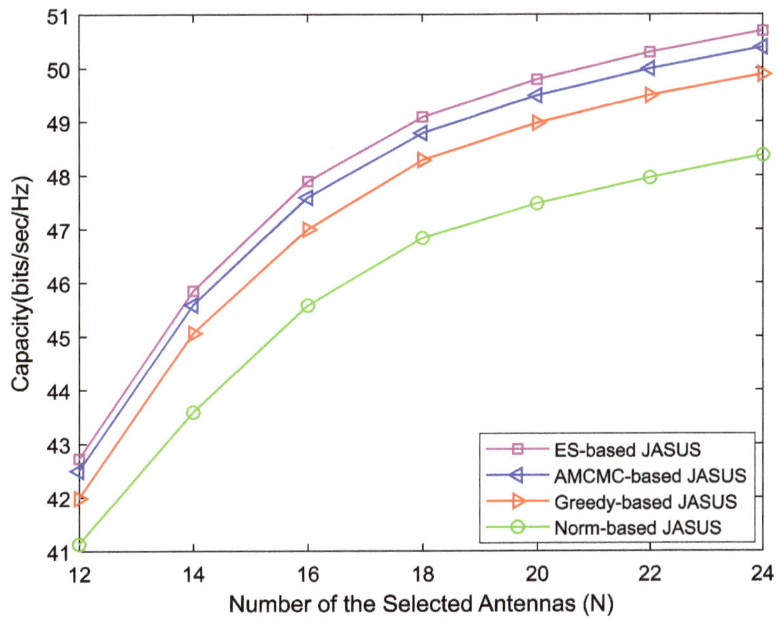

Fig. 6 Ergodic capacity of different algorithms with different numbers of the selected antennas with $B = 7$, $M = 64$, $U = 50$, $K = 10$, and SINR = 20 dB

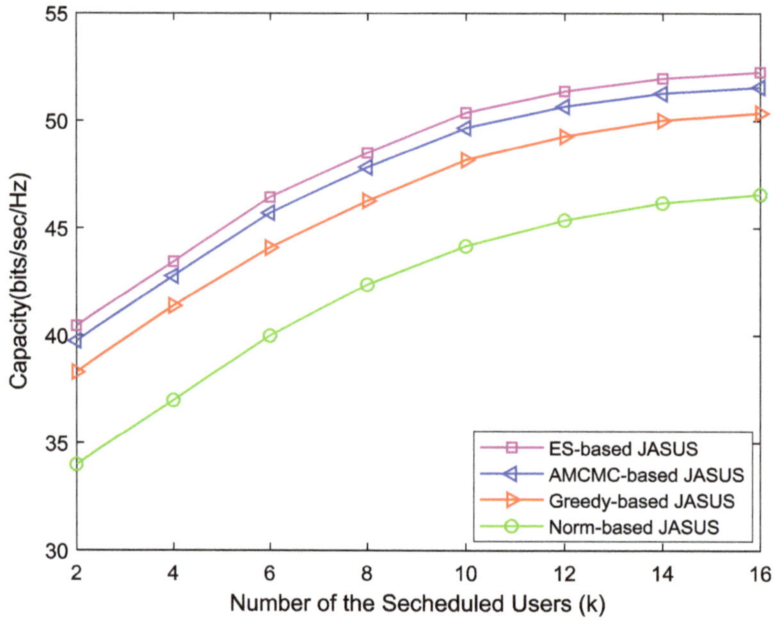

Fig. 7 Ergodic capacity of different algorithms with different scheduled number of users with $B = 7$, $M = 64$, $U = 50$, $N = 16$, and SINR = 20 dB

the ES-based JASUS algorithm. At the same time, we find that ES-based JASUS and AMMC-based JASUS have better SER performance than greedy-based JASUS and norm-based JASUS, especially the SINR to high. When the SINR is 20 dB, the SER performances of the ES-based JASUS, AMCMC-based JASUS, greedy-based JASUS, and norm-based JASUS algorithms are approximately 3.5×10^{-2}, 3.9×10^{-2}, 4.9×10^{-2}, and 6.8×10^{-2}, respectively.

5.2 Computational complexity analysis

The computational complexities of the introduced different JASUS algorithm are analyzed in this section. Table 2 summarizes the computational complexity of our proposed algorithm along with the complexity of another algorithm. The asymptotic notations, which reflect the computational complexity, was used to evaluation how the scheme responds to changes of parameters which are M, N, U, and K. From Table 2, we can easily observe the

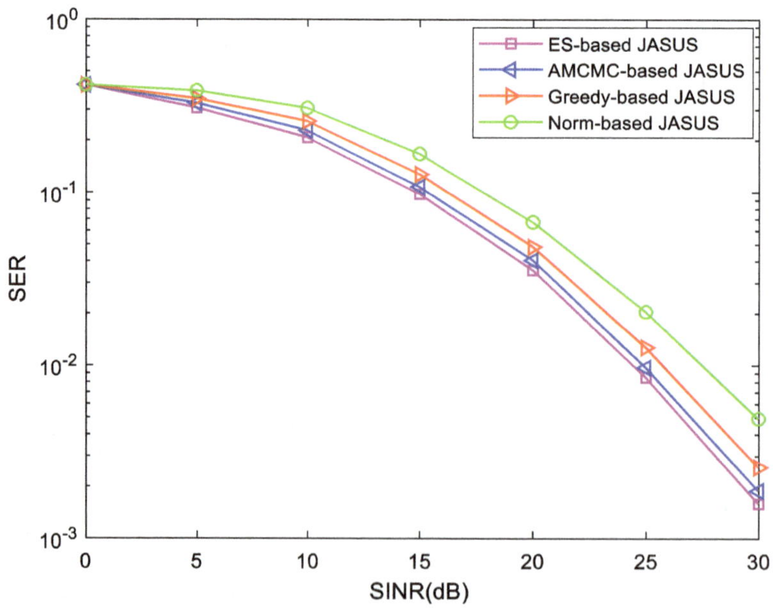

Fig. 8 SER versus SINR with $B = 7$, $M = 64$, $U = 50$, $N = 16$, and $K = 10$

Table 2 Computational complexity analysis

Algorithms	General case	$M = 64$, $N = 16$, $U = 50$, $k = 10$
ES-based JASUS	$O(C_M^N C_U^K N^3)$	2.1×10^{28}
AMCMC-based JASUS	$O(N^3 t L_{MCMC})$	2.8×10^7
Greedy-based JASUS	$tO(MUN^3)$	3.9×10^8
Norm-based JASUS	$O(MU) + tO(N^3)$	1.3×10^5
Number of iterations t	$1 \leq t \leq 50$	$t = 30$
Constant δ	0.5/1/1.5/2/2.5	$\delta = 2$
Number of promising samples L_{MCMC}	$L_{MCMC} = \delta \times (M + U)$	$L_{MCMC} = 228$

Notes: $\delta = 2$ is the best value in our experiments, because when δ is larger than 2, the cell capacity performance improvement is no longer significant

computational complexity gap between the four methods. Note that C_M^N denotes the binomial coefficient, and the matrix inverse operation [29–32] makes the computational complexity of per sample up to $O(N^3)$. Thus, the overall complexity of our proposed Algorithm 1 for the problem of (11) is $O(N^3 t L_{MCMC})$, where $t \times L_{MCMC}$ is the total number of target function evaluations. Ultimately, we can be observed from Table 2 that our proposed algorithm has a very low computational complexity compared to ES-based JASUS and greedy-based JASUS algorithms. However, the norm-based JASUS algorithm has a very low computational complexity compared to our proposed algorithm, but it also has a very low cell capacity. This result shows that our proposed algorithm is suitable for practical multi-cell multi-user massive MIMO system.

6 Conclusion

In this paper, we studied the problem of JASUS in a multi-cell multi-user massive MIMO downlink system operating with TDD mode. Considering the trade-off between network performance and computational complexity, we proposed a low-complexity algorithm for JASUS method based on AMCMC algorithm in the downlink multi-cell multi-user massive MIMO systems. AMCMC algorithm has been proven helpful for selecting combination subset of antennas and users to approach the maximum cell capacity with consideration of the inter-cell interference. In our algorithm, the updating rules of the selection probability of each base station antenna and scheduling probability for each user are proposed. In addition, we proposed a new projection strategy to satisfy the constraints of selection. Performance analysis and simulation results show that our proposed algorithm can produce promising results and achieve a good trade-off between complexity and performance. Compared with ES-based JASUS algorithm, the proposed algorithm achieved comparable performance with very low complexity. In addition, we demonstrate that our proposed algorithm outperforms greedy-based JASUS and norm-based JASUS methods in terms of cell capacity and SER performance with under poorly conditioned channels. At the same time, the computational complexity is reduced significantly by combining with the proposed algorithm.

Abbreviations

AMCMC: Adaptive Markov chain Monte Carlo; BS: Base station; CSI: Channel state information; ES: Exhaustive search; JASUS: Joint antenna selection and user scheduling; MIMO: Multiple-input multiple-output; MU: Multi-user; RF: Radio frequency; SINR: Signal-to-interference-plus-noise ratio; TDD: Time division duplexing; ZF: Zero-forcing

Acknowledgements

This work described in this paper was supported by the National Science and Technology Major Project: No. 2018ZX03001029-004.

Authors' contributions

WG conceived and designed the study. SM and KZ performed the simulation experiments. SM and KZ wrote the paper. XL and ZS reviewed and edited the manuscript. All authors read and approved the final manuscript.

Authors' information

Saidiwaerdi Maimaiti received the M.Sc. degree in signal and information processing from Southwest Jiaotong University, Chengdu, China, in 2014. He is currently working towards the Ph.D. degree in Information and Communications Engineering, Key Laboratory of Universal Wireless Communications, Ministry of Education, Beijing University of Posts and Telecommunications Beijing, China. His research interests include massive MIMO, interference management, radio network planning, resource management, and intelligent network optimization in 5G network systems.

Funding

The funding for the research reported is provided by the National Science and Technology Major Project: No. 2018ZX03001029-004. The funds are mainly used for simulation hardware support.

Competing interests

The authors declare that they have no competing interests.

References

1. T.A. Sheikh, J. Bora, A. Hussain, in *CCTCEEC*. A survey of antenna and user scheduling techniques for massive MIMO-5G wireless system (Mysore, 2017), pp. 578–583. https://doi.org/10.1109/CTCEEC.2017.8455177
2. T.L. Marzetta, Noncooperative cellular wireless with unlimited numbers of base station antennas. IEEE Trans. Wirel. Commun. **9**(11), 3590–3600 (2010)
3. E.G. Larsson, O. Edfors, F. Tufvesson, T.L. Marzetta, Massive MIMO for next generation wireless systems. IEEE Comun. Mag. **52**(2), 186–195 (2014)
4. S. Han, I. Chih-Lin, Z. Xu, C. Rowell, Large-scale antenna systems with hybrid analog and digital beamforming for millimeter wave 5G. IEEE Comun. Mag. **53**(1), 186–194 (2015)
5. Y. Dong, Y. Tang, K.Z. Shenzhen, in *IEEE/ACIS International Conference on Computer and Information Science IEEE*. Improved joint antenna selection and user scheduling for massive MIMO systems (2017)
6. M. Torabi, D. Haccoun, Performance analysis of joint user scheduling and antenna selection over MIMO fading channels. IEEE Signal Process. Lett. **18**(4), 235–238 (2011)
7. C. Bandeira, D. Moreira, P. Normando, et al., in *Proc. IEEE Vehicular Technology Conf. (VTC-Spring)*. Performance of joint antenna and user selection schemes for interference alignment in MIMO interference channels (2013)
8. S. Xue, Z. Qian, W. Shao, et al., in *Proc. IEEE International Conf. Computational Electromagnetics (ICCEM)*. Distributed joint antenna and user selection for MIMO broadcast channels (2016)
9. J.-C. Chen, "Joint Antenna Selection and User Scheduling for Massive Multiuser MIMO Systems With Low-Resolution ADCs," IEEE Trans. Veh. Technol. **68**(1) 1019–1024 (2019)

10. X. Gao, L. Dai, A.M. Sayeed, Low RF-complexity technologies to enable millimeter-wave MIMO with large antenna array for 5G wireless communications. IEEE Commun. Mag. **56**(4), 211–217 (2018)
11. T.-H. Tai, W.-H. Chung, T.-S. Lee, "A low complexity antenna selection algorithm for energy efficiency in massive MIMO systems", Proc. IEEE Int. Conf. Data Sci. Data Intensive Syst. 284-289 (2015)
12. X. Gao , O. Edfors, F. Tufvesson, and E. G. Larsson, "Massive MIMO in real propagation environments: Do all antennas contribute equally?," IEEE Trans. Commun. **63**(11) 3917–3928 (2015)
13. B. Makki, A. Ide, T. Svensson, T. Eriksson, M.-S. Alouini, A genetic algorithm-based antenna selection approach for large-but-finite MIMO networks. IEEE Trans. Veh. Technol. **66**(7), 6591–6595 (2016)
14. M. Benmimoune, E. Driouch, W. Ajib, D. Massicotte, in *Proc. IEEE WCNC*. Joint transmit antenna selection and user scheduling for massive MIMO systems (2015), pp. 381–386
15. M. Olyaee, M. Eslami, J. Haghighat, An energy-efficient joint antenna and user selection algorithm for multi-user massive MIMO downlink. IET Commun. **12**(3), 255–260 (2018)
16. G. Xu, A. Liu, W. Jiang, H. Xiang, W. Luo, Joint user scheduling and antenna selection in distributed massive MIMO systems with limited backhaul capacity. China Commun. **11**(5), 17–30 (2014)
17. B.M. Lee, J. Choi, J. Bang, B.-C. Kang, in *IEEE Int. Symp. On Circuits and Systems (ISCAS)*. An energy efficient antenna selection for large scale green MIMO systems (2013), pp. 950–953
18. Q.H. Spencer, A.L. Swindlehurst, M. Haardt, Zero-forcing methods for downlink spatial multiplexing in multiuser MIMO channels. IEEE Trans. Signal Process. **52**(2), 461–471 (2004)
19. G. Dimic, N.D. Sidiropoulos, On downlink beamforming with greedy user selection: performance analysis and a simple new algorithm. IEEE Trans. Signal Process. **53**(10), 3857–3868 (2005)
20. M. Al-Saedy, M. Al-Imari, M. Al-Shuraifi, et al. Joint User Selection and Multi-Mode Scheduling in Multicell MIMO Cellular Networks[J]. IEEE Transactions on Vehicular Technology, 1-1 (2017)
21. J.C. Spall, Estimation via Markov chain Monte Carlo. Control Syst. IEEE **23**(2), 34–45 (2003)
22. W.R. Gilks, S. Richardson, D.J. Spiegelhalter, *Markov Chain Monte Carlo in Practice* (Chapman and Hall, London, 1996)
23. J. Liu, *Monte Carlo Strategies in Scientific Computing* (Springer-Verlag, New York, 2001)
24. S. Kullback, R. Leibler, On information and sufficiency. Ann. Math. Stat. **22**(1), 79–86 (1951)
25. R.Y. Rubinstein, D.P. Kroese, *The Cross-Entropy Method: A Unified Approach to Combinatorial Optimization, Monte-Carlo Simulation and Machine Learning* (Springer-Verlag, Berlin, 2004)
26. H. Robbins, S. Monro, A stochastic approximation method. Ann. Math. Stat. **22**(3), 400–407 (1951)
27. H. Kushner, G. Yin, *Stochastic Approximation Algorithms and Applications*, 2nd edn. (Springer-Verlag, New York, 2003)
28. Y. Liu, Y.Y. Zhang, C.L. Ji, W.Q. Malik, D.J. Edwards, A low complexity receive-antenna-selection algorithm for MIMO-OFDM wireless systems. IEEE Trans. Veh. Technol. **58**(6), 2793–2802 (2009)
29. X. Gao, L. Dai, Y. Ma, Z. Wang, Low-complexity near-optimal signal detection for uplink large-scale MIMO systems. Electron. Lett. **50**(18), 1326–1328 (2014)
30. X. Gao, L. Dai, Y. Hu, Y. Zhang, Z. Wang, Low-complexity signal detection for large-scale MIMO in optical wireless communications. IEEE J. Sel. Areas Commun. **33**(9), 1903–1912 (2015)
31. L. Dai, X. Gao, X. Su, S. Han, I. Chih-Lin, Z. Wang, Low-complexity soft-output signal detection based on Gauss-Seidel method for uplink multiuser large-scale MIMO systems. IEEE Trans. Veh. Technol. **64**(10), 4839–4845 (2015)
32. Z. Wu, C. Zhang, Y. Xue, S. Xu, Z. You, "Efficient architecture for soft-output massive MIMO detection with gauss-seidel method", Proc. IEEE Int. Conf. Circuits Syst. (ISCAS), 1886-1889 (2016)

8

Improving MIMO relay compressed sensing-based channel estimation and pilot allocation

Abbas Akbarpour-Kasgari[1] and Mehrdad Ardebilipour[1*]

Abstract

Increasing the data rate of communication systems together with the performance are main goals of the communication networks in the future. Multiple-input multiple-output (MIMO)-orthogonal frequency division multiplexing (OFDM) relay networks as a key technology is one the main techniques to maintain the performance and data rate. Being influenced by fading channels, the MIMO-OFDM relay networks need some reliable approaches to estimate the channel response. Compressed sensing (CS) is one of the critical approaches to maintain the accuracy together with the spectral efficiency. In this paper, we have utilized CS-based approaches to estimate the MIMO-OFDM relay channel. Specifically, forward backward prediction (FBP) is used to estimate the fading channels. This approach benefits from backward correction which distinguishes it from other iterative approaches and helps interestingly in channel estimation accuracy. Moreover, in order to improve the channel estimation, pilot allocation approaches are proposed based on the system model and probability function. Furthermore, a cross entropy-based approach is utilized to propose two different approaches called sequential cross entropy self-coherence (SCE-SC) and parallel cross entropy self-coherence. Actually, mutual coherence is divided into two parts namely cross coherence and self coherence. It is demonstrated that in FBP-based approach self-coherence is important in mutual coherence. Consequently, the computations are interestingly decreased. The superiority of the method is represented using comparing simulations with other well-known approaches.

Keywords: Cross entropy, Compressed sensing, Forward-backward prediction, MIMO-OFDM relay channel estimation, Pilot allocation

1 Introduction

By increasing demand of data rate and coverage area in modern communication systems, utilizing of relay communication networks is necessitated. In order to combat with the frequency selectivity of the channel and long-distance impairments, multiple-input multiple-output (MIMO)-orthogonal frequency division multiplexing (OFDM) is used as the pioneer technology [1–3]. Supplying some advanced features in MIMO-OFDM relays such as beamforming, relay selection, and power minimization required the active nodes in the network to be aware of channel state information (CSI) [4, 5]. Pilot-aided channel estimation suffers from bandwidth efficiency which is very essential for the high-rate communication. To provide bandwidth-efficient pilot-aided channel estimation, compressed sensing (CS) is emerged, recently [6]. Wireless channels could be modeled by sparse signals, since there are considerable diffusion sources in the wireless environments. Hence, CS is very precious in sparse channel estimation together with the increase of accuracy and bandwidth efficiency. Furthermore, block sparse behavior of MIMO communication channels which is resulted by the existence of common scatterers, is the critical characteristic which is utilized by researchers in recent MIMO-OFDM compressed channel estimation [7]. As a consequence of joint sparsity, the support of the different channel ensembles between transmit-receive antenna pairs in MIMO nodes is identically distributed.

*Correspondence: mehrdad@eetd.kntu.ac.ir
[1]Department of Electrical Engineering, K. N. Toosi University of Technology, Shariati St., Tehran, Iran

Thus, channel estimation could be extended to block-sparse signal processing in MIMO communications [8, 9]. At first, distributed CS (DCS)-simultaneous orthogonal matching pursuit (SOMP) was developed for channel estimation in single-input single-output (SISO)-OFDM channel estimation in [10]. Then, the authors of [11] utilized DCS-SOMP-based channel estimation for MIMO channel identification. Subsequently, a joint-OMP algorithm has been proposed to estimate the CSI of massive MIMO (mMIMO) in [12]. Recently, MIMO-OFDM relay channel estimation has been developed using DCS-based approach and a novel algorithm has been proposed using compressive sampling matching pursuit (CoSaMP) called block-verified CoSaMP (B-vCoSaMP) [13]. Furthermore, CS-based channel estimation is getting more attractive in massive MIMO communications [14, 15] which utilize angular domain sparsity and time domain sparsity.

Designing appropriate pilot sequences to improve channel estimation performance is the other key obstacle in front of researchers. In DCS-based channel estimation, the trivial pilot pattern is the random pilot allocation; while utilizing restricted isometry property (RIP), the measurement matrix could be designed to develop channel estimation. Of course, mutual coherence is optimized instead of RIP since there is no polynomial time [16]. In DCS-based channel estimation, the pilot sequences are translated to the measurement matrix identification and one may optimize the mutual coherence to design appropriate pilot sequence. It is shown that designing pilot sequences in terms of mutual coherence optimization is a combinatorial optimization. Hence, evolutionary algorithms are used to design pilot sequences. Specifically, the authors in [17] proposed a genetic algorithm (GA)-based pilot allocation algorithm in CS-based channel estimation. Furthermore, in [18] a suboptimal pilot allocation algorithm is designed based on GA for DCS-based channel estimation. Moreover, Qi et al. in [19] utilize the estimation of distribution algorithm (EDA) to define optimized pilot positions in SISO-OFDM systems. Moreover, He et al. [18] and Akbarpour-Kasgari and Ardebilipour [13] generate optimized pilot sequences for MIMO-OFDM compressed channel estimation using a GA and cross entropy (CE)-based approach.

In this paper, we have proposed to utilize a forward backward prediction (FBP) algorithm in order to increase the estimation accuracy. We proposed a FBP-based channel estimation algorithm where the algorithm further to forward collecting steps consists of backward elimination steps. In forward steps, the best atom is gathered. Further, exploiting backward steps makes it possible to omit bad atoms gathered in current and previous iterations. Hence, it would increase the accuracy of estimation rather than OMP and CoSaMP. In OMP and CoSaMP, the backward elimination steps are absent; thus, the previously unsuitable atoms would increase the error in channel estimation and decrease the accuracy as well. To exploit the proposed FBP, we design the channel matrix and measurement matrix to exploit the common sparsity using ζ-norm which is introduced. The system model is formulated using matrix representation to exploit the common sparsity of the channel. Then an FBP-based channel estimation is developed to estimate the channel coefficients in a time domain. By simulation results, it is validated that the proposed method is superior than the other existing methods. The superiority is caused by the backward stages which omit the evil atoms for accuracy amplification.

Moreover, we have introduced two probability-based pilot allocation called sequential cross entropy self-coherence (SCE-SC) and parallel cross entropy self-coherence (PCE-SC). As mentioned, in order to exploit the FBP-based channel estimation and ζ-norm, we have designed appropriate measurement matrix using pilot symbols. It is represented in [13] that mutual coherence is related to the self-coherence of the sub-matrices and cross coherence of the sub-matrices related to each of the transmitter antenna, simultaneously. Here, we have changed the measurement matrix and we have shown that mutual coherence is only related to the self-coherence of the transmitter's sub-matrices. Consequently, the computation burden is considerably reduced rather than the previous works in [9, 13]. Using self-coherence phenomena, we have designed more simpler mutual coherence fitness function to be optimized. To optimize the introduced fitness function, we have proposed the probability-based pilot allocation algorithms, specifically SCE-SC and PCE-SC, where we have tried to optimize the new fitness function using a probability approach. SCE-SC performs optimization based on sampling the probability density function (pdf) of the available subcarriers. The pdf is updated in each iteration of the algorithm and then the fitness function is calculated. Afterward, the updated pdf is sampled for the next iteration. The iterations are continued till the steady state is resulted in the fitness function. In SCE-SC, in each iteration, only one pdf is followed. It will reduce the speed of convergence. In order to increase the speed of convergence, PCE-SC is developed where the number of pdfs are followed, simultaneously.

The contributions of the paper are summarized as follows:

- At first, we have proposed a novel approach to estimate the channel impulse response. In the proposed method which is based on FBP, the extracted appropriate atoms are reconsidered in the backward stages to be omitted if they are not suitable enough for estimation.
- In order to exploit the joint sparsity in a FBP approach, the channel measurement matrix is

developed utilizing the introduced norm. Hence, the Gram matrix is in diagonal matrix form, and as a consequence, mutual coherence could be defined using self-coherence metrics.

- A novel approach called SCE-SC is designed using a cross entropy-based method to optimize the self-coherence of the developed measurement matrix.
- To improve the pilot sequences and considering multiple probability density functions, PCE-SC is introduced which is more rapid and accurate in convergence rather than SCE-SC.

The remainder of the paper is as follows. The methods are demonstrated in Section 2. The system model is represented in Section 3. Section 4 covers the channel estimation approach using the proposed FBP. The pilot allocation scheme for SISO and MIMO systems is described in Section 5. Moreover, the proposed pilot allocation methods are represented in Section 6. Eventually, numerical results are expressed in Section 7 and concluding remarks are demonstrated in Section 8.

Notations: Matrices and vectors are denoted by uppercase and lowercase boldfaced letters, respectively. $|.|$ and $(.)^*$ denote the complex modulus and the conjugate of a complex number. For a given matrix $\mathbf{A}$, $\mathbf{A}^T$, $\mathbf{A}^H$, and Trace($\mathbf{A}$) denote its transpose, conjugate transpose, and trace, respectively, and $\mathbf{A}_{i,j}$ denotes the (i,j)th element of $\mathbf{A}$. For a given vector $\mathbf{x}$ with its element denoted by x_l, $\|\mathbf{x}\|_2 = \sqrt{\mathbf{x}^H\mathbf{x}}$ represents the Euclidean norm, $\|\mathbf{x}\|_1 = \sum_l |x_l|$ is the l_1-norm, and diag($\mathbf{x}$) denotes a diagonal matrix with $\mathbf{x}$ on its main diagonal. For two vectors $\mathbf{x}$ and $\mathbf{y}$, $< \mathbf{x}, \mathbf{y} >$ denotes their inner product. For a given set $\mathbf{\Lambda}$, $n(\mathbf{\Lambda})$ is the number of elements in $\mathbf{\Lambda}$. $\mathbb{C}^{m\times n}$ stands for the set of all complex-valued $m \times n$ matrices, and $\emptyset$ denotes the null set.

2 Methods

In this paper, the channel estimation is addressed using CS perspective. In order to increase the channel estimation accuracy together with decreasing the spectrum utilization, we have proposed the FBP-based channel estimation and SCE-SC and PCE-SC pilot allocation algorithms. To compare the proposed method with the existing ones, we have considered a GA-based approach and B-vCoSaMP-based channel estimation.

3 System model

Consider an amplify-and-forward (AF) relay network consists of MIMO terminals (Fig. 1). The network consists of a source node ($\mathbb{S}$), a relay node ($\mathbb{R}$), and a destination node ($\mathbb{D}$). Besides, each terminal is equipped with $N_\mathbb{S}$, $N_\mathbb{R}$, and $N_\mathbb{D}$ transceiver antennas, respectively. Without loss of generality, we consider OFDM transmission in MIMO terminals. Consider an OFDM system with N subcarriers where N_p of them are selected as pilot subcarriers. Besides, to omit the interference of other antenna's pilots on each other, we consider to utilize the orthogonal pilot allocation, i.e., not only N_p subcarriers are allocated to pilot subcarriers, but also pilot subcarriers assigned to other $N_\mathbb{S} - 1$ transmit antennas are reserved to be zero [20]. Thus, the number of data subcarriers on each of the transmit antenna is equal to $N - N_\mathbb{S}N_p$. Assuming $\mathbf{x}_m \in \mathbb{C}^{N\times 1}$ as the OFDM symbol before cyclic prefix (CP) adding to be transmitted on the mth transmit antenna of $\mathbb{S}$. In the first time slot of the time-division-duplexing (TDD), the received signal $\mathbf{r}_q$ for $q = 1, 2, \ldots, N_\mathbb{R}$ is formulated as

$$\mathbf{r}_q = \sum_{m=1}^{N_\mathbb{S}} \mathbf{X}_m \mathbf{F}_m^L \mathbf{f}_{mq} + \mathbf{v}_q \tag{1}$$

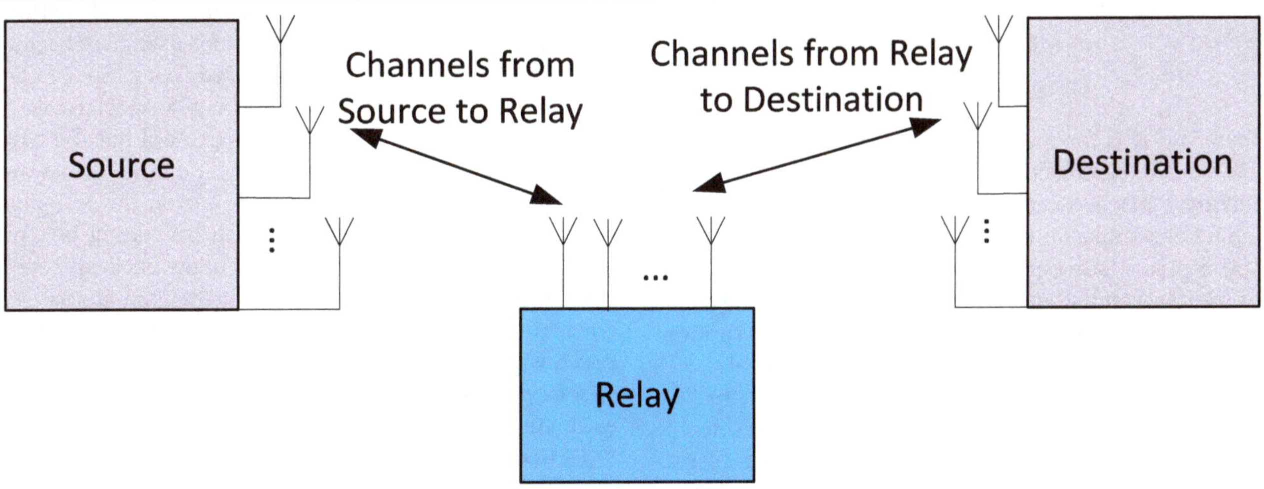

Fig. 1 Conceptual system model for a MIMO-OFDM relay network and joint sparsity

where $\mathbf{X}_m$ is the diagonal matrix with $\mathbf{x}_m$ as its main diagonal, $\mathbf{F}_m^L$ is the partial discrete Fourier transform (DFT) matrix with N_p rows corresponding to the N_p pilot subcarriers of mth transmit antenna and first L columns of $N \times N$ DFT matrix, $\mathbf{f}_{mq} = \left[f_{mq}(0), f_{mq}(1), \ldots, f_{mq}(L-1)\right]^T$ is the channel vector between source node $\mathbb{S}$ and relay node $\mathbb{R}$ with length L which K of them are non-zero to represent the channel sparsity. Besides, $\mathbf{v}_q$ is the additive white Gaussian noise (AWGN) vector in the qth antenna of relay $\mathbb{R}$. In the second time slot, relay $\mathbb{R}$ amplifies and retransmits the received signal to the destination node $\mathbb{D}$. As a consequence, the received pilots in the nth antenna of destination $\mathbb{D}$ could be formulated as

$$\mathbf{y}_n = \sum_{q=1}^{N_{\mathbb{R}}} \mathbf{X}_m \mathbf{F}_m^{2L-1} \mathbf{h}_{mn} + \mathbf{z}_n \tag{2}$$

where $\mathbf{h}_{mn} = \sum_{m=1}^{N_{\mathbb{S}}} \beta \mathbf{f}_{mq} * \mathbf{g}_{qn}$ where $*$ stands for convolution, is the overall channel between $\mathbb{S}$ and $\mathbb{D}$ passing by $\mathbb{R}$ and $\mathbf{g}_{qn}$ is the channel vector between qth antenna of $\mathbb{R}$ and nth antenna of $\mathbb{D}$ and $\mathbf{F}_m^{2L-1}$ is the partial DFT matrix with N_p rows corresponding to the N_p pilot subcarriers of mth transmit antenna and first $2L-1$ columns of $N \times N$ DFT matrix. Collecting all the received pilot sequences, we can represent the $N_p \times N_{\mathbb{D}}$ received pilot matrix as $\mathbf{Y} = [\mathbf{y}_{00}, \mathbf{y}_{01}, \ldots, \mathbf{y}_{0N_{\mathbb{D}}-1}]$ and collecting all the channel ensembles, we can represent the $2L-1 \times N_{\mathbb{D}}$ as $\mathbf{H} = [\mathbf{h}_{00}, \mathbf{h}_{01}, \ldots, \mathbf{h}_{0N_{\mathbb{D}}-1}]$. Hence, the extension of (2) could be represented in matrix form for single-input multiple-output (SIMO) case as

$$\mathbf{Y} = \mathbf{\Phi}\mathbf{H} + \mathbf{Z} \tag{3}$$

where $\mathbf{Z}$ is the AWGN matrix with corresponding columns according to the $\mathbf{z}_{0v}$. $\mathbf{\Phi}$ is the measurement matrix with size of $N_p \times L$. Defining ζ norm for matrices as $\zeta(\mathbf{H}) = \text{card}\{\|\mathbf{H}_v\|_2 \neq 0\}$ where $\mathbf{H}_v$ is the vth column of $\mathbf{H}$, estimating the channel could be accomplished by following optimization criterion.

$$\begin{aligned} &\min_{\mathbf{H}} \ \|\mathbf{Y} - \mathbf{\Phi}\mathbf{H}\|_2^2 \\ &s.t. \ \ \zeta(\mathbf{H}) \leq K \end{aligned} \tag{4}$$

Obviously, K is the maximum sparsity of the columns of $\mathbf{H}$. We called the objective function as $\mathbf{F}(\mathbf{H}) = \|\mathbf{Y} - \mathbf{\Phi}\mathbf{H}\|_2^2$. Utilizing $\zeta(\mathbf{H})$, we can exploit the joint sparsity of the channel ensembles in the system. The objective function in (4) represents the error of channel estimation method and the constraint controls the sparsity order of the channel ensembles in $\mathbf{H}$. Moreover, using ζ-norm definition, the block sparsity of the channels are exploited.

As mentioned, the channel between transmit-receive pair which is denoted by $\mathbf{h}_{uv}$ consists of L resolvable paths. These resolvable paths are resulted from L scatterers which are encountered by the signal conveying from the transmitter antenna u to the receiver antenna v. K of these L scatterers are significant scatterers where $K << L$. Consequently, the channel could be modeled using sparse vectors. Moreover, the signal conveyed distance is large relative to the transmit-receive antenna spacing in each terminal. Hence, the encountered scatterers in each chip period is identical between different antennas. In other words, the delays of different paths are the same in all the channel ensembles between two terminals. Thus, the sparsity pattern of different channel pairs could be assumed to be the same while the channel attenuation is different. Each path consists of different subpaths which are scattered from different scatterers which are zero-mean and identically independent distributed (i.i.d.). Thus, each path's attenuation is assumed to be $\mathcal{CN}\left(0, \sigma^2\right)$. Hence, the channel coefficient could be represented as

$$h_{uv}(l) = \sum_{i=0}^{I-1} \alpha_{uv}(i) g(lT - \tau(i)) \tag{5}$$

where $\tau(I-1) \geq \cdots \geq \tau(1) \geq \tau(0)$ are the respective paths' delay and $g(.)$ is the shaping pulse in continuous domain. The shaping pulse is zero outside the the interval $[0, T_g]$, where T_g is the integer multiple of chip time T. Without loss of generality, we assumed that $\tau(i)$ are integer multiples of T. Thus, the number of channel paths, caused by the channel itself and shaping filter is derived by $L = \tau(I-1)/T + T_g/T + 1$. Furthermore, we assume that L is lower than T_g/T. Using the mentioned notations, we can represent the channel impulse response using $\mathbf{h}_{uv} = [h_{uv}(0), h_{uv}(1), \ldots, h_{uv}(L-1)]^T$.

4 Forward-backward pursuit channel estimation

In order to handle the optimization in (4), we have proposed a forward-backward pursuit (FBP) based on [21] where ℓ_0 norm was used. The algorithm is represented in details in Algorithm 1. Equation 4 could be solved using three different methods as convex relaxation, greedy methods, and message-passing (MP) algorithms. FBP which is based on MP algorithm is used here because of its forward selection and backward fixing. Specifically, OMP which is a greedy algorithm is a special case of FBP which the forward selection is present but the backward fixing is absent; consequently, it cannot fix its own mistakes in the previous steps. Moreover, the FBP algorithm constructs the new subspace by adding just one atom to the previous subspace, and in the backward steps, it reconstructs the subspace by omitting bad atoms. As a consequence, the proposed FBP algorithm could be compared with its greedy one called OMP, where OMP is the special case of FBP without backward steps to increase the estimation accuracy.

At the end of forward stages, a metric called $\delta_F^{(t)}$ is defined which represents the difference by considering the new added atom. In the backward stages, we first consider

Algorithm 1 FBP-based channel estimation

Input: the measurement matrix $\mathbf{\Phi}$, the received pilot matrix $\mathbf{Y}$ and sparsity order K
Output: the estimation of channel impulse responses $\mathbf{H}$
$\mathbf{R}^{(0)} = \mathbf{Y}, \boldsymbol{\lambda}^0 = \phi$ and $t = 0$
while stop criterion not met **do**
  $i^{(t)} = \arg\max_{j \notin \boldsymbol{\lambda}^{(t-1)}} \|(\mathbf{R}^{(t)})^H \phi_j\|_2$
  Merging supports of the pervious iteration and the present one $\boldsymbol{\lambda}^{(t)} = \boldsymbol{\lambda}^{(t-1)} \cup i^{(t)}$
  Define $\mathbf{H}^{(t)} = \mathbf{\Phi}^{\dagger}_{\boldsymbol{\lambda}^{(t)}} \mathbf{Y}$
  $\delta_F^{(t)} = \mathbf{F}(\mathbf{H}^{(t-1)}) - \mathbf{F}(\mathbf{H}^{(t)})$
  while 1 **do**
    $j^{(t)} = \arg\min_{j \in \boldsymbol{\lambda}^{(t)}} \mathbf{F}(\mathbf{H}^{(t)} - \mathbf{H}_{j/}^{(t)})$
    $\delta_B^{(t)} = \mathbf{F}(\mathbf{H}_{j/}^{(t)}) - \mathbf{F}(\mathbf{H}^{(t)})$
    if $\delta_B^{(t)} \geq 0.5\delta_F^{(t)}$ **then**
      Update residual $\mathbf{R}^{(t)} = \mathbf{Y} - \mathbf{\Phi}_{\boldsymbol{\lambda}^{(t)}} \mathbf{H}^{(t)}$
      break
    end if
    Omit bad atom $\boldsymbol{\lambda}^{(t)} = \boldsymbol{\lambda}^{(t)} - j^{(t)}$
    Update $\mathbf{H}^{(t)} = \mathbf{\Phi}^{\dagger}_{\boldsymbol{\lambda}^{(t)}} \mathbf{Y}$
    Update residual $\mathbf{R}^{(t)} = \mathbf{Y} - \mathbf{\Phi}_{\boldsymbol{\lambda}^{(t)}} \mathbf{H}^{(t)}$
  end while
end while

all the collected atoms individually and their effect on the residual is considered by calculating $\arg\min_{j \in \boldsymbol{\lambda}^{(t)}} \mathbf{F}(\mathbf{H}^{(t)} - \mathbf{H}_{j/}^{(t)})$ and subsequently $\delta_B^{(t)}$ is computed to determine the worst collected atom. These calculation is absent in other known algorithms which makes this algorithm an efficient one in the CS-based channel estimation approaches.

In Algorithm 1, $|i^{(t)}|$ denotes the number of selected atoms in forward step, and $\mathbf{H}_{j/}^{(t)}$ represents the $\mathbf{H}^{(t)}$ while $j^{(t)}$th column is omitted.

In order to extend the proposed method to the MIMO case, one can extend the measurement matrix and channel matrix as follows. In order to extend the channel matrix, we add the other transmitting antenna caused channels row-wise to the each other and represent channel matrix $\mathbf{H} \in \mathbb{C}^{(2L-1)N_{\mathbb{S}} \times N_{\mathbb{D}}}$ as

$$\mathbf{H} = \begin{pmatrix} \mathbf{h}_{00} & \mathbf{h}_{01} & \cdots & \mathbf{h}_{0(N_{\mathbb{D}}-1)} \\ \mathbf{h}_{10} & \mathbf{h}_{11} & \cdots & \mathbf{h}_{1(N_{\mathbb{D}}-1)} \\ \vdots & \vdots & \ddots & \vdots \\ \mathbf{h}_{(N_{\mathbb{S}}-1)0} & \mathbf{h}_{(N_{\mathbb{S}}-1)1} & \cdots & \mathbf{h}_{(N_{\mathbb{S}}-1)(N_{\mathbb{D}}-1)} \end{pmatrix}. \tag{6}$$

Moreover, the measurement matrix $\mathbf{\Phi} \in \mathbb{C}^{N_{\mathbb{S}}N_p \times (2L-1)N_{\mathbb{S}}}$ is extended as

$$\mathbf{\Phi} = \begin{pmatrix} \mathbf{\Phi}_0 & \mathbf{0} & \cdots & \mathbf{0} \\ \mathbf{0} & \mathbf{\Phi}_1 & \cdots & \mathbf{0} \\ \vdots & \vdots & \ddots & \vdots \\ \mathbf{0} & \mathbf{0} & \cdots & \mathbf{\Phi}_{N_{\mathbb{S}}-1} \end{pmatrix} \tag{7}$$

where $\mathbf{0}$ is $N_p \times (2L-1)$ zeros matrix. Furthermore, the received pilots are gathered in $\mathbf{Y} \in \mathbb{C}^{N_{\mathbb{S}}N_p \times N_{\mathbb{D}}}$ as

$$\mathbf{Y} = \begin{pmatrix} \mathbf{y}_{00} & \mathbf{y}_{01} & \cdots & \mathbf{y}_{0(N_{\mathbb{D}}-1)} \\ \mathbf{y}_{10} & \mathbf{y}_{11} & \cdots & \mathbf{y}_{1(N_{\mathbb{D}}-1)} \\ \vdots & \vdots & \ddots & \vdots \\ \mathbf{y}_{(N_{\mathbb{S}}-1)0} & \mathbf{y}_{(N_{\mathbb{S}}-1)1} & \cdots & \mathbf{y}_{(N_{\mathbb{S}}-1)(N_{\mathbb{D}}-1)} \end{pmatrix}. \tag{8}$$

Using the introduced matrices, one can utilize the proposed FBP algorithm to estimate MIMO-OFDM relay system channels. One of the main advantages of the proposed approach is the advantage of the measurement matrix in designing optimal pilot subcarriers to improve the channel estimation accuracy which will be discussed in the following sections.

5 Pilot allocation for compressed channel estimation

In FBP, jointly sparse channels are estimated altogether. The space existing between adjacent antennas in MIMO nodes is close to each other where the sparsity pattern between transmit-receive pairs is the same. Moreover, the channel coefficients in each of the non-zero paths are not the same and are rayleigh random variable since they are a consequence of the number of normal distributed paths. Utilizing the FBP, the jointly sparse channels could be estimated altogether. The measurement matrix which is represented in (7) could be generated using random pilot subcarriers and optimized pilot subcarriers. In order to improve the accuracy of estimated channels, it is mandatory to select pilot subcarriers to optimize the estimation metric. In CS, RIP is used as the key metric in designing appropriate measurement matrices. But, there is no polynomial time approach to calculate RIP; thus, we have used mutual coherence to design optimal measurement matrix.

Mutual coherence is defined as

$$\mu\{\mathbf{\Phi}\} = \max_{\lambda_i, \lambda_j; \lambda_i \neq \lambda_j} \left| \sum_{l=0}^{2L-2} e^{-j2\pi(\lambda_i - \lambda_j)l/N} \right| \tag{9}$$

where λ_i and λ_j are pilot subcarriers among N available subcarriers. Consequently, mutual coherence is defined as maximum off-diagonal entries of Gram matrix $\boldsymbol{G}\{\mathbf{\Phi}\} = \mathbf{\Phi}^H\mathbf{\Phi}$ if $\mathbf{\Phi}$ is orthonormal [13, 22]. Accordingly, $\mu\{\mathbf{\Phi}\}$ is related to the positions of pilots λ_i and λ_j. Since, they are positions, the problem can be deduced which is defined as

$$\begin{aligned} \min_{\lambda_i, \lambda_j} \quad & \mu\{\mathbf{\Phi}\} \\ & \lambda_i \neq \lambda_j \\ & \lambda_i, \lambda_j \in \mathcal{A} \end{aligned} \tag{10}$$

where $\mathcal{A}$ is the set of all the available subcarriers. Obviously, the selection of subcarriers is a combinatorial optimization. In order to generate optimal pilot sequences, we have used probability-based approaches which will be discussed in the following sections. Here, we consider the measurement matrix Φ.

Theorem 1 *Assuming* $\mathbf{\Phi}$ *as an orthonormal measurement matrix in Eq. (7), then* $\mu\{\mathbf{\Phi}\}$ *could be defined by*

$$\max_{i\in\{1,2,\ldots,N_{\mathbb{S}}\}} \mu\{\mathbf{\Phi}_i, \mathbf{\Phi}_i\} \tag{11}$$

Considering Gram matrix

$$\mathbf{G}\{\mathbf{\Phi}\} = \begin{pmatrix} \mathbf{\Phi}_1^H\mathbf{\Phi}_1 & 0 & \cdots & 0 \\ 0 & \mathbf{\Phi}_2^H\mathbf{\Phi}_2 & \cdots & 0 \\ \cdots & \cdots & \ddots & \cdots \\ 0 & 0 & \cdots & \mathbf{\Phi}_{N_{\mathbb{S}}}^H\mathbf{\Phi}_{N_{\mathbb{S}}} \end{pmatrix} \tag{12}$$

where $\mathbf{G}\{\mathbf{\Phi}\} \in \mathbb{C}^{N_{\mathbb{S}}(2L-1)\times N_{\mathbb{S}}(2L-1)}$, we define two concepts. The first one is self-coherence and the other one is the cross-coherence. $\mathbf{\Phi}$ is consist of different submatrices from different antennas called $\mathbf{\Phi}_i$ for $i = 0, 1, \ldots, N_{\mathbb{S}}-1$. We define self-coherence as $\mu\{\mathbf{\Phi}_i, \mathbf{\Phi}_i\}$ and cross-coherence as $\mu\{\mathbf{\Phi}_i, \mathbf{\Phi}_j\}$ where $i \neq j$ where

$$\mu\{\mathbf{\Phi}_i, \mathbf{\Phi}_i\} = \max_{k\neq t} \phi_{ik}^H\phi_{it} \tag{13}$$

$$\mu\{\mathbf{\Phi}_i, \mathbf{\Phi}_j\} = \max_{k\neq t} \phi_{ik}^H\phi_{jt} \tag{14}$$

and ϕ_{ik} is the kth column of $\mathbf{\Phi}_i$. Obviously, according to (12), the mutual coherence could be defined as the maximum of self-coherence between all the antennas. ■

According to the above, among other jointly sparse estimation, our proposed method and formulation leads to self-coherence while others lead to self-coherence and cross-coherence [11, 13]. Consequently, it could decrease the number of computations since the number of matrix multiplications is $N_{\mathbb{S}}$, while in other approaches the number of matrix multiplications is $N_{\mathbb{S}}N_{\mathbb{D}}$. Hence, the number of matrix multiplications is outstandingly minimized.

In order to omit the interference of antennas on each other, we have considered orthogonal pilot sequences which are demonstrated in [13].

6 Proposed pilot allocation algorithms

Here, we will demonstrate two numerical algorithms to optimize the mutual coherence of measurement matrix and defining optimized pilot sequences. The optimization is performed over the search space $\mathcal{S} \in \binom{N}{N_p}$. Hence, the exhaustive search is intractable and computationally inefficient by increased number of N and N_p. Consequently, the combinatorial optimization problem in Eq. (10) is solved using pdf sampling by the proposed algorithms. The sampling is continued until convergence of the pdf. The steady state is achieved while the pdf consists of zero or $1/N_pN_{\mathbb{S}}$ values. The pdf in each of the iteration is updated utilizing appropriate population in each generation. In order to detect the appropriate individuals, we use the fitness function

$$\max_{i\in\{1,2,\ldots,N_{\mathbb{S}}\}} \mu\{\mathbf{\Phi}_i, \mathbf{\Phi}_i\} \tag{15}$$

To decrease the computational complexity in fitness function, we used self-coherence of different measurement matrices. Evidently, there are lots of zero elements in matrix $\mathbf{\Phi}$ which are unnecessary to be multiplied. Moreover, the cross-coherence is not used in mutual coherence. Since, in Eq. (12) only self-coherence is included. Hence, we used self-coherence in Eq. (13) to optimize the computations in fitness function evaluation. In each generation, the updated pdf is utilized to generate the new individuals.

6.1 Sequential cross entropy self-coherence (SCE-SC) pilot allocation approach

In this approach, there are some definitions which are very important to understand the approach. *Generation* ($\mathcal{G}$) is the set of pilot sequences which are sampled from the pdf. Each generation is constructed from (I) *individuals* which are referred to each of the sample pilot sequence. Furthermore, each individual consists of N *elements*. The probability of each element is represented in a *pdf* called $\mathcal{P}$ which demonstrates the probability of the element to be selected as an appropriate selection for the pilot sequence. Moreover, *elite individuals* in each generation are J individuals with best fitness function. Hence, *fitness function* is the metric for detecting elite individuals in each generation which is demonstrated in Eq. (15). The initial state of the pdf $\mathcal{P}$ is the uniform pdf with elements equal to $1/N$. Gradually, during different generations, the pdf is updated using J elite individuals and its elements converge to the steady state. Since the generation of individuals consists of sampling from the probability function, this method is more robust than other mutation-based approaches in trapping to local minima.

The step-by-step representation of the SCE-SC algorithm is demonstrated in Algorithm 2. Considering $p_g(\mathcal{G})$ as the pdf vector of gth iteration, we represent the pdf vector by

$$p_g(\mathcal{G}) = [p_g(\mathcal{G}_1), p_g(\mathcal{G}_2), \ldots, p_g(\mathcal{G}_N)]^T \tag{16}$$

where $p_g(\mathcal{G}_i)$ refers to the probability of obtaining a value of 1 in the ith element of $\mathcal{G}$. In each generation, the pdf vector is utilized to generate I individuals. Among these individuals, J elite individuals are selected. Then, pdf is

Algorithm 2 Sequential cross entropy self-coherence (SCE-SC) pilot allocation

Initialize the remaining subcarrier set $\mathcal{A} = [1 : N]$
for $n = 1$ **to** N_s **do**
  Initialize the population $\mathcal{G}^0$ using I random individuals with length N_p from remaining subcarrier set $\mathcal{A}$
  while stop criteria not met **do**
    Sort the individuals in $\mathcal{G}^g$ in ascending order according to corresponding self-coherence
    Select J of the individuals in $\mathcal{G}^g$ with the best fitness function and establish subset $\mathcal{G}^g_{j:I}$ for $j = 1, 2, \ldots, J$
    Construct pdf $p_{g+1}(\mathcal{G})$ using $\mathcal{G}^g_{j:I}$ and Eq. (17)
    Sample the probability function and generate the new individual without repeat and generate the new generation $\mathcal{G}_{g+1}$
  end while
  Update the remaining subcarrier set $\mathcal{A}$
end for

updated using these elite individuals according to

$$p_{g+1}(\mathcal{G}) = (1 - \tau)p_g(\mathcal{G}) + \frac{\tau}{J}\sum_{j=1}^{J} \mathcal{G}^g_{j:M} \tag{17}$$

where $\mathcal{G}^g_{j:I}$ is the jth elite individual in gth generation and τ is the regulation parameter of the algorithm which demonstrate the dependency of the update procedure to the current generation. Furthermore, τ is non-zero positive lower than 1. According to the previous discussions, the algorithm is performed till the steady state is met. In the steady state, the pdf vector is only included for zero or $1/N_{\mathbb{S}}N_p$ elements.

6.2 Parallel cross entropy self-coherence (PCE-SC) pilot allocation algorithm

In the SCE-SC algorithm, the initial condition is set just one time. Consequently, iteratively the pdf is updated based on the initial generation. Since the initial generation is randomly selected, it would be helpful to revise the algorithm by generating some generations as the initial generation. Consequently, by selecting different initial conditions, we are pursuing the optimal point in some parallel avenues. Parallelism is helpful as it would decrease the probability of local minima trapping and the resulting sun-optimal point would be more robust and closer to the optimal point. Hence, in PCE-SC algorithm which is represented stage-by-stage in Algorithm 3, α initial generation from the search space $\mathcal{S}$ are considered and iteratively the pdf of these generations are converged to the steady state, simultaneously. Considering multiple pdfs at the same time will increase the computation burden which

Algorithm 3 Parallel cross entropy self-coherence (PCE-SC) pilot allocation

Initialize the remaining subcarrier set $\mathcal{A} = [1 : N]$
for $n = 1$ **to** $N_{\mathbb{S}}$ **do**
  Initialize the population $\mathcal{G}^0$ using I random individuals with length N_p from remaining subcarrier set $\mathcal{A}$
  while stop criteria not met **do**
    for $\alpha = 1$ **to** M **do**
      Sort the individuals in $\mathcal{G}^g_\alpha$ in ascending order
      Select J of the individuals in $\mathcal{G}^g_\alpha$ with the best objective value and establish subset $\mathcal{G}^g_{\alpha,j:I}$ for $j = 1, 2, \ldots, J$
      Construct pdf $p^\alpha_g(\mathcal{G})$ using $\mathcal{G}^g_{\alpha,j:I}$ and Eq. (17)
      Sample the probability function and generate the new individual without repeat and generate the new generation $\mathcal{G}^\alpha_{g+1}$
    end for
  end while
  Update the remaining subcarrier set $\mathcal{A}$
end for

can be handled by the parallel processing units. However, the proposed method will decrease the local minima trapping more than the SCE-SC approach and increase the accuracy of the resulted steady state point. Consequently, utilizing PCE-SC will increase two substantial perspective of pilot allocation algorithm in channel estimation.

7 Numerical results

Here, we demonstrate the performance of the proposed algorithms in channel estimation and pilot allocations called FBP, SCE-SC, and PCE-SC by simulation results. The simulation parameters are listed in Table 1. Sparse Reyleigh channel is modeled using finite impulse response (FIR) filters where non-zero taps (significant taps) are independently and identically distributed (i.i.d.) utilizing zero mean and unit variance complex Gaussian variables. Moreover, Monte Carlo simulations are performed over 1000 independent runs and the results are averaged over them.

Table 1 Simulation parameters

Simulation parameter	Value
Modulation	16-QAM
Number of subcarriers (N)	512
Number of transmit antenna ($N_{\mathbb{S}}$)	4
Number of receive antenna ($N_{\mathbb{D}}$)	4
Number of relay antenna ($N_{\mathbb{R}}$)	4
Channel taps (L)	30
Channel significant taps (K)	3

Here, we will consider the applicability of the proposed algorithm and the obtained pilot sequences utilizing two main characteristics of channel estimator called NMSE and BER. NMSE is calculated using

$$\mathrm{NMSE} = \frac{1}{N_{MC}} \sum_{i=1}^{N_{MC}} \frac{\|\mathbf{h}_i - \hat{\mathbf{h}}_i\|_2^2}{\|\mathbf{h}_i\|_2^2} \tag{18}$$

where $\mathbf{h}$ is the complete channel vector and $\hat{\mathbf{h}}$ is its estimation. Moreover, N_{MC} is the Monte Carlo iterations which is 1000. BER is evaluated using Monte Carlo simulation using N_{MC} individual simulation according to the following equation.

$$\mathrm{BER} = \sum_{i=1}^{N_{MC}} \frac{N_b(i)}{N_t(i)} \tag{19}$$

where $N_b(i)$ is the number of the correct received bits in ith Monte Carlo iteration and $N_t(i)$ is the number of transmitted bits in each Monte Carlo iteration.

7.1 Comparison of channel estimation performance

The results of the proposed channel estimation method are represented in Fig. 2 considering MSE and BER, respectively. For comparison, we have utilized BStOMP algorithm developed in [23] and B-vCoSaMP developed in [13]. Here, in Fig. 3a, from the NMSE point of view, the

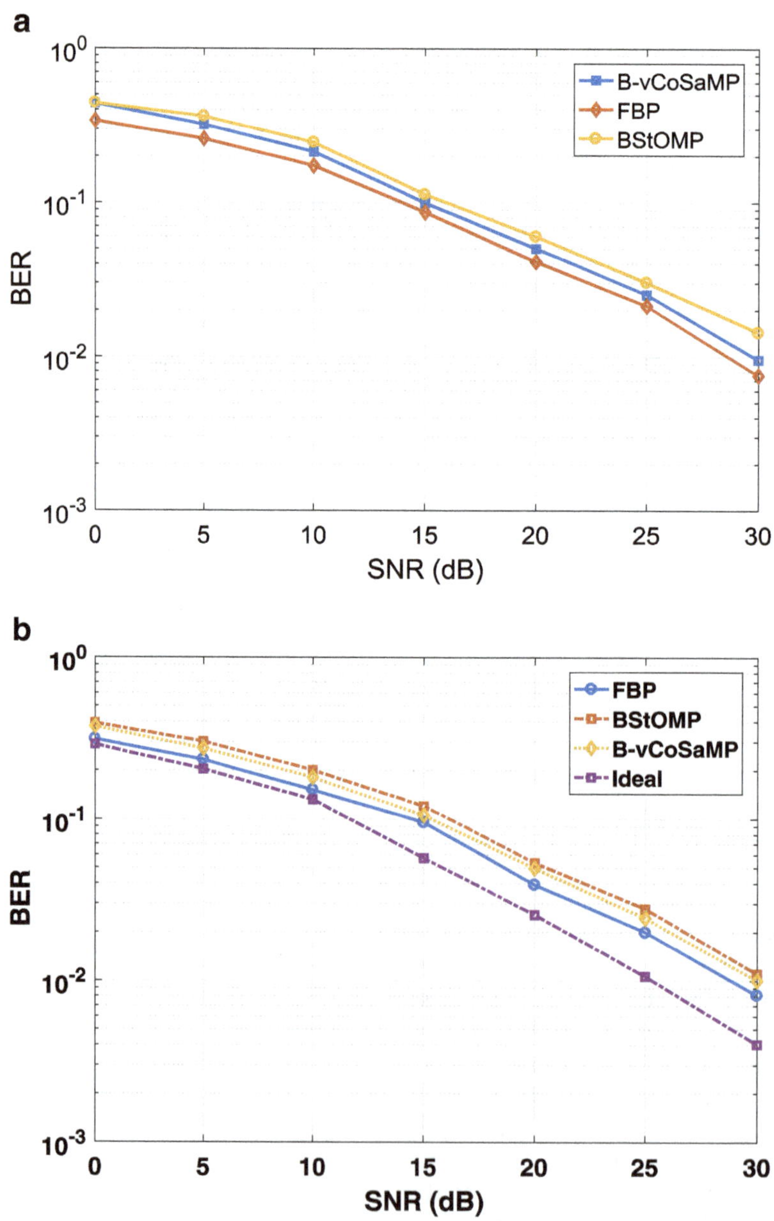

Fig. 2 Comparison of the proposed FBP-based method, B-vCoSaMP, and BStOMP: **a** MSE and **b** BER

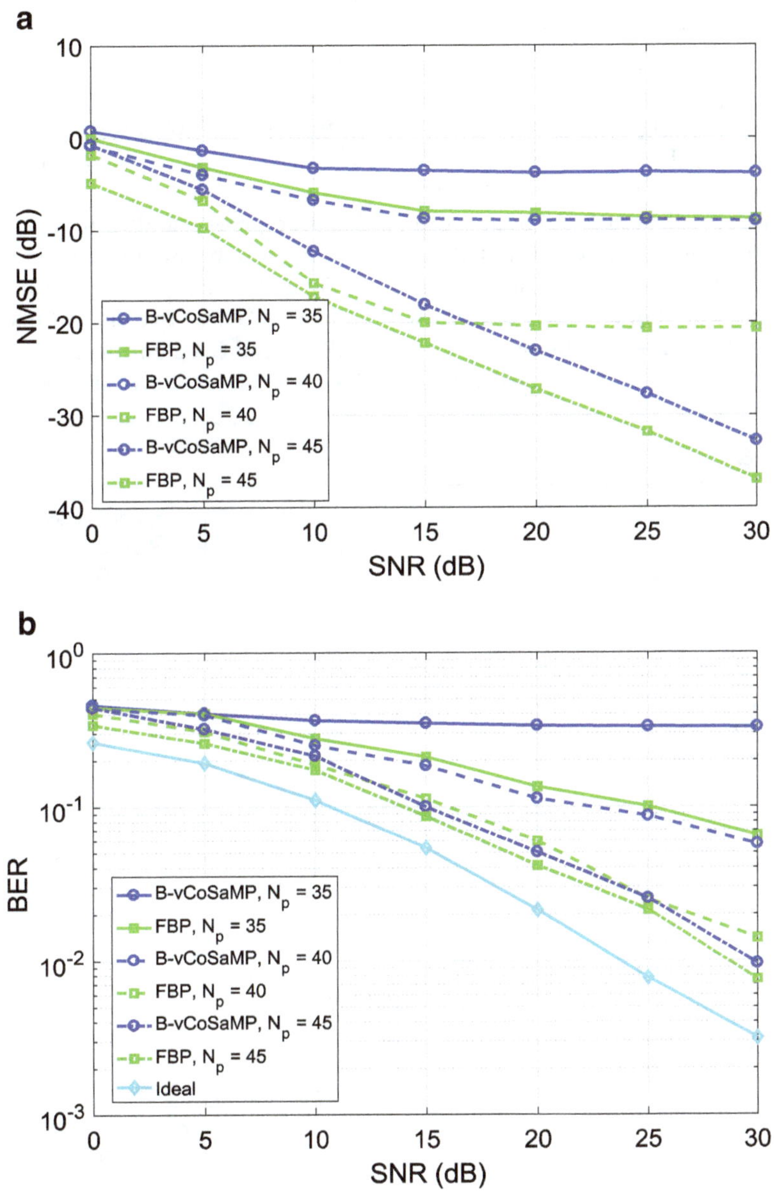

Fig. 3 The effect of pilot numbers in channel estimation: **a** MSE and **b** BER

proposed approach is more efficient than others. Moreover, the comparison of BER is demonstrated and, as it is shown, the BER of the proposed method is superior almost 2 dB better than BStOMP approach and 1.8 dB better than B-vCoSaMP. Moreover, using only $N_p = 45$ subcarriers as pilot, we are almost 3 dB away from the ideal case. This means that more than 65% spectral efficiency is maintained.

7.2 Comparison of the number of pilots

In order to compare the effect of the number of pilots, we have considered the proposed method and B-vCoSaMP developed in [13], since these two methods where performed better than BStOMP. Moreover, the number of pilots is changed and selected to be 35, 40, and 45. Illustratively, the NMSE and BER comparisons are represented in Fig. 3. Obviously, in case of 35 pilots, both methods are irreducible while the proposed FBP-based approach is irreducible in lower NMSE. By increasing the number of pilots to 40 and 45, the estimation accuracy is getting more applicable. Furthermore, in all the cases of pilots, the proposed FBP-based approach is superior than the B-vCoSaMP. This superiority is obvious in BER, too.

7.3 Comparison of proposed pilot allocation algorithms

Here, we consider pilot allocation algorithms proposed in this paper called PCE-SC and SCE-SC. In order to compare these proposed algorithms, we have considered NMSE and BER, separately. These figures are represented in Fig. 4. NMSE is compared in Fig. 4a and BER is compared in Fig. 4b. Obviously, using optimized pilot placement creates approximately 6 dB and 7 dB superiority than random placement utilizing SCE-based and PCE-based algorithms, respectively. Moreover, this superiority is encountered in BER represented in Fig. 4b almost 2 dB and 3 dB in higher SNRs. The BER gap between PCE-SC pilot placement using 45 pilots and the ideal case is approximately 1 dB which represented the superiority of the proposed approach in terms of bandwidth efficiency. Among 512 subcarriers, only 100 of them is utilized as the pilots and others are used as data subcarriers. Hence, more than 65% of the subcarriers are used as the data subcarriers which can be used by transmitting antennas.

7.4 Comparison of the proposed pilot allocation algorithm with the existing one

In Fig. 5, two different algorithms for pilot allocation are compared. These two algorithms are defined by proposed PCE-SC and GA algorithm developed in [18]. To estimate the channel, we have utilized the proposed FBP-based

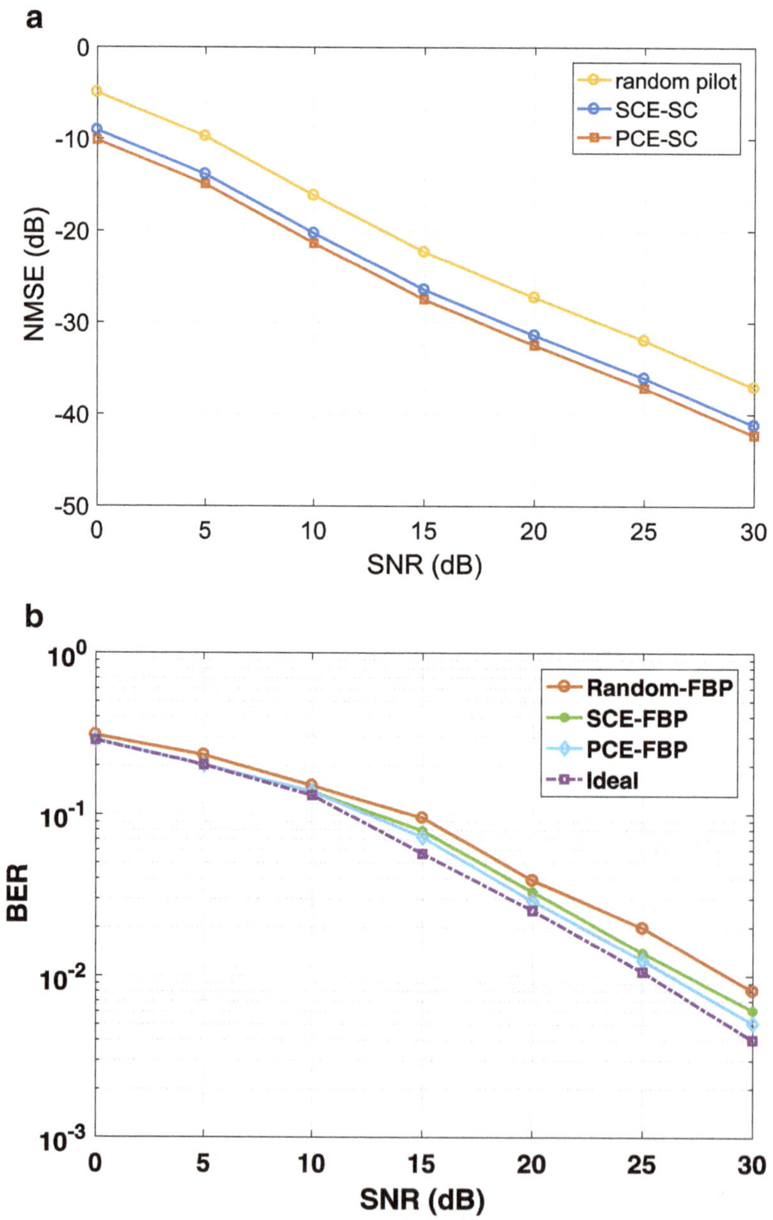

Fig. 4 Comparison of different proposed pilot allocation algorithms utilizing (**a**) NMSE and (**b**) BER performance metric

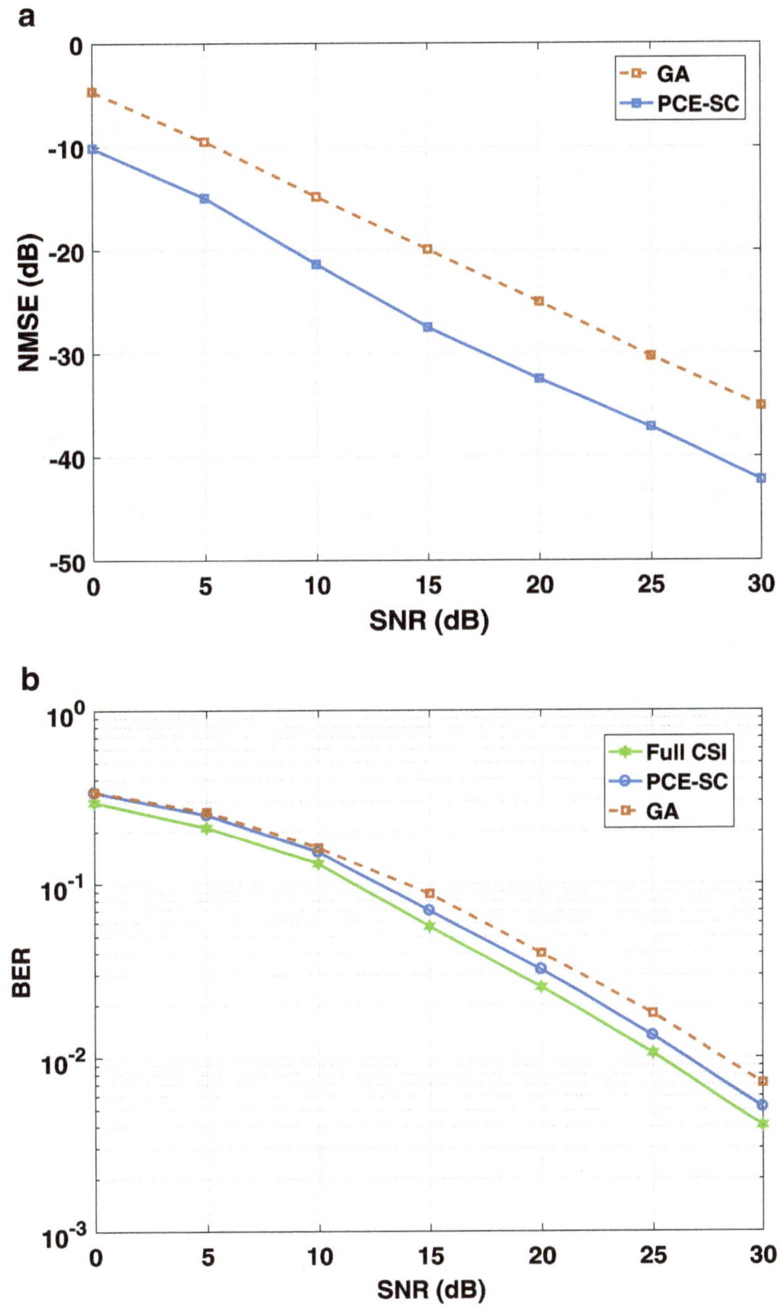

Fig. 5 Performance comparison of proposed method and GA-based pilot allocation algorithm: **a** NMSE performance metric and **b** BER performance metric

channel estimation approach. The number of the pilots for each of the transmitting antenna is $N_p = 45$. Illustratively, the results of the NMSE and BER are represented in Fig. 5. Obviously, NMSE of the proposed method is better than a GA-based approach approximately 3 dB. Moreover, concerning BER, this superiority is almost 1.5 dB. Evidently, by utilizing $N_p = 45$ optimized pilots on each antenna, the performance is almost 1 dB away from the optimal performance which could be compensated using other facilities. Actually, using optimized pilots will lead to 65% spectral efficiency which is one of the most critical characteristics of the system.

In Fig. 6, the proposed pilot allocation algorithms are compared with the optimized least squares (LS)-based channel estimation as the state-of-the-art approach in channel estimation of MIMO-OFDM systems in [24]. Obviously, the LS-based channel estimation is performed using 256 number of pilots where the spectral efficiency

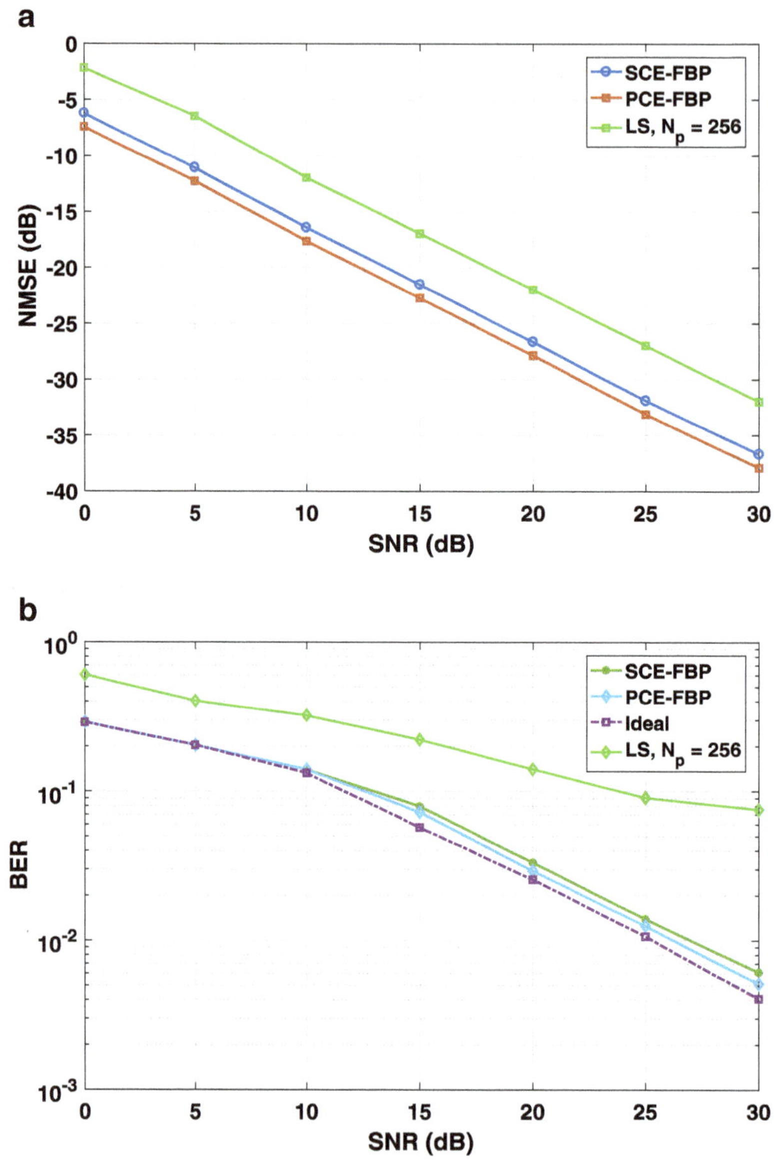

Fig. 6 Performance comparison of proposed method and optimized LS-based pilot allocation algorithm: (**a**) NMSE performance metric, (**b**) BER performance metric

is extremely decreased. The simulation parameters for the proposed pilot allocation is as before.

8 Results and discussion

In this paper, channel estimation of AF MIMO relay is considered and CS-based approaches is utilized due to their spectral efficiency and accuracy improvement. Here, we have proposed FBP-based channel estimation algorithm for forward selection and backward elimination. The proposed method benefits from backward elimination to improve the accuracy of estimation. Moreover, the measurement matrix is introduced and its Gram matrix is developed to minimize the mutual coherence. As discussed, mutual coherence is related to the self-coherence of sub-matrices related to each of the antennas. Hence, two pilot allocation algorithms are proposed based on the cross entropy where the number of multiplications are decreased rather than other approaches since they utilize cross-coherence and self-coherence, simultaneously. Accordingly, PCE-SC and SCE-SC algorithms are proposed for pilot generation and channel estimation accuracy improvement. They are compared with other approaches and their superiority validation is represented using simulation results.

Abbreviations

BER: Bit error rate; B-vCoSaMP: Block-verified compressive sampling; CS : Compressed sensing; CSI: Channel state information; DCS: Distributed compressed sensing; EDA: Estimating of distribution algorithm; FBP: Forward backward prediction; MIMO: Multiple-input multiple-output; NMSE: Normalized mean square error; OFDM: Orthogonal frequency division multiplexing; OMP: Orthogonal matching pursuit; PCE-SC: Parallel cross entropy self-coherence; RIP: Restricted isometry property; SCE-SC: Sequential cross entropy self-coherence

Funding
There is no source of funding for this paper.

Authors' contributions
Both authors carried out the mathematical proofs and developed the algorithms. AA participated in simulation part. Both authors read and approved the final manuscript.

Competing interests
This paper is written in order to fulfill the Ph.D. of telecommunication engineering and there are no other competing interests.

References
1. D. Hu, X. Wang, L. He, A new sparse channel estimation and tracking method for time-varying ofdm systems. IEEE Trans. Veh. Technol. **62**(9), 4648–4653 (2013)
2. V. R. Asghari, M. Ardebilipour, Spread spectrum code estimation by genetic algorithm. Int. J. Signal Proc. **1**(4), 301–304 (2004)
3. M. A. Mohammadi, M. Ardabilipour, B. Moussakhani, Z. Mobini, in *2008 5th IFIP International Conference on Wireless and Optical Communications Networks (WOCN'08)*. Performance comparison of RLS and LMS channel estimation techniques with optimum training sequences for MIMO-OFDM systems (IEEE, 2008), pp. 1–5
4. C. Qi, G. Yue, L. Wu, N. Arumugam, Pilot design for sparse channel estimation in OFDM-based cognitive radio systems. IEEE Trans. Veh. Technol. **63**(2), 982–987 (2014)
5. F. Azami, S. M. Safavi Hemami, A. Akbarpour-Kasgari, Joint power allocation and beamforming in amplify-and-forward relay networks under per-node power constraint. Wirel. Commun. Mob. Comput. **2017**(5681236), 9 (2017)
6. M. F. Duarte, Y. Eldar, Structured compressed sensing: From theory to applications. IEEE Tran. Signal Process. **59**(9), 4053–4085 (2011)
7. Z. Gao, L. Dai, W. Dai, S. B, Z. Wang, Structured compressive sensing-based spatio-temporal joint channel estimation for FDD massive MIMO. IEEE Trans. Commun. **64**(2), 601–617 (2016)
8. S. Sarvotham, D. Baron, M. Wakin, M. F. Duarte, R. G. Baraniuk, in *Asilomar conference on signals, systems, and computers*. Distributed compressed sensing of jointly sparse signals, (2005), pp. 1537–1541
9. A. Akbarpour-Kasgari, M. Ardebilipour, in *2017 Iranian Conference on Electrical Engineering (ICEE)*. MIMO relay compressed channel estimation using optimized pilot placements (IEEE, 2017), pp. 2028–2031
10. P. Cheng, Z. Chen, Y. Rui, Y. J. Guo, L. Gui, M. Tao, Q. T. Zhang, Channel estimation for OFDM systems over doubly selective channels: A distributed compressive sensing based approach. IEEE Trans. Commun. **61**(10), 4173–4185 (2013)
11. L. Xu, K. Niu, Z. He, W. Xu, Z. Zheng, in *2013 9th International Conference on Information, Communications & Signal Processing*. MIMO channel estimation based on distributed compressed sensing for LTE-advanced (IEEE, 2013), pp. 1–5
12. R. X, V. K. N. Lau, Distributed compressive CSIT estimation and feedback for FDD multi-user massive MIMO systems. IEEE Trans. Signal Process. **62**(12), 3261–3271 (2014)
13. A. Akbarpour-Kasgari, M. Ardebilipour, Probability-based pilot allocation for MIMO relay distributed compressed sensing based channel estimation. Eurasip J. on Advances in Signal Process. **2018**(1), 18 (2018)
14. Y. Han, J. Lee, D. J. Love, Compressed sensing-aided downlink channel training for FDD massive MIMO systems. IEEE Trans. Commun. **65**(7), 2852–2862 (2017)
15. L. Chen, A. Liu, X. Yuan, Structured turbo compressed sensing for massive MIMO channel estimation using a Markov prior. IEEE Trans. Veh. Technol. **67**(5), 4635–4639 (2018)
16. J. Tropp, Greed is good: Algorithmic results for sparse approximation. IEEE Trans. Inf. Theory. **50**(10), 2231–2242 (2004)
17. X. He, R. Song, W. P. Zhu, Pilot allocation for sparse channel estimation in MIMO-OFDM systems. IEEE Trans. Circuits Syst. II, Exp. Briefs. **60**(9), 612–616 (2013)
18. X. He, R. Song, W. P. Zhu, Pilot allocation for distributed-compressed-sensing-based sparse channel estimation in MIMO-OFDM systems. IEEE Trans. Veh. Technol. **65**(5), 2990–3004 (2016)
19. C. Qi, G. Yue, L. Wu, Y. Huang, A. Nallathan, Pilot design schemes for sparse channel estimation in OFDM systems. IEEE Trans. Veh. Technol. **64**(4), 1493–1505 (2015)
20. A. Akbarpour-Kasgari, M. Ardebilipour, F. Afghah, in *2015 49th Annual Conference on Information Sciences and Systems (CISS)*. MIMO-OFDM non-regenerative relay channel estimation using compressed sensing, (2015), pp. 1–5. IEEE
21. T. Zhang, Adaptive forward-backward greedy algorithm for learning sparse representations. IEEE Trans. Inf. Theory. **57**(7), 4689–4708 (2011)
22. J. M. Duarte-Carvajalino, G. Sapiro, Learning to sense sparse signals: Simultaneous sensing matrix and sparsifying dictionary optimization. IEEE Trans. Image Proc. **18**(7), 1395–1408 (2009)
23. D. Lee, MIMO-OFDM channel estimation via block stagewise orthogonal matching pursuit. IEEE Commun. Lett. **20**(10), 2115–2118 (2016)
24. I. Barhumi, G. Leus, M. Moonen, Optimal training design for MIMO OFDM systems in mobile wireless channels. IEEE Trans. Signal proc. **51**(6), 1615–1624 (2003)

9

Channel estimation for spatially/temporally correlated massive MIMO systems with one-bit ADCs

Hwanjin Kim and Junil Choi*

Abstract

This paper considers the channel estimation problem for massive multiple-input multiple-output (MIMO) systems that use one-bit analog-to-digital converters (ADCs). Previous channel estimation techniques for massive MIMO using one-bit ADCs are all based on single-shot estimation without exploiting the inherent temporal correlation in wireless channels. In this paper, we propose an adaptive channel estimation technique taking the spatial and temporal correlations into account for massive MIMO with one-bit ADCs. We first use the Bussgang decomposition to linearize the one-bit quantized received signals. Then, we adopt the Kalman filter to estimate the spatially and temporally correlated channels. Since the quantization noise is not Gaussian, we assume the effective noise as a Gaussian noise with the same statistics to apply the Kalman filtering. We also implement the truncated polynomial expansion-based low-complexity channel estimator with negligible performance loss. Numerical results reveal that the proposed channel estimators can improve the estimation accuracy significantly by using the spatial and temporal correlations of channels.

Keywords: Massive MIMO, Channel estimation, One-bit ADC, Kalman filter, Spatial and temporal correlations, Truncated polynomial expansion

1 Introduction

Massive multiple-input multiple-output (MIMO) systems are one of the promising techniques for next-generation wireless communication systems [2–5]. By using a large number of antennas at base stations (BSs), it is possible to support multiple users simultaneously to boost network throughput and improve the energy-efficiency by beamforming techniques [4]. Due to the large number of antennas at the BS, high implementation cost and power consumption could be major problems for implementing massive MIMO in practice.

Using low-resolution analog-to-digital converters (ADCs) is an effective way of mitigating the power consumption problem in massive MIMO systems because the ADC power consumption exponentially decreases as its resolution level [6]. However, symbol detection and channel estimation in massive MIMO systems with low-resolution ADCs become difficult tasks because the quantization process using low-resolution ADCs becomes highly non-linear. Recent works have revealed that it is possible to implement practical symbol detectors and channel estimators for massive MIMO even with low-resolution ADCs. For the symbol detection, a massive spatial modulation MIMO approach based on sum-product-algorithm was developed in [7], a convex optimization-based multiuser detection for massive MIMO with low-resolution ADC was considered in [8], a mixed-ADC massive MIMO detector was proposed in [9], and a blind detection technique was developed by exploiting supervised learning [10]. Also, an iterative detection and decoding scheme based on the message-passing algorithm and low-resolution aware (LRA) minimum mean square error (MMSE) receive filter was presented in [11], a low-complexity maximum likelihood detection (MLD) algorithm called one-bit-sphere-decoding was developed in [12], and a successive cancelation soft-output detector by exploiting a previous decoded message was proposed in [13].

*Correspondence: junil@kaist.ac.kr
This paper was presented in part at the IEEE Global Communications Conference, 2018 [1].
School of Electrical Engineering, Korea Advanced Institute of Science and Technology, Daejeon, Korea

For the channel estimation, a near-maximum likelihood channel estimator based on the convex optimization was developed in [14], and a Bayes-optimal joint channel and data estimator was proposed in [15]. To reduce the complexity, the generalized approximate message-passing algorithm was applied in [16], and the hybrid architectures were considered in [17]. Moreover, an oversampling based LRA-MMSE channel estimator that exploits the correlation of filtered noise for a given channel was proposed in [18]. However, up to the authors' knowledge, the previous channel estimators with low-resolution ADCs have not considered the temporal correlation in channels, which is inherent in communication channels.

In this paper, we develop a channel estimator taking both spatial and temporal correlations into consideration for massive MIMO systems with one-bit ADCs. We first discuss how to estimate the spatial correlation matrix for the channel estimation. Then, we reformulate the non-linear one-bit quantizer to the linear operator based on the Bussgang decomposition[19]. To exploit both the spatial and temporal correlations, we implement the Kalman filter-based (KFB) estimator [20] assuming the statistically equivalent quantization noise after the Bussgang decomposition follows a Gaussian distribution with the same mean and covariance matrix. The numerical results demonstrate that the normalized mean square error (NMSE) of the proposed KFB estimator decreases as the time slot increases. Moreover, as channels are more correlated with space and time, it is possible to track the channels more accurately. To reduce the complexity of KFB estimator, which comes from the large size matrix inversion, we also exploit a truncated polynomial expansion (TPE) approximation for the matrix inversion in the Kalman gain matrix. We analytically show that, with some moderate assumptions, the NMSE of the TPE-based estimator also keeps decreasing with the time slots. The numerical results show that the low-complexity TPE-based estimator gives approximately the same performance as the KFB estimator even with low approximation orders.

The rest of the paper is organized as follows. In Section 2, we explain a system model with one-bit ADCs. In Section 3, we first discuss how to estimate the spatial correlation matrix. Then, we review the single-shot channel estimator based on the Bussgang decomposition [21]. After, we explain our proposed successive channel estimator based on the Bussgang decomposition and the Kalman filter. We also propose the low-complexity TPE-based channel estimator and analyze the complexities of competing estimators. After explaining the data transmission with one-bit ADCs in Section 4, we evaluate numerical results in Section 5. Finally, we conclude the paper in Section 6.

Notation: Lower and upper boldface letters represent column vectors and matrices. $\mathbf{A}^{\mathrm{T}}$, $\mathbf{A}^*$, $\mathbf{A}^{\mathrm{H}}$, and $\mathbf{A}^\dagger$ denote the transpose, conjugate, conjugate transpose, and pseudo inverse of the matrix $\mathbf{A}$. $\mathbb{E}\{\cdot\}$ represents the expectation, and $\Re\{\cdot\}$, $\Im\{\cdot\}$ denote the real part and imaginary part of the variable. $\mathbf{0}_m$ is used for the $m \times 1$ all zero vector, and $\mathbf{I}_m$ denotes the $m \times m$ identity matrix. $\otimes$ denotes the Kronecker product. diag$(\cdot)$ returns the diagonal matrix. vec$(\cdot)$ denotes the column-wise vectorization. $\mathbb{C}^{m\times n}$ and $\mathbb{R}^{m\times n}$ represent the set of all $m \times n$ complex and real matrices. $|\cdot|$ denotes the amplitude of the scalar, and $\|\cdot\|$ represents the ℓ_2-norm of the vector. $\mathcal{CN}(m, \sigma^2)$ denotes the complex normal distribution with mean m and variance σ^2. tr$(\cdot)$ represents the trace operator. $\mathcal{O}$ denotes the Big-O notation.

2 System model

As in Fig. 1, we assume a single-cell massive MIMO system with M BS antennas and K single-antenna users with $M \gg K$. Each BS antenna is connected to two one-bit ADCs, one for the in-phase component and the other for the quadrature component of received signals. We consider the block-fading channel with the coherence time of T. The received signal at the i-th fading block is given by

$$\mathbf{y}_i = \sqrt{\rho}\mathbf{H}_i\mathbf{s}_i + \mathbf{n}_i, \tag{1}$$

where ρ is the signal-to-noise ratio (SNR), $\mathbf{H}_i = [\mathbf{h}_{i,1}, \mathbf{h}_{i,2}, ..., \mathbf{h}_{i,K}] \in \mathbb{C}^{M\times K}$ is the channel matrix, $\mathbf{h}_{i,k}$ is the channel between the BS and the k-th user in the i-th fading block, $\mathbf{s}_i$ is the transmit signal, and $\mathbf{n}_i \sim \mathcal{CN}(\mathbf{0}_M, \mathbf{I}_M)$ is the complex Gaussian noise. We consider the spatially and temporally correlated channels by assuming $\mathbf{h}_{i,k}$ follows the first-order Gauss-Markov process,

$$\begin{aligned}\mathbf{h}_{0,k} &= \mathbf{R}_k^{\frac{1}{2}}\mathbf{g}_{0,k},\\ \mathbf{h}_{i,k} &= \eta_k\mathbf{h}_{i-1,k} + \sqrt{1-\eta_k^2}\mathbf{R}_k^{\frac{1}{2}}\mathbf{g}_{i,k}, \quad i \geq 1,\end{aligned} \tag{2}$$

where η_k is the temporal correlation coefficient, $\mathbf{R}_k = \mathbb{E}\left\{\mathbf{h}_{i,k}\mathbf{h}_{i,k}^{\mathrm{H}}\right\}$ is the spatial correlation matrix, and $\mathbf{g}_{i,k} \sim \mathcal{CN}(\mathbf{0}_M, \mathbf{I}_M)$ is the innovation process of the k-th user in the i-th fading block. Note that η_k and $\mathbf{R}_k$ do not have the time index i since both are long-term statistics that are static for multiple coherence blocks.

Although other models are also possible, to have concrete analyses, we adopt the exponential model for the spatial correlation matrix $\mathbf{R}_k$,

$$\mathbf{R}_k = \begin{bmatrix} 1 & r_k & \cdots & r_k^{M-1} \\ r_k^* & 1 & \cdots & \vdots \\ \vdots & \vdots & \ddots & \vdots \\ r_k^{*(M-1)} & \cdots & \cdots & 1 \end{bmatrix}, \tag{3}$$

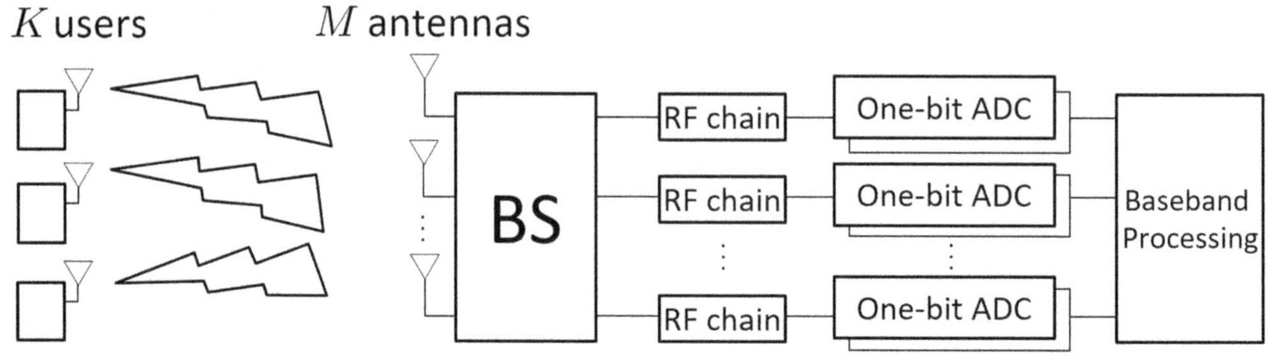

Fig. 1 Massive MIMO systems with M BS antennas and K single-antenna users. Each RF chain is equipped with two one-bit ADCs for the in-phase and quadrature component, respectively

where $r_k = re^{j\theta_k}$ satisfying $0 \leq r < 1$ and $0 \leq \theta_k < 2\pi$. We assume all users experience the same spatial correlation coefficient r since it is dominated by the BS antenna spacing while each user has an indifferent phase θ_k since it is more related to the user position [22].

The quantized signal by the one-bit ADCs is

$$\mathbf{r}_i = \mathcal{Q}(\mathbf{y}_i) = \mathcal{Q}(\sqrt{\rho}\mathbf{H}_i\mathbf{s}_i + \mathbf{n}_i), \tag{4}$$

where $\mathcal{Q}(\cdot)$ is the element-wise one-bit quantization operator, i.e., $\mathcal{Q}(\cdot) = \frac{1}{\sqrt{2}}(\text{sign}(\Re\{\cdot\}) + j\,\text{sign}(\Im\{\cdot\}))$.

3 Channel estimation with one-bit ADCs

In this section, we first discuss how to estimate the spatial correlation matrix. Then, we explain the Bussgang linear minimum mean square error (BLMMSE) estimator, which is the baseline of the proposed estimator. The BLMMSE estimator is a single-shot channel estimator based on the Bussgang decomposition without exploiting any temporal correlation [21]. Then, we propose the KFB estimator, which is a successive channel estimator, for massive MIMO with one-bit ADCs exploiting the temporal correlation. Also, we propose the low-complexity TPE-based estimator to reduce the complexity of the proposed KFB estimator.

3.1 Spatial correlation matrix estimation

In this subsection, we discuss how to estimate the spatial correlation matrix since all the channel estimators in this paper exploit the spatial correlation of channel. We omit the user index k since the BS can estimate the spatial correlation of each user separately.

When the BS does not have any prior channel information, it can use the least square (LS) estimate of the quantized signal $\mathbf{r}_i$, which is given by

$$\mathbf{h}_i^{\text{LS}} = \mathbf{\Phi}_i^{\dagger}\mathbf{r}_i, \tag{5}$$

where $\mathbf{\Phi}_i$ is the pilot matrix. The LS estimator for one-bit quantized signal performs well when the number of antennas at the BS is large, as shown in [23]. The BS then can obtain a sampled spatial correlation matrix as

$$\hat{\mathbf{R}} = \frac{1}{N_s}\sum_{n=1}^{N_s}\mathbf{h}_n^{\text{LS}}\left(\mathbf{h}_n^{\text{LS}}\right)^{\text{H}}, \tag{6}$$

where N_s is the number of samples. We evaluate the performance loss by using the sampled correlation matrix in Fig. 3 in Section 5. After this subsection, we assume that the true spatial correlation matrices and the temporal correlation coefficients of all users are known to the BS to derive analytical results.

3.2 BLMMSE estimator

In this subsection, we omit the time slot index i since the single-shot channel estimator does not use any temporal correlation. To estimate the channel at the BS, K users transmit the length τ pilot sequences to the BS,

$$\mathbf{Y} = \sqrt{\rho}\mathbf{H}\mathbf{\Phi}^{\text{T}} + \mathbf{N}, \tag{7}$$

where $\mathbf{Y} \in \mathbb{C}^{M\times\tau}$ is the received signal, $\mathbf{\Phi} \in \mathbb{C}^{\tau\times K}$ is the pilot matrix, and $\mathbf{N} = [\mathbf{n}_1, \mathbf{n}_2, ..., \mathbf{n}_\tau] \in \mathbb{C}^{M\times\tau}$ is the complex Gaussian noise. We assume that the pilot sequences are column-wise orthogonal, i.e., $\mathbf{\Phi}^{\text{T}}\mathbf{\Phi}^* = \tau\mathbf{I}_K$, and all the elements of the pilot matrix have the same magnitude. For the sake of simplicity, the receive signal is vectorized as

$$\text{vec}(\mathbf{Y}) = \underline{\mathbf{y}} = \bar{\mathbf{\Phi}}\underline{\mathbf{h}} + \underline{\mathbf{n}}, \tag{8}$$

where $\bar{\mathbf{\Phi}} = (\mathbf{\Phi} \otimes \sqrt{\rho}\mathbf{I}_M)$, $\underline{\mathbf{h}} = \text{vec}(\mathbf{H})$, and $\underline{\mathbf{n}} = \text{vec}(\mathbf{N})$. The quantized signal by one-bit ADCs is

$$\underline{\mathbf{r}} = \mathcal{Q}(\underline{\mathbf{y}}). \tag{9}$$

Assuming independent spatial correlations across the users, the aggregated spatial correlation matrix $\mathbf{R} = \mathbb{E}\{\underline{\mathbf{h}}\underline{\mathbf{h}}^{\text{H}}\}$ is given by

$$\mathbf{R} = \begin{bmatrix} \mathbf{R}_1 & \cdots & 0 & 0 \\ \vdots & \mathbf{R}_2 & \cdots & 0 \\ 0 & \vdots & \ddots & \vdots \\ 0 & 0 & \cdots & \mathbf{R}_K \end{bmatrix}. \tag{10}$$

The Bussgang decomposition of quantized signal is given by

$$\underline{\mathbf{r}} = \mathcal{Q}(\underline{\mathbf{y}}) = \mathbf{A}\underline{\mathbf{y}} + \mathbf{q}, \tag{11}$$

where $\mathbf{A}$ denotes the linear operator and $\mathbf{q}$ represents the statistically equivalent quantization noise. The linear operator $\mathbf{A}$ is obtained from [21],

$$\begin{aligned} \mathbf{A} &= \sqrt{\frac{2}{\pi}}\mathrm{diag}(\mathbf{C}_{\underline{\mathbf{y}}})^{-\frac{1}{2}} \\ &= \sqrt{\frac{2}{\pi}}\mathrm{diag}\left(\bar{\mathbf{\Phi}}\mathbf{R}\bar{\mathbf{\Phi}}^{\mathrm{H}} + \mathbf{I}_{M\tau}\right)^{-\frac{1}{2}} \\ &\overset{(a)}{=} \sqrt{\frac{2}{\pi}}\sqrt{\frac{1}{K\rho+1}}\mathbf{I}_{M\tau}, \end{aligned} \tag{12}$$

where $\mathbf{C}_{\underline{\mathbf{y}}}$ is the auto-covariance matrix of the received signal. In (12), (a) is derived in Appendix 1. Substituting (8) into (11), $\underline{\mathbf{r}}$ is represented as

$$\underline{\mathbf{r}} = \mathcal{Q}(\underline{\mathbf{y}}) = \tilde{\mathbf{\Phi}}\underline{\mathbf{h}} + \tilde{\underline{\mathbf{n}}}, \tag{13}$$

where $\tilde{\mathbf{\Phi}} = \mathbf{A}\bar{\mathbf{\Phi}} \in \mathbb{C}^{M\tau\times MK}$ and $\tilde{\underline{\mathbf{n}}} = \mathbf{A}\underline{\mathbf{n}} + \mathbf{q} \in \mathbb{C}^{M\tau\times 1}$.

After adopting the Bussgang decomposition, a linear MMSE estimator, which is denoted as the BLMMSE channel estimator [21], is given as

$$\hat{\underline{\mathbf{h}}}^{\mathrm{BLM}} = \mathbf{C}_{\underline{\mathbf{h}}}\tilde{\mathbf{\Phi}}^{\mathrm{H}}\mathbf{C}_{\underline{\mathbf{r}}}^{-1}\underline{\mathbf{r}}, \tag{14}$$

where $\mathbf{C}_{\underline{\mathbf{h}}}$ is the auto-covariance matrix of the channel $\underline{\mathbf{h}}$, and $\mathbf{C}_{\underline{\mathbf{r}}}$ is the auto-covariance matrix of the quantized signal $\underline{\mathbf{r}}$. In (14), $\mathbf{C}_{\underline{\mathbf{r}}}$ is obtained by the arcsin law [24],

$$\begin{aligned} \mathbf{C}_{\underline{\mathbf{r}}} = \frac{2}{\pi}\Big(&\arcsin\left(\Sigma_{\underline{\mathbf{y}}}^{-1/2}\Re\{\mathbf{C}_{\underline{\mathbf{y}}}\}\Sigma_{\underline{\mathbf{y}}}^{-1/2}\right) \\ &+ j\arcsin\left(\Sigma_{\underline{\mathbf{y}}}^{-1/2}\Im\{\mathbf{C}_{\underline{\mathbf{y}}}\}\Sigma_{\underline{\mathbf{y}}}^{-1/2}\right)\Big), \end{aligned} \tag{15}$$

where $\Sigma_{\underline{\mathbf{y}}} = \mathrm{diag}(\mathbf{C}_{\underline{\mathbf{y}}})$.

3.3 Proposed KFB estimator

Although effective, the BLMMSE estimator does not exploit any inherent temporal correlation in wireless channels. We now propose a simple, yet effective, channel estimator based on the Bussgang decomposition and the Kalman filtering. We recover the time slot index i to explicitly use the temporal correlation. We first reformulate the channel model in (2) by vectorization,

$$\begin{aligned} \underline{\mathbf{h}}_0 &= \mathbf{R}^{\frac{1}{2}}\underline{\mathbf{g}}_0, \\ \underline{\mathbf{h}}_i &= \boldsymbol{\eta}\underline{\mathbf{h}}_{i-1} + \boldsymbol{\zeta}\mathbf{R}^{\frac{1}{2}}\underline{\mathbf{g}}_i, \quad i \ge 1, \end{aligned} \tag{16}$$

where $\underline{\mathbf{g}}_i$ is the vectorized innovation process, which is expressed as

$$\underline{\mathbf{g}}_i = \left[\mathbf{g}_{i,1}^{\mathrm{T}}, \mathbf{g}_{i,2}^{\mathrm{T}}, ..., \mathbf{g}_{i,K}^{\mathrm{T}}\right]^{\mathrm{T}}, \quad i \ge 0. \tag{17}$$

The temporal correlation matrices $\boldsymbol{\eta}$ and $\boldsymbol{\zeta}$ in (16) are given by the Kronecker product,

$$\begin{aligned} \boldsymbol{\eta} &= \mathrm{diag}(\eta_1, \eta_2, ..., \eta_K) \otimes \mathbf{I}_M, \\ \boldsymbol{\zeta} &= \mathrm{diag}(\zeta_1, \zeta_2, ..., \zeta_K) \otimes \mathbf{I}_M, \end{aligned} \tag{18}$$

where η_k denotes the k-th user temporal correlation coefficient and $\zeta_k = \sqrt{1-\eta_k^2}$.

Following the same steps as in Section 3.2, the one-bit quantized signal can be represented using the Bussgang decomposition as

$$\underline{\mathbf{r}}_i = \mathcal{Q}(\underline{\mathbf{y}}_i), \tag{19}$$

$$= \mathbf{A}_i\underline{\mathbf{y}}_i + \mathbf{q}_i, \tag{20}$$

$$= \tilde{\mathbf{\Phi}}_i\underline{\mathbf{h}}_i + \tilde{\underline{\mathbf{n}}}_i, \tag{21}$$

where $\mathbf{A}_i$ is the linear operator, $\mathbf{q}_i$ is the statistically equivalent quantization noise, $\tilde{\mathbf{\Phi}}_i = \mathbf{A}_i\bar{\mathbf{\Phi}}_i \in \mathbb{C}^{M\tau\times MK}$, and $\tilde{\underline{\mathbf{n}}}_i = \mathbf{A}_i\underline{\mathbf{n}}_i + \mathbf{q}_i \in \mathbb{C}^{M\tau\times 1}$.

Algorithm 1 Kalman Filter-Based Channel Estimator

1: Initialization:

$$\hat{\underline{\mathbf{h}}}_{0|0} = \mathbf{0}_{MK}, \ \mathbf{M}_{0|0} = \mathbf{R} = \mathbb{E}\left\{\underline{\mathbf{h}}_0\underline{\mathbf{h}}_0^{\mathrm{H}}\right\}$$

2: Prediction:

$$\hat{\underline{\mathbf{h}}}_{i|i-1} = \boldsymbol{\eta}\hat{\underline{\mathbf{h}}}_{i-1|i-1}$$

3: Minimum prediction MSE matrix ($MK \times MK$):

$$\mathbf{M}_{i|i-1} = \boldsymbol{\eta}\mathbf{M}_{i-1|i-1}\boldsymbol{\eta}^{\mathrm{H}} + \boldsymbol{\zeta}\mathbf{R}\boldsymbol{\zeta}^{\mathrm{H}}$$

4: Kalman gain matrix ($MK \times M\tau$):

$$\mathbf{K}_i = \mathbf{M}_{i|i-1}\tilde{\mathbf{\Phi}}_i^{\mathrm{H}}\left(\mathbf{C}_{\tilde{\underline{\mathbf{n}}}_i} + \tilde{\mathbf{\Phi}}_i\mathbf{M}_{i|i-1}\tilde{\mathbf{\Phi}}_i^{\mathrm{H}}\right)^{-1}$$

5: Correction:

$$\hat{\underline{\mathbf{h}}}_{i|i} = \hat{\underline{\mathbf{h}}}_{i|i-1} + \mathbf{K}_i\left(\underline{\mathbf{r}}_i - \tilde{\mathbf{\Phi}}_i\hat{\underline{\mathbf{h}}}_{i|i-1}\right)$$

6: Minimum MSE matrix ($MK \times MK$):

$$\mathbf{M}_{i|i} = \left(\mathbf{I}_{MK} - \mathbf{K}_i\tilde{\mathbf{\Phi}}_i\right)\mathbf{M}_{i|i-1}$$

The Kalman filter guarantees the optimality when the noise is Gaussian distributed [20]; however, the effective noise $\tilde{\underline{\mathbf{n}}}_i$ in (21) is not Gaussian because of the one-bit quantization noise $\mathbf{q}_i$. Although the noise is not Gaussian, it is still possible to apply the Kalman filter using the same covariance matrix $\mathbf{C}_{\tilde{\underline{\mathbf{n}}}_i}$. The proposed KFB channel estimator is summarized in Algorithm 1.

Remark 1 *Assuming the effective noise is Gaussian distributed may result in inaccurate channel estimation. This effect becomes more dominant as SNR increases, which is shown in Fig. 4 in Section 5. In the high SNR regime, the noise* $\underline{\mathbf{n}}_i$ *in (1) becomes negligible, and the effective noise* $\tilde{\underline{\mathbf{n}}}_i$ *in (21) is dominated by the quantization noise* $\mathbf{q}_i$*, which would severely violate the Gaussian assumption of* $\tilde{\underline{\mathbf{n}}}_i$*. In the low SNR regime, however, the effective noise* $\tilde{\underline{\mathbf{n}}}_i$ *is more like Gaussian, and the proposed KFB estimator is nearly optimal.*

3.4 Low-complexity TPE-based estimator

The BLMMSE estimator is a single-shot estimator, which returns a new channel estimate while the KFB estimator is a successive channel estimator, which tracks the channel based on a previous channel estimate at each time slot. Thus, the complexity of both channel estimators is the same at each time slot.

The matrix inversion has the most dominant computation complexity among matrix operations. The large channel dimensions in massive MIMO systems even exacerbate the complexity of matrix inversion. Therefore, when comparing the complexity of algorithms, we only consider the complexity of the matrix inversion. To reduce the complexity of KFB estimator, the truncated polynomial expansion [25] can be used to approximate the matrix inversion at the Kalman gain matrix $\mathbf{K}_i$ in Step 4 of Algorithm 1.

The Lth-order TPE approximation of the inversion of $N \times N$ matrix $\mathbf{X}$ is expressed as

$$\mathbf{X}^{-1} \approx \alpha \sum_{l=0}^{L} (\mathbf{I} - \alpha\mathbf{X})^l. \tag{22}$$

In (22), α is the convergence coefficient, which can be set as $0 < \alpha < \frac{2}{\max_n \lambda_n(\mathbf{X})}$ where $\lambda_n(\mathbf{X})$ is the n-th eigenvalue of the matrix $\mathbf{X}$[25].

The complexity of TPE approximation in (22) is $\mathcal{O}(LN^2)$ since it has only the matrix multiplication with the Lth-order. This is a large complexity reduction as compared to $\mathcal{O}(N^3)$ for the complexity of the $N \times N$ matrix inversion when L is much smaller than N. In Table 1, we summarize the complexity of three competing estimators. The TPE-based estimator has much lower computational complexity than the other estimators because $L \ll M\tau$ in practice.

To verify the effectiveness of TPE approximation, we evaluate the minimum NMSE of TPE-based estimator. For a tractable analysis, we assume $\mathbf{R} = \mathbf{I}_{MK}$ and $\tau = K$ as in [21], which results in $\mathbf{C}_{\underline{\mathbf{r}}} = \mathbf{I}_{MK}$ in (15) since $\mathbf{C}_{\underline{\mathbf{y}}} = (K\rho + 1)\mathbf{I}_{MK}$. Then the NMSE of BLMMSE estimator in [21], which is a performance baseline of the proposed estimators, is represented as

Table 1 Computational complexity of BLMMSE estimator, KFB estimator, and TPE-based estimator

Channel estimator	Computational complexity
BLMMSE estimator	$\mathcal{O}(M^3\tau^3)$
KFB estimator	$\mathcal{O}(M^3\tau^3)$
TPE-based estimator	$\mathcal{O}(LM^2\tau^2)$

M number of antennas, *τ* pilot symbol length, *L* approximation order

$$\begin{aligned}\text{NMSE}_{\text{BLM}} &= \frac{1}{MK}\mathbb{E}\left\{\left\|\hat{\underline{\mathbf{h}}}^{\text{BLM}} - \underline{\mathbf{h}}\right\|_2^2\right\} \\ &= 1 - \frac{2}{\pi}\frac{K\rho}{K\rho + 1} \\ &= 1 - \beta,\end{aligned} \tag{23}$$

where $\beta = \frac{2}{\pi}\frac{K\rho}{K\rho+1}$.

To derive the NMSE of TPE-based estimator, we first expand the covariance matrix of $\mathbf{q}_i$ as

$$\begin{aligned}\mathbf{C}_{\mathbf{q}_i} &= \mathbf{C}_{\underline{\mathbf{r}}_i} - \mathbf{A}_i\mathbf{C}_{\underline{\mathbf{y}}_i}\mathbf{A}_i^{\text{H}} \\ &\overset{(a)}{=} \mathbf{C}_{\underline{\mathbf{r}}_i} - \frac{2}{\pi}(\mathbf{I}_{MK}) \\ &\overset{(b)}{=} \frac{2}{\pi}(\arcsin(\mathbf{X}_i) + j\arcsin(\mathbf{Y}_i)) - \frac{2}{\pi}(\mathbf{I}_{MK}) \\ &\overset{(c)}{=} \left(1 - \frac{2}{\pi}\right)\mathbf{I}_{MK},\end{aligned} \tag{24}$$

where we define

$$\begin{aligned}\mathbf{X}_i &= \Sigma_{\underline{\mathbf{y}}_i}^{-1/2}\Re\{\mathbf{C}_{\underline{\mathbf{y}}_i}\}\Sigma_{\underline{\mathbf{y}}_i}^{-1/2}, \\ \mathbf{Y}_i &= \Sigma_{\underline{\mathbf{y}}_i}^{-1/2}\Im\{\mathbf{C}_{\underline{\mathbf{y}}_i}\}\Sigma_{\underline{\mathbf{y}}_i}^{-1/2}.\end{aligned} \tag{25}$$

In (24), (a) comes from $\mathbf{A}_i = \sqrt{\frac{2}{\pi}}\sqrt{\frac{1}{K\rho+1}}\mathbf{I}_{MK}$ and $\mathbf{C}_{\underline{\mathbf{y}}_i} = (K\rho + 1)\mathbf{I}_{MK}$, (b) is from the arcsin law in [24], and (c) is derived by substituting $\mathbf{C}_{\underline{\mathbf{y}}_i} = (K\rho + 1)\mathbf{I}_{MK}$ into (25).

The first-order TPE approximation of matrix inversion in Kalman gain matrix is given by

$$\mathbf{X}^{-1} \approx \alpha(\mathbf{I} + (\mathbf{I} - \alpha\mathbf{X})). \tag{26}$$

Thus, the Kalman gain matrix is approximated as

$$\begin{aligned}\mathbf{K}_i &= \mathbf{M}_{i|i-1}\tilde{\mathbf{\Phi}}_i^{\text{H}}\left(\mathbf{C}_{\tilde{\underline{\mathbf{n}}}_i} + \tilde{\mathbf{\Phi}}_i\mathbf{M}_{i|i-1}\tilde{\mathbf{\Phi}}_i^{\text{H}}\right)^{-1} \\ &\approx \mathbf{M}_{i|i-1}\tilde{\mathbf{\Phi}}_i^{\text{H}}\left(2\alpha\mathbf{I}_{MK} - \alpha^2\left(\mathbf{C}_{\tilde{\underline{\mathbf{n}}}_i} + \tilde{\mathbf{\Phi}}_i\mathbf{M}_{i|i-1}\tilde{\mathbf{\Phi}}_i^{\text{H}}\right)\right).\end{aligned} \tag{27}$$

We define the normalized trace of $\mathbf{M}_{i|i-1}$ and $\mathbf{M}_{i|i}$ as

$$\begin{aligned}m_{i|i-1} &\triangleq \frac{1}{MK}\text{tr}(\mathbf{M}_{i|i-1}), \\ m_{i|i} &\triangleq \frac{1}{MK}\text{tr}(\mathbf{M}_{i|i}),\end{aligned} \tag{28}$$

where we denote $m_{i|i-1}$ as the prediction NMSE and $m_{i|i}$ as the minimum NMSE.

We assume that the temporal correlation coefficient is identical for all users, i.e., $\eta_k = \eta$ for all k. With this assumption, we can further expand $m_{i|i-1}$ and $m_{i|i}$ as

$$\begin{aligned} m_{i|i-1} &= \frac{1}{MK}\mathrm{tr}(\mathbf{M}_{i|i-1}) \\ &= \frac{1}{MK}\mathrm{tr}\left(\eta\mathbf{M}_{i-1|i-1}\eta^{\mathrm{H}} + \zeta\mathbf{R}\zeta^{\mathrm{H}}\right) \\ &= \eta^2 m_{i-1|i-1} + (1-\eta^2), \end{aligned} \tag{29}$$

and

$$\begin{aligned} &m_{i|i} \\ &= \frac{1}{MK}\mathrm{tr}(\mathbf{M}_{i|i}) \\ &= \frac{1}{MK}\mathrm{tr}\left(\left(\mathbf{I}_{MK} - \mathbf{K}_i\tilde{\mathbf{\Phi}}_i\right)\mathbf{M}_{i|i-1}\right) \\ &\overset{(a)}{\approx} \frac{1}{MK}\mathrm{tr}\Big(\Big(\mathbf{I}_{MK} - \mathbf{M}_{i|i-1}\tilde{\mathbf{\Phi}}_i^{\mathrm{H}}\Big(2\alpha\mathbf{I}_{MK} \\ &\qquad -\alpha^2\Big(\mathbf{C}_{\tilde{\mathbf{n}}_i} + \tilde{\mathbf{\Phi}}_i\mathbf{M}_{i|i-1}\tilde{\mathbf{\Phi}}_i^{\mathrm{H}}\Big)\Big)\tilde{\mathbf{\Phi}}_i\Big)\mathbf{M}_{i|i-1}\Big) \\ &\overset{(b)}{=} \frac{1}{MK}\mathrm{tr}\Big(\Big(\mathbf{I}_{MK} - \mathbf{M}_{i|i-1}\Big((2\alpha-\alpha^2(1-\beta))\tilde{\mathbf{\Phi}}_i^{\mathrm{H}}\tilde{\mathbf{\Phi}}_i \\ &\qquad -\alpha^2\tilde{\mathbf{\Phi}}_i^{\mathrm{H}}\tilde{\mathbf{\Phi}}_i\mathbf{M}_{i|i-1}\tilde{\mathbf{\Phi}}_i^{\mathrm{H}}\tilde{\mathbf{\Phi}}_i\Big)\Big)\mathbf{M}_{i|i-1}\Big) \\ &\overset{(c)}{=} \frac{1}{MK}\mathrm{tr}\Big(\Big(\mathbf{I}_{MK} - \mathbf{M}_{i|i-1}\Big((2\alpha-\alpha^2(1-\beta))\beta \\ &\qquad -\alpha^2\beta\mathbf{M}_{i|i-1}\beta\Big)\Big)\mathbf{M}_{i|i-1}\Big) \\ &\overset{(d)}{=} \Big(1 - m_{i|i-1}\Big((2\alpha-\alpha^2(1-\beta))\beta \\ &\qquad -\alpha^2\beta m_{i|i-1}\beta\Big)\Big)m_{i|i-1} \\ &= (1-(2\alpha-\alpha^2(1-\beta)-\alpha^2\beta m_{i|i-1})\beta m_{i|i-1})m_{i|i-1}, \end{aligned} \tag{30}$$

where (a) is derived by the Kalman gain matrix approximation in (27), (b) is from

$$\begin{aligned} \mathbf{C}_{\tilde{\mathbf{n}}_i} &= \mathbf{A}_i\mathbf{A}_i^{\mathrm{H}} + \mathbf{C}_{\mathbf{q}_i} \\ &= \left(\frac{2}{\pi}\frac{1}{K\rho+1} + \left(1-\frac{2}{\pi}\right)\right)\mathbf{I}_{MK} \\ &= (1-\beta)\mathbf{I}_{MK}, \end{aligned} \tag{31}$$

(c) comes from

$$\begin{aligned} \tilde{\mathbf{\Phi}}_i^{\mathrm{H}}\tilde{\mathbf{\Phi}}_i &= \bar{\mathbf{\Phi}}_i^{\mathrm{H}}\mathbf{A}_i^{\mathrm{H}}\mathbf{A}_i\bar{\mathbf{\Phi}}_i \\ &= \frac{2}{\pi}\frac{1}{K\rho+1}\bar{\mathbf{\Phi}}_i^{\mathrm{H}}\bar{\mathbf{\Phi}}_i \\ &= \frac{2}{\pi}\frac{K\rho}{K\rho+1}\mathbf{I}_{MK} \\ &= \beta\mathbf{I}_{MK}, \end{aligned} \tag{32}$$

and (d) is derived by the fact that $\mathbf{M}_{i|i-1}$ and $\mathbf{M}_{i|i}$ are diagonal matrix based on the mathematical induction with $\mathbf{M}_{0|0} = \mathbf{R} = \mathbf{I}_{MK}$.

Now, we will show that $m_{i|i} < m_{i-1|i-1}$, i.e., the minimum NMSE decreases as the time slot index i increases. It is enough to show that $m_{i|i-1}$ is a monotonic decreasing sequence since $m_{i|i}$ and $m_{i|i-1}$ has a linear relationship in (29),

$$m_{i+1|i} < m_{i|i-1} \Leftrightarrow m_{i|i} < m_{i-1|i-1}. \tag{33}$$

First, we can reformulate (29)

$$\begin{aligned} &m_{i+1|i} \\ &= \eta^2 m_{i|i} + (1-\eta^2) \\ &= \eta^2(1-(2\alpha-\alpha^2(1-\beta)-\alpha^2\beta m_{i|i-1})\beta m_{i|i-1})m_{i|i-1} \\ &\quad +(1-\eta^2). \end{aligned} \tag{34}$$

We define $f(x)$ as

$$f(x) \triangleq \eta^2(1-(2\alpha-\alpha^2(1-\beta)-\alpha^2\beta x)\beta x)x + (1-\eta^2). \tag{35}$$

Then, in Appendix 2, we prove

$$f(x) < x, \; 0 < \gamma < x < 1, \tag{36}$$

where γ is the root of $f(x) = x$. In (36), we exploited the condition $0 < \alpha < 2$ that is proved in Appendix 3. Thus, we conclude

$$m_{i+1|i} = f(m_{i|i-1}) < m_{i|i-1}, \tag{37}$$

which is equivalent to $m_{i|i} < m_{i-1|i-1}$. Furthermore, we prove

$$\lim_{i\to\infty} m_{i|i-1} = \gamma, \tag{38}$$

in Appendix 4. Therefore, the prediction NMSE $m_{i|i-1}$ decreases as the time slot index i increases and converges to γ. Also, we can easily check that $m_{1|1} = 1-(2\alpha-\alpha^2)\beta = 1-\beta = \mathrm{NMSE_{BLM}}$ with $m_{1|0} = 1$ and $\alpha = 1$. After many time instances, we will have

$$\mathrm{NMSE_{BLM}} = m_{1|1} \gg m_{i|i}, \tag{39}$$

and the TPE-based estimator would outperform the BLMMSE estimator.

So far, we assume that $\mathbf{R} = \mathbf{I}_{MK}$, i.e., spatially uncorrelated channels, to derive the NMSE of the TPE estimator. Even for spatially correlated channels, the numerical results in Section 5 show that the TPE-based estimator outperforms the BLMMSE estimator.

4 Uplink data transmission

In this section, we derive the achievable sum-rate of massive MIMO with one-bit ADCs following similar steps as in [21] for the sake of completeness. The K users transmit data symbols to the BS. Based on the Bussgang decomposition, the quantized signal in the i-th time slot can be represented as

$$\begin{aligned}\mathbf{r}_{d,i} &= \mathcal{Q}(\sqrt{\rho_{d,i}}\mathbf{H}_i\mathbf{s}_i + \mathbf{n}_{d,i}) \\ &= \sqrt{\rho_{d,i}}\mathbf{A}_{d,i}\mathbf{H}_i\mathbf{s}_i + \mathbf{A}_{d,i}\mathbf{n}_{d,i} + \mathbf{q}_{d,i},\end{aligned} \tag{40}$$

where $\mathbf{s}_i$ is the transmit signal satisfying $\mathbb{E}\{|s_{i,k}|^2\} = 1$, and the subscript d denotes the data transmission. The linear operator in (40) can be approximated as

$$\begin{aligned}\mathbf{A}_{d,i} &= \sqrt{\frac{2}{\pi}}\mathrm{diag}(\mathbf{C}_{\mathbf{y}_{d,i}})^{-\frac{1}{2}} \\ &= \sqrt{\frac{2}{\pi}}\mathrm{diag}\left(\rho_{d,i}\mathbf{H}_i\mathbf{H}_i^{\mathrm{H}} + \mathbf{I}_M\right)^{-\frac{1}{2}} \\ &\overset{(a)}{\approx} \sqrt{\frac{2}{\pi}}\sqrt{\frac{1}{K\rho_{d,i}+1}}\mathbf{I}_M.\end{aligned} \tag{41}$$

In (41), (a) is from the channel hardening effect in massive MIMO systems as in [21]. After applying the receive combiner for the quantized signal, we have

$$\begin{aligned}\hat{\mathbf{s}}_i &= \mathbf{W}_i^{\mathrm{T}}\mathbf{r}_{d,i} \\ &= \sqrt{\rho_{d,i}}\mathbf{W}_i^{\mathrm{T}}\mathbf{A}_{d,i}(\hat{\mathbf{H}}_i\mathbf{s}_i + \boldsymbol{\mathcal{E}}_i\mathbf{s}_i) \\ &\quad + \mathbf{W}_i^{\mathrm{T}}\mathbf{A}_{d,i}\mathbf{n}_{d,i} + \mathbf{W}_i^{\mathrm{T}}\mathbf{q}_{d,i},\end{aligned} \tag{42}$$

where $\mathbf{W}_i$ is the receive combining matrix, $\hat{\mathbf{H}}_i = \mathrm{unvec}(\hat{\underline{\mathbf{h}}}_i)$ is the unvectorized channel estimation matrix, and $\boldsymbol{\mathcal{E}}_i = \mathbf{H}_i - \hat{\mathbf{H}}_i$ is the estimation error matrix. The k-th element of $\hat{\mathbf{s}}_i$ can be represented as

$$\begin{aligned}&\hat{s}_{i,k} \\ &= \sqrt{\rho_{d,i}}\mathbf{w}_{i,k}^{\mathrm{T}}\mathbf{A}_{d,i}\hat{\mathbf{h}}_{i,k}s_{i,k} + \sqrt{\rho_{d,i}}\mathbf{w}_{i,k}^{\mathrm{T}}\sum_{j\neq k}^{K}\mathbf{A}_{d,i}\hat{\mathbf{h}}_{i,j}s_{i,j} \\ &+ \sqrt{\rho_{d,i}}\mathbf{w}_{i,k}^{\mathrm{T}}\sum_{j=1}^{K}\mathbf{A}_{d,i}\boldsymbol{\epsilon}_{i,j}s_{i,j} + \mathbf{w}_{i,k}^{\mathrm{T}}\mathbf{A}_{d,i}\mathbf{n}_{d,i} + \mathbf{w}_{i,k}^{\mathrm{T}}\mathbf{q}_{d,i},\end{aligned} \tag{43}$$

where $\mathbf{w}_{i,k}$, $\hat{\mathbf{h}}_{i,k}$ and $\boldsymbol{\epsilon}_{i,k}$ represent the k-th columns of $\mathbf{W}_i$, $\hat{\mathbf{H}}_i$, and $\boldsymbol{\mathcal{E}}_i$, respectively.

We can obtain a lower bound on the achievable rate of the k-th user by treating the uncorrelated inter-user interference (IUI) and the quantization noise (QN) $\mathbf{q}_{d,i}$ as a Gaussian noise [26], and assuming the Gaussian channel input as in [21],

$$R_{i,k} = \mathbb{E}\left\{\log_2\left(1 + \frac{\mathrm{S}_{i,k}}{\mathrm{IUI}_{i,k} + \mathrm{QN}_{i,k}}\right)\right\}, \tag{44}$$

where

$$\begin{aligned}\mathrm{S}_{i,k} &= \rho_{d,i}\left|\mathbf{w}_{i,k}^{\mathrm{T}}\mathbf{A}_{d,i}\hat{\mathbf{h}}_{i,k}\right|^2, \\ \mathrm{IUI}_{i,k} &= \rho_{d,i}\sum_{j\neq k}^{K}\left|\mathbf{w}_{i,k}^{\mathrm{T}}\mathbf{A}_{d,i}\hat{\mathbf{h}}_{i,j}\right|^2, \\ \mathrm{QN}_{i,k} &= \rho_{d,i}\sum_{j=1}^{K}\left|\mathbf{w}_{i,k}^{\mathrm{T}}\mathbf{A}_{d,i}\boldsymbol{\epsilon}_{i,j}\right|^2 + \|\mathbf{w}_{i,k}^{\mathrm{T}}\mathbf{A}_{d,i}\|^2 \\ &\quad + \mathbf{w}_{i,k}^{\mathrm{T}}\mathbf{C}_{\mathbf{q}_{d,i}}\mathbf{w}_{i,k}^{*}.\end{aligned} \tag{45}$$

The auto-covariance matrix of $\mathbf{q}_{d,i}$ is given by

$$\begin{aligned}\mathbf{C}_{\mathbf{q}_{d,i}} &= \mathbf{C}_{\mathbf{r}_{d,i}} - \mathbf{A}_{d,i}\mathbf{C}_{\mathbf{y}_{d,i}}\mathbf{A}_{d,i}^{\mathrm{H}} \\ &= \frac{2}{\pi}(\arcsin(\mathbf{X}_{d,i}) + j\arcsin(\mathbf{Y}_{d,i})) \\ &\quad - \frac{2}{\pi}(\mathbf{X}_{d,i} + j\mathbf{Y}_{d,i}) \\ &\overset{(a)}{\approx} (1 - 2/\pi)\mathbf{I}_M,\end{aligned} \tag{46}$$

where we define

$$\begin{aligned}\mathbf{X}_{d,i} &= \Sigma_{\mathbf{y}_{d,i}}^{-1/2}\Re\{\mathbf{C}_{\mathbf{y}_{d,i}}\}\Sigma_{\mathbf{y}_{d,i}}^{-1/2}, \\ \mathbf{Y}_{d,i} &= \Sigma_{\mathbf{y}_{d,i}}^{-1/2}\Im\{\mathbf{C}_{\mathbf{y}_{d,i}}\}\Sigma_{\mathbf{y}_{d,i}}^{-1/2}.\end{aligned} \tag{47}$$

In (46), $\mathbf{C}_{\mathbf{r}_{d,i}}$ can be obtained by the arcsin law in (15), and (a) comes from the approximation of the low SNR as in [21]. This approximation holds even in correlated channels, which is different from (24) that is based on the assumption $\mathbf{R} = \mathbf{I}_{MK}$. We define the achievable sum-rate as

$$R_i = \sum_{k=1}^{K} R_{i,k}. \tag{48}$$

To reduce the interference, we adopt the zero-forcing (ZF) combiner,

$$\mathbf{W}_{i,\mathrm{ZF}}^{\mathrm{T}} = \left(\hat{\mathbf{H}}_i^{\mathrm{H}}\hat{\mathbf{H}}_i\right)^{-1}\hat{\mathbf{H}}_i^{\mathrm{H}}, \tag{49}$$

for numerical studies.

5 Results and discussion

In this section, we verify the proposed channel estimator by Monte-Carlo simulation. We define the NMSE as the performance metric,

$$\mathrm{NMSE} = \frac{1}{MK}\mathbb{E}\left\{\left\|\hat{\underline{\mathbf{h}}} - \underline{\mathbf{h}}\right\|_2^2\right\}, \tag{50}$$

where $\hat{\underline{\mathbf{h}}}$ is the channel estimate and $\underline{\mathbf{h}}$ is the true channel. We adopt the pilot matrix $\boldsymbol{\Phi}$ by the discrete Fourier transform (DFT) matrix, which satisfies the assumptions in Section 3.2, and select K columns of $\tau \times \tau$ DFT matrix with $\tau \geq K$ to obtain the pilot sequences. We adopt the Jakes' model for the temporal correlation, which is given as $\eta_k = J_0(2\pi f_{D,k} t)$ where $J_0(\cdot)$ denotes the 0-th order Bessel function, $f_{D,k}$ is the Doppler frequency, and t is the channel instantiation interval. For simulations, we set $f_{D,k} = v_k f_c/c$ with the user speed v_k, the carrier frequency $f_c = 2.5$ GHz, and the speed of light $c = 3 \times 10^8$ m $\cdot$ s^{-1}. We also set $t = 5$ ms [27]. We denote NMSE($\mathbf{h}_i$) as the NMSE of KFB estimator at the i-th time slot and NMSE($\mathbf{M}$) $= \frac{1}{MK}\text{tr}(\mathbf{M}_{i|i})$ as the theoretical NMSE of Kalman filtering with the Gaussian noise, not the quantization noise. Therefore, NMSE($\mathbf{M}$) gives the performance limit of Kalman filtering with the Gaussian noise. We depict the "BLMMSE" as the NMSE performance of the single-shot channel estimator discussed in Section 3.2.

In Fig. 2, we compare the NMSEs of BLMMSE estimator and KFB estimator with the time slot i for $r = 0.5$ or $r = 0.8$ with SNR = -5 dB. We assume the BS antennas $M = 128$, the users $K = 8$, and the symbols $\tau = 8$. We set the temporal correlation coefficient $\eta_k = 0.988$, which corresponds to $v = 3$ km/h. As the time slot increases, the proposed KFB estimator outperforms the BLMMSE estimator. By comparing NMSE($\mathbf{h}_i$) and NMSE($\mathbf{M}$), the loss from using one-bit ADCs is around 1.5 dB. As the amount of spatial correlation increases from 0.5 to 0.8, all estimators perform better since it becomes easier to estimate channels as the channels become more correlated in space [28–30].

In Fig. 3, we compare the NMSEs of the channel estimators with and without the perfect spatial correlation knowledge. All the parameters are the same as in Fig. 2 with $r = 0.8$. Without the spatial correlation knowledge, we use N_s samples to estimate the spatial channel correlation by the LS estimates; then, we estimate the channel. When we use $N_s = 500, 1000$, the performance loss is about 4, 2 dB compared to the case of perfect correlation knowledge. Although the performance degradation due to the imperfect knowledge of spatial correlation is non-negligible, the loss is inevitable for the channel estimators, including the BLMMSE estimator, that exploit the spatial correlation. The KFB estimator outperforms the BLMMSE estimator even with the sample correlation matrix, and as time slot increases, the KFB estimator using the sample correlation matrix achieves lower NMSE than the BLMMSE estimator using the true spatial correlation matrix.

Figure 4 depicts the NMSEs of the KFB estimator when each user experiences different temporal fading. We set $r = 0.8$ and the temporal correlation coefficient of user 1 to 4 as $\eta_k = 0.872, 0.936, 0.967$, and 0.988, which correspond to $v_k = 10$ km/h, 7 km/h, 5 km/h, and 3 km/h. All other settings are the same as in Fig. 2. As expected, the users with high temporal correlations benefit more from the KFB estimator. Even the user with the moderate velocity of 10 km/h also has the gain more than 1 dB.

In Figs. 5 and 6, we compare the NMSEs of the KFB estimator according to SNR with different time slots when $M = 128, K = 8, \tau = 8, \eta_k = 0.988, 0.724$ (correspond to $v_k = 5, 15$ km/h), and $r = 0.5$. When the temporal correlation is high, the NMSEs of the KFB estimator decreased as the SNR increased in low SNR regime. In the low SNR

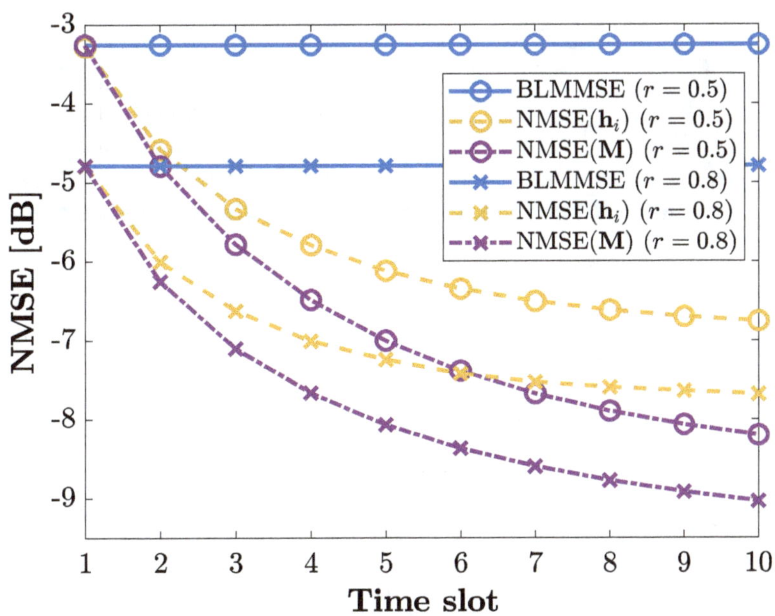

Fig. 2 The NMSEs of BLMMSE estimator, KFB estimator, and theoretical limit of Kalman filtering according to time slot i for different spatial correlation coefficient r when $M = 128, K = 8, \tau = 8, \eta_k = 0.988$, and SNR = -5 dB

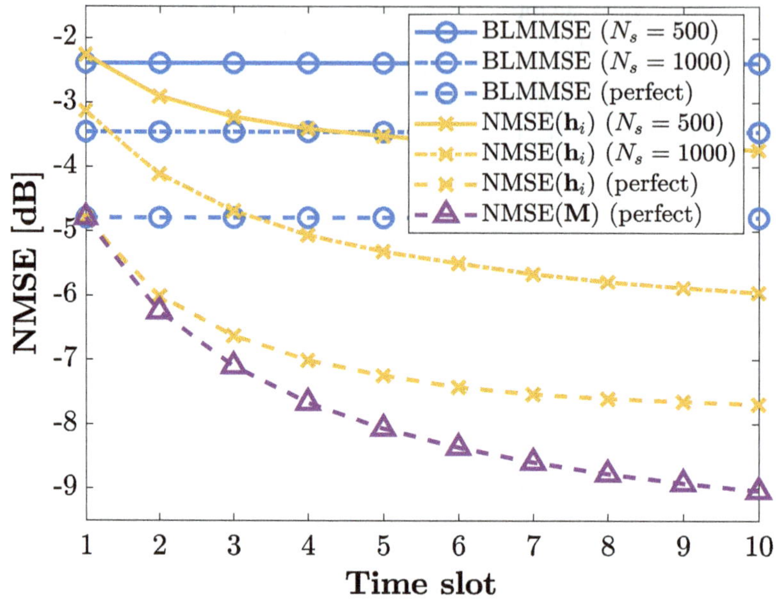

Fig. 3 The NMSEs of BLMMSE estimator, KFB estimator, and theoretical limit of Kalman filtering according to time slot i with and without the perfect spatial correlation knowledge when $M = 128$, $K = 8$, $\tau = 8$, $\eta_k = 0.988$, $r = 0.8$, and SNR $= -5$ dB

regime, NMSE($\mathbf{h}_i$) is almost the same as the theoretical NMSE of NMSE(**M**) after 10 successive estimations. In the high SNR, however, the NMSE of KFB estimator suffers from the saturation effect, which is referred as the stochastic resonance due to one-bit quantization noise [31]. In the proposed KFB estimator, the loss also comes from the Gaussian model mismatch in the one-bit quantization as explained in Remark 1 in Section 3.3. When the temporal correlation is low, the NMSEs of the KFB estimator decreased as the SNR increased in all SNR regime. This is because the channel estimation error comes mostly from the large temporal channel variation, not from the one-bit quantization.

In Figs. 7 and 8, we compare the achievable sum-rates of the BLMMSE estimator and KFB estimator according to the time slot when $M = 128$, $K = 8$, $\tau = 8$, r=0.8, and

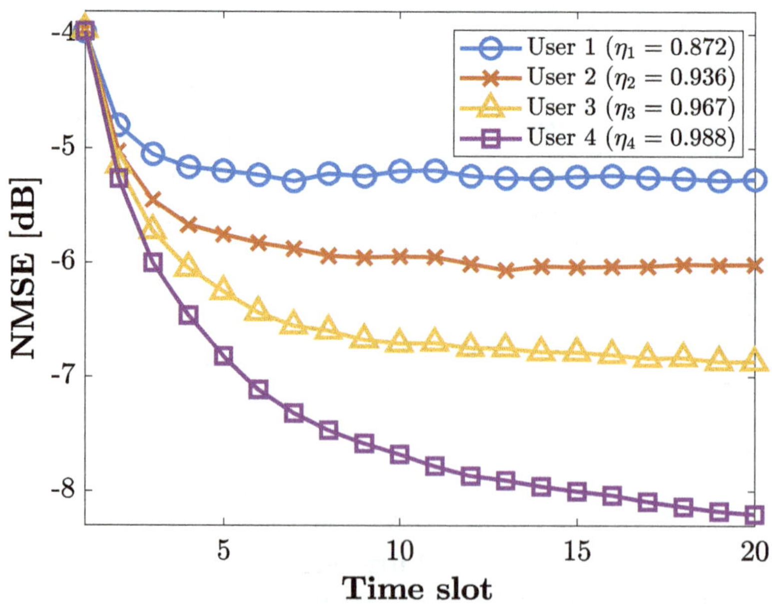

Fig. 4 The NMSEs of KFB estimator according to time slot i for different temporal fading users when $M = 128$, $K = 4$, $\tau = 4$, $r = 0.8$, and SNR $= -5$ dB

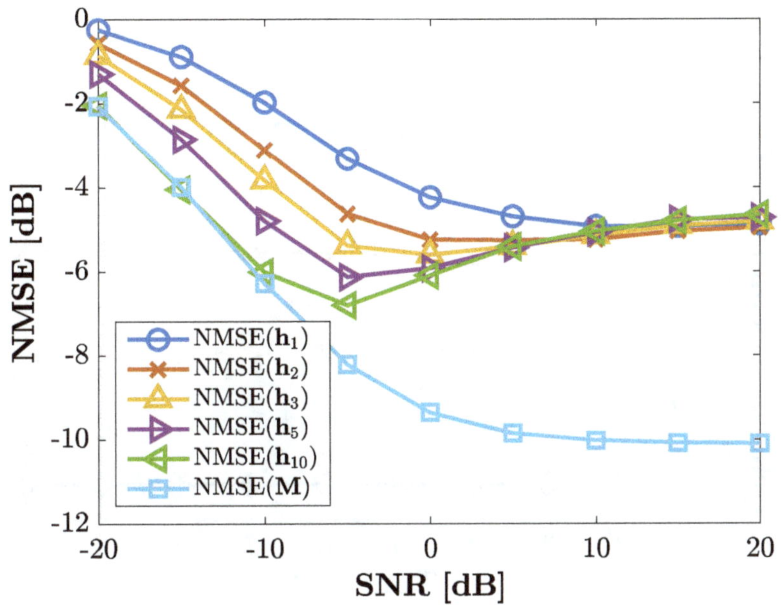

Fig. 5 The NMSEs of KFB estimator according to SNR with different time slots when $M = 128$, $K = 8$, $\tau = 8$, $\eta_k = 0.988$, and $r = 0.5$

SNR = 0 and 10 dB. We assume all users experience the same η. In both scenarios, the achievable sum-rate of the KFB estimator outperforms the BLMMSE estimator as the time slot increases.

In Figs. 9 and 10, we compare the NMSEs of the KFB estimator and the low-complexity TPE-based estimator with the time slot. We set $M = 128, K = 8, \tau = 8, r = 0.5$, $\eta_k = 0.988, 0.872$, and SNR = $-5, 10$ dB. We numerically optimize $\alpha = 0.5$ for the TPE-based estimator. In the high temporal correlation and low SNR case (Fig. 9), the NMSE gap between the KFB and TPE-based estimators is negligible and already quite small even with $L = 1$. In the low temporal correlation and high SNR case (Fig. 10), the performance is degraded but the gap becomes small with

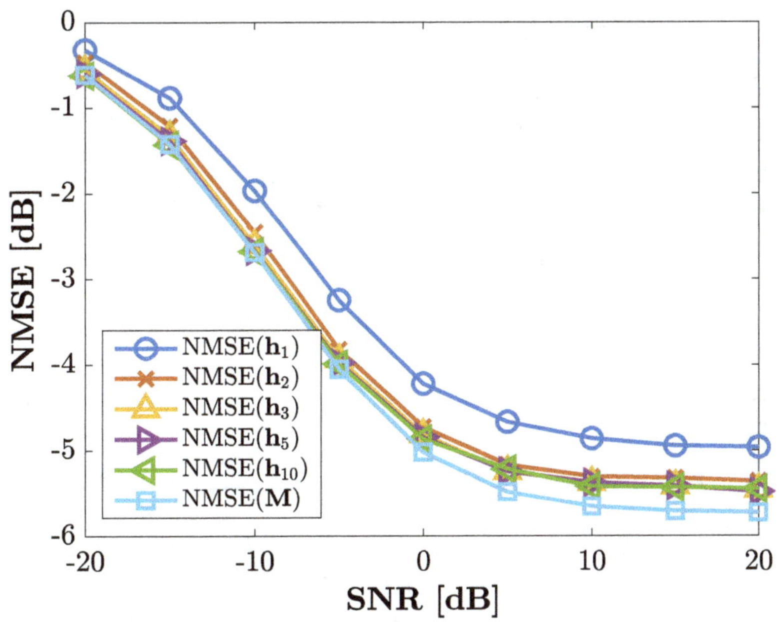

Fig. 6 The NMSEs of KFB estimator according to SNR with different time slots when $M = 128$, $K = 8$, $\tau = 8$, $\eta_k = 0.724$, and $r = 0.5$

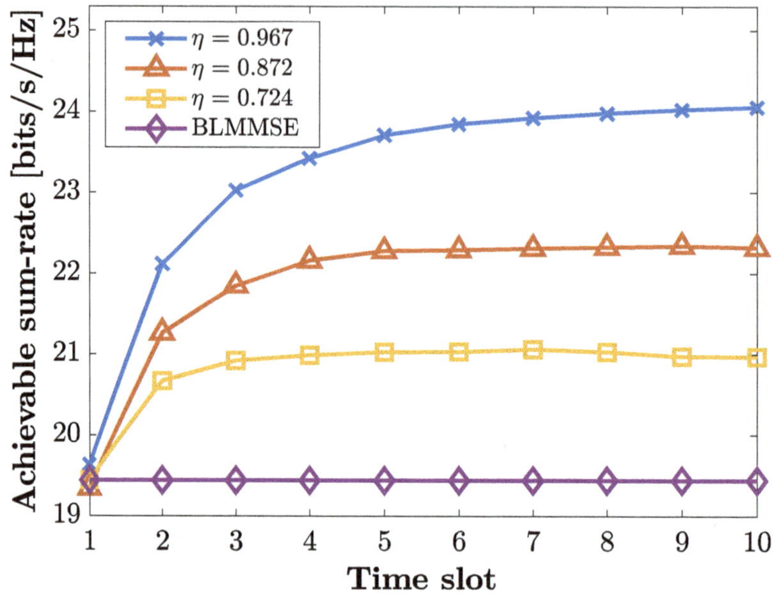

Fig. 7 The achievable sum-rates of BLMMSE estimator and KFB estimator according to time slot i with different temporal correlations when $M = 128$, $K = 8$, $\tau = 8$, $r = 0.8$, and SNR = 0 dB

$L = 2$. Therefore, in practice, the low-complexity TPE-based estimator can be used with negligible performance loss.

6 Conclusion

In this paper, we proposed the Kalman filter-based (KFB) channel estimators that exploit both the spatial and temporal correlations of channels for massive MIMO systems using one-bit ADCs. We adopted the Bussgang decomposition to linearize the non-linear effect from one-bit quantization. Based on the linearized model and assuming the effective noise as Gaussian, we exploited the Kalman filter to estimate the channel successively. The proposed KFB estimator has a remarkable gain compared to the previous estimator in [21], which does not exploit any temporal correlation in channels. To resolve the complexity issue of

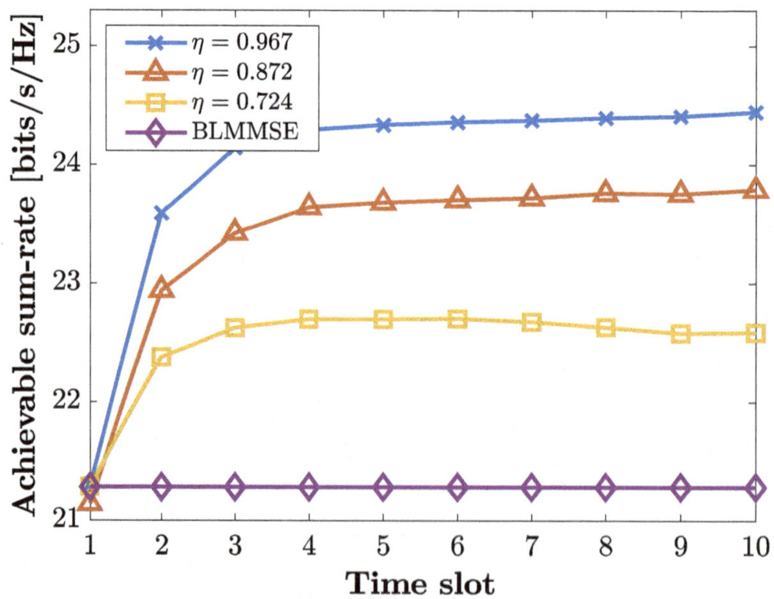

Fig. 8 The achievable sum-rates of BLMMSE estimator and KFB estimator according to time slot i with different temporal correlations when $M = 128$, $K = 8$, $\tau = 8$, $r = 0.8$, and SNR = 10 dB

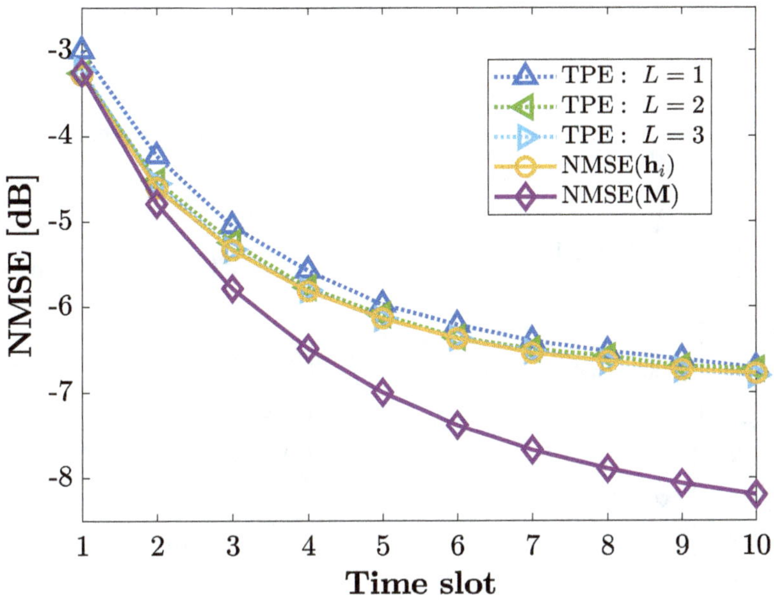

Fig. 9 The NMSEs of KFB estimator and TPE-based estimator according to time slot i when $M = 128$, $K = 8$, $\tau = 8$, $\eta_k = 0.988$, $r = 0.5$, $\alpha = 0.5$, and SNR $= -5$ dB

the KFB estimator due to the large-scale matrix inversion, we also implemented the truncated polynomial expansion (TPE)-based estimator. We analytically derived the minimum NMSE based on the first-order TPE approximation, and the numerical results showed that the low-complexity TPE-based estimator gives nearly the same accuracy as the KFB estimator even with lower approximation orders.

7 Methods

This paper studies the channel estimation problem for massive MIMO systems using one-bit ADCs. The performance of proposed algorithm was evaluated by different settings and metrics, which are NMSE and achievable sum-rate. We use MATLAB R2018a to simulate the algorithm.

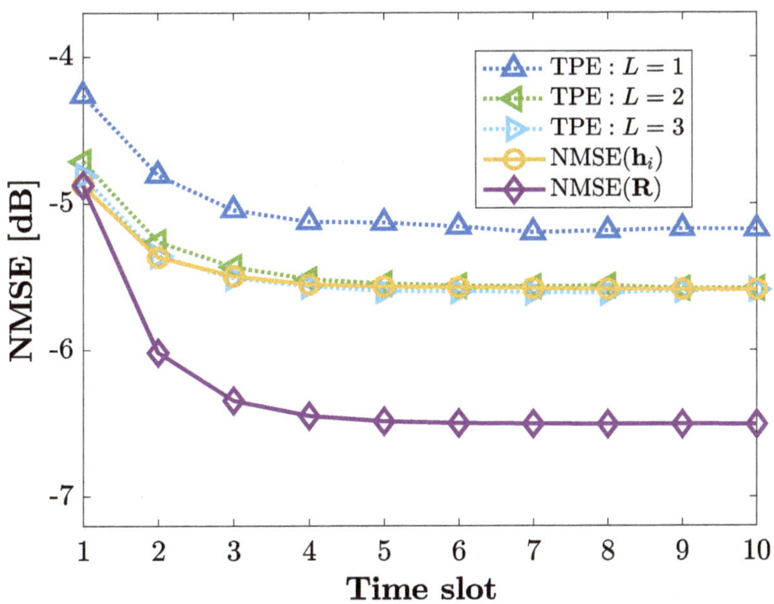

Fig. 10 The NMSEs of KFB estimator and TPE-based estimator according to time slot i when $M = 128$, $K = 8$, $\tau = 8$, $\eta_k = 0.872$, $r = 0.5$, $\alpha = 0.5$, and SNR $= 10$ dB

Appendix 1: Proof of (12)

We first expand $\bar{\mathbf{\Phi}}\mathbf{R}\bar{\mathbf{\Phi}}^{\mathrm{H}}$ as,

$$\begin{aligned}
&\bar{\mathbf{\Phi}}\mathbf{R}\bar{\mathbf{\Phi}}^{\mathrm{H}} \\
&= (\mathbf{\Phi} \otimes \sqrt{\rho}\mathbf{I}_M)\mathbf{R}(\mathbf{\Phi} \otimes \sqrt{\rho}\mathbf{I}_M)^{\mathrm{H}} \\
&= \rho \left(\begin{bmatrix} \phi_{1,1} & \cdots & \phi_{1,K} \\ \vdots & \ddots & \vdots \\ \phi_{\tau,1} & \cdots & \phi_{\tau,K} \end{bmatrix} \otimes \mathbf{I}_M \right) \mathbf{R}(\mathbf{\Phi} \otimes \mathbf{I}_M)^{\mathrm{H}} \\
&= \rho \begin{bmatrix} \phi_{1,1}\mathbf{I}_M & \cdots & \phi_{1,K}\mathbf{I}_M \\ \vdots & \ddots & \vdots \\ \phi_{\tau,1}\mathbf{I}_M & \cdots & \phi_{\tau,K}\mathbf{I}_M \end{bmatrix} \mathbf{R}(\mathbf{\Phi} \otimes \mathbf{I}_M)^{\mathrm{H}} \\
&\overset{(a)}{=} \rho \begin{bmatrix} \phi_{1,1}\mathbf{R}_1 & \cdots & \phi_{1,K}\mathbf{R}_K \\ \vdots & \ddots & \vdots \\ \phi_{\tau,1}\mathbf{R}_1 & \cdots & \phi_{\tau,K}\mathbf{R}_K \end{bmatrix} (\mathbf{\Phi} \otimes \mathbf{I}_M)^{\mathrm{H}} \\
&= \rho \begin{bmatrix} \phi_{1,1}\mathbf{R}_1 & \cdots & \phi_{1,K}\mathbf{R}_K \\ \vdots & \ddots & \vdots \\ \phi_{\tau,1}\mathbf{R}_1 & \cdots & \phi_{\tau,K}\mathbf{R}_K \end{bmatrix} \begin{bmatrix} \phi_{1,1}^*\mathbf{I}_M & \cdots & \phi_{\tau,1}^*\mathbf{I}_M \\ \vdots & \ddots & \vdots \\ \phi_{1,K}^*\mathbf{I}_M & \cdots & \phi_{\tau,K}^*\mathbf{I}_M \end{bmatrix} \\
&\overset{(b)}{=} \rho \begin{bmatrix} \sum_{k=1}^{K} \mathbf{R}_k & \cdots & \\ \vdots & \ddots & \vdots \\ & \cdots & \sum_{k=1}^{K} \mathbf{R}_k \end{bmatrix},
\end{aligned} \tag{51}$$

where (a) comes from the independent spatial correlation matrix $\mathbf{R}$ in (10) and (b) follows by assuming that the pilot sequences are column-wise orthogonal with the same magnitude for all elements. Since the diagonal term of $\mathbf{R}_k$ is 1 for all k, we have

$$\operatorname{diag}\left(\bar{\mathbf{\Phi}}\mathbf{R}\bar{\mathbf{\Phi}}^{\mathrm{H}}\right) = K\rho\mathbf{I}_{M\tau}, \tag{52}$$

which finishes the proof.

Appendix 2: Proof of (36)

First, we define $g(x)$ as

$$\begin{aligned}
g(x) &\triangleq f(x) - x \\
&= \eta^2(\alpha^2\beta^2x^3 - (2\alpha - \alpha^2 + \alpha^2\beta)\beta x^2 + x) \\
&\quad + 1 - \eta^2 - x.
\end{aligned} \tag{53}$$

Then, we have

$$\begin{aligned}
g(-\infty) &< 0, \\
g(0) &= 1 - \eta^2 > 0, \\
g(1) &= \eta^2\beta(-2\alpha + \alpha^2) < 0, \\
g(\infty) &> 0,
\end{aligned} \tag{54}$$

where the inequality of $g(1)$ is due to $0 < \alpha < 2$. The roots of $g(x)$ are $\gamma_{0^-} \in (-\infty, 0)$, $\gamma \in (0, 1)$, $\gamma_{1^+} \in (1, \infty)$ since $g(x)$ is the third-order polynomial, and $g(-\infty)g(0) < 0$, $g(0)g(1) < 0$, $g(1)g(\infty) < 0$ based on the intermediate value theorem. Therefore, γ is the unique solution of $g(x) = 0$ on $x \in (0, 1)$.

Now, we will show $g(x) < 0$ for $0 < \gamma < x < 1$. The derivative of $g(x)$ is given by

$$\begin{aligned}
g'(x) &= f'(x) - 1 \\
&= \eta^2\left(3\alpha^2\beta^2x^2 - 2(2\alpha - \alpha^2 + \alpha^2\beta)\beta x + 1\right) - 1.
\end{aligned} \tag{55}$$

Then, we have

$$\begin{aligned}
g'(0) &= \eta^2 - 1 < 0, \\
g'(1) &= \eta^2(\alpha^2\beta^2 - 4\alpha\beta + 2\alpha^2\beta + 1) - 1 \\
&= \eta^2\alpha\beta(\alpha(\beta + 2) - 4) + \eta^2 - 1 \\
&< \eta^2\alpha\beta(\alpha(\beta + 2) - 4) \\
&< \eta^2\alpha\beta(2\alpha - 4) \\
&< 0,
\end{aligned} \tag{56}$$

since $0 < \alpha < 2$. This result implies $g'(x) < 0$ for $0 < x < 1$ because $g'(x)$ is the second-order polynomial with the positive leading coefficient $3\eta^2\alpha^2\beta^2$. Since $g'(x) < 0$ for $0 < x < 1$ and $g(\gamma) = 0$, then $g(x) < g(\gamma) = 0$ for $0 < \gamma < x < 1$, which finishes the proof.

Appendix 3: Proof of $0 < \alpha < 2$

We first reformulate $\max_n \lambda_n\left(\mathbf{C}_{\tilde{\underline{\mathbf{n}}}_i} + \tilde{\mathbf{\Phi}}_i\mathbf{M}_{i|i-1}\tilde{\mathbf{\Phi}}_i^{\mathrm{H}}\right)$ as,

$$\begin{aligned}
\max_n \lambda_n\left(\mathbf{C}_{\tilde{\underline{\mathbf{n}}}_i} + \tilde{\mathbf{\Phi}}_i\mathbf{M}_{i|i-1}\tilde{\mathbf{\Phi}}_i^{\mathrm{H}}\right) &\overset{(a)}{=} 1 - \beta + \beta m_{i|i-1} \\
&= 1 - \beta(1 - m_{i|i-1}) \\
&\overset{(b)}{\leq} 1,
\end{aligned} \tag{57}$$

where (a) comes from the fact that $\mathbf{M}_{i|i-1}$ is a diagonal matrix, and (b) is from $0 < \gamma \leq m_{i|i-1} \leq 1$. By plugging (57) into the bound on α,

$$0 < \alpha < \frac{2}{\max_n \lambda_n\left(\mathbf{C}_{\tilde{\underline{\mathbf{n}}}_i} + \tilde{\mathbf{\Phi}}_i\mathbf{M}_{i|i-1}\tilde{\mathbf{\Phi}}_i^{\mathrm{H}}\right)}, \tag{58}$$

we have the tightened bound

$$0 < \alpha < 2 \leq \frac{2}{\max_n \lambda_n\left(\mathbf{C}_{\tilde{\underline{\mathbf{n}}}_i} + \tilde{\mathbf{\Phi}}_i\mathbf{M}_{i|i-1}\tilde{\mathbf{\Phi}}_i^{\mathrm{H}}\right)}. \tag{59}$$

Appendix 4: Proof of (38)

Based on the mathematical induction, we assume

$$\begin{aligned}
m_{1|0} &= 1 > \gamma, \\
m_{i|i-1} &> \gamma.
\end{aligned} \tag{60}$$

First, we proof that $f(x)$ is the increasing function on $x \in (\gamma, 1)$. The derivative of $f(x)$ is

$$f'(x) = \eta^2\Big(3\alpha^2\beta^2x^2 - 2(2\alpha - \alpha^2 + \alpha^2\beta)\beta x + 1\Big). \quad (61)$$

Then, we have

$$\begin{aligned} f'(0) &= \eta^2 > 0, \\ f'(1) &= \eta^2(\alpha^2\beta^2 - 4\alpha\beta + 2\alpha^2\beta + 1) \\ &= \eta^2((\beta^2 + 2\beta)\alpha^2 - 4\beta\alpha + 1) \\ &= \eta^2\left((\beta^2 + 2\beta)\alpha - \frac{2\beta}{\beta^2 + 2\beta}\right)^2 + \frac{\beta(2 - 3\beta)}{\beta^2 + 2\beta} \\ &> 0, \end{aligned} \quad (62)$$

which comes from $0 < \beta < \frac{2}{\pi} < \frac{2}{3}$. Thus, $f(x)$ is the increasing function on $x \in (\gamma, 1)$. Finally, we get $f(x) > f(\gamma) = \gamma$, which implies $m_{i+1|i} = f(m_{i|i-1}) > f(\gamma) = \gamma$. Thus, $m_{i|i-1} > \gamma$ for all $i > 0$ due to the mathematical induction. Since $m_{i|i-1}$ is the monotonic decreasing and bounded sequence, $m_{i|i-1}$ converges by the monotone convergence theorem [32].

Thus, we can define $L_m = \lim_{i\to\infty} m_{i|i-1}$,

$$\begin{aligned} L_m &= \lim_{i\to\infty} m_{i+1|i} \\ &= \lim_{i\to\infty} f(m_{i|i-1}) \\ &= f(L_m), \end{aligned} \quad (63)$$

which implies L_m is also a root of $f(x) = x$. Since $m_{i|i-1}$ converges L_m, and γ is the unique solution of $f(x) = x$ on $x \in (0, 1)$, $L_m = \lim_{i\to\infty} m_{i|i-1} = \gamma$, which finishes the proof.

Abbreviations

ADC: Analog-to-digital converter; BLMMSE: Bussgang linear minimum mean square error; BS: Base station; DFT: Discrete Fourier transform; IUI: Inter-user interference; KFB: Kalman filter-based; LRA-MMSE: Low-resolution aware minumum mean square error; LS: Least square; MIMO: Multiple-input multiple-output; MLD: maximum likelihood detection; NMSE: Normalized mean square error; QN: Quantization noise; SNR: Signal-to-noise ratio; TPE: Truncated polynomial expansion

Acknowledgements

This work was partly supported by Institute for Information & communications Technology Promotion(IITP) grant funded by the Korea government(MSIT) (No. 2016-0-00123, Development of Integer-Forcing MIMO Transceivers for 5G & Beyond Mobile Communication Systems) and by the National Research Foundation (NRF) grant funded by the MSIT of the Korea government (2019R1C1C1003638).

Authors' contributions

All authors read and approved the final manuscript.

Competing interests

The authors declare that they have no competing interests.

References

1. H. Kim, J. Choi, in *2018 IEEE Global Communications Conference (GLOBECOM)*. Channel estimation for one-bit massive MIMO systems exploiting spatio-temporal correlations, (2018), pp. 1–6. https://doi.org/10.1109/GLOCOM.2018.8647574
2. T. L. Marzetta, Noncooperative cellular wireless with unlimited numbers of base station antennas. IEEE Trans. Wirel. Commun. **9**(11), 3590–3600 (2010). https://doi.org/10.1109/TWC.2010.092810.091092
3. F. Rusek, D. Persson, B. K. Lau, E. G. Larsson, T. L. Marzetta, O. Edfors, F. Tufvesson, Scaling up MIMO: opportunities and challenges with very large arrays. IEEE Signal Proc. Mag. **30**(1), 40–60 (2013). https://doi.org/10.1109/MSP.2011.2178495
4. J. Hoydis, S. ten Brink, M. Debbah, Massive MIMO in the UL/DL of cellular networks: how many antennas do we need? IEEE J. Sel. Areas Commun. **31**(2), 160–171 (2013). https://doi.org/10.1109/JSAC.2013.130205
5. E. G. Larsson, O. Edfors, F. Tufvesson, T. L. Marzetta, Massive MIMO for next generation wireless systems. IEEE Commun. Mag. **52**(2), 186–195 (2014). https://doi.org/10.1109/MCOM.2014.6736761
6. R. H. Walden, Analog-to-digital converter survey and analysis. IEEE J. Sel. Areas Commun. **17**(4), 539–550 (1999). https://doi.org/10.1109/49.761034
7. S. Wang, Y. Li, J. Wang, Multiuser detection in massive spatial modulation MIMO with low-resolution ADCs. IEEE Trans. Wirel. Commun. **14**(4), 2156–2168 (2015). https://doi.org/10.1109/TWC.2014.2382098
8. S. Wang, Y. Li, J. Wang, in *2014 IEEE International Conference on Communications (ICC)*. Convex optimization based multiuser detection for uplink large-scale MIMO under low-resolution quantization, (2014), pp. 4789–4794. https://doi.org/10.1109/ICC.2014.6884078
9. T. Zhang, C. Wen, S. Jin, T. Jiang, Mixed-ADC massive MIMO detectors: performance analysis and design optimization. IEEE Trans. Wirel. Commun. **15**(11), 7738–7752 (2016). https://doi.org/10.1109/TWC.2016.2606592
10. Y. Jeon, S. Hong, N. Lee, in *2017 IEEE International Conference on Communications (ICC)*. Blind detection for MIMO systems with low-resolution ADCs using supervised learning, (2017), pp. 1–6. https://doi.org/10.1109/ICC.2017.7997434
11. Z. Shao, R. C. de Lamare, L. T. N. Landau, Iterative detection and decoding for large-scale multiple-antenna systems with 1-bit ADCs. IEEE Wirel. Commun. Lett. **7**(3), 476–479 (2018). https://doi.org/10.1109/LWC.2017.2787159
12. Y. Jeon, N. Lee, S. Hong, R. W. Heath, in *2018 IEEE 87th Vehicular Technology Conference (VTC Spring)*. A low complexity ML detection for uplink massive MIMO systems with one-bit ADCs, (2018), pp. 1–5. https://doi.org/10.1109/VTCSpring.2018.8417580
13. Y. Cho, S. Hong, One-bit successive-cancellation soft-output (OSS) detector for uplink MU-MIMO systems with one-bit ADCs. IEEE Access. **7**, 27172–27182 (2019). https://doi.org/10.1109/ACCESS.2019.2901942
14. J. Choi, J. Mo, R. W. Heath, Near maximum-likelihood detector and channel estimator for uplink multiuser massive MIMO systems with one-bit ADCs. IEEE Trans. Commun. **64**(5), 2005–2018 (2016). https://doi.org/10.1109/TCOMM.2016.2545666
15. C. K. Wen, C. J. Wang, S. Jin, K. K. Wong, P. Ting, Bayes-optimal joint channel-and-data estimation for massive MIMO with low-precision ADCs. IEEE Trans. Signal Process. **64**(10), 2541–2556 (2016). https://doi.org/10.1109/TSP.2015.2508786
16. J. Mo, P. Schniter, R. W. Heath, Channel estimation in broadband millimeter wave MIMO systems with few-bit ADCs. IEEE Trans. Signal Process. **66**(5), 1141–1154 (2018). https://doi.org/10.1109/TSP.2017.2781644
17. J. Mo, A. Alkhateeb, S. Abu-Surra, R. W. Heath, Hybrid architectures with few-bit ADC receivers: achievable rates and energy-rate tradeoffs. IEEE Trans. Wirel. Commun. **16**(4), 2274–2287 (2017). https://doi.org/10.1109/TWC.2017.2661749
18. Z. Shao, L. T. N. Landau, R. C. de Lamare, in *ICASSP 2019 - 2019 IEEE International Conference on Acoustics, Speech and Signal Processing (ICASSP)*. Channel estimation using 1-bit quantization and oversampling for large-scale multiple-antenna systems, (2019), pp. 4669–4673. https://doi.org/10.1109/ICASSP.2019.8683750
19. J. J. Bussgang, Crosscorrelation functions of amplitude-distorted Gaussian signals. MIT Res. Lab. Elec. Tech. Rep. **216**, 1–14 (1952)
20. S. M. Kay, *Fundamentals of Statistical Signal Processing: Estimation Theory, 1st edn.* (Prentice Hall, New Jersey, 2000)

21. Y. Li, C. Tao, G. Seco-Granados, A. Mezghani, A. L. Swindlehurst, L. Liu, Channel estimation and performance analysis of one-bit massive MIMO systems. IEEE Trans. Signal Process. **65**(15), 4075–4089 (2017). https://doi.org/10.1109/TSP.2017.2706179
22. B. Clerckx, G. Kim, S. Kim, in *2008 IEEE Global Telecommunications Conference*. Correlated fading in broadcast MIMO channels: curse or blessing? (2008), pp. 1–5. https://doi.org/10.1109/GLOCOM.2008.ECP.735
23. J. Choi, D. J. Love, D. R. Brown, in *2014 48th Asilomar Conference on Signals, Systems and Computers*. Channel estimation techniques for quantized distributed reception in MIMO systems, (2014), pp. 1066–1070. https://doi.org/10.1109/ACSSC.2014.7094619
24. G. Jacovitti, A. Neri, Estimation of the autocorrelation function of complex Gaussian stationary processes by amplitude clipped signals. IEEE Trans. Inf. Theory. **40**(1), 239–245 (1994). https://doi.org/10.1109/18.272490
25. N. Shariati, E. Björnson, M. Bengtsson, M. Debbah, Low-complexity polynomial channel estimation in large-scale MIMO with arbitrary statistics. IEEE J. Sel. Top. Signal Process. **8**(5), 815–830 (2014). https://doi.org/10.1109/JSTSP.2014.2316063
26. S. N. Diggavi, T. M. Cover, The worst additive noise under a covariance constraint. IEEE Trans. Inf. Theory. **47**(7), 3072–3081 (2001). https://doi.org/10.1109/18.959289
27. J. Choi, B. Clerckx, N. Lee, G. Kim, A new design of polar-cap differential codebook for temporally/spatially correlated MISO channels. IEEE Trans. Wirel. Commun. **11**(2), 703–711 (2012). https://doi.org/10.1109/TWC.2011.122211.110313
28. J. H. Kotecha, A. M. Sayeed, Transmit signal design for optimal estimation of correlated MIMO channels. IEEE Trans. Signal Process. **52**(2), 546–557 (2004). https://doi.org/10.1109/TSP.2003.821104
29. E. Bjornson, B. Ottersten, A framework for training-based estimation in arbitrarily correlated Rician MIMO channels with Rician disturbance. IEEE Trans. Signal Process. **58**(3), 1807–1820 (2010). https://doi.org/10.1109/TSP.2009.2037352
30. J. Choi, D. J. Love, P. Bidigare, Downlink training techniques for FDD massive MIMO systems: open-loop and closed-loop training with memory. IEEE J. Sel. Top. Signal Process. **8**(5), 802–814 (2014). https://doi.org/10.1109/JSTSP.2014.2313020
31. U. Gustavsson, C. Sanchéz-Perez, T. Eriksson, F. Athley, G. Durisi, P. Landin, K. Hausmair, C. Fager, L. Svensson, in *2014 IEEE Globecom Workshops (GC Wkshps)*. On the impact of hardware impairments on massive MIMO, (2014), pp. 294–300. https://doi.org/10.1109/GLOCOMW.2014.7063447
32. J. Bibby, Axiomatisations of the average and a further generalisation of monotonic sequences. Glasg. Math. J. **15**(1), 63–65 (1974). https://doi.org/10.1017/S0017089500002135

10

Pilot allocation scheme based on coalition game for TDD massive MIMO systems

Hui Zhi[1*] and Xiaoguang Ding[2]

Abstract

Pilot contamination is a major factor that restricts the system performance of time division duplex (TDD) massive multiple-input multiple-output (MIMO) systems. Reasonable pilot allocation is an effective method to reduce pilot contamination. However, most of the existing pilot allocation methods focus on cells of the same size, a regular shape, an equal number of users per cell, and/or independent fading channels. When they are generalized or modified to actual scenes (in which cells have irregular shapes, users are randomly distributed, and channels are correlated fading channels), they are inflexible and cannot guarantee the target performance requirement. Considering both flexibility of pilot allocation and system performance, this paper introduces the idea of a coalition game into pilot allocation and proposes a pilot allocation scheme based on the coalition game. Different utility functions are analyzed under correlated fading channels; the adjustment principle of coalition structure and the coalition formation algorithm are given. Simulation results show that the proposed scheme can be used in any actual scene as well as guarantee the target performance requirement.

Keywords: Massive MIMO, Pilot allocation scheme, Coalition game

1 Introduction

1.1 The research background of this paper

With the appearance of smartphones and tablets, wireless communication requires increasingly higher communication rates. Therefore, many technologies have been proposed to improve the communication rate of 5G, and massive multiple-input multiple-output (MIMO) is one of these technologies. Compared with traditional wireless networks, massive MIMO significantly enhances the spectral and power efficiency [1].

However, the excellent performance of massive MIMO depends on the accuracy of channel estimation. For conventional channel estimation in time division duplex (TDD) massive MIMO systems, users transmit pilot signals in the uplink, and base stations (BSs) receive the pilot signal and estimate channel state information (CSI). Then, BSs can use the CSI for uplinking received signal detection and downlinking transmission precoding. Therefore, the accuracy of channel estimation is one of the key factors that affect the system performance of uplink and downlink. In pilot-based channel estimation, the number of pilots is finite and limited by the channel coherence time, so mobile users in different cells might have to reuse the same pilot. Pilot reuse seriously affects the channel estimation accuracy at BSs and further influences the system performance of uplink and downlink, where the influence is known as pilot contamination [2]. Recent studies have shown that pilot contamination is one of the key factors that affect the performance of TDD massive MIMO systems [1–4]. Therefore, many methods have been proposed to reduce pilot contamination, such as pilot contamination precoding [4], time-shifted pilots [5, 6], spatial domain-based pilot allocation [7], design and transmit power control of pilot sequences [8, 9], and pilot reuse [10, 11].

For instance, reference [4] shows that the performance of TDD massive MIMO system is limited by pilot contamination. Furthermore, reference [4] considers the uplink precoding and downlink precoding (called pilot contamination precoding) based on slow fading coefficients to reduce pilot contamination. Papers [5, 6]

* Correspondence: zhihui_0902@163.com
[1]Key Lab of Intelligent Computing and Signal Processing (Ministry of Education), Anhui University, Hefei 230601, China
Full list of author information is available at the end of the article

discuss the time-shifted pilot method in hexagonal cells, which means cells are divided into different groups, and cells in different groups transmit pilots in different time slots; analyses show that this method can effectively reduce pilot contamination. However, this method assumes that the number of users in each cell is fixed and equal. In addition, it requires strict system synchronization, making implementation complex and difficult. Paper [7] proposes a spatial domain method with the main idea that users with high channel orthogonality can use the same pilot. This method also assumes that cells have a regular shape and that the number of users in each cell is fixed and equal. Furthermore, this method is highly complex because BSs need to know the CSIs between all users and all BSs to carry out the search for highly orthogonal channels. Paper [8] gives a design for non-orthogonal pilot signals, and paper [9] proposes a pilot power allocation method based on cell grouping to alleviate pilot contamination. The methods in [8, 9] both assume that cells have a regular shape and the number of users per cell is fixed. Paper [10] presents a pilot reuse method based on cell grouping and analyzes the system performance of different pilot reuse factors. In paper [11], a fractional pilot reuse method is proposed in which users are classified into cell-center users and cell-edge users, where the cell-center users reuse the same pilot subset and the cell-edge users use the remaining pilots according to cell grouping.

Studies [12, 13] try to introduce game theory into pilot allocation to reduce pilot contamination. However, the pilot allocation methods in [12, 13] both consider the games between cells. Simply because of the games between cells, the suitable application scene of [12, 13] is still fixed and has an equal number of users in each cell, although reference [13] attempts to discuss irregularly shaped cells.

Reviewing the studies in [4–13], reasonable pilot allocation can effectively reduce pilot contamination and improve system performance. However, in order to facilitate theoretical analysis and simulation, most of these pilot allocation methods assume that cells have a regular shape (most are hexagons), the number of users per cell is equal, and/or channels are independent. However, in reality, the cell shapes are affected by various factors, such as landforms and buildings, and do not always have a regular shape. Moreover, mobile users are randomly distributed in reality, so it is impossible for the number of users in each cell to be equal or fixed. When the number of base station antennas is very large, it is difficult to ensure the distance between the antennas, and it is difficult to ensure that all channels are independent.

Most of these pilot allocation methods (in references [4–13]) assume that the number of users per cell is equal or fixed for two main reasons. First, when the number of users in each cell is equal or fixed, it is easy to allocate pilots. For example, in references [4–13], in order to avoid the use of the same pilot in adjacent cells (pilot contamination is large in this case), these pilot allocation methods [4–13] always reserve a certain number of pilots for each cell. If the number of users in each cell is equal or fixed, it is convenient to estimate the number of pilots per cell, so that when pilots are allocated, the pilots of adjacent cells can be differentiated. Second, when the number of users in each cell is equal, the dimension of the channel matrix in each cell is equal, which is convenient for merging in mathematical operation in theoretical analysis.

If we modify these methods [4–13] to the actual scenes in which all users are randomly distributed and the number of users per cell is unequal, the system performance will not be guaranteed. For example, if the number of users is large in a cell, using these methods [4–13], pre-allocated pilots are insufficient, and pilot reuse will cause large pilot contamination. On the other hand, if the number of users in a cell is small, using these methods [4–13], the pilot resource cannot be fully utilized, and this will cause the waste of pilot resources and then affect overall performance. Obviously, this point reflects that these methods [4–13] lack flexibility.

The assumption of a regular cell shape is also beneficial to pilot allocation. When the cell shape is regular, cells can be grouped first. For example, as shown in Fig. 1a, taking pilot reuse factor 3 as an example, the cells in the picture have three types of background pattern. The cells with the same background pattern are grouped together and reuse the same pilot subset (the total pilot set has been divided into three orthogonal pilot subsets). Cells with different background patterns use different orthogonal pilot subsets. References [9–11] use similar cell grouping pilot allocation methods. However, the actual situation is that the cell shape is irregular. The pilot multiplexing in 11 cells shown in Fig. 1b is based on the pilot allocation method in Fig. 1a. It can be seen that in Fig. 1b, because of the irregular cell shape, the same pilot subset is used in adjacent cells, which will cause serious pilot contamination; thus, the system performance cannot be guaranteed.

So we can conclude that if we modify these methods [4–13] to the actual scenes (in which cells have irregular shapes, users are randomly distributed), the system performance will not be guaranteed.

Papers [14, 15] propose different greedy algorithms to allocate pilots in the scene of correlated fading channels. When compared with the methods in [4–13], greedy algorithms in papers [14, 15] are flexible and can be easily

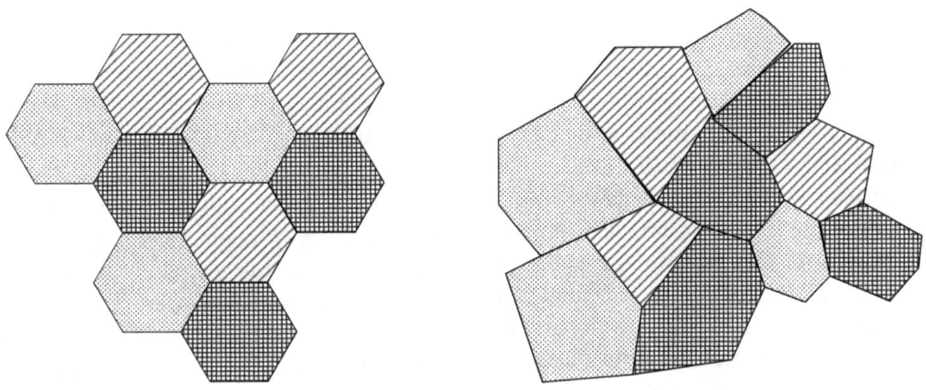

Fig. 1 Pilot allocation based on cell grouping

generalized or modified to the actual scenes, but they still cannot guarantee the target performance requirement. Paper [14] presents a statistical greedy pilot scheduling algorithm, which focuses on users in a single cell and searches users with small channel correlations to use the same pilot. This algorithm does not consider the application scenario of multi-cells and the correlation of the channels between different users and different BSs, so it cannot reduce the interference among different cells and thus cannot guarantee system performance in multi-cell scenarios. That is to say, although this algorithm can be generalized to multi-cell scenarios of arbitrary BS locations and an unequal number of users per cell, it cannot guarantee target performance satisfaction. Thus, it is essentially different from our work; the proposed pilot allocation scheme in our work can be used in arbitrary scenarios as well as guarantee system performance.

Paper [15] proposes a greedy approach; its main idea is selecting one user from each cell (there are P users per cell) to form a user set during each search, and users in the same set use the same pilot. The principle of user selection is to minimize the sum mean square error of channel estimation for the users using the same pilot. The greedy approach assumes the number of users per cell is equal; it cannot be used in the case of different numbers of users per cell, so it is different from our work and we have to make modifications in simulations. In addition, it only considers the sum-mean-square error of users with the same pilot and does not consider the total performance of all users; furthermore, it is a one-time search for all users and cannot achieve global optimization, whereas the proposed pilot allocation scheme in our work considers the total performance of all users and performs a cyclic search for all users. Therefore, in comparison, the greedy approach in [15] cannot meet the needs of total performance, whereas our scheme can meet overall performance requirements through cyclic search.

1.2 The main contents of this paper

In order to find the flexible pilot allocation method that can be used in any actual scenes and provide performance guarantee, we introduce the idea of a coalition game into pilot allocation and propose a pilot allocation scheme based on the coalition game in this paper. We consider the game between users and divide users into different sub-coalitions. Because this coalition game is among users, the proposed pilot allocation scheme is more flexible and can be used in any actual scene. Because the purpose of each coalition adjustment is to improve the target performance (average utility function), the target performance can be satisfied through the cyclic search of coalition adjustment. The contributions of this paper are as follows.

1) To be applied in arbitrary actual scene (with irregular shape cells, unequal number of users per cell, and correlated fading channels) and provide target performance guarantee, a novel pilot allocation scheme based on the coalition game is proposed.
   For the game between users, the coalition structure is defined, different utility functions are analyzed, and the coalition formation algorithm is given.
2) The implementation method of the proposed pilot allocation scheme is presented.
3) Simulations are developed to show that the proposed scheme can be used in any actual scene, effectively reduce pilot contamination, and provide a certain target performance guarantee. Moreover, the comparisons of computational complexity between different schemes are discussed.

The remainder of this paper is organized as follows. The system model for the massive MIMO system is described in Section 2. Different utility functions are analyzed under correlated fading channels in Section 3, including mean square error of channel estimation (MSE-CE), mean square error of signal detection (MSE-SD), received signal-to-interference-plus-noise ratio (SINR), and spectrum efficiency (SE). In Section 4, the definition of the coalition game is given, and then the coalition structure, the adjustment principle of coalition structure, and the condition of final stability are defined. After that, the coalition formation algorithm is given, a pilot allocation scheme based on the coalition game is proposed, and the implementation method of the proposed scheme is discussed. In Section 5, simulations are developed with respect to different utility functions, and the proposed pilot allocation scheme is compared with the greedy approach in reference [15] and the random coalition scheme. Section 6 concludes this paper. The proof of theorem is given in the Appendices 1, 2, and 3.

2 System model

Consider the massive MIMO system as shown in Fig. 2. There are L cells and N users in a fixed area, where we focus on the uplink. Each cell is assigned an index in the set $\mathscr{L}=\{1,2,\cdots,L\}$. The kth user in the jth cell is denoted by user (j,k). Assume cell j has K_j active users, thus $\sum_{j=1}^{L} K_j = N$. All users form the set $\mathcal{N}$ with $\mathcal{N}=\{(1,1),\cdots(1,K_1),\cdots(j,1),\cdots(j,K_j),\cdots(L,1),\cdots(L,K_L)\}$. Each cell has one BS, and one cell and its BS use the same index number, i.e., BS j is in the jth cell. Each BS is equipped with an array of M antennas, the shape of antennas at BSs is uniformly spaced linear array (ULA) [15], and each user is equipped with a single antenna. Assume that BSs are subject to a random distribution and that the users are subject to another random distribution, the distributions of BSs and users are independent. Each user selects and accesses its nearest BS according to the strength of the received BS signals, so the number of users in each cell (K_j, $j\in\{1,2,\cdots,L\}$) is not equal.

Assume $z_{jk}^{u}\in\mathrm{R}^2$ and $z_l^{b}\in\mathrm{R}^2$ are the locations of user (j,k) and BS l respectively, and z_{jk} is known at all BSs. $d_l(z_{jk})$ is the distance from user (j,k) to BS l. The channel between user (j,k) and BS l is defined as $\boldsymbol{h}_{ljk}\in\mathbb{C}^{M\times 1}$, and we consider correlated Rayleigh fading channels [15–17]. (For other types of correlated channel models and independent channel models, our pilot allocation scheme is also applicable.) The channel $\boldsymbol{h}_{ljk}$ is the superposition of F arriving paths,

$$\boldsymbol{h}_{ljk}=\frac{1}{\sqrt{F}}\sum_{f=1}^{F} a\left(\theta_{ljk}^{(f)}\right)\alpha_{ljk}^{(f)} \tag{1}$$

where $a(\theta_{ljk}^{(f)})\in\mathbb{C}^{M\times 1}$ is the antenna steering vector corresponding to angle-of-arrivals (AoAs) $\theta_{ljk}\in[0,2\pi)$, f is the path index, and $\alpha_{ljk}^{(f)}\overset{i.i.d.}{\sim}\mathcal{CN}(0,\ \beta_{ljk})$ is the channel coefficient of the fth path, β_{ljk} denotes the large-scale fading coefficient and $\beta_{ljk}=\delta\|z_{jk}^{u}-z_l^{b}\|_2^{-v}$, where v is the path loss exponent, δ is a constant of channel coefficient, and we let δ be a constant for arbitrary path f and arbitrary channel $\boldsymbol{h}_{ljk}$ in our simulations.

The expression of $a(\theta)$ is

$$a(\theta)=\begin{bmatrix} 1 \\ e^{-j2\pi\frac{D}{\lambda}\cos(\theta)} \\ \vdots \\ e^{-j2\pi(M-1)\frac{D}{\lambda}\cos(\theta)} \end{bmatrix} \tag{2}$$

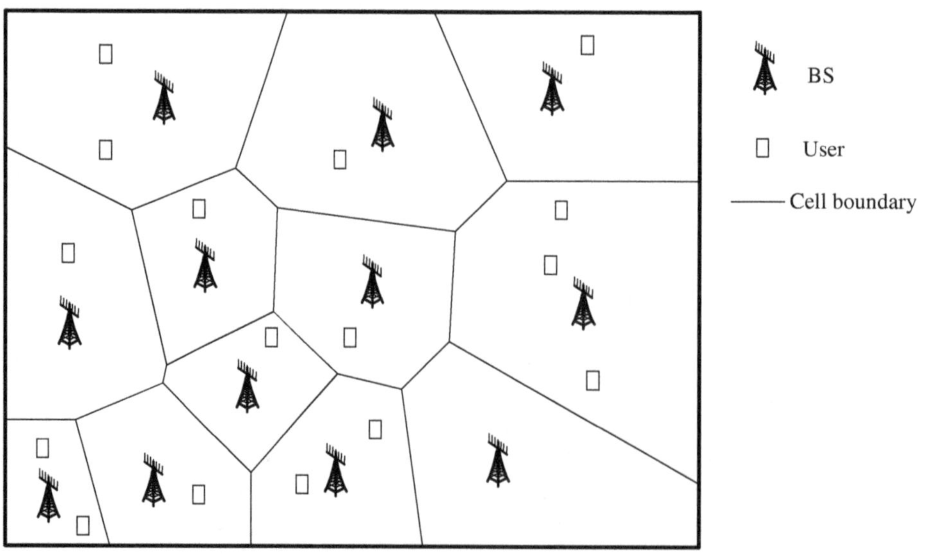

Fig. 2 A massive MIMO system with irregularly shaped cells and randomly located users

where D is the interval of antenna spacing at the BS, and λ is the signal wavelength.

As an approximation, we consider a ring of radius r_s comprising many scatters around the user. In this case [16], θ_{ljk} obeys a uniform distribution on $(\theta_{ljk}^{\min}, \theta_{ljk}^{\max})$, where $\theta_{ljk}^{\min} < \theta_{ljk}^{\max} \in [0, \pi)$, and

$$\theta_{ljk}^{\min} = \arctan\left(\frac{\left[z_{jk}^{u}\right]_2 - \left[z_l^b\right]_2}{\left[z_{jk}^{u}\right]_1 - \left[z_l^b\right]_1}\right) - \arcsin\left(\frac{r_s}{\left\|z_{jk}^{u} - z_l^b\right\|_2}\right) \tag{3}$$

$$\theta_{ljk}^{\max} = \arctan\left(\frac{\left[z_{jk}^{u}\right]_2 - \left[z_l^b\right]_2}{\left[z_{jk}^{u}\right]_1 - \left[z_l^b\right]_1}\right) + \arcsin\left(\frac{r_s}{\left\|z_{jk}^{u} - z_l^b\right\|_2}\right) \tag{4}$$

The channel covariance matrix is $\mathbf{R}_{ljk} = \mathbb{E}\{\boldsymbol{h}_{ljk}\boldsymbol{h}_{ljk}^H\} = \frac{1}{F}\beta_{ljk}\sum_{f=1}^{F}\int_{\theta_{ljk}^{\min}}^{\theta_{ljk}^{\max}} a(\theta_{ljk}^{(f)})[a(\theta_{ljk}^{(f)})]^H d\theta_{ljk}^{(f)}$, where $\mathbb{E}\{\cdot\}$ denotes the expectation.

3 Analysis of different utility functions

The uplink transmission time-frequency resources are divided into frames consisting of T_c seconds and B_c Hz; thus, each frame contains $Q = T_cB_c$ transmission symbols (P pilot symbols and Q-P data symbols, pilot symbols and data symbols are transmitted in different time slots). Assume channels between BSs and users have constant channel response within a frame but vary between frames. Therefore, the number of available orthogonal pilot sequences is P.

Assume that the coalition game divides all users into P sub-coalitions ($N > P$). P pilots are allocated to the P sub-coalitions respectively. That is, users in the same sub-coalition use the same pilot, users in different sub-coalitions use different pilots, and P pilots are mutually orthogonal. The analysis of utility function must be carried out under a certain coalition structure, so we first give the coalition structure definition in definition 1.

Definition 1 *(Coalition structure). User set $\mathcal{N}$ is divided into P disjoint sub-coalitions $\mathcal{S}_1, \mathcal{S}_2, \cdots, \mathcal{S}_P$, and $\cup_{n=1}^{P}\mathcal{S}_n = \mathcal{N}$, $\cap_{n=1}^{P}\mathcal{S}_n = \varnothing$. Coalition structure $\mathcal{S}$ is the coalitional division state, i.e., $\mathcal{S} = \{\mathcal{S}_1, \mathcal{S}_2, \cdots, \mathcal{S}_P\}$.*

3.1 Analysis of MSE-CE

MSE-CE means the mean square error of channel estimation. During the uplink pilot transmission, all users transmit pilots at the same time, thus the received pilot signal at the BS l is

$$\boldsymbol{y}_l^{\text{pilot}} = \sum_{j=1}^{L}\sum_{k=1}^{K_j}\sqrt{t_{jk}}\ \boldsymbol{h}_{ljk}\boldsymbol{w}_{jk}^H + \boldsymbol{N}_l^{\text{pilot}} \tag{5}$$

where $\boldsymbol{w}_{jk} \in \mathbb{C}^{P\times 1}$ is the pilot sequence used by user (j, k), and $\mathbb{E}\{\boldsymbol{w}_{jk}^H\boldsymbol{w}_{jk}\} = 1$. t_{jk} is the pilot transmit power of user (j, k). In order to avoid the near-far issue in the uplink, here we consider the transmit power control and let $t_{jk} = T/\beta_{jjk}$, where T is a design parameter of transmit power. $\boldsymbol{N}_l^{\text{pilot}} \in \mathbb{C}^{M\times P}$ is the equivalent additive white Gaussian noise (AWGN) at the receiver with independent and identically distributed (i.i.d.) elements of $\mathcal{CN}(0, \sigma^2)$, where σ^2 is the variance of each element in noise .

After decorrelation and power normalization of $\boldsymbol{y}_l^{\text{pilot}}$, BS l can obtain the channel observation of $\boldsymbol{h}_{ljk}$, which is

$$\bar{\bar{\boldsymbol{h}}}_{ljk} = \frac{\boldsymbol{y}_l^{\text{pilot}}}{\sqrt{t_{jk}}}\boldsymbol{w}_{jk} = \boldsymbol{h}_{ljk} + \sum_{\substack{(m,n)\in\Lambda_{\mathcal{S}}(j,k)\\(m,n)\neq(j,k)}}\frac{\sqrt{t_{mn}}}{\sqrt{t_{jk}}}\boldsymbol{h}_{lmn} + \frac{1}{\sqrt{t_{jk}}}\boldsymbol{N}_l^{\text{pilot}}\boldsymbol{w}_{jk} \tag{6}$$

where $\Lambda_{\mathcal{S}}(j,k)$ is one of the sub-coalitions in coalition structure $\mathcal{S}$, and user (j, k) belongs to the sub-coalition $\Lambda_{\mathcal{S}}(j,k)$.

According to the definition of minimum mean square error (MMSE) estimation and expressions (14), (15) in reference [14], the MMSE estimation of channel $\boldsymbol{h}_{ljk}$ at BS l, which is based on the channel observation $\bar{\bar{\boldsymbol{h}}}_{ljk}$, is given by

$$\hat{\boldsymbol{h}}_{ljk} = \mathbf{R}_{ljk}\mathbf{C}_{ljk}^{-1}\bar{\bar{\boldsymbol{h}}}_{ljk} \tag{7}$$

where $\mathbf{C}_{ljk} = \sum_{(m,n)\in\Lambda_{\mathcal{S}}(j,k)}\frac{t_{mn}}{t_{jk}}\mathbf{R}_{lmn} + \frac{\sigma^2}{t_{jk}}\mathbf{I}_M$ is the covariance matrix of $\bar{\bar{\boldsymbol{h}}}_{ljk}$.

Similar to expression (7), we can obtain the channel estimation of $\boldsymbol{h}_{jjk}$ (the channel between user (j, k) and its own BS j), which is

$$\begin{aligned}\hat{\boldsymbol{h}}_{jjk} &= \mathbf{R}_{jjk}\mathbf{C}_{jjk}^{-1}\bar{\bar{\boldsymbol{h}}}_{jjk}\\ &= \mathbf{R}_{jjk}\mathbf{C}_{jjk}^{-1}\left(\boldsymbol{h}_{jjk} + \sum_{\substack{(m,n)\in\Lambda_{\mathcal{S}}(j,k)\\(m,n)\neq(j,k)}}\frac{\sqrt{t_{mn}}}{\sqrt{t_{jk}}}\boldsymbol{h}_{jmn} + \frac{1}{\sqrt{t_{jk}}}\boldsymbol{N}_j^{\text{pilot}}\boldsymbol{w}_{jk}\right)\end{aligned} \tag{8}$$

with covariance $\mathbf{R}_{\hat{h}_{jjk}} = \mathbf{R}_{jjk}\mathbf{C}_{jjk}^{-1}\mathbf{R}_{jjk}$. Thus, we can determine that the channel estimation error is $\tilde{\boldsymbol{h}}_{jjk} = \boldsymbol{h}_{jjk} - \hat{\boldsymbol{h}}_{jjk}$. From the orthogonality principle of MMSE estimation [14, 18], channel estimation error $\tilde{\boldsymbol{h}}_{jjk}$ is independent of $\hat{\boldsymbol{h}}_{jjk}$ and the covariance of $\tilde{\boldsymbol{h}}_{jjk}$ is $\mathbf{R}_{\tilde{h}_{jjk}} = \mathbf{R}_{jjk} - \mathbf{R}_{\hat{h}_{jjk}} = \mathbf{R}_{jjk} - \mathbf{R}_{jjk}\mathbf{C}_{jjk}^{-1}\mathbf{R}_{jjk}$. Note that $\hat{\boldsymbol{h}}_{jjk}$ and $\mathbf{R}_{\tilde{h}_{jjk}}$ are also mean and covariance of $\boldsymbol{h}_{jjk}$ conditioned on $\bar{\bar{\boldsymbol{h}}}_{ljk}$, respectively [18].

The mean square error of channel estimation (MSE-CE) $\hat{\boldsymbol{h}}_{jjk}$ means the sum of squares estimation errors for each element of $\boldsymbol{h}_{jjk}$ (you can refer to expression (18) in [14]); so the MSE-CE of $\hat{\boldsymbol{h}}_{jjk}$ is $\text{MSE-CE}_{jk} = \varepsilon_{\tilde{h}_{jjk}}^{\text{MSE-CE}} = \mathbb{E}\{(\tilde{\boldsymbol{h}}_{jjk})^H \tilde{\boldsymbol{h}}_{jjk}\} = \mathbb{E}\{\text{tr}\{\tilde{\boldsymbol{h}}_{jjk}\tilde{\boldsymbol{h}}_{jjk}^H\}\} = \text{tr}\{\mathbf{R}_{\tilde{h}_{jjk}}\}$. We define the sum MSE-CE of all users as the sum of the estimation mean square error (MSE) of the channels between each user and its own BS. Thus, the average MSE-CE per user is

$$\text{MSE-CE}_{\text{ave}} = \frac{1}{N}\sum_{j=1}^{L}\sum_{k=1}^{K_j} \text{tr}\left\{\mathbf{R}_{\tilde{h}_{jjk}}\right\} \tag{9}$$

3.2 Analysis of MSE-SD and received SINR

MSE-SD means the mean square error of signal detection. During the uplink data transmission, all users send their data to all BSs at the same time, and the received signal at BS j is

$$\begin{aligned} \boldsymbol{y}_j^{\text{data}} &= \sum_{m=1}^{L}\sum_{n=1}^{K_m} \sqrt{t_{mn}^d}\, \boldsymbol{h}_{jmn} x_{mn} + \boldsymbol{n}_j^{\text{data}} \\ &= \sum_{m=1}^{L}\sum_{n=1}^{K_m} \boldsymbol{g}_{jmn}\, x_{mn} + \boldsymbol{n}_j^{\text{data}} \end{aligned} \tag{10}$$

where $x_{mn} \in \mathbb{C}^{1\times 1}$ represents the symbol transmitted from user (m, n), $\mathbb{E}\{|x_{mn}|^2\} = 1$, $\boldsymbol{y}_j^{\text{data}} = [y_{j1}, \cdots, y_{jM}]^T \in \mathbb{C}^{M\times 1}$, $\boldsymbol{n}_j^{\text{data}}$ is AWGN, and $\boldsymbol{n}_j^{\text{data}} \sim \mathcal{CN}(0, \sigma^2\mathbf{I}_M)$, $\boldsymbol{g}_{jmn} = \sqrt{t_{mn}^d}\boldsymbol{h}_{jmn}$.

Let $\boldsymbol{H}_{jm} = [\boldsymbol{h}_{jm1}, \boldsymbol{h}_{jm2}, \cdots, \boldsymbol{h}_{jmK_m}] \in \mathbb{C}^{M\times K_m}$, and $\boldsymbol{T}_m^d = \text{diag}(\sqrt{t_{m1}^d}, \cdots, \sqrt{t_{mK_m}^d})$, $\boldsymbol{x}_m = [x_{m1}, \cdots, x_{mK_m}]^T$, $\boldsymbol{G}_{jm} = \boldsymbol{H}_{jm}\boldsymbol{T}_m^d = [\boldsymbol{g}_{jm1}, \cdots, \boldsymbol{g}_{jmK_m}]$; thus, the received signal at BS j can be rewritten as $\boldsymbol{y}_j^{\text{data}} = \sum_{m=1}^{L} \boldsymbol{H}_{jm}\, \boldsymbol{T}_m^d \boldsymbol{x}_m + \boldsymbol{n}_j^{\text{data}} = \sum_{m=1}^{L} \boldsymbol{G}_{jm}\boldsymbol{x}_m + \boldsymbol{n}_j^{\text{data}}$.

To distinguish the symbol of each user, the BS j selects the combination matrix $\boldsymbol{W}_{jj}^{\text{scheme}} \in \mathbb{C}^{M\times K_j}$, which is directly multiplied with the received signal, and the processed data of all K_j users in the jth cell can be written as

$$\hat{\boldsymbol{x}}_j = \left(\boldsymbol{T}_j^d\right)^{-1}\left(\boldsymbol{W}_{jj}^{\text{scheme}}\right)^H \boldsymbol{y}_j^{\text{data}} \tag{11}$$

Considering the maximum ratio combining (MRC), zero-forcing combining (ZFC), and MMSE schemes of the received signal $\boldsymbol{y}_j^{\text{data}}$, thus

$$\boldsymbol{W}_{jj}^{\text{scheme}} = \begin{cases} \hat{\boldsymbol{H}}_{jj}\boldsymbol{D}_{jj}, & \text{for MRC} \\ \hat{\boldsymbol{H}}_{jj}\left(\hat{\boldsymbol{H}}_{jj}^H\hat{\boldsymbol{H}}_{jj}\right)^{-1}, & \text{for ZRC} \\ \hat{\boldsymbol{H}}_{jj}\left(\hat{\boldsymbol{H}}_{jj}^H\hat{\boldsymbol{H}}_{jj} + \sigma^2\left(\boldsymbol{T}_j^d\right)^{-2}\right)^{-1}, & \text{for MMSE} \end{cases} \tag{12}$$

where $\hat{\boldsymbol{H}}_{jj} = [\hat{\boldsymbol{h}}_{jj1}, \hat{\boldsymbol{h}}_{jj2}, \cdots, \hat{\boldsymbol{h}}_{jjK_j}] \in \mathbb{C}^{M\times K_j}$, and $\boldsymbol{T}_j^d = \text{diag}(\sqrt{t_{j1}^d}, \cdots, \sqrt{t_{jK_j}^d})$, $\boldsymbol{D}_{jj} = \text{diag}(\frac{1}{\delta_{jj1}}, \cdots, \frac{1}{\delta_{jjK_j}})$ is the diagonal matrix parameter that is set for $\mathbb{E}\{(\boldsymbol{w}_{jjk}^{MRC})^H \boldsymbol{h}_{jjk}\} = 1$, and $\delta_{jjk} = \text{tr}\{\mathbf{R}_{\hat{h}_{jjk}}\}$.

Theorem 1 *For a given coalition structure $\mathcal{S}$, the mean square error of signal detection (MSE-SD) for user (j, k) at BS j for MRC, ZFC, and MMSE schemes are given by the following expressions* (13), (14) *and* (15) *respectively,*

$$\begin{aligned} \text{MSE-SD}_{jk}^{\text{mrc}} = &\frac{1}{\delta_{jjk}^2}\text{sum}\left\{\mathbf{R}_{\tilde{h}_{jjk}} \circ \mathbf{R}_{\hat{h}_{jjk}}^T\right\} + \sum_{(l,m)\notin\Lambda(j,k)} \frac{t_{lm}^d}{t_{jk}^d}\frac{1}{\delta_{jjk}^2}\text{sum}\left\{\mathbf{R}_{jlm} \circ \mathbf{R}_{\hat{h}_{jjk}}^T\right\} \\ &+ \sum_{(l,m)\in\Lambda(j,k)} \frac{t_{lm}^d}{t_{jk}^d}\frac{1}{\delta_{jjk}^2}\left\{\frac{t_{lm}^d}{t_{jk}^d}\left|\text{sum}\left\{\left(\mathbf{R}_{jlm}\mathbf{R}_{jjk}^{-1}\right)\circ\mathbf{R}_{\hat{h}_{jjk}}^T\right\}\right|^2 \right. \\ &\left. + \text{sum}\left\{\mathbf{R}_{\tilde{h}_{jlm}} \circ \mathbf{R}_{\hat{h}_{jjk}}^T\right\}\right\} + \frac{\sigma^2}{t_{jk}^d\delta_{jjk}} - 1 \end{aligned} \tag{13}$$

$$\begin{aligned} \text{MSE-SD}_{jk}^{\text{zfc}} = &\sum_{(l,m)\notin\Lambda(j,k)} \frac{t_{lm}^d}{t_{jk}^d}\text{sum}\left\{\mathbf{R}_{jlm}\circ\boldsymbol{Q}_1^T\right\} \\ &+ \sum_{(l,m)\in\Lambda(j,k)} \frac{t_{lm}^d}{t_{jk}^d}\left\{\frac{t_{lm}^d}{t_{jk}^d}\left|\text{sum}\left\{\left(\mathbf{R}_{jlm}\mathbf{R}_{jjk}^{-1}\right)\circ\boldsymbol{Q}_2^T\right\}\right|^2 \right. \\ &\left. + \text{sum}\left\{\mathbf{R}_{\tilde{h}_{jlm}}\circ\boldsymbol{Q}_1^T\right\}\right\} + \frac{\sigma^2}{t_{jk}^d}\cdot[\mathbf{Q}_3]_{k,k} - 1 \end{aligned} \tag{14}$$

$$\begin{aligned} \text{MSE-SD}_{jk}^{\text{mmse}} = &\sum_{(l,m)\notin\Lambda(j,k)} \frac{t_{lm}^d}{t_{jk}^d}\text{sum}\left\{\mathbf{R}_{jlm}\circ\boldsymbol{Q}_5^T\right\} \\ &+ \sum_{(l,m)\in\Lambda(j,k)} \frac{t_{lm}^d}{t_{jk}^d}\left\{\frac{t_{lm}^d}{t_{jk}^d}\left|\text{sum}\left\{\left(\mathbf{R}_{jlm}\mathbf{R}_{jjk}^{-1}\right)\circ\boldsymbol{Q}_6^T\right\}\right|^2 \right. \\ &\left. + \text{sum}\left\{\mathbf{R}_{\tilde{h}_{jlm}}\circ\boldsymbol{Q}_5^T\right\}\right\} + \frac{\sigma^2}{t_{jk}^d}\cdot[\mathbf{Q}_7]_{k,k} - 2[\mathbf{Q}_4]_{k,k} - 1 \end{aligned} \tag{15}$$

where $\boldsymbol{Q}_1 = \mathbb{E}_{\boldsymbol{h}}\{\boldsymbol{w}_{jjk}^{zfc}(\boldsymbol{w}_{jjk}^{zfc})^H\}$, $\boldsymbol{Q}_2 = \mathbb{E}_{\boldsymbol{h}}\{\hat{\boldsymbol{h}}_{jjk}(\boldsymbol{w}_{jjk}^{zfc})^H\}$, $\boldsymbol{Q}_3 = \mathbb{E}_{\boldsymbol{h}}\{(\hat{\boldsymbol{H}}_{jj}^H\hat{\boldsymbol{H}}_{jj})^{-1}\}$, $\boldsymbol{Q}_4 = \mathbb{E}_{\boldsymbol{h}}\{\sigma^2(\hat{\boldsymbol{H}}_{jj}^H\hat{\boldsymbol{H}}_{jj} + \sigma^2(\boldsymbol{T}_j^d)^{-2})^{-1}\}$,

$\boldsymbol{Q}_5 = \mathbb{E}_{\boldsymbol{h}}\{\boldsymbol{w}_{jjk}^{\mathrm{mmse}}(\boldsymbol{w}_{jjk}^{\mathrm{mmse}})^H\}$, $\boldsymbol{Q}_6 = \mathbb{E}_{\boldsymbol{h}}\{\hat{\boldsymbol{h}}_{jjk}(\boldsymbol{w}_{jjk}^{\mathrm{mmse}})^H\}$, $\boldsymbol{Q}_7 = \mathbb{E}_{\boldsymbol{h}}\{(\boldsymbol{W}_{jj}^{\mathrm{mmse}})^H\boldsymbol{W}_{jj}^{\mathrm{mmse}}\}$, $\boldsymbol{w}_{jjk}^{\mathrm{zfc}}$ is the kth column of the matrix $\boldsymbol{W}_{jj}^{\mathrm{zfc}}=\hat{\boldsymbol{H}}_{jj}(\hat{\boldsymbol{H}}_{jj}^H\hat{\boldsymbol{H}}_{jj})^{-1}$, and $\boldsymbol{w}_{jjk}^{\mathrm{mmse}}$ is the kth column of the matrix $\boldsymbol{W}_{jj}^{\mathrm{mmse}}=\hat{\boldsymbol{H}}_{jj}(\hat{\boldsymbol{H}}_{jj}^H\hat{\boldsymbol{H}}_{jj} + \sigma^2(\boldsymbol{T}_j^d)^{-2})^{-1}$.

For a given coalition structure $\mathcal{S}$, the uplink received SINR for user (j, k) at BS j for MRC, ZFC, and MMSE schemes are given by the expressions (17), (18) and (19) respectively.

Proof The proofs of Theorem 1 for MRC, ZFC, and MMSE schemes are given in Appendices 1, 2, and 3, respectively.

3.3 Analysis of spectrum efficiency

Based on previous theoretical analyses, according to Shannon capacity and referring to expression (11) in [19], the uplink spectrum efficiency (SE) of user (j,k) is

$$\mathrm{SE}_{jk}^{\mathrm{scheme}} = \left(1-\frac{P}{Q}\right)\cdot\log_2\left(1+\mathrm{SINR}_{jk}^{\mathrm{scheme}}\right) \tag{16}$$

where the $\mathrm{SINR}_{jk}^{\mathrm{scheme}}$ for MRC, ZFC, and MMSE schemes are given by expressions (17), (18), and (19), respectively.

4 Pilot allocation scheme based on a coalition game

As mentioned in Section 3, we use the idea of partition form [20] in the coalition game, and all users will be divided into P sub-coalitions ($N > P$). P pilots are allocated to the P sub-coalitions respectively. That is, users are players in the coalition game. Users in the same sub-coalition use the same pilot and interact with each other in channel estimation. Users in different sub-coalitions use different pilots, and they do not affect each other in channel estimation. Therefore, there are mutually influencing and mutually constraining relationships between users, which coincides with the idea of the coalition game. For this reason, we introduce the idea of the coalition game into pilot allocation. The definition of the coalition game is given as follows.

Definition 2 *(Coalition game). The coalition game is* $\mathcal{G} = \langle\mathcal{N}, \{u_{j,k}\}_{(j,k)\in\mathcal{N}}\rangle$, *where* $\mathcal{N}$ *is the set of game players; that is to say, users are players in the coalition game.* $u_{j,k}$ *is the utility function of player* (j, k) *(user* (j, k)*), and it is also called the payoff function in game theory.*

In this coalition game, each player must belong to a sub-coalition, and the definition of coalition structure is given in definition 1.

The purpose of the coalition game is to optimize utility function, so different performance metrics can be used as the utility function in the coalition game. In this paper, we use the MSE-CE ($u_{j,k}(\mathcal{S}) = \text{MSE-CE}_{jk}(\mathcal{S})$), MSE-SD ($u_{j,k}(\mathcal{S}) = \text{MSE-SD}_{jk}^{\mathrm{scheme}}(\mathcal{S})$), received SINR ($u_{j,k}(\mathcal{S}) = \mathrm{SINR}_{jk}^{\mathrm{scheme}}(\mathcal{S})$), and SE ($u_{j,k}(\mathcal{S}) = \mathrm{SE}_{jk}^{\mathrm{scheme}}(\mathcal{S})$) as the utility functions in the coalition game when we want to satisfy the corresponding system performance. The optimization of a certain utility function is also the minimization of pilot contamination corresponding to this performance.

According to the partition form definition in the coalition game theory [20], we need to describe three elements: (1) the coalition structure, (2) the adjustment principle of the coalition structure, and (3) the final stability of the coalition structure.

During each adjustment of the coalition structure, one user leaves its sub-coalition and joins another sub-coalition. This adjustment reduces interference of the original sub-coalition of the user and increases the interference of the newly joined sub-coalition.

The definition of coalition structure is given in definition 1. To benefit each user, the coalition structure needs to continuously adjust until final stability. For a certain adjustment, assume player (j, k) leaves its sub-coalition $\Lambda_{\mathcal{S}}(j, k)$ and joins another sub-coalition $\mathcal{S}_i$, where $\mathcal{S}_i \in \{\{\mathcal{S}_1, \mathcal{S}_2, \cdots, \mathcal{S}_P\}\backslash\Lambda_{\mathcal{S}}(j, k)\}$. That is, the entire coalition structure changes from the original $\mathcal{S}$ into a new structure $\mathcal{S}^{\Phi}$. We describe this adjustment as $\mathcal{S} \overset{(j,k)}{\rightarrow} \mathcal{S}^{\Phi}$. According to the individual stability concept in the coalition game [21], the adjustment of a user (j, k) leaving its sub-coalition $\Lambda_{\mathcal{S}}(j, k)$ and joining another sub-coalition $\mathcal{S}_i$ is permissible if this adjustment can

$$\mathrm{SINR}_{jk}^{\mathrm{mrc}} = \frac{1}{\sum\limits_{(l,m)\notin\Lambda(j,k)}^{L}\frac{t_{lm}^d}{t_{jk}^d}\frac{1}{\delta_{jjk}^2}\mathrm{sum}\left\{\mathbf{R}_{jlm}\circ\mathbf{R}_{\hat{h}_{jjk}}^T\right\} + \sum\limits_{(l,m)\in\Lambda(j,k)}^{L}\frac{t_{lm}^d}{t_{jk}^d}\frac{1}{\delta_{jjk}^2}\left\{\frac{t_{lm}^d}{t_{jk}^d}\left|\mathrm{sum}\left\{\left(\mathbf{R}_{jlm}\mathbf{R}_{jjk}^{-1}\right)\circ\mathbf{R}_{\hat{h}_{jjk}}^T\right\}\right|^2 + \mathrm{sum}\left\{\mathbf{R}_{\tilde{h}_{jlm}}\circ\mathbf{R}_{\hat{h}_{jjk}}^T\right\}\right\} + \frac{\sigma^2}{t_{jk}^d\delta_{jjk}} - 1} \tag{17}$$

$$\mathrm{SINR}_{jk}^{\mathrm{zfc}} = \frac{1}{\sum_{(l,m)\notin\Lambda(j,k)}^{L} \frac{t_{lm}^d}{t_{jk}^d} \mathrm{sum}\{\mathbf{R}_{jlm} \circ \mathbf{Q}_1^T\} + \sum_{(l,m)\in\Lambda(j,k)}^{L} \frac{t_{lm}^d}{t_{jk}^d} \left\{ \frac{t_{lm}^d}{t_{jk}^d} \left| \mathrm{sum}\left\{ \left(\mathbf{R}_{jlm}\mathbf{R}_{jjk}^{-1}\right) \circ \mathbf{Q}_2^T \right\} \right|^2 + \mathrm{sum}\left\{ \mathbf{R}_{\tilde{h}_{jlm}} \circ \mathbf{Q}_1^T \right\} \right\} + \frac{\sigma^2}{t_{jk}^d} \cdot [\mathbf{Q}_3]_{k,k} - 1} \tag{18}$$

$$\mathrm{SINR}_{jk}^{\mathrm{mmse}} = \frac{1-[\mathbf{Q}_4]_{k,k}}{\sum_{(l,m)\notin\Lambda(j,k)}^{L} \frac{t_{lm}^d}{t_{jk}^d} \mathrm{sum}\{\mathbf{R}_{jlm} \circ \mathbf{Q}_5^T\} + \sum_{(l,m)\in\Lambda(j,k)}^{L} \frac{t_{lm}^d}{t_{jk}^d} \left\{ \frac{t_{lm}^d}{t_{jk}^d} \left| \mathrm{sum}\left\{ \left(\mathbf{R}_{jlm}\mathbf{R}_{jjk}^{-1}\right) \circ \mathbf{Q}_6^T \right\} \right|^2 + \mathrm{sum}\left\{ \mathbf{R}_{\tilde{h}_{jlm}} \circ \mathbf{Q}_5^T \right\} \right\} + \frac{\sigma^2}{t_{jk}^d} \cdot [\mathbf{Q}_7]_{k,k} - 1 + [\mathbf{Q}_4]_{k,k}} \tag{19}$$

strictly improve its utility function and does not reduce the average utility function of all users. Therefore, we have definition 3.

Definition 3 *(Principle of adjustment). The adjustment $\mathcal{S} \overset{(j,k)}{\rightarrow} \mathcal{S}^{\Phi}$ is permissible if $u_{j,k}(\mathcal{S}^{\Phi})$ is better than $u_{j,k}(\mathcal{S})$ and $\frac{1}{N}\sum_{(l,m)\in\mathcal{N}} u_{l.m}(\mathcal{S}^{\Phi})$ is not worse than $\frac{1}{N}\sum_{(l,m)\in\mathcal{N}} u_{l.m}(\mathcal{S})$.*

In definition 3, that $u_{j,k}(\mathcal{S}^{\Phi})$ is better than $u_{j,k}(\mathcal{S})$ means $u_{j,k}(\mathcal{S}^{\Phi}) < u_{j,k}(\mathcal{S})$ for MSE-CE and MSE-SD, and $u_{j,k}(\mathcal{S}^{\Phi}) > u_{j,k}(\mathcal{S})$ for SINR and SE. That $\frac{1}{N}\sum_{(l,m)\in\mathcal{N}} u_{l.m}(\mathcal{S}^{\Phi})$ is not worse than $\frac{1}{N}\sum_{(l,m)\in\mathcal{N}} u_{l.m}(\mathcal{S})$ means $\frac{1}{N}\sum_{(l,m)\in\mathcal{N}} u_{l.m}(\mathcal{S}^{\Phi}) \le \frac{1}{N}\sum_{(l,m)\in\mathcal{N}} u_{l.m}(\mathcal{S})$ for MSE-CE and MSE-SD, and $\frac{1}{N}\sum_{(l,m)\in\mathcal{N}} u_{l.m}(\mathcal{S}^{\Phi}) \ge \frac{1}{N}\sum_{(l,m)\in\mathcal{N}} u_{l.m}(\mathcal{S})$ for SINR and SE.

From definition 3, we can see that any user that wants to join another sub-coalition should be beneficial to itself without damaging the average interest of all users. Thus, the average utility function will be optimal at the final stability.

This adjustment comes to an end as defined in definition 4, which is similar to the definition of final stability in [21].

Definition 4 *(Final stability). The whole coalition structure S is stable if there is no permission of adjustment $\mathcal{S} \overset{(j,k)}{\rightarrow} \mathcal{S}^{\Phi}$ for all users $(j,k)\in\mathcal{N}$ and all sub-coalitions, or $\alpha \ge \tau$, or the average utility function ($\frac{1}{N}\sum_{(l,m)\in\mathcal{N}} u_{l.m}(\mathcal{S})$) is not worse than U for a given performance index U (where $U \ge U_{\min}$ for MSE-CE and MSE-SD, $U \le U_{\min}$ for SINR and SE).*

In definition 4, assume the average utility function is equal to $U_{\min}$ when there is no permission of adjustment $\mathcal{S} \overset{(j,k)}{\rightarrow} \mathcal{S}^{\Phi}$ for all users $(j,k)\in\mathcal{N}$ and all sub-coalitions. That is to say, $U_{\min}$ is the global optimal value of average utility function. In addition, for the given performance index of the average utility function, $\frac{1}{N}\sum_{(l,m)\in\mathcal{N}} u_{l.m}(\mathcal{S})$ is not worse than U means $\frac{1}{N}\sum_{(l,m)\in\mathcal{N}} u_{l.m}(\mathcal{S}^{\Phi}) \le U$ for MSE-CE and MSE-SD, and $\frac{1}{N}\sum_{(l,m)\in\mathcal{N}} u_{l.m}(\mathcal{S}^{\Phi}) \ge U$ for SINR and SE.

According to the discussions above, the coalition formation algorithm is given in algorithm 1. In definition 4 and algorithm 1, α is the current number of searches and τ is the limitation of total number of searches, which is used to prevent the coalition formation algorithm from falling into an endless loop.

We can see from algorithm 1 that each adjustment of the coalition structure further optimizes the average utility function. Taking the utility function MSE-CE as an example, the purpose of every searching in algorithm 1 is to reduce the average MSE-CE, and every adjustment of the coalition structure can reduce the average MSE-CE. According to the conditions of final stability in definition 4, algorithm 1 is convergent. Therefore, when the condition of final stability is that $\frac{1}{N}\sum_{(l,m)\in\mathcal{N}} u_{l.m}(\mathcal{S})$ is not worse than U for a given performance index U (where $U \ge U_{\min}$ for MSE-CE), this algorithm can search continuously until the performance U is satisfied, it is a state of finite convergence. Therefore, this pilot allocation scheme can provide performance guarantee.

Algorithm 1 Coalition formation algorithm

Initialize: random coalition structure $\mathcal{S}=\{\mathcal{S}_1,\mathcal{S}_2,\cdots,\mathcal{S}_P\}$, and $\alpha=0$

Repeat

1. For all users $(j,k)\in\mathcal{N}$, do cyclic search

 2. For all $\mathcal{S}_i\in\{\{\mathcal{S}_1,\mathcal{S}_2,\cdots,\mathcal{S}_P\}\backslash\Lambda_{\mathcal{S}}(j,k)\}$ ($i=1,2,\cdots,P$), do search

 Ask for the permission of adjustment $\mathcal{S}\xrightarrow{(j,k)}\mathcal{S}^{\Phi}$, i.e., judge $u_{j,k}(\mathcal{S}^{\Phi})$ is better than $u_{j,k}(\mathcal{S})$ and $\frac{1}{N}\sum_{(l,m)\in\mathcal{N}}u_{l.m}(\mathcal{S}^{\Phi})$ is not worse than $\frac{1}{N}\sum_{(l,m)\in\mathcal{N}}u_{l.m}(\mathcal{S})$.

 If adjustment $\mathcal{S}\xrightarrow{(j,k)}\mathcal{S}^{\Phi}$ is permissible, (j,k) leaves its sub-coalition and joins sub-coalition $\mathcal{S}_i$, update the coalition structure ($\mathcal{S}=\mathcal{S}^{\Phi}$).

 Else if adjustment $\mathcal{S}\xrightarrow{(j,k)}\mathcal{S}^{\Phi}$ is not permissible, coalition structure remains unchanged ($\mathcal{S}=\mathcal{S}$), end if.

 End (Corresponding to 2)

 Increase the number of searches, i.e., $\alpha=\alpha+1$.

End (Corresponding to 1)

Return to **repeat**, until there is no permission of adjustment $\mathcal{S}\xrightarrow{(j,k)}\mathcal{S}^{\Phi}$, or $\alpha\geq\tau$, or $\frac{1}{N}\sum_{(l,m)\in\mathcal{N}}u_{l.m}(\mathcal{S})$ is not worse than U.

We go on to discuss the implementation method of the proposed pilot allocation scheme. The utility function calculations require the distance of each user to each BS. If algorithm 1 operates at BSs, all BSs should cooperate and exchange many messages. Therefore, both the calculations of utility functions and the processes of algorithm 1 can be set at the upper layer of BSs, i.e., completed by the base station controller (BSC). Each BS sends its own location and user locations to BSC. BSC calculates distances, utility functions, and the coalition formation algorithm, and then BSC sends the results of the coalition formation algorithm (i.e., the results of pilot allocation) to each user through BS forwarding. This can greatly reduce the complexity of the whole system. Although the computational complexity of this scheme is still somewhat high, with the general improvement of computing power, processing delays of BSC can be reduced. In addition, BSC needs to send the results of pilot allocation to each user; this will increase the signaling delays of the system. However, this is similar to the case that the receiver transmits the channel estimation back to the sender, only the transmission delay between BSs and BSC is added; however, the transmission delay of this kind of optical fiber line is very small, so we think this signal transmission can be hopefully achievable. Stated thus, the proposed scheme can be hopefully applied in actual situations.

5 Simulation results and discussions

In this section, simulation results are provided to illustrate the system performance of the proposed pilot allocation scheme based on the coalition game. All experiments are Monte Carlo simulations (based on the results of theoretical analysis of different utility functions) under correlated Rayleigh fading channels. Then, we compare the proposed pilot allocation scheme with the pilot allocation scheme based on the greedy approach in [15] and the random coalition scheme (i.e., randomized pilot allocation) in terms of different performance and computational complexity.

The greedy approach in [15] assumes that there are P users per cell and there are P orthogonal pilots. The greedy approach selects one user from each cell to form a user set during each search, all users are grouped into P sets by P searches. Users in the same set use the same pilot, and users in the different sets use different pilots. The principle of user selection is to minimize the sum mean square error of channel estimation for the users using the same pilot.

The greedy approach in [15] assumes the number of users per cell is equal, and it cannot be used in the case of different number of users per cell, so we have to make modifications to facilitate the comparison. When simulating this greedy approach, $\eta=\lceil\frac{N}{P}\rceil$ (η represents the average number of users in each set) users are selected to form a set during each search, regardless of whether the selected users are in different cells. In addition, in all the following simulations, this greedy approach uses the MMSE combining scheme. Taking the MSE-CE as an example, the performance comparison of the greedy approach before and after the modification is shown in Fig. 3 (when using parameters in Table 1). We can see that the values of MSE-CE are very close before and after modification. As to other performance (like MSE-SD, SINR, SE), the results are similar according to our simulations; thus, we omit these results to save space. So, we can compare our scheme with the modified greedy approach, and the performance improvement can be seen in Figs. 4, 5, 6, 7, 8, 9, 10, 11, and 12. The greedy approach in Figs. 4, 5, 6, 7, 8, 9, 10, 11, and 12 is referring to the greedy approach after modification.

Simulation parameters are given in Table 1. We use a fixed area of 2000 m × 2000 m; BSs are subject to a uniform distribution, and the users are subject to a uniform distribution; and the distributions of BSs and users are independent. Each user selects and accesses its nearest BS. About the simulation parameters in Table 1, we refer to the simulation parameters in references [7, 16, 17]. For the setting of the ring radius of scatters, we consider a non-line-of-sight (non-LOS) scenario with 50 scattering paths as in references [7, 16, 17].

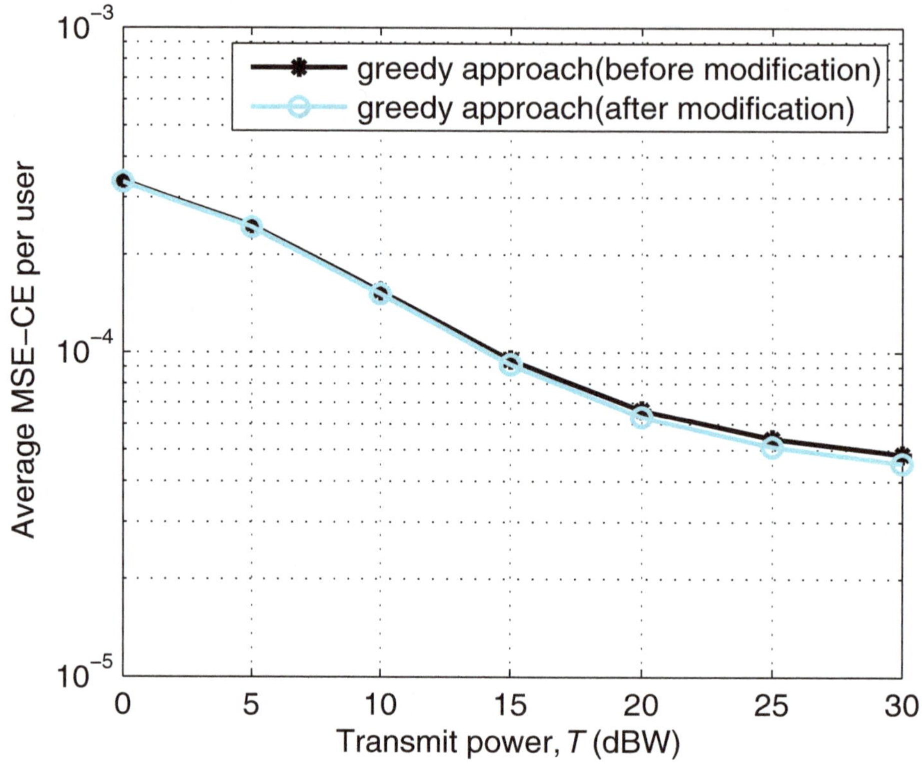

Fig. 3 Comparison of the greedy approach before and after the modification

In the following simulations, for all coalition game pilot allocation schemes, we show the optimal results of the coalition formation algorithm, i.e., the condition of the final stability is the no permission of adjustment $\mathcal{S} \xrightarrow{(j,k)} \mathcal{S}^{\Phi}$. However, we know that the proposed pilot allocation scheme based on the coalition game can provide a certain performance guarantee through a user cyclic search. For a given performance (average utility function) index U, the proposed scheme can search continuously until the performance index U is satisfied. Taking the utility function of MSE-CE as an example, Fig. 4 shows the relationship between the average MSE-CE and the number of searches (α) for a certain location of BSs and users. We can see that, with the increase of α, the average MSE decreases, and when$\alpha \geq 22$, the average MSE reaches its minimum value. To satisfy the average MSE-CE performance index of 10^{-4}, we should set the total number of search limitation τ to greater than or equal to 14.

Table 1 Simulation parameters

Parameter	Symbol	Value
Number of cells	L	4
Number of users	N	12
Path-loss exponent	v	2.5
Number of antennas	M	100
Number of paths	F	50
Number of symbols in each frame	Q	15
Ring radius of scatters around users	r_s	100 m
Antenna spacing at BS	D	$\lambda/2$
Channel coefficient	δ	10
Standard deviation of noise power	σ^2	10 W
Wave length	λ	0.1 m

Figure 5 shows the average MSE-CE per user versus the transmit power T with $P = 4$. The unit of transmission power T is the watt in decibels (i.e., dBW). The "coalition game, MSE-CE" means the pilot allocation scheme based on the coalition game when utility function is MSE-CE. The "greedy approach" means the greedy approach in [15]. We can see that the coalition game for MSE-CE can achieve lower average MSE than the greedy approach and random coalition. Although the object function of the greedy approach in [15] is also MSE-CE, the greedy approach is worse than the coalition game for MSE-CE, because the greedy approach in [15] is a one-time search for users, but the coalition

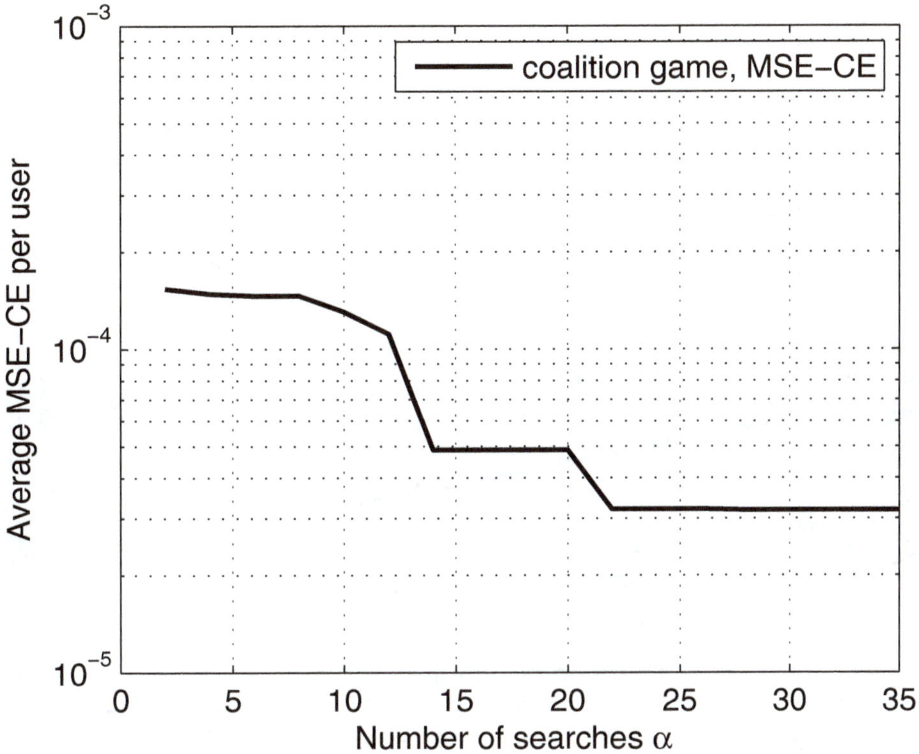

Fig. 4 Average MSE-CE versus the number of searches (α) with $P = 4$ and $T = 20$ dBW

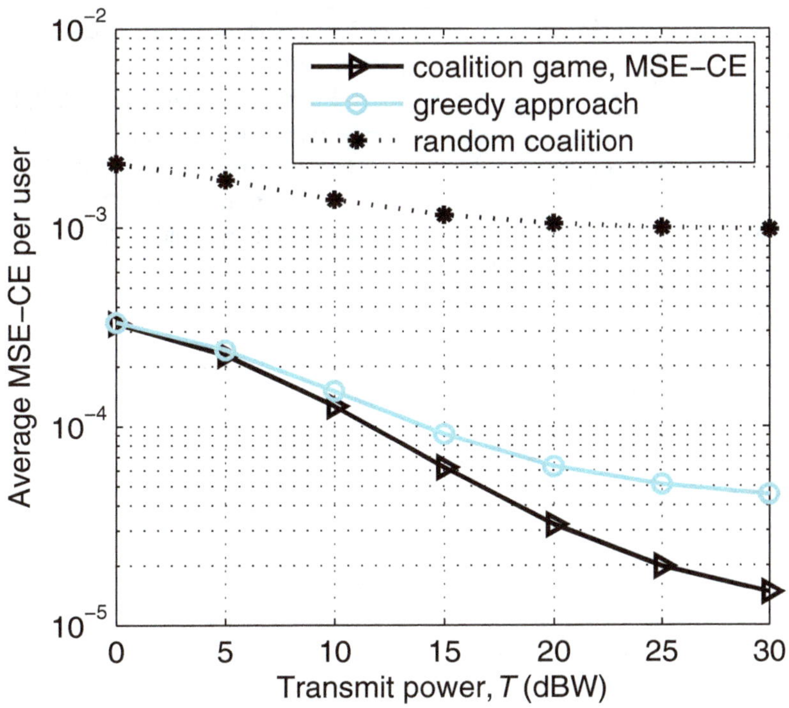

Fig. 5 Average MSE-CE versus transmit power T with $P = 4$

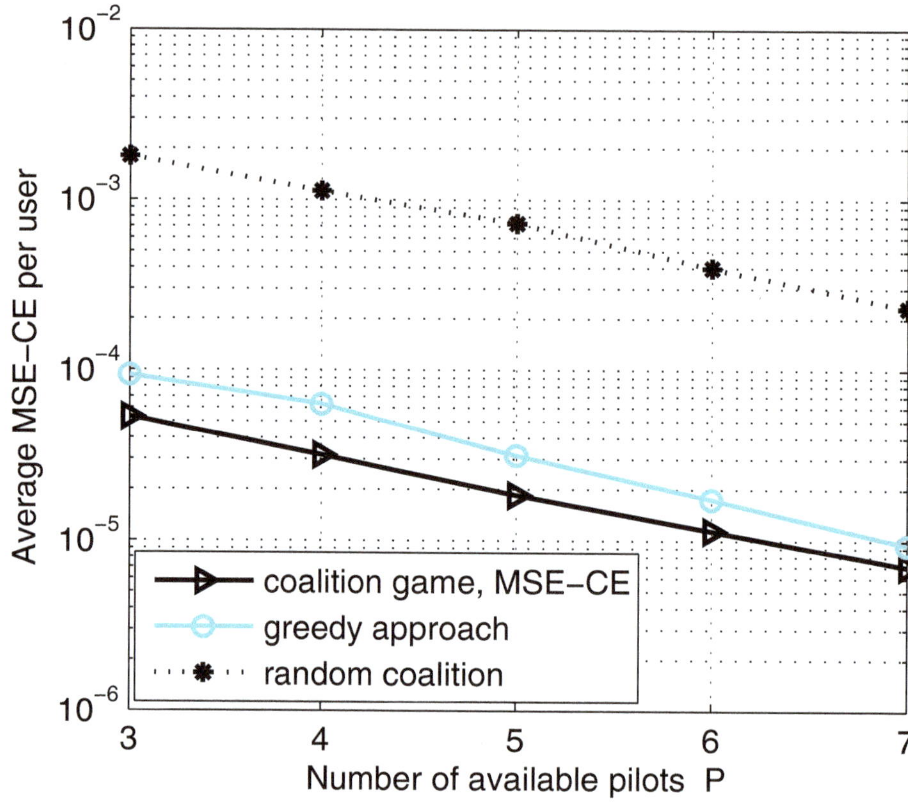

Fig. 6 Average MSE-CE versus the number of available pilots P with $T = 20$ dBW

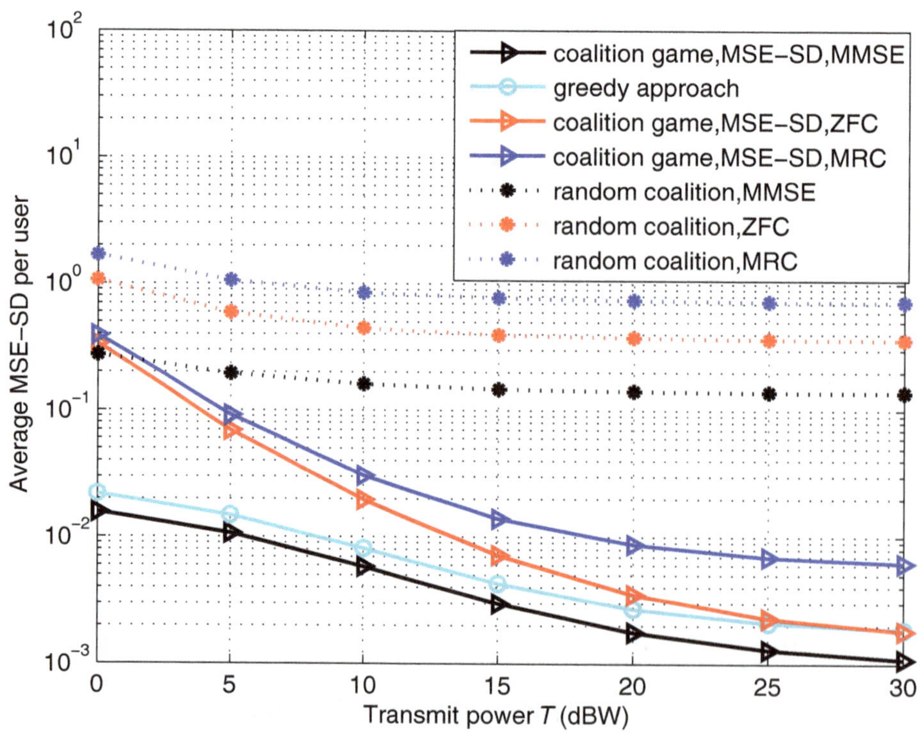

Fig. 7 Average MSE-SD versus transmit power T with $P = 4$

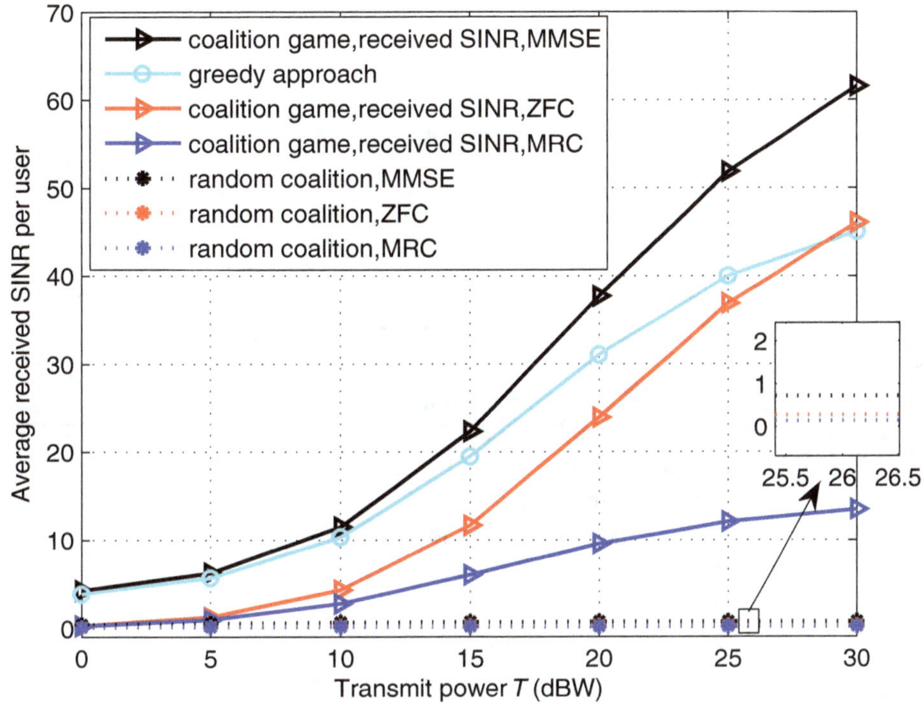

Fig. 8 The average received SINR versus transmit power T with $P = 4$

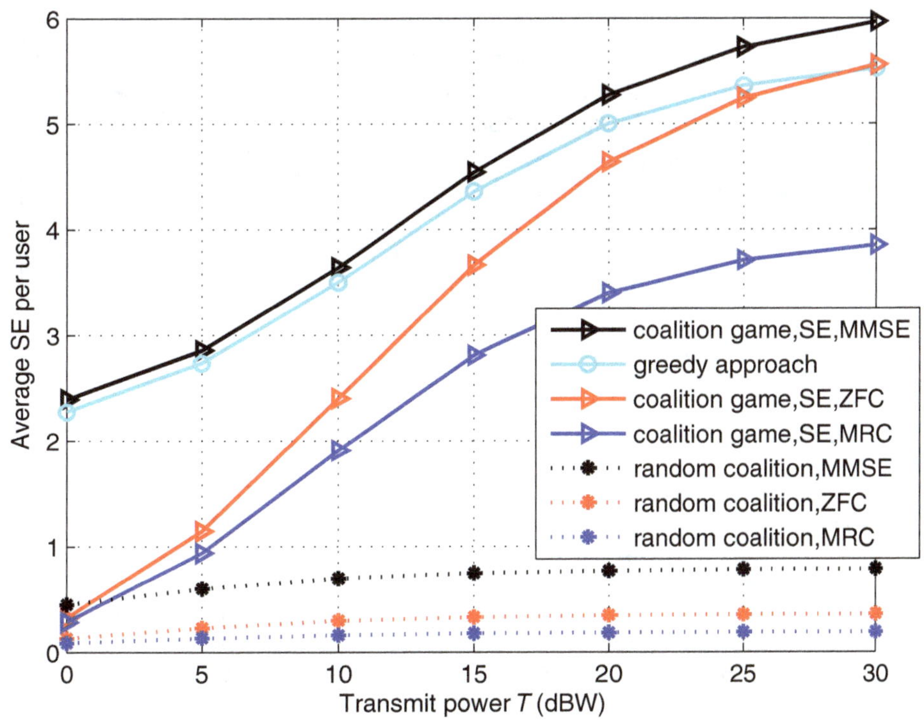

Fig. 9 Average spectrum efficiency (SE) versus transmit power T with $P = 4$

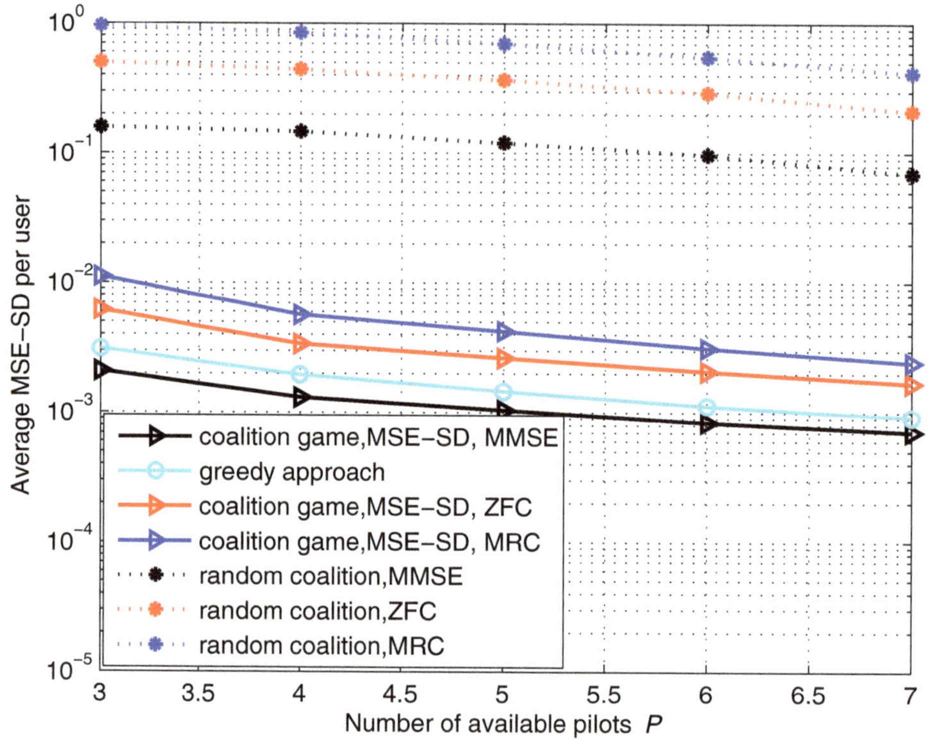

Fig. 10 Average MSE-SD versus the number of pilots with $T = 20$ dBW

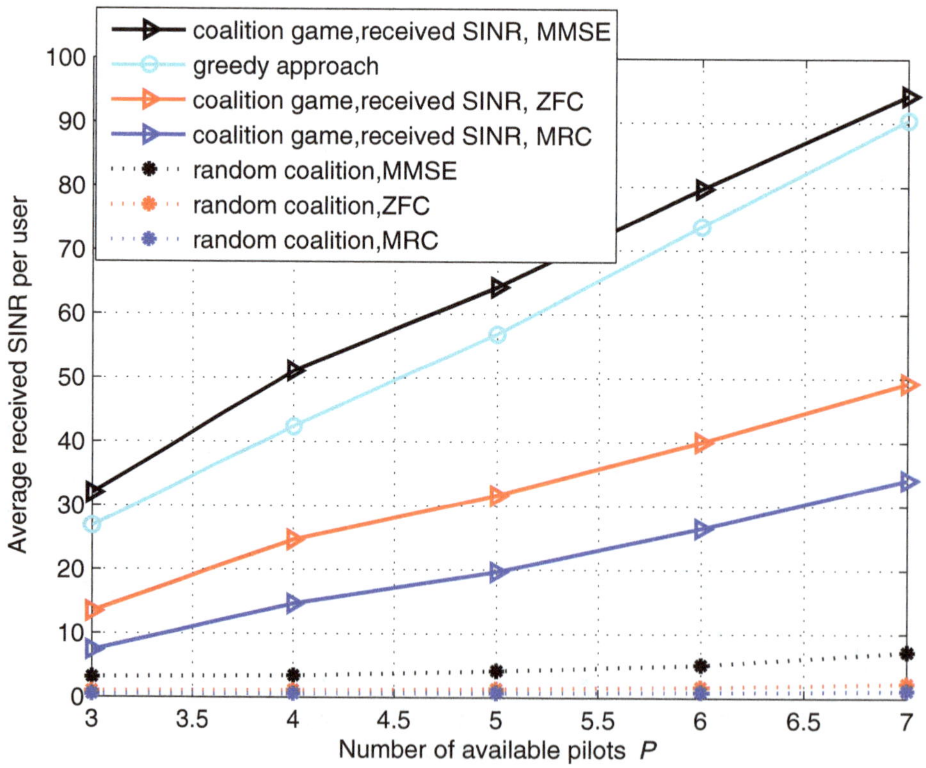

Fig. 11 Average received SINR versus the number of pilots with $T = 20$ dBW

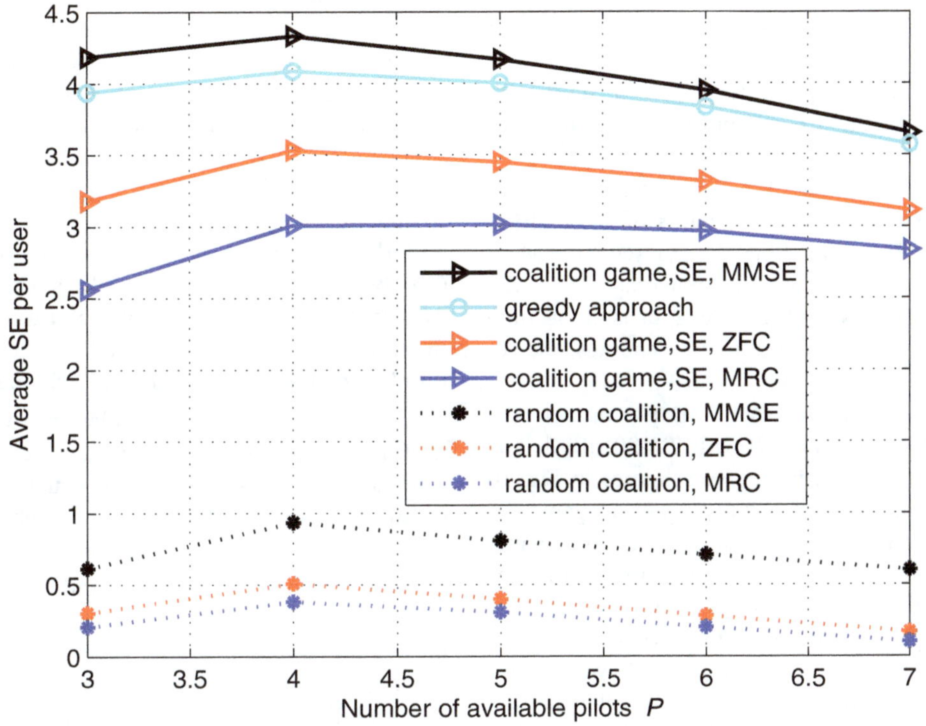

Fig. 12 Average SE versus the number of pilots with $T = 20$ dBW

game for MSE-CE is a cyclic search for users. In addition, the greater the transmit power, the more obvious the performance improvements.

Figure 6 shows the average MSE-CE per user versus the number of available pilots P with $T = 20$ dBW. We can see that the coalition game for MSE-CE can achieve lower average MSE-CE than the greedy approach. The performance of random alliance is the worst because of random allocation of pilots. When the number of pilots becomes larger, the average MSE-CE of all schemes becomes smaller. This is because the larger the number of pilots, the larger the number of sub-coalitions and the smaller the interference between users.

Figure 7 shows the average MSE-SD per user versus the transmit power T with $P = 4$. The "coalition game, MSE-SD, MMSE," "coalition game, MSE-SD, ZFC," and "coalition game, MSE-SD, MRC" means the pilot allocation scheme based on the coalition game when utility function is MSE-SD with the MMSE scheme, when utility function is MSE-SD with the ZFC scheme, and when utility function is MSE-SD with the MRC scheme, respectively. It can be seen that the coalition game for MSE-SD with MMSE is better than the coalition game for MSE-SD with ZFC, and the coalition game for MSE-SD with ZFC is better than the coalition game for MSE-SD with MRC. That is because the MRC scheme cannot reduce interference and noise, and the ZFC scheme can reduce interference, but cannot reduce noise, whereas the MMSE scheme can reduce both interference and noise. When the transmit power is small, the performance of the greedy approach is slightly worse than that of the coalition game for MSE-SD with MMSE and better than that of the coalition game for MSE-SD with ZFC and the coalition game for MSE-SD with MRC. Because the purpose of the greedy approach in [15] is to minimize the sum of normalized MSE-CE of all users, its performance is better than the coalition game for MSE-SD with ZFC and the coalition game for MSE-SD with MRC. The greedy approach is worse than the coalition game for MSE-SD with MMSE because the greedy approach is a one-time search for users, but the coalition game for MSE-SD with MMSE is a cyclic search for users. With the increase of transmit power T, the advantage of the coalition game for MSE-SD with MMSE relative to the greedy approach is increasingly obvious. In addition, for each combining scheme (MMSE, ZFC, or MRC), the coalition game is better than random coalition.

Figure 8 shows the average received SINR per user versus the transmit power T with $P = 4$. The utility functions of all coalition game schemes in Fig. 8 are received SINR. When transmit power T becomes larger, the average received SINR of the coalition game schemes and greedy approach become larger. But random coalition schemes change little with the increase of T, that is because the interference between all users is increasing when transmit power T becomes larger. From Fig. 8, we can see that the coalition game for received SINR with MMSE is better than the greedy approach, and other results are similar to Fig. 7.

Figure 9 shows the average SE per user versus transmit power T with $P = 4$. The utility functions of all coalition game schemes in Fig. 9 are SE. We can see that when transmit power T becomes larger, the average SE of the coalition game schemes and greedy approach become larger. When comparing these different schemes, the results are similar to Figs. 7 and 8.

Figure 10 shows the average MSE-SD per user versus the number of available pilots P with $T = 20$ dBW. The utility functions of all coalition game schemes in Fig. 10 are MSE-SD. It can be seen that the coalition game for MSE-SD with MMSE is better than the coalition game for MSE-SD with ZFC, and the coalition game for MSE-SD with ZFC is better than the coalition game for MSE-SD with MRC. All these results are the same as Fig. 7. In addition, the larger the number of available pilots, the smaller the average MSE-SD of all schemes. This is because when the number of pilots becomes larger, fewer users use the same pilot; thus, the interference between users can be reduced, and the system performance can be improved.

Figure 11 shows the average received SINR per user versus the number of available pilots with $T = 20$ dBW. The utility functions of all coalition game schemes in Fig. 11 are received SINR. We can get the similar results as in Fig. 8.

Figure 12 shows the average SE per user versus the number of available pilots with $T = 20$ dBW. The utility functions of all coalition game schemes in Fig. 12 are SE. The results in Fig. 12 are different from those in Fig. 9. When the number of pilots becomes larger, the average SE of all schemes first becomes larger and then becomes smaller. That is because when the number of pilots is very small, the number of sub-coalitions is small and the interference between users is large, so the SE is small. On the other hand, when the number of pilots becomes very large, the number of data symbols in each frame becomes small and the coefficient in front of the log in expression (16) becomes small, so the SE is also small. That is to say, there exists the optimal number of pilots to maximize SE. We can see in Fig. 12 that the optimal number of pilots is four for all schemes.

Table 2 shows the complexity (in terms of the multiplication times) during each search for different schemes. It is assumed that calculating the inversion of an M-dimensional matrix needs about M^3 multiplications, $\eta = \lceil \frac{N}{P} \rceil$ represents the average number of users in each sub-coalition, and $\xi = \lceil \frac{N}{L} \rceil$ represents the average number of users in each cell, where $\lceil * \rceil$ means the round up number of *. The expressions in Table 2 are obtained according to the multiplication times of Eqs. (9),(13)~(19), and we have assumed that $\mathbf{R}_{jlm}$ is known according to large-scale channel fading and the locations of users and BSs and that $\mathbf{R}_{\hat{h}_{jlm}}$, $\boldsymbol{Q}_1$, $\boldsymbol{Q}_2$, $\boldsymbol{Q}_3$, $\boldsymbol{Q}_4$, $\boldsymbol{Q}_5$, $\boldsymbol{Q}_6$, and $\boldsymbol{Q}_7$ are obtained through statistical average of channel estimation and calculated once during each search. $o(NM^2)$ represents higher order infinitesimal of NM^2.

Table 3 shows the average number of searches (i.e., $\overline{\alpha}$) for each scheme, and also means the calculation times of average utility function in coalition formation algorithm. The values in Table 3 are obtained through Monte Carlo simulations, and they are the average number of searches for 10,000 user and BS locations. For all

Table 2 The multiplication times during each search for different schemes

	Different combining schemes		
	MMSE	ZFC	MRC
Coalition game for different utility function			
MSE-SD	$(3N+2\eta)NM^3+$ $[(N+\eta)\xi^2+3\xi^2+2(N+\eta)]NM^2+o(NM^2)$	$(3N+2\eta)NM^3+$ $[(N+\eta)\xi^2+\xi^2+2(N+\eta)]NM^2+o(NM^2)$	$(3N+2\eta)NM^3+$ $[N+\eta+1]NM^2+o(NM^2)$
Received SINR	$(3N+2\eta)NM^3+$ $[(N+\eta)\xi^2+3\xi^2+2(N+\eta)]NM^2+o(NM^2)$	$(3N+2\eta)NM^3+$ $[(N+\eta)\xi^2+\xi^2+2(N+\eta)]NM^2+o(NM^2)$	$(3N+2\eta)NM^3+$ $[N+\eta+1]NM^2+o(NM^2)$
SE	$(3N+2\eta)NM^3+$ $[(N+\eta)\xi^2+3\xi^2+2(N+\eta)]NM^2+o(NM^2)$	$(3N+2\eta)NM^3+$ $[(N+\eta)\xi^2+\xi^2+2(N+\eta)]NM^2+o(NM^2)$	$(3N+2\eta)NM^3+$ $[N+\eta+1]NM^2+o(NM^2)$
MSE-CE	$3NM^3+o(NM^2)$ (without combining schemes)		
Random coalition	0 (for all combining schemes)		
Greedy approach in [15]			
MSE-CE	$3NM^3+o(NM^2)$ (for MMSE combining scheme)		

Table 3 The average number of searches ($\overline{a}$) for different schemes

	Different combining schemes		
	MMSE	ZFC	MRC
Coalition game for different utility function			
MSE-SD	36.1	35.2	34.2
Received SINR	36.5	35.8	34.2
SE	36.5	35.8	34.2
MSE-CE	24 (without combining schemes)		
Random coalition	0 (for all combining schemes)		
Greedy approach in [15]			
MSE-CE	12 (for MMSE combining scheme)		

coalition game pilot allocation schemes in Table 3, the condition of the final stability is the no permission of adjustment $\mathcal{S} \overset{(j,k)}{\rightarrow} \mathcal{S}^{\Phi}$.

In Table 3, the average number of search of random coalition is 0 because random coalition means randomized pilot allocation and there is no adjustment of the coalition structure. The average number of search of the greedy approach in [15] is 12 because the greedy approach is a one-time search for all users and the number of users is 12. The average number of search of the coalition game for different utility functions (MSE-SD, received SINR, or SE) is very close, and they are all close to 36, which means each user is searched about 3 times on average. The average number of searches of the coalition game for MSE-CE is 24, which means each user is searched about 2 times on average.

The total complexity of each scheme is the result of the multiplication of the corresponding items in Tables 2 and 3. Combining Tables 2 and 3, when comparing the total multiplication times, we can see that random coalition has the lowest complexity, the greedy approach is higher than random coalition, the coalition game for MSE-CE is higher than the greedy approach, and the coalition game for MSE-SD, received SINR, and SE are very close and has the highest complexity. The total multiplication times of the coalition game for different utility functions (MSE-SD, received SINR, or SE) are also very close, and they are higher than the coalition game for MSE-CE and the greedy approach, although their average search time does not differ much. That is because the multiplication times during each search of the coalition game for different utility functions (MSE-SD, received SINR, or SE) are larger than the coalition game for MSE-CE and the greedy approach. In a word, the total multiplication times of the coalition game for MSE-CE and the greedy approach is much higher than that of random coalition, but lower than that of the coalition game for MSE-SD, received SINR, and SE.

6 Conclusions

A pilot allocation scheme based on the coalition game is proposed for TDD massive MIMO systems. This scheme can be suitable for any actual scene in which cells are irregularly shaped (BSs distribute randomly), numbers of users per cell are not equal (users distribute randomly), and channels are correlated fading channels, and this scheme can provide performance guarantee. The definition of the coalition structure is given, and different utility functions are analyzed. To benefit each user, the adjustment principle of the coalition structure is set, and the coalition formation algorithm is provided. According to the coalition formation algorithm, the pilot allocation scheme based on the coalition game (for different utility functions) is simulated and compared with the greedy approach in [15] and random coalition. And then, the computational complexity in terms of multiplication times for different pilot allocation schemes is compared. Simulation results show that when considering the performance comparison of different pilot allocation schemes, the results for different performance metrics (different utility functions) are similar. The coalition game for each utility function with MMSE achieves the best performance with high computational complexity. For each utility function and for each combining scheme (MMSE, ZFC, or MRC), the coalition game is better than random coalition. When comparing the performance of different combining schemes, with regard to the coalition game for each utility function, MMSE is better than ZFC and ZFC is better than MRC. The performance of the greedy approach in [15] is between the coalition game for each utility function with MMSE and the coalition game for each utility function with ZFC. In addition, we obtain the optimal number of pilots to maximize SE through simulations.

7 Appendix 1

According to expression (11), the mean square error of signal detection (MSE-SD) for user (j,k) is

$$\begin{aligned}\text{MSE-SD}_{jk}^{\text{scheme}} &= \mathbb{E}\Bigg\{\Bigg|\frac{1}{\sqrt{t_{jk}^{d}}}\left(\boldsymbol{w}_{jjk}^{\text{scheme}}\right)^{H}\left(\sum_{m=1}^{L}\sum_{n=1}^{K_m}\sqrt{t_{mn}^{d}}\ \boldsymbol{h}_{jmn}x_{mn}+\boldsymbol{n}_{j}^{\text{data}}\right)\\ &-x_{jk}\Bigg|^{2}\Bigg\} = \mathbb{E}_{\boldsymbol{h}}\left\{\left|\left(\boldsymbol{w}_{jjk}^{\text{scheme}}\right)^{H}\hat{\boldsymbol{h}}_{jjk}-1\right|^{2}\right\}+\mathbb{E}_{\boldsymbol{h}}\left\{\left|\left(\boldsymbol{w}_{jjk}^{\text{scheme}}\right)^{H}\tilde{\boldsymbol{h}}_{jjk}\right|^{2}\right\}\\ &+\sum_{l=1}^{L}\sum_{m=1}^{K_l}\frac{t_{lm}^{d}}{t_{jk}^{d}}\mathbb{E}_{\boldsymbol{h}}\left\{\left|\left(\boldsymbol{w}_{jjk}^{\text{scheme}}\right)^{H}\boldsymbol{h}_{jlm}\right|^{2}\right\}-\mathbb{E}_{\boldsymbol{h}}\left\{\left|\left(\boldsymbol{w}_{jjk}^{\text{scheme}}\right)^{H}\boldsymbol{h}_{jjk}\right|^{2}\right\}\\ &+\frac{1}{t_{jk}^{d}}\mathbb{E}_{\boldsymbol{h}}\left\{\left|\left(\boldsymbol{w}_{jjk}^{\text{scheme}}\right)^{H}\boldsymbol{n}_{j}^{\text{data}}\right|^{2}\right\}\end{aligned} \tag{20}$$

Then, according to reference ([19], Lemma 2), the received SINR for user (j,k) at BS j is

$$\mathrm{SINR}_{jjk}^{\mathrm{scheme}} = \frac{\left|\mathbb{E}_{\boldsymbol{h}}\left\{\left(\boldsymbol{w}_{jjk}^{\mathrm{scheme}}\right)^{H}\boldsymbol{h}_{jjk}\right\}\right|^{2}}{\sum_{l=1}^{L}\sum_{m=1}^{K_l}\frac{t_{lm}^{d}}{t_{jk}^{d}}\mathbb{E}_{\boldsymbol{h}}\left\{\left|\left(\boldsymbol{w}_{jjk}^{\mathrm{scheme}}\right)^{H}\boldsymbol{h}_{jlm}\right|^{2}\right\}-\left|\mathbb{E}_{\boldsymbol{h}}\left\{\left(\boldsymbol{w}_{jjk}^{\mathrm{scheme}}\right)^{H}\boldsymbol{h}_{jjk}\right\}\right|^{2}+\frac{1}{t_{jk}^{d}}\mathbb{E}_{\boldsymbol{h}}\left\{\left|\left(\boldsymbol{w}_{jjk}^{\mathrm{scheme}}\right)^{H}\boldsymbol{n}_{j}^{\mathrm{data}}\right|^{2}\right\}} \quad (21)$$

For the MRC scheme, we can obtain $\mathbb{E}_{\boldsymbol{h}}\{|(\boldsymbol{w}_{jjk}^{\mathrm{mrc}})^{H}\boldsymbol{h}_{jjk}-1|^{2}\} = \mathbb{E}_{\boldsymbol{h}}\{|(\boldsymbol{w}_{jjk}^{\mathrm{mrc}})^{H}\hat{\boldsymbol{h}}_{jjk}-1|^{2}\} + \mathbb{E}_{\boldsymbol{h}}\{|(\boldsymbol{w}_{jjk}^{\mathrm{mrc}})^{H}\tilde{\boldsymbol{h}}_{jjk}|^{2}\}$. That is the first and the second items of expression (20).

According to the definition of the MRC scheme, we obtain

$$\mathbb{E}_{\boldsymbol{h}}\left\{\left(\boldsymbol{w}_{jjk}^{\mathrm{mrc}}\right)^{H}\boldsymbol{h}_{jjk}\right\} = \mathbb{E}_{\boldsymbol{h}}\left\{\left(\boldsymbol{w}_{jjk}^{\mathrm{mrc}}\right)^{H}\hat{\boldsymbol{h}}_{jjk}\right\} = \frac{1}{\delta_{jjk}}\mathbb{E}_{\boldsymbol{h}}\left\{\hat{\boldsymbol{h}}_{jjk}{}^{H}\hat{\boldsymbol{h}}_{jjk}\right\} = 1 \quad (22)$$

Therefore, the first item of expression (20) is

$$\mathbb{E}_{\boldsymbol{h}}\left\{\left|\left(\boldsymbol{w}_{jjk}^{\mathrm{mrc}}\right)^{H}\hat{\boldsymbol{h}}_{jjk}-1\right|^{2}\right\} = 0. \quad (23)$$

When $M\rightarrow\infty$, $\mathbb{E}_{\boldsymbol{h}}\{|(\boldsymbol{w}_{jjk}^{\mathrm{mrc}})^{H}\boldsymbol{h}_{jjk}|^{2}\} \approx |\mathbb{E}_{\boldsymbol{h}}\{(\boldsymbol{w}_{jjk}^{\mathrm{mrc}})^{H}\boldsymbol{h}_{jjk}\}|^{2} = 1$. This is the result of the fourth item of expression (20).

Because $\tilde{\boldsymbol{h}}_{jjk}$ is independent of $\boldsymbol{w}_{jjk}^{\mathrm{wrc}}$, the second item of expression (20) is

$$\mathbb{E}_{\boldsymbol{h}}\left\{\left|\left(\boldsymbol{w}_{jjk}^{\mathrm{mrc}}\right)^{H}\tilde{\boldsymbol{h}}_{jjk}\right|^{2}\right\} = \frac{1}{\delta_{jjk}^{2}}\mathbb{E}_{\boldsymbol{h}}\left\{\hat{\boldsymbol{h}}_{jjk}^{H}\tilde{\boldsymbol{h}}_{jjk}\tilde{\boldsymbol{h}}_{jjk}^{H}\hat{\boldsymbol{h}}_{jjk}\right\} = \frac{1}{\delta_{jjk}^{2}}\mathrm{sum}\left\{\mathbf{R}_{\tilde{h}_{jjk}}\circ\mathbf{R}_{\hat{h}_{jjk}}^{T}\right\} \quad (24)$$

where $\boldsymbol{A}\circ\boldsymbol{B}$ is the Hadamard product of $\boldsymbol{A}$ and $\boldsymbol{B}$.

For the third item of expression (20) and the first item of the denominator in expression (21), two cases (i.e., $(l,m)\notin\Lambda_{S}(j,k)$ and $(l,m)\in\Lambda_{S}(j,k)$) are discussed.

When $(l,m)\notin\Lambda_{S}(j,k)$,

$$\begin{aligned}\sum_{(l,m)\notin\Lambda(j,k)}\frac{t_{lm}^{d}}{t_{jk}^{d}}\mathbb{E}_{\boldsymbol{h}}\left\{\left|\left(\boldsymbol{w}_{jjk}^{\mathrm{mrc}}\right)^{H}\boldsymbol{h}_{jlm}\right|^{2}\right\} &= \sum_{(l,m)\notin\Lambda(j,k)}\frac{t_{lm}^{d}}{t_{jk}^{d}}\frac{1}{\delta_{jjk}^{2}}\mathbb{E}_{\boldsymbol{h}}\left\{\hat{\boldsymbol{h}}_{jjk}^{H}\boldsymbol{h}_{jlm}\boldsymbol{h}_{jlm}^{H}\hat{\boldsymbol{h}}_{jjk}\right\}\\ &= \sum_{(l,m)\notin\Lambda(j,k)}\frac{t_{lm}^{d}}{t_{jk}^{d}}\frac{1}{\delta_{jjk}^{2}}\mathrm{sum}\left\{\mathbf{R}_{jlm}\circ\mathbf{R}_{\hat{h}_{jjk}}^{T}\right\}\end{aligned} \quad (25)$$

where the second equation exists because of the independence of $\hat{\boldsymbol{h}}_{jjk}$ and $\boldsymbol{h}_{jlm}$ for $(l,m)\notin\Lambda_{S}(j,k)$.

When $(l,m)\in\Lambda_{S}(j,k)$,

$$\begin{aligned}&\sum_{(l,m)\in\Lambda(j,k)}\frac{t_{lm}^{d}}{t_{jk}^{d}}\mathbb{E}_{\boldsymbol{h}}\left\{\left|\left(\boldsymbol{w}_{jjk}^{\mathrm{mrc}}\right)^{H}\boldsymbol{h}_{jlm}\right|^{2}\right\}\\ &= \sum_{(l,m)\in\Lambda(j,k)}\frac{t_{lm}^{d}}{t_{jk}^{d}}\frac{1}{\delta_{jjk}^{2}}\left(\mathbb{E}_{\boldsymbol{h}}\left\{\hat{\boldsymbol{h}}_{jjk}^{H}\hat{\boldsymbol{h}}_{jlm}\hat{\boldsymbol{h}}_{jlm}^{H}\hat{\boldsymbol{h}}_{jjk}\right\}+\mathbb{E}_{\boldsymbol{h}}\left\{\hat{\boldsymbol{h}}_{jjk}^{H}\tilde{\boldsymbol{h}}_{jlm}\tilde{\boldsymbol{h}}_{jlm}^{H}\hat{\boldsymbol{h}}_{jjk}\right\}\right)\end{aligned} \quad (26)$$

According to expression (7) and the analysis above, when $(l,m)\in\Lambda_{S}(j,k), \hat{\boldsymbol{h}}_{jlm} = \sqrt{\frac{t_{lm}^{d}}{t_{jk}^{d}}}\mathbf{R}_{jlm}\mathbf{R}_{jjk}^{-1}\hat{\boldsymbol{h}}_{jjk}$. Substituting it into expression (26), we can get (A-8) for $(l,m)\in\Lambda_{S}(j,k)$

$$\begin{aligned}\mathbb{E}_{\boldsymbol{h}}\left\{\hat{\boldsymbol{h}}_{jjk}^{H}\hat{\boldsymbol{h}}_{jlm}\hat{\boldsymbol{h}}_{jlm}^{H}\hat{\boldsymbol{h}}_{jjk}\right\} &= \frac{t_{lm}^{d}}{t_{jk}^{d}}\mathbb{E}_{\boldsymbol{h}}\left\{\hat{\boldsymbol{h}}_{jjk}^{H}\mathbf{R}_{jlm}\mathbf{R}_{jjk}^{-1}\hat{\boldsymbol{h}}_{jjk}\hat{\boldsymbol{h}}_{jjk}^{H}\left(\mathbf{R}_{jlm}\mathbf{R}_{jjk}^{-1}\right)^{H}\hat{\boldsymbol{h}}_{jjk}\right\}\\ &= \frac{t_{lm}^{d}}{t_{jk}^{d}}\mathbb{E}_{\boldsymbol{h}}\left\{\left|\hat{\boldsymbol{h}}_{jjk}^{H}\mathbf{R}_{jlm}\mathbf{R}_{jjk}^{-1}\hat{\boldsymbol{h}}_{jjk}\right|^{2}\right\}\\ &\overset{(a)}{\approx} \frac{t_{lm}^{d}}{t_{jk}^{d}}\left|\mathbb{E}_{\boldsymbol{h}}\left\{\hat{\boldsymbol{h}}_{jjk}^{H}\mathbf{R}_{jlm}\mathbf{R}_{jjk}^{-1}\hat{\boldsymbol{h}}_{jjk}\right\}\right|^{2}\\ &= \frac{t_{lm}^{d}}{t_{jk}^{d}}\left|\mathrm{sum}\left\{\left(\mathbf{R}_{jlm}\mathbf{R}_{jjk}^{-1}\right)\circ\mathbf{R}_{\hat{h}_{jjk}}^{T}\right\}\right|^{2}\end{aligned} \quad (27)$$

where equation (a) exists for $\mathbf{R}_{jlm}\mathbf{R}_{jjk}^{-1}$ is a Hermitian matrix, and $\hat{\boldsymbol{h}}_{jjk}^{H}\mathbf{R}_{jlm}\mathbf{R}_{jjk}^{-1}\hat{\boldsymbol{h}}_{jjk} = \mathrm{sum}\{(\mathbf{R}_{jlm}\mathbf{R}_{jjk}^{-1})\circ(\hat{\boldsymbol{h}}_{jjk}\hat{\boldsymbol{h}}_{jjk}^{H})^{T}\}$ is a real number. When $M\rightarrow\infty$, $\mathbb{E}_{\boldsymbol{h}}\{|\hat{\boldsymbol{h}}_{jjk}^{H}\mathbf{R}_{jlm}\mathbf{R}_{jjk}^{-1}\hat{\boldsymbol{h}}_{jjk}|^{2}\} \approx |\mathbb{E}_{\boldsymbol{h}}\{\hat{\boldsymbol{h}}_{jjk}^{H}\mathbf{R}_{jlm}\mathbf{R}_{jjk}^{-1}\hat{\boldsymbol{h}}_{jjk}\}|^{2}$. In addition, due to the mutual independence of $\hat{\boldsymbol{h}}_{jjk}$ and $\boldsymbol{h}_{jlm}$, $\mathbb{E}_{\boldsymbol{h}}\{\hat{\boldsymbol{h}}_{jjk}^{H}\tilde{\boldsymbol{h}}_{jlm}\tilde{\boldsymbol{h}}_{jlm}^{H}\hat{\boldsymbol{h}}_{jjk}\} = \mathrm{sum}\{\mathbf{R}_{\tilde{h}_{jlm}}\circ\mathbf{R}_{\hat{h}_{jjk}}^{T}\}$. Therefore, the third item of expression (20) (the first item of the denominator in expression (21)) is

$$\begin{aligned}&\sum_{l=1}^{L}\sum_{m=1}^{K_l}\frac{t_{lm}^{d}}{t_{jk}^{d}}\mathbb{E}_{\boldsymbol{h}}\left\{\left|\left(\boldsymbol{w}_{jjk}^{\mathrm{scheme}}\right)^{H}\boldsymbol{h}_{jlm}\right|^{2}\right\}\\ &= \sum_{(l,m)\notin\Lambda(j,k)}\frac{t_{lm}^{d}}{t_{jk}^{d}}\mathbb{E}_{\boldsymbol{h}}\left\{\left|\left(\boldsymbol{w}_{jjk}^{\mathrm{mrc}}\right)^{H}\boldsymbol{h}_{jlm}\right|^{2}\right\} + \sum_{(l,m)\in\Lambda(j,k)}\frac{t_{lm}^{d}}{t_{jk}^{d}}\mathbb{E}_{\boldsymbol{h}}\left\{\left|\left(\boldsymbol{w}_{jjk}^{\mathrm{mrc}}\right)^{H}\boldsymbol{h}_{jlm}\right|^{2}\right\}\\ &= \sum_{(l,m)\notin\Lambda(j,k)}\frac{t_{lm}^{d}}{t_{jk}^{d}}\frac{1}{\delta_{jjk}^{2}}\mathrm{sum}\left\{\mathbf{R}_{jlm}\circ\mathbf{R}_{\hat{h}_{jjk}}^{T}\right\}\\ &+ \sum_{(l,m)\in\Lambda(j,k)}\frac{t_{lm}^{d}}{t_{jk}^{d}}\frac{1}{\delta_{jjk}^{2}}\left\{\frac{t_{lm}^{d}}{t_{jk}^{d}}\left|\mathrm{sum}\left\{\left(\mathbf{R}_{jlm}\mathbf{R}_{jjk}^{-1}\right)\circ\mathbf{R}_{\hat{h}_{jjk}}^{T}\right\}\right|^{2} + \mathrm{sum}\left\{\mathbf{R}_{\tilde{h}_{jlm}}\circ\mathbf{R}_{\hat{h}_{jjk}}^{T}\right\}\right\}\end{aligned} \quad (28)$$

For the last item of expression (20) and the last item of the denominator in expression (21),

$$\frac{1}{t_{jk}^{d}}\mathbb{E}_{\boldsymbol{h}}\left\{\left|\left(\boldsymbol{w}_{jjk}^{\mathrm{mrc}}\right)^{H}\boldsymbol{n}_{j}^{\mathrm{data}}\right|^{2}\right\}=\frac{1}{t_{jk}^{d}}\frac{1}{\delta_{jjk}^{2}}\mathbb{E}_{\boldsymbol{h}}\left\{\hat{\boldsymbol{h}}_{jjk}^{H}\boldsymbol{n}_{j}^{\mathrm{data}}\left(\boldsymbol{n}_{j}^{\mathrm{data}}\right)^{H}\hat{\boldsymbol{h}}_{jjk}\right\}$$
$$=\frac{\sigma^{2}}{t_{jk}^{d}\delta_{jjk}^{2}}\mathbb{E}_{\boldsymbol{h}}\left\{\hat{\boldsymbol{h}}_{jjk}^{H}\hat{\boldsymbol{h}}_{jjk}\right\}=\frac{\sigma^{2}}{t_{jk}^{d}\delta_{jjk}}\tag{29}$$

Combining expressions (20), (23), (24), (28), and (29), the MSE-SD for user (*j*, *k*) at BS *j* is

$$\begin{aligned}\text{MSE-SD}_{jk}^{\mathrm{mrc}}&=\frac{1}{\delta_{jjk}^{2}}\mathrm{sum}\left\{\mathbf{R}_{\tilde{h}_{jjk}}\circ\mathbf{R}_{\hat{h}_{jjk}}^{T}\right\}\\&+\sum_{(l,m)\notin\Lambda(j,k)}\frac{t_{lm}^{d}}{t_{jk}^{d}}\frac{1}{\delta_{jjk}^{2}}\mathrm{sum}\left\{\mathbf{R}_{jlm}\circ\mathbf{R}_{\hat{h}_{jjk}}^{T}\right\}\\&+\sum_{(l,m)\in\Lambda(j,k)}\frac{t_{lm}^{d}}{t_{jk}^{d}}\frac{1}{\delta_{jjk}^{2}}\left\{\frac{t_{lm}^{d}}{t_{jk}^{d}}\left|\mathrm{sum}\left\{\left(\mathbf{R}_{jlm}\mathbf{R}_{jjk}^{-1}\right)\circ\mathbf{R}_{\hat{h}_{jjk}}^{T}\right\}\right|^{2}\right.\\&\left.+\mathrm{sum}\left\{\mathbf{R}_{\tilde{h}_{jlm}}\circ\mathbf{R}_{\hat{h}_{jjk}}^{T}\right\}\right\}+\frac{\sigma^{2}}{t_{jk}^{d}\delta_{jjk}}-1\end{aligned}\tag{30}$$

Combining expressions (21), (22), (28), and (29), the received SINR for user (*j*, *k*) at BS *j* is

$$\mathrm{SINR}_{jk}^{\mathrm{mrc}}=\frac{1}{\begin{aligned}&\sum_{(l,m)\notin\Lambda(j,k)}\frac{t_{lm}^{d}}{t_{jk}^{d}}\frac{1}{\delta_{jjk}^{2}}\mathrm{sum}\left\{\mathbf{R}_{jlm}\circ\mathbf{R}_{\hat{h}_{jjk}}^{T}\right\}\\&+\sum_{(l,m)\in\Lambda(j,k)}\frac{t_{lm}^{d}}{t_{jk}^{d}}\frac{1}{\delta_{jjk}^{2}}\left\{\frac{t_{lm}^{d}}{t_{jk}^{d}}\left|\mathrm{sum}\left\{\left(\mathbf{R}_{jlm}\mathbf{R}_{jjk}^{-1}\right)\circ\mathbf{R}_{\hat{h}_{jjk}}^{T}\right\}\right|^{2}\right.\\&\left.+\mathrm{sum}\left\{\mathbf{R}_{\tilde{h}_{jlm}}\circ\mathbf{R}_{\hat{h}_{jjk}}^{T}\right\}\right\}+\frac{\sigma^{2}}{t_{jk}^{d}\delta_{jjk}}-1\end{aligned}}\tag{31}$$

8 Appendix 2

We consider the ZFC scheme. For ZFC, the $\boldsymbol{w}_{jjk}^{\mathrm{zfc}}$ is the kth column of the matrix $\boldsymbol{W}_{jj}^{\mathrm{zfc}}=\hat{\boldsymbol{H}}_{jj}(\hat{\boldsymbol{H}}_{jj}^{H}\hat{\boldsymbol{H}}_{jj})^{-1}$, and thus, $(\boldsymbol{W}_{jj}^{\mathrm{zfc}})^{H}\hat{\boldsymbol{H}}_{jj}=\mathbf{I}_{K_j}$ and $(\boldsymbol{W}_{jj}^{\mathrm{zfc}})^{H}\boldsymbol{W}_{jj}^{\mathrm{zfc}}=(\hat{\boldsymbol{H}}_{jj}^{H}\hat{\boldsymbol{H}}_{jj})^{-1}$. Therefore

$$\mathbb{E}_{\boldsymbol{h}}\left\{\left(\boldsymbol{w}_{jjk}^{\mathrm{zfc}}\right)^{H}\boldsymbol{h}_{jjk}\right\}=\mathbb{E}_{\boldsymbol{h}}\left\{\left(\boldsymbol{w}_{jjk}^{\mathrm{zfc}}\right)^{H}\left(\hat{\boldsymbol{h}}_{jjk}+\tilde{\boldsymbol{h}}_{jjk}\right)\right\}$$
$$=\mathbb{E}_{\boldsymbol{h}}\left\{\left(\boldsymbol{w}_{jjk}^{\mathrm{zfc}}\right)^{H}\hat{\boldsymbol{h}}_{jjk}\right\}=1\tag{32}$$

Therefore, the first item of expression (20) is

$$\mathbb{E}_{\boldsymbol{h}}\left\{\left|\left(\boldsymbol{w}_{jjk}^{\mathrm{zfc}}\right)^{H}\hat{\boldsymbol{h}}_{jjk}-1\right|^{2}\right\}=0\tag{33}$$

Because $\tilde{\boldsymbol{h}}_{jjk}$ is independent of $\boldsymbol{w}_{jjk}^{\mathrm{zfc}}$, the second item of expression (20) is

$$\mathbb{E}_{\boldsymbol{h}}\left\{\left|\left(\boldsymbol{w}_{jjk}^{\mathrm{zfc}}\right)^{H}\tilde{\boldsymbol{h}}_{jjk}\right|^{2}\right\}=\mathbb{E}_{\boldsymbol{h}}\left\{\left(\boldsymbol{w}_{jjk}^{\mathrm{zfc}}\right)^{H}\tilde{\boldsymbol{h}}_{jjk}\tilde{\boldsymbol{h}}_{jjk}^{H}\boldsymbol{w}_{jjk}^{\mathrm{zfc}}\right\}$$
$$=\mathrm{sum}\left\{\mathbf{R}_{\tilde{h}_{jjk}}\circ\boldsymbol{Q}_{1}^{T}\right\}\tag{34}$$

Here, we assume that $\boldsymbol{Q}_{1}=\mathbb{E}_{\boldsymbol{h}}\{\boldsymbol{w}_{jjk}^{\mathrm{zfc}}(\boldsymbol{w}_{jjk}^{\mathrm{zfc}})^{H}\}$. The fourth item of expression (20) is

$$\begin{aligned}&\mathbb{E}_{\boldsymbol{h}}\left\{\left|\left(\boldsymbol{w}_{jjk}^{\mathrm{zfc}}\right)^{H}\boldsymbol{h}_{jjk}\right|^{2}\right\}\\&=\mathbb{E}_{\boldsymbol{h}}\left\{\left(\boldsymbol{w}_{jjk}^{\mathrm{zfc}}\right)^{H}\hat{\boldsymbol{h}}_{jjk}\hat{\boldsymbol{h}}_{jjk}^{H}\boldsymbol{w}_{jjk}^{\mathrm{zfc}}\right\}+\mathbb{E}_{\boldsymbol{h}}\left\{\left(\boldsymbol{w}_{jjk}^{\mathrm{zfc}}\right)^{H}\tilde{\boldsymbol{h}}_{jjk}\tilde{\boldsymbol{h}}_{jjk}^{H}\boldsymbol{w}_{jjk}^{\mathrm{zfc}}\right\}\\&=1+\mathrm{sum}\left\{\mathbf{R}_{\tilde{h}_{jjk}}\circ\boldsymbol{Q}_{1}^{T}\right\}\end{aligned}\tag{35}$$

Using a similar analysis method as MRC in Appendix 1, we can divide the third item of expression (20) (the first item of the denominator in expression (21)) into two cases.

When $(l,m)\notin\Lambda_{S}(j,k)$,

$$\begin{aligned}&\sum_{(l,m)\notin\Lambda(j,k)}\frac{t_{lm}^{d}}{t_{jk}^{d}}\mathbb{E}_{\boldsymbol{h}}\left\{\left|\left(\boldsymbol{w}_{jjk}^{\mathrm{zfc}}\right)^{H}\boldsymbol{h}_{jlm}\right|^{2}\right\}\\&=\sum_{(l,m)\notin\Lambda(j,k)}\frac{t_{lm}^{d}}{t_{jk}^{d}}\mathbb{E}_{\boldsymbol{h}}\left\{\left(\boldsymbol{w}_{jjk}^{\mathrm{zfc}}\right)^{H}\boldsymbol{h}_{jlm}\boldsymbol{h}_{jlm}^{H}\boldsymbol{w}_{jjk}^{\mathrm{zfc}}\right\}\\&=\sum_{(l,m)\notin\Lambda(j,k)}\frac{t_{lm}^{d}}{t_{jk}^{d}}\mathrm{sum}\{\mathbf{R}_{jlm}\circ\boldsymbol{Q}_{1}^{T}\}\end{aligned}\tag{36}$$

When $(l,m)\in\Lambda_{S}(j,k)$,

$$\begin{aligned}&\sum_{(l,m)\in\Lambda(j,k)}\frac{t_{lm}^{d}}{t_{jk}^{d}}\mathbb{E}_{\boldsymbol{h}}\left\{\left|\left(\boldsymbol{w}_{jjk}^{\mathrm{zfc}}\right)^{H}\boldsymbol{h}_{jlm}\right|^{2}\right\}\\&=\sum_{(l,m)\in\Lambda(j,k)}\frac{t_{lm}^{d}}{t_{jk}^{d}}\left(\mathbb{E}_{\boldsymbol{h}}\left\{\left(\boldsymbol{w}_{jjk}^{\mathrm{zfc}}\right)^{H}\hat{\boldsymbol{h}}_{jlm}\hat{\boldsymbol{h}}_{jlm}^{H}\boldsymbol{w}_{jjk}^{\mathrm{zfc}}\right\}\right.\\&\left.+\mathbb{E}_{\boldsymbol{h}}\left\{\left(\boldsymbol{w}_{jjk}^{\mathrm{zfc}}\right)^{H}\tilde{\boldsymbol{h}}_{jlm}\tilde{\boldsymbol{h}}_{jlm}^{H}\boldsymbol{w}_{jjk}^{\mathrm{zfc}}\right\}\right)\\&=\sum_{(l,m)\in\Lambda(j,k)}\frac{t_{lm}^{d}}{t_{jk}^{d}}\left\{\frac{t_{lm}^{d}}{t_{jk}^{d}}\left|\mathrm{sum}\left\{\left(\mathbf{R}_{jlm}\mathbf{R}_{jjk}^{-1}\right)\circ\boldsymbol{Q}_{2}^{T}\right\}\right|^{2}\right.\\&\left.+\mathrm{sum}\left\{\mathbf{R}_{\tilde{h}_{jlm}}\circ\boldsymbol{Q}_{1}^{T}\right\}\right\}\end{aligned}\tag{37}$$

where $\boldsymbol{Q}_{2}=\mathbb{E}_{\boldsymbol{h}}\{\hat{\boldsymbol{h}}_{jjk}(\boldsymbol{w}_{jjk}^{\mathrm{zfc}})^{H}\}$.

In addition, the last item of expression (20) (the last item of the denominator in expression (21)) is

$$\frac{1}{t_{jk}^d}\mathbb{E}_h\left\{\left|\left(w_{jjk}^{\text{zfc}}\right)^H n_j^{\text{data}}\right|^2\right\} = \frac{1}{t_{jk}^d}\mathbb{E}_h\left\{\left(w_{jjk}^{\text{zfc}}\right)^H n_j^{\text{data}}\left(n_j^{\text{data}}\right)^H w_{jjk}^{\text{zfc}}\right\}$$
$$= \frac{\sigma^2}{t_{jk}^d}\mathbb{E}_h\left\{\left(w_{jjk}^{\text{zfc}}\right)^H w_{jjk}^{\text{zfc}}\right\} = \frac{\sigma^2}{t_{jk}^d}\cdot[Q_3]_{k,k} \quad (38)$$

where $Q_3 = \mathbb{E}_h\{(\hat{H}_{jj}^H\hat{H}_{jj})^{-1}\}$ and $[Q_3]_{k,\,k}$ means the kth row and kth column element of matrix Q_3.

Combining expressions (20), (33), (34), (35), (36), (37), and (38), the MSE-SD for user (j, k) at BS j is

$$\text{MSE-SD}_{jk}^{\text{zfc}} = \sum_{(l,m)\notin\Lambda(j,k)}\frac{t_{lm}^d}{t_{jk}^d}\,\text{sum}\left\{\mathbf{R}_{jlm}\circ Q_1^T\right\} + \sum_{(l,m)\in\Lambda(j,k)}\frac{t_{lm}^d}{t_{jk}^d}\left\{\frac{t_{lm}^d}{t_{jk}^d}\left|\text{sum}\left\{\left(\mathbf{R}_{jlm}\mathbf{R}_{jjk}^{-1}\right)\circ Q_2^T\right\}\right|^2 + \text{sum}\left\{\mathbf{R}_{\tilde{h}_{jlm}}\circ Q_1^T\right\}\right\} + \frac{\sigma^2}{t_{jk}^d}\cdot[Q_3]_{k,k} - 1 \quad (39)$$

Combining expressions (21), (32), (36), (37) and (38), the received SINR for user (j, k) at BS j is

$$\text{SINR}_{jk}^{\text{zfc}} = \frac{1}{\sum_{(l,m)\notin\Lambda(j,k)}\frac{t_{lm}^d}{t_{jk}^d}\text{sum}\left\{\mathbf{R}_{jlm}\circ Q_1^T\right\} + \sum_{(l,m)\in\Lambda(j,k)}\frac{t_{lm}^d}{t_{jk}^d}\left\{\frac{t_{lm}^d}{t_{jk}^d}\left|\text{sum}\left\{\left(\mathbf{R}_{jlm}\mathbf{R}_{jjk}^{-1}\right)\circ Q_2^T\right\}\right|^2 + \text{sum}\left\{\mathbf{R}_{\tilde{h}_{jlm}}\circ Q_1^T\right\}\right\} + \frac{\sigma^2}{t_{jk}^d}\cdot[Q_3]_{k,k} - 1} \quad (40)$$

9 Appendix 3

We consider the MMSE scheme, the w_{jjk}^{mmse} is the kth column of the matrix $W_{jj}^{\text{mmse}} = \hat{H}_{jj}(\hat{H}_{jj}^H\hat{H}_{jj} + \sigma^2(T_j^d)^{-2})^{-1}$. Because $(W_{jj}^{\text{zfc}})^H\hat{H}_{jj} + \sigma^2(\hat{H}_{jj}^H\hat{H}_{jj} + \sigma^2(T_j^d)^{-2})^{-1} = \mathbf{I}_{K_j}$, $(w_{jjk}^{\text{mmse}})^H\hat{h}_{jjk} = [(W_{jj}^{\text{zfc}})^H\hat{H}_{jj}]_{k,k} = 1 - [\sigma^2(\hat{H}_{jj}^H\hat{H}_{jj} + \sigma^2(T_j^d)^{-2})^{-1}]_{k,k}$, thus

$$\mathbb{E}_h\left\{\left(w_{jjk}^{\text{mmse}}\right)^H\hat{h}_{jjk}\right\} = 1 - \mathbb{E}_h\left\{\left[\sigma^2\left(\hat{H}_{jj}^H H_{jj} + \sigma^2\left(T_j^d\right)^{-2}\right)^{-1}\right]_{k,k}\right\} = 1 - [Q_4]_{k,k} \quad (41)$$

Therefore, the first item of expression (20) is

$$\mathbb{E}_h\left\{\left|\left(w_{jjk}^{\text{mmse}}\right)^H\hat{h}_{jjk} - 1\right|^2\right\} = \left|[Q_4]_{k,k}\right|^2 \quad (42)$$

where $Q_4 = \mathbb{E}_h\{\sigma^2(\hat{H}_{jj}^H\hat{H}_{jj} + \sigma^2(T_j^d)^{-2})^{-1}\}$. Because $\tilde{h}_{jjk}$ is independent of w_{jjk}^{zfc}, the second item of expression (20) is

$$\mathbb{E}_h\left\{\left|\left(w_{jjk}^{\text{mmse}}\right)^H\tilde{h}_{jjk}\right|^2\right\} = \mathbb{E}_h\left\{\left(w_{jjk}^{\text{mmse}}\right)^H\tilde{h}_{jjk}\tilde{h}_{jjk}^H w_{jjk}^{\text{mmse}}\right\} = \text{sum}\left\{\mathbf{R}_{\tilde{h}_{jjk}}\circ Q_5^T\right\} \quad (43)$$

where $Q_5 = \mathbb{E}_h\{w_{jjk}^{\text{mmse}}(w_{jjk}^{\text{mmse}})^H\}$. The fourth item of expression (20) is

$$\mathbb{E}_h\left\{\left|\left(w_{jjk}^{\text{mmse}}\right)^H h_{jjk}\right|^2\right\} = \mathbb{E}_h\left\{\left(w_{jjk}^{\text{mmse}}\right)^H\hat{h}_{jjk}\hat{h}_{jjk}^H w_{jjk}^{\text{mmse}}\right\} + \mathbb{E}_h\left\{\left(w_{jjk}^{\text{mmse}}\right)^H\tilde{h}_{jjk}\tilde{h}_{jjk}^H w_{jjk}^{\text{mmse}}\right\} \approx \left|1 - [Q_4]_{k,k}\right|^2 + \text{sum}\left\{\mathbf{R}_{\tilde{h}_{jjk}}\circ Q_5^T\right\} \quad (44)$$

where the last approximate equality exists when $M \to \infty$.

Using a similar analysis method as MRC in Appendix 1, we can divide the third item of expression (20) (the first item of the denominator in expression (21)) into two cases. The results is

$$\sum_{l=1}^{L}\sum_{m=1}^{K_l}\frac{t_{lm}^d}{t_{jk}^d}\mathbb{E}_h\left\{\left|\left(w_{jjk}^{\text{mmse}}\right)^H h_{jlm}\right|^2\right\} = \sum_{(l,m)\notin\Lambda(j,k)}\frac{t_{lm}^d}{t_{jk}^d}\,\text{sum}\left\{\mathbf{R}_{jlm}\circ Q_5^T\right\} + \sum_{(l,m)\in\Lambda(j,k)}\frac{t_{lm}^d}{t_{jk}^d}\left\{\frac{t_{lm}^d}{t_{jk}^d}\left|\text{sum}\left\{\left(\mathbf{R}_{jlm}\mathbf{R}_{jjk}^{-1}\right)\circ Q_6^T\right\}\right|^2 + \text{sum}\left\{\mathbf{R}_{\tilde{h}_{jlm}}\circ Q_5^T\right\}\right\} \quad (45)$$

where $Q_6 = \mathbb{E}_h\{\hat{h}_{jjk}(w_{jjk}^{\text{mmse}})^H\}$.

In addition, the last item of expression (20) (the last item of the denominator in expression (21)) is5

$$\frac{1}{t_{jk}^d}\mathbb{E}_h\left\{\left|\left(w_{jjk}^{\text{mmse}}\right)^H n_j^{\text{data}}\right|^2\right\} = \frac{1}{t_{jk}^d}\mathbb{E}_h\left\{\left(w_{jjk}^{\text{mmse}}\right)^H n_j^{\text{data}}\left(n_j^{\text{data}}\right)^H w_{jjk}^{\text{mmse}}\right\} = \frac{\sigma^2}{t_{jk}^d}\mathbb{E}_h\left\{\left(w_{jjk}^{\text{mmse}}\right)^H w_{jjk}^{\text{mmse}}\right\} = \frac{\sigma^2}{t_{jk}^d}\cdot[Q_7]_{k,k} \quad (46)$$

where $Q_7 = \mathbb{E}_h\{(W_{jj}^{\text{mmse}})^H W_{jj}^{\text{mmse}}\} = \mathbb{E}_h\{(\hat{H}_{jj}^H\hat{H}_{jj} + \sigma^2(T_j^d)^{-2})^{-1}\hat{H}_{jj}^H\hat{H}_{jj}(\hat{H}_{jj}^H\hat{H}_{jj} + \sigma^2(T_j^d)^{-2})^{-1}\}$.

Combining expressions (20), (42), (43), (44), (45), and (46), the MSE-SD for user (j, k) at BS j is

$$
\begin{aligned}
\text{MSE-SD}_{jk}^{\text{mmse}} = & \sum_{(l,m)\notin\Lambda(j,k)} \frac{t_{lm}^{d}}{t_{jk}^{d}} \text{sum}\{\mathbf{R}_{jlm}\circ\boldsymbol{Q}_5^T\} \\
& + \sum_{(l,m)\in\Lambda(j,k)} \frac{t_{lm}^{d}}{t_{jk}^{d}} \left\{ \frac{t_{lm}^{d}}{t_{jk}^{d}} \left|\text{sum}\left\{ \left(\mathbf{R}_{jlm}\mathbf{R}_{jjk}^{-1}\right)\circ\boldsymbol{Q}_6^T \right\}\right|^2 \right. \\
& \left. +\text{sum}\left\{\mathbf{R}_{\tilde{h}_{jlm}}\circ\boldsymbol{Q}_5^T\right\} \right\} + \frac{\sigma^2}{t_{jk}^{d}} \cdot [\boldsymbol{Q}_7]_{k,k} - 2[\boldsymbol{Q}_4]_{k,k} - 1
\end{aligned}
\tag{47}
$$

Combining expressions (21), (41), (45), and (46), the received SINR for user (j,k) at BS j is

$$
\text{SINR}_{jk}^{\text{mmse}} = \frac{1-[\boldsymbol{Q}_4]_{k,k}}{I_{jk}^{\text{mmse}}} \tag{48}
$$

where
$$
\begin{aligned}
I_{jk}^{\text{mmse}} = & \sum_{(l,m)\notin\Lambda(j,k)} \frac{t_{lm}^{d}}{t_{jk}^{d}} \text{sum}\{\mathbf{R}_{jlm}\circ\boldsymbol{Q}_5^T\} \\
& + \sum_{(l,m)\in\Lambda(j,k)} \frac{t_{lm}^{d}}{t_{jk}^{d}} \left\{ \frac{t_{lm}^{d}}{t_{jk}^{d}} \left|\text{sum}\left\{ \left(\mathbf{R}_{jlm}\mathbf{R}_{jjk}^{-1}\right)\circ\boldsymbol{Q}_6^T \right\}\right|^2 \right. \\
& \left. +\text{sum}\left\{\mathbf{R}_{\tilde{h}_{jlm}}\circ\boldsymbol{Q}_5^T\right\} \right\} + \frac{\sigma^2}{t_{jk}^{d}} \cdot [\boldsymbol{Q}_7]_{k,k} - 1 + [\boldsymbol{Q}_4]_{k,k}.
\end{aligned}
$$

Abbreviations

AoA: Angle-of-arrival; AWGN: Additive white Gaussian noise; BS: Base station; BSC: Base station controller; CSI: Channel state information; MIMO: Multiple-input multiple-output; MMSE: Minimum mean square error; MRC: Maximum ratio combining; MSE: Mean square error; MSE-CE: Mean square error of channel estimation; MSE-SD: Mean square error of signal detection; SE: Spectrum efficiency; SINR: Signal-to-interference-plus-noise ratio; TDD: Time division duplex; ULA: Uniformly spaced linear array; ZFC: Zero-forcing combining

Funding

This research is supported by the Natural Science Foundation of Anhui Province, the Startup Research Foundation of Anhui University (01001770-10117700011), the College Natural Science Research Project of Anhui Province (KJ2016A042).

Authors' contributions

HZ has contributed 70% of the work, and the rest is contributed by XD. Both authors read and approved the final manuscript.

Competing interests

The authors declare that they have no competing interests.

Author details

[1]Key Lab of Intelligent Computing and Signal Processing (Ministry of Education), Anhui University, Hefei 230601, China. [2]QuantumCTek Co., Ltd., Hefei 230088, China.

References

1. L. Lu, G.Y. Li, A.L. Swindlehurst, et al., An overview of massive MIMO: benefits and challenges. IEEE J. Sel. Top. Sign. Proces. **8**(5), 742–758 (2014).
2. Z. Gong, C. Li, F. Jiang, Pilot contamination mitigation strategies in massive MIMO systems. IET Commun. **11**(16), 2403–2409 (2017).
3. T.L. Marzetta, Noncooperative cellular wireless with unlimited numbers of base station antennas. IEEE Trans. Wirel. Commun. **9**(11), 3590–3600 (2010).
4. A. Ashikhmin, T. Marzetta, *Pilot Contamination Precoding in Multi-Cell Large Scale Antenna Systems. Proceedings International Symposium on Information Theory (ISIT)* (2012), pp. 1137–1141.
5. M.H.C. Garcia, J. Luo, in *Proceedings of 2018 IEEE Wireless Communications and Networking Conference (WCNC)*. Time-shifted pilots multiplexed with uplink data and unequal power allocation (IEEE, Barcelona, 2018), pp. 1–6.
6. W. Chang, Y.K. Hua, S.F. Liao, Partial overlapped time-shifted pilots for massive MIMO systems. IEEE Commun. Lett. **21**(11), 2480–2483 (2017).
7. H. Echigo, T. Ohtsuki, W. Jiang, Y. Takatori, in *Proceedings of 2017 IEEE Global Communications Conference (GLOBECOM)*. Fair pilot assignment based on AOA and pathloss with location information in massive MIMO (IEEE, Singapore, 2017), pp. 1–6.
8. H. Wang, W. Zhang, Y. Liu, Q. Xu, P. Pan, On design of non-orthogonal pilot signals for a multi-cell massive MIMO system. IEEE Wireless Commun. Lett. **4**(2), 129–132 (2015).
9. P. Liu, S. Jin, T. Jiang, Q. Zhang, M. Matthaiou, Pilot power allocation through user grouping in multi-cell massive MIMO systems. IEEE Trans. Commun. **65**(4), 1561–1574 (2017).
10. H. Zhi, Q. Yuan, J. Zhu, Y. Hu, *A pilot allocation scheme for TDD massive MIMO system enabled HetNet*, 9th IEEE International Conference on Communication Software and Networks (ICCSN, GuangZhou, 2017), pp. 443–447.
11. J. Fan, W. Li, Y. Zhang, J. Deng, in *Proceedings of IEEE 85th Vehicular Technology Conference (VTC Spring)*. Fractional pilot reuse with vertical sectorization in massive MIMO systems (IEEE, Sydney, 2017), pp. 1–5.
12. H. Ahmadi, A. Farhang, N. Marchetti, A. MacKenzie, A game theoretic approach for pilot contamination avoidance in massive MIMO. IEEE Wireless Commun. Lett. **5**(1), 12–15 (2016).
13. R. Mochaourab, E. Björnson, M. Bengtsson, Adaptive pilot clustering in heterogeneous massive MIMO networks. IEEE Trans. Wirel. Commun. **15**(8), 5555–5568 (2016).
14. L. You, X. Gao, X. Xia, N. Ma, Y. Peng, Pilot reuse for massive MIMO transmission over spatially correlated rayleigh fading channels. IEEE Trans. Wirel. Commun. **14**(6), 3352–3366 (2015).
15. H. Yin, D. Gesbert, M. Filippou, Y. Liu, A coordinated approach to channel estimation in large-scale multiple-antenna systems. IEEE J. Sel. Areas Commun. **31**(2), 264–273 (2013).
16. L.S. Muppirisetty, H. Wymeersch, J. Karout, G. Fodor, in *Proceedings of 2015 IEEE Global Communications Conference (GLOBECOM)*. Location-aided pilot contamination elimination for massive MIMO systems (IEEE, San Diego, 2015), pp. 1–5.
17. L.S. Muppirisetty, T. Charalambous, J. Karout, G. Fodor, H. Wymeersch, Location-aided pilot contamination avoidance for massive MIMO systems. IEEE Trans. Wirel. Commun. **17**(4), 2662–2674 (2018).
18. T. Kailath, A.H. Sayed, B. Hassibi, *Linear Estimation* (Prentice-Hall, Upper Saddle River, 2000).
19. E. Björnson, E.G. Larsson, M. Debbah, Massive MIMO for maximal spectral efficiency: How many users and pilots should be allocated? IEEE Trans. Wirel. Commun. **15**(2), 1293–1308 (2016).
20. M.J. Osborne, A. Rubinstein, *A Course in Game Theory* (MIT Press, Cambridge, 1994).
21. A. Bogomolnaia, M.O. Jackson, The stability of hedonic coalition structures. Games Econ. Behavior. **38**(2), 201–230 (2002).

11

Robust full-dimension MIMO transmission based on limited feedback angular-domain CSIT

Stefan Schwarz

Abstract

In this paper, we propose robust multi-user beamforming and precoding techniques for full-dimension MIMO transmission based on limited feedback. We propose to employ an over-complete basis decomposition in the angular domain to approximate the channel matrices as sums of few dominant specular components, facilitating efficient channel state information (CSI) quantization at the users. The selected expansion vectors of such a sparse approximation, parametrized by azimuth and elevation angles, are relatively robust with respect to channel estimation errors as well as channel variations over time. Based on this CSI feedback, we propose incoherent beamforming/precoding methods that make use only of the azimuth and elevation angles as well as the norm of the expansion coefficients and do not rely on coherent multipath interference to eliminate inter-user interference. Our optimization aims at maximizing the signal power or the achievable rate of a user, while limiting the amount of interference leakage caused to other users. To further improve robustness, we account for uncertainty in the angular channel decomposition in the proposed precoder optimization.

Keywords: Full-dimension MIMO, 3D beamforming, Robust precoding, Interference leakage, High mobility, Limited feedback

1 Introduction

Full-dimension multiple-input multiple-output (FD-MIMO) transmission refers to wireless transmission systems that support two-dimensional antenna arrays with a large number of antenna elements. This enables high-resolution beamforming in both, the elevation and the azimuth domain, in order to achieve space-division multiple access gains through spatial separation of users, as well as to enhance the energy efficiency of wireless data transmission by concentrating the radiated energy towards intended users [1–3]. The great potential of FD-MIMO systems has been recognized by standardization bodies, such as the third generation partnership project (3GPP); correspondingly, development and standardization efforts are ongoing within long-term evolution (LTE) Release 14 and 5G new radio (NR) to further enhance the initial realization of FD-MIMO in Release 13. Currently, most implementation proposals of FD-MIMO transceivers are based on hybrid architectures, where part of the signal processing is performed in base band and part in the analog radio frequency (RF) domain in order to limit the number of required RF chains. This approach is especially important in the millimeter wave band, where hardware complexity is a limiting factor [4–8]. When designing hybrid precoding systems, several challenges need to be addressed [9–11]: the design freedom of RF precoders is commonly restricted by hardware limitations, such as discrete angular resolution and constant modulus of the phase-shifting elements, as well as partial connectivity of the phase-shift network to limit insertion loss. An alternative to hybrid architectures, which potentially provides even higher spectral and energy efficiency in certain operating regimes, is to employ cheap low (even single bit) resolution analog to digital (AD)/digital to analog (DA) converters in FD-MIMO transceivers.

Correspondence: stefan.schwarz@nt.tuwien.ac.at
Christian Doppler Laboratory for Dependable Wireless Connectivity for the Society in Motion, Institute of Telecommunications, Technische Universität (TU) Wien, Vienna, Austria

At the receive-side of most existing MIMO transceivers, high-resolution AD converters are employed to minimize the signal distortions introduced by the receiver. Such high-resolution AD converters are, however, among the most power hungry devices of the receiver chain. This motivates the use of low-resolution energy efficient AD converters in FD-MIMO systems to achieve energy-efficient operation [12–15]. At the transmit side, on the other hand, power expenditure is dominated by power amplifiers, which are usually required to operate within the high linearity regime to avoid signal distortion. Employing low-resolution DA converters relaxes these linearity requirements, allowing the amplifiers to operate closer to saturation thereby increasing their efficiency [16, 17].

In this work, our focus is on an enhancing the robustness of downlink three-dimensional FD-MIMO point to multi-point transmission with respect to imperfections of the channel state information at the transmitter (CSIT), especially due to high user mobility, without specifically addressing the restrictions of hybrid precoding or low-resolution AD/DA implementations. In MIMO transmission in general, and especially when considering multi-point communication, CSIT is an essential ingredient. CSIT for downlink transmission is commonly obtained either from uplink measurements using dedicated uplink pilot signals [18, 19] or from limited channel state information (CSI) feedback from the users [20–22]. We consider frequency division duplex (FDD) transmission in this paper, where CSIT is provided via feedback from the users over dedicated limited capacity feedback links. For this purpose, the users have to quantize the required CSI utilizing a proper quantization codebook. Two approaches are most common in current mobile communication systems [23]: (1) Implicit CSI feedback, where the users select preferred transmit beamformers/precoders from a predefined codebook given the current channel conditions. This has the advantage that the users can accurately estimate the expected performance during transmission, provided the channel conditions are quasi-static. (2) Explicit CSI feedback, where the users directly quantize the MIMO channel matrix. This approach is better suited for multi-user MIMO transmission because it facilitates calculation of efficient multi-user precoders at the transmitter on-the-fly. Either way, providing CSI via limited feedback is always prone to inaccuracies, on the one hand, caused by the quantization process, which is inevitable due to the limited capacity of the feedback links. On the other hand, CSI inaccuracies also result from the delay of the feedback path, leading to a mismatch of the CSI experienced during the quantization phase and the CSI valid during transmission. Several theoretical studies exist that evaluate the performance degradation due to imperfect and outdated CSIT for a variety of transmission scenarios [24–27].

Popular coherent multi-user beamforming/precoding methods, such as zero forcing (ZF) beamforming [28], block diagonalization (BD) precoding [29] and interference alignment (IA) [30, 31], utilize the phase difference between the multipath components contributing to the channel between the transmitter and the receiver to achieve constructive/destructive signal interference. The phases of multipath components, however, can change very quickly over time, especially for mobile users, and hence coherent beamforming/precoding methods react very sensitive to non-static scenarios [32, 33]. The corresponding performance degradation can partly be mitigated by robust coherent beamforming/precoding techniques that account for the CSIT imperfection. One prominent example of robust beamforming techniques is minimum variance robust adaptive beamforming [34], which accounts for imperfections in the estimation of signal covariance matrices. In [35], the authors achieve robustness w.r.t. CSI quantization by utilizing multiple receive antennas to improve system performance without the need for full channel feedback information; this idea is in essence similar to [36], where subspace-selection strategies for CSI feedback are discussed that enable mitigating the impact of CSI quantization. In [37], robust single-user MIMO precoding methods are proposed that maximize the worst-case received signal-to-noise ratio or minimize the worst-case error probability with imperfect CSIT. Furthermore, hybrid transceiver architectures that incorporate robustness against channel estimation errors are proposed in [38].

Contribution In this paper, we take a different approach to achieve robustness of multi-user beamforming/precoding w.r.t. CSIT imperfections due to limited feedback. Specifically, we employ an over-complete basis decomposition of the MIMO channel matrix at each receiver to obtain a sparse approximation of the CSI, which can efficiently be quantized and fed back to the transmitter. The over-complete basis decomposition is based on an angular-domain quantization codebook (also called dictionary following the nomenclature of sparse approximation literature), which is parametrized by azimuth and elevation signal departure angles. Under line of sight (LOS) conditions, the selected expansion vectors of the sparse approximation are robust w.r.t. channel estimation errors as well as channel variations over time. We demonstrate this robustness by means of numerical simulations employing state-of-the-art three-dimensional stochastic-geometric channel models. However, the complex-valued coefficients of the over-complete basis expansion are still significantly impacted by channel estimation errors and channel variations over time, which can impair the performance of coherent precoding techniques, such as, as ZF beamforming, BD precoding and IA, that rely on destructive multipath interference to eliminate inter-user interference. We therefore propose to apply incoherent beamforming/precoding techniques that are inherently robust w.r.t. phase uncertainties of multipath components, similar to space-division multiple access (SDMA). Specifically, we propose leakage-based angular beamforming/precoding methods that make use only of the azimuth and elevation angles as well as the norm of the expansion coefficients and do not rely on the phase of the expansion coefficients to achieve destructive multipath interference. This approach also improves robustness of beamforming/precoding solutions at high mobility, where the relative phases of multipath components change very quickly over time. To further improve robustness we additionally account for uncertainty in the angular channel decomposition in the proposed precoder optimization. Naturally, this approach cannot compete with coherent precoding methods in case of highly accurate CSIT; however, our numerical simulations demonstrate significant gains in terms of achievable rate and symbol error ratio (SER) in case of imperfect CSIT under LOS conditions. The present work is an extension of our two conference publications [39, 40] to multi-user beamforming and multi-stream multi-user precoding. Specifically, in [39] we propose and approximately solve the FD-MIMO transmit beam pattern optimization problem of worst-case signal to leakage ratio (SLR) maximization, where we maximize the ratio of signal power radiated into an angular region of interest to leakage power radiated into an unintended angular region. A similar design and optimization principle guides us also in the present paper, yet extended to adaptive operation with limited feedback from the users and transmission of multiple data streams per user. In [40], we consider so-called double-sided FD-MIMO systems, where both transmitter and receiver are equipped with FD-MIMO antenna arrays, and we apply a similar over-complete basis decomposition as in the present paper to achieve efficient limited feedback operation. However, we do not consider robust multi-user beamforming/precoding in [40]. Hence, the present paper combines results of [39, 40] and extends them to adaptive limited-feedback based multi-user precoding with multi-stream transmission per user.

Organization In Section 2, we introduce our system model and state the underlying assumptions of this work. In Section 3, we present the proposed CSI feedback approach that is based on a sparse approximation of the channel utilizing an over-complete basis decomposition. Based on this CSI feedback, we then propose in Section 4 our robust leakage-bounded angular beamforming and precoding methods. We briefly summarize for the readers convenience in Section 5 the benchmark methods employed in our simulations, in order to clarify how these methods are applied in our considered scenario. Finally, in Section 6, we investigate the performance of the proposed beamformers and precoders by means of numerical simulations and we provide concluding remarks in Section 7.

Notation We denote scalars by upper and lower case italic letters (e.g., N, n), vectors by lower case bold letters (e.g., $\mathbf{x}$), and matrices by upper case bold letters (e.g., $\mathbf{H}$). To address the element in the i-th row and the j-th column of matrix $\mathbf{H}$, we use the notation $[\mathbf{H}]_{i,j}$. We write the trace of a matrix as $\mathrm{tr}\,(\mathbf{H})$, the transpose as $\mathbf{H}^{\mathrm{T}}$, the conjugate-transpose as $\mathbf{H}^{\mathrm{H}}$, the Frobenius norm as $\|\mathbf{H}\|$, the pseudo-inverse as $\mathrm{pinv}(\mathbf{H})$, and the vector obtained by stacking the columns of the matrix below each other as $\mathrm{vec}(\mathbf{H})$. We denote the logarithm of the determinant of a quadratic matrix $\mathbf{H}$ as $\log|\mathbf{H}|$. We denote sets by calligraphic letters (e.g., $\mathcal{A}$); the size of set $\mathcal{A}$ is $|\mathcal{A}|$. To calculate the expected value of a random variable r, we employ the notation $\mathbb{E}r$. The Dirac delta function is $\delta(t)$; the discrete Kronecker delta is δ_{ij}, i.e., $\delta(t) \to \infty \mathrm{iff} t = 0$ and $\delta_{ij} = 1 \mathrm{iff} i = j$. We denote the real part of a complex number z as $\Re(z)$, the imaginary part as $\Im(z)$, and the complex conjugate as z^*. We denote the uniform distribution on the interval $[a, b]$ as $\mathcal{U}(a, b)$ and the vector-valued complex-Gaussian distribution with mean vector $\boldsymbol{\mu}$ and covariance matrix $\mathbf{C}$ as $\mathcal{CN}(\boldsymbol{\mu}, \mathbf{C})$.

2 System model

2.1 Channel model

In wireless communications, the channel between transmitter and receiver can be characterized by a double-directional time-variant channel impulse response [41]:

$$h(t, \tau, \boldsymbol{\Psi}_t, \boldsymbol{\Psi}_r) = \sum_{\ell=1}^{N_L} \alpha_\ell \delta(\tau - \tau_\ell) \delta(\boldsymbol{\Psi}_t - \boldsymbol{\Psi}_{t,\ell}) \delta(\boldsymbol{\Psi}_r - \boldsymbol{\Psi}_{r,\ell}), \tag{1}$$

with N_L denoting the number of multipath components and α_ℓ, τ_ℓ, respectively, representing the complex-valued amplitude and the propagation delay associated to path ℓ. The vectors $\boldsymbol{\Psi}_{t,\ell} = [\phi_{t,\ell}, \theta_{t,\ell}]^{\mathrm{T}}$ and $\boldsymbol{\Psi}_{r,\ell} = [\phi_{r,\ell}, \theta_{r,\ell}]^{\mathrm{T}}$ denote the azimuth and elevation angles of departure and arrival. Although not explicitly shown in (1), the parameters on the right-hand side may all implicitly depend on the absolute time t due to movement of users and scattering objects. Incorporating the responses of the transmit and receive antenna elements w.r.t. a plane wave from angular direction $\boldsymbol{\Psi}$, i.e., $a_t(\boldsymbol{\Psi})$ and $a_r(\boldsymbol{\Psi})$, we obtain the time-variant channel impulse response as:

$$h(t,\tau) = \int \cdots \int_{-\pi}^{\pi} a_t(\boldsymbol{\Psi}_t) a_r(\boldsymbol{\Psi}_r) h(t,\tau,\boldsymbol{\Psi}_t,\boldsymbol{\Psi}_r) d\boldsymbol{\Psi}_r d\boldsymbol{\Psi}_t = \sum_{\ell=1}^{N_L} \alpha_\ell \delta(\tau - \tau_\ell) a_r(\boldsymbol{\Psi}_{r,\ell}) a_t(\boldsymbol{\Psi}_{t,\ell}). \tag{2}$$

If we furthermore incorporate the overall impulse response of the transmit and receive filters $f(\tau)$, we obtain the transceiver-dependent time-variant channel impulse response as:

$$\tilde{h}(t,\tau) = \int_{\tau'} h(t,\tau') f(\tau - \tau') d\tau' = \sum_{\ell=1}^{N_L} \alpha_\ell f(\tau - \tau_\ell) a_r(\boldsymbol{\Psi}_{r,\ell}) a_t(\boldsymbol{\Psi}_{t,\ell}), \tag{3}$$

where again the time dependence of the parameters on the right-hand side is implicit. Under the well-known planar wave and narrowband array assumptions [42], we can obtain the time-variant impulse response of a MIMO system simply by concatenating the individual impulse responses of antenna element pairs:

$$\tilde{\mathbf{H}}(t,\tau) = \sum_{\ell=1}^{N_L} \alpha_\ell f(\tau - \tau_\ell) \mathbf{a}_r(\boldsymbol{\Psi}_{r,\ell}) \mathbf{a}_t(\boldsymbol{\Psi}_{t,\ell})^{\mathrm{T}} \tag{4}$$

with $\mathbf{a}_r(\boldsymbol{\Psi}) \in \mathbb{C}^{N_r \times 1}, \mathbf{a}_t(\boldsymbol{\Psi}) \in \mathbb{C}^{N_t \times 1}$ denoting the receive and transmit antenna array response vectors. For example, consider a uniform planar array (UPA) with N_v rows and N_h columns at the transmitter, with vertical antenna element spacing d_v and horizontal antenna element spacing d_h in multiples of the wavelength λ. Then, the response of the antenna element in the ℓ-th row and the k-th column with respect to a plane wave from direction $\boldsymbol{\Psi} = [\phi, \theta]^{\mathrm{T}}$ is [43]:

$$\left[\mathbf{a}_t^{\mathrm{UPA}}(\phi,\theta)\right]_{(\ell-1)N_h+k} = \frac{1}{\sqrt{N_t}} \exp\left(j2\pi(d_v(\ell-1)\cos\theta) + d_h(k-1)\sin\phi\sin\theta\right) g_e(\phi,\theta), \tag{5}$$

$$\ell \in \{1,\ldots,N_v\},\ k \in \{1,\ldots,N_h\},\ N_t = N_v N_h, \tag{6}$$

where angles are measured as illustrated in Fig. 1 in relation to the antenna array and the individual antenna element responses are arranged row-by-row in the vector $\mathbf{a}_t^{\mathrm{UPA}}(\phi,\theta) \in \mathbb{C}^{N_t \times 1}$. In (5), the function $g_e(\phi,\theta) \in \mathbb{C}$ denotes the complex-valued angle-dependent antenna element gain pattern, e.g., a Hertzian dipole pattern. We do not focus on a specific antenna array geometry in this work nor do we restrict the antenna element gain pattern $g_e(\phi,\theta)$. Notice, in (4), we do not account for coupling in-between antenna elements, which may be incorporated with additional coupling matrices [43].

2.2 Transmission model

We consider downlink transmission from a base station that is equipped with an antenna array of N_t antenna elements, e.g., a UPA of size $N_t = N_h \cdot N_v$ as illustrated in Fig. 1. The base station serves U users. The users are each equipped with N_r receive antennas; we assume $N_t \geq N_r$. We consider orthogonal frequency division multiplexing (OFDM) transmission; the MIMO-OFDM channel matrix $\mathbf{H}[n,k] \in \mathbb{C}^{N_r \times N_t}$ at OFDM symbol n and subcarrier k is obtained by sampling the Fourier transform of the time-variant impulse response $\tilde{\mathbf{H}}(t,\tau)$ of (4). We assume that time variations within OFDM symbols are small enough such that the inter-carrier interference due to Doppler shifts is negligible. In the following, we focus on a single subcarrier and omit the subcarrier index k for brevity. To

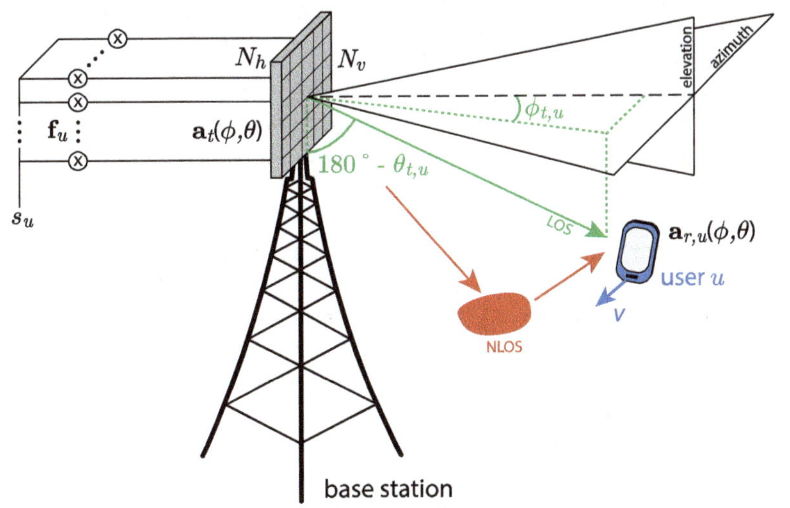

Fig. 1 Illustration of the considered FD-MIMO transmission system employing three-dimensional downlink beamforming

refer to the channel matrix associated to a specific user u, we employ a subscript, such as $\mathbf{H}_u[n]$.

The base station selects at each time instant a set $\mathcal{S}[n] \subseteq \{1,\ldots,U\}$ of users to be served in parallel, employing spatial multiplexing by means of linear beamforming/precoding. Assuming transmission of L_u streams to user $u \in \mathcal{S}[n]$, the base station applies the precoding matrix $\mathbf{F}_u[n] \in \mathbb{C}^{N_t \times L_u}$ to transmit the data symbol vector $\mathbf{s}_u[n] \in \mathbb{C}^{L_u \times 1}$ to user u. We assume $\mathbb{E}(\mathbf{s}_u[n]\,\mathbf{s}_u[n]^{\mathrm{H}}) = \mathbf{I}_{L_u}$ and account for the power allocation in $\mathbf{F}_u[n]$, such that $\operatorname{tr}(\mathbf{F}_u[n]\,\mathbf{F}_u[n]^{\mathrm{H}}) = 1/|\mathcal{S}[n]|$. The input-output relationship of user u hence is

$$\mathbf{y}_u[n] = \mathbf{H}_u[n] \sum_{j \in \mathcal{S}[n]} \mathbf{F}_j[n]\,\mathbf{s}_j[n] + \mathbf{n}_u[n] \in \mathbb{C}^{N_r \times 1}, \quad (7)$$

with $\mathbf{n}_u[n]$ denoting zero-mean complex Gaussian receiver noise of variance $\sigma^2_{n,u}$.

2.3 Targeted deployment scenarios

The beamforming and precoding methods proposed below are basically enhanced interference-aware SDMA techniques. As such, the methods require a certain degree of spatial separability of users in the angular domain. Such separability can mainly be guaranteed in limited scattering environments with a small number of dominant specular components per user and relatively weak diffuse background scattering, as common in LOS conditions. Even though this might sound restrictive for mobile communications, many scenarios envisioned for deployment of FD-MIMO systems are expected to exhibit dominant LOS transmission. Example scenarios where LOS propagation could dominate are the urban micro and macro scenarios illustrated in ([2], Fig. 2) and ([44], Fig. 2), where users on different floors of a building are served in parallel by means of spatial beamforming from base stations mounted on neighboring buildings. Another interesting deployment scenario is multi-user multiplexing in large indoor open spaces, e.g., within stadiums, airport terminals, shopping malls, where high-ceiling-mounted small cells and remote radio heads can provide LOS connectivity to users. Furthermore of interest are vehicular communication scenarios on highways and motorways, where FD-MIMO beamforming can be employed for multi-user transmission between road-side access points and vehicles as described in [45]. Notice also that channel fading models that contain only two dominant specular components in addition to weak diffuse background scattering, such as the two-wave diffuse power (TWDP), generalized two-ray (GRT), and fluctuating two-ray (FRT) fading models [46–50], are well suited to capture the behavior of channel measurements conducted in the millimeter wave band [49, 51], suggesting that millimeter wave transmissions are in many cases well characterized by few specular components. Importantly, in the millimeter wave band, even relatively low mobility of users (pedestrians) causes significant time selectivity of the channel due to the small wavelength; hence, CSIT becomes outdated even faster.

3 Channel state information feedback

We assume that the transmitter inserts pilot symbols in the OFDM time-frequency grid to support channel estimation for CSI feedback calculation at the receivers. Notice, these CSI pilots may be provided sparsely in the time-frequency grid to facilitate coarse channel estimation only for CSI calculation; additional more densely populated precoded pilot symbols can be employed to support accurate channel estimation for data detection, similar to LTE's CSI and UE-specific pilot symbols [52]. We consider channel quantization and feedback in the frequency domain; alternatively, it is also possible to quantize the channel impulse response [53]. We write the MIMO-OFDM channel matrix $\mathbf{H}_u[n]$ as a function of the estimated matrix $\hat{\mathbf{H}}_u[n]$ as:

$$\mathbf{H}_u[n] = \hat{\mathbf{H}}_u[n] + \mathbf{E}_u[n], \quad (8)$$

with $\mathbb{E}\left(\operatorname{vec}(\mathbf{E}_u[n])\operatorname{vec}(\mathbf{E}_u[n])^{\mathrm{H}}\right) = \sigma^2_{e,u}\mathbf{I}_{N_r N_t}$. If we assume that downlink CSI pilot signals are provided relatively sparse in the time-frequency domain, to keep the pilot signal overhead small, the channel estimation error variance $\sigma^2_{e,u}$ will be comparatively large.

In order to attain more robust behavior w.r.t. channel estimation errors and outdated CSIT, we propose to employ an over-complete basis decomposition of $\hat{\mathbf{H}}_u[n]$ to obtain a sparse approximation of $\mathbf{H}_u[n]$.[1] Choosing a proper quantization codebook/dictionary $\mathcal{D}$ for the decomposition, we can achieve a signal denoising gain that provides the wanted robustness. More specifically, we apply the following decomposition:

$$\hat{\mathbf{H}}_u[n] = \sum_{s=1}^{S} \mathbf{c}_u^s[n]\,(\mathbf{d}_u^s[n])^{\mathrm{T}} + \mathbf{R}_u[n], \quad (9)$$

with $\mathbf{d}_u^s[n] \in \mathcal{D}$ representing the s-th selected entry/atom from the dictionary $\mathcal{D}$ and $\mathbf{c}_u^s[n] \in \mathbb{C}^{N_r \times 1}$ being the corresponding expansion coefficient vector. Matrix $\mathbf{R}_u[n] \in \mathbb{C}^{N_r \times N_t}$ is the residual decomposition error. The dictionary $\mathcal{D} = \{\mathbf{d}_1,\ldots,\mathbf{d}_{|\mathcal{D}|}\}$, $\mathbf{d}_i \in \mathbb{C}^{N_t \times 1}$, $|\mathcal{D}| \geq N_t$ is chosen to span the entire N_t-dimensional complex Euclidean space, in order to be able to represent in principle any matrix error-free with sufficiently many expansion coefficients. Our goal is to obtain a good sparse approximation of $\hat{\mathbf{H}}_u[n]$ with a very small number S of expansion terms by means of a dictionary that supports easy parametrization and therefore efficient CSI quantization/feedback. This can be achieved if the channel matrix is compressible in the chosen dictionary [54], which basically means that the norm of the expansion coefficient vectors $\|\mathbf{c}_u^s[n]\|$

decreases exponentially with the index s (assuming sorting in decreasing order).

If we assume transmission under LOS conditions, the channel impulse response $\tilde{\mathbf{H}}(t,\tau)$ is dominated by the LOS specular component. This dominant component is then also present in the Fourier transform $\mathbf{H}_u[n]$. Correspondingly, we propose to utilize a dictionary that is matched to the transmit antenna array response vector $\mathbf{a}_t(\boldsymbol{\Psi})$, since this will result in a compressible decomposition of the channel matrix.[2] We therefore propose to set the dictionary as:

$$\mathcal{D} = \left\{ \mathbf{d}_i = \mathbf{a}_t(\boldsymbol{\Psi}_i) \middle| \boldsymbol{\Psi}_i \in \left\{\phi_1, \ldots, \phi_{N_\phi}\right\} \times \left\{\theta_1, \ldots, \theta_{N_\theta}\right\} \right\}. \tag{10}$$

With $|\mathcal{D}| = N_\phi N_\theta \geq N_t$, this dictionary spans the N_t-dimensional complex Euclidean space and can thus be used to decompose in principle any matrix; however, for a general unstructured matrix, such decomposition will not be compressible. Notice, as the codebook is matched to the transmit antenna array geometry, the users must be aware of this geometry which requires additional downlink signaling. However, since mainly UPAs are employed in practice, we expect this to be a minor issue. The size of the quantization codebook in the azimuth and elevation domains N_ϕ and N_θ should be chosen according to the spatial resolution of the antenna array, i.e., with a larger number of antenna elements in the horizontal/vertical dimension also the resolution of the quantization codebook in the corresponding angular domain should be increased. In order to be able to closely match the angular directions of the dominant specular components, we employ a dictionary with $|\mathcal{D}| \gg N_t$ in our simulations.

To obtain a sparse channel decomposition, we consider the following optimization problem:

$$\min_{\{\mathbf{c}_i \in \mathbb{C}^{N_r \times 1}\}_{i=1}^{S}, S \leq |\mathcal{D}|} \sum_{i=1}^{S} \|\mathbf{c}_i\| \tag{11}$$

$$\text{subject to: } \left\| \hat{\mathbf{H}}_u[n] - \sum_{i=1}^{S} \mathbf{c}_i \mathbf{d}_i^{\mathrm{T}} \right\|^2 \leq \epsilon^2, \; \mathbf{d}_i \in \mathcal{D}$$

We determine an approximate solution of this optimization problem by applying a variation of the orthogonal matching pursuit (OMP) algorithm [55], as detailed in Algorithm 1. Notice, to approximately solve (11), we would actually have to adapt the number S of expansion terms in Algorithm 1 according to the desired approximation accuracy ϵ^2; however, in our simulations, we employ a fixed S. Notice the objective function in (11) can also be written as an $\ell 1$-norm $\|[\|\mathbf{c}_1\|, \ldots, \|\mathbf{c}_S\|]\|_1$ justifying its use for sparse decomposition. Given the sparse approximation, the CSI feedback is set equal to the corresponding codebook indices of the optimal azimuth and elevation angles $\left(\phi_u^s[n], \theta_u^s[n]\right)$ of the expansion vectors $\mathbf{d}_u^s[n]$. Additionally, we provide the norms of the expansion coefficient vectors $\left\|\mathbf{c}_u^s[n]\right\|$ as feedback information; the reason for this will become clear in Section 4. In our simulations in Section 6, we consider unquantized feedback of the expansion coefficient vector norms; we have investigated the impact of quantization of these norms for $N_r = 1$ in [39], showing that coarse quantization is possible without significant performance degradation.

Algorithm 1 Channel decomposition for CSI feedback calculation utilizing a variation of OMP.

Initialize the decomposed channel matrix $\mathbf{H}_d = \mathbf{0}$
Initialize the expansion matrix $\mathbf{D} = [\,]$
for $s = 1$ to S **do**
  Calculate the intermediate decomposition residual

$$\mathbf{R} = \hat{\mathbf{H}}_u[n] - \mathbf{H}_d \tag{12}$$

  Calculate the decomposition metric

$$m_i = \left\|\mathbf{R}\mathbf{d}_i^*\right\|^2, \; \forall \mathbf{d}_i \in \mathcal{D} \tag{13}$$

  Determine the s-th expansion vector

$$\mathbf{d}_u^s[n] = \arg\max_{\mathbf{d}_i \in \mathcal{D}} m_i \tag{14}$$

  Update the expansion matrix

$$\mathbf{D} = \left[\mathbf{D}, \mathbf{d}_u^s[n]\right] \in \mathbb{C}^{N_t \times s} \tag{15}$$

  Calculate the intermediate expansion coefficient matrix

$$\mathbf{C} = \hat{\mathbf{H}}_u[n] \; \mathrm{pinv}(\mathbf{D})^{\mathrm{T}} \in \mathbb{C}^{N_r \times s} \tag{16}$$

  Update the decomposed channel matrix

$$\mathbf{H}_d = \mathbf{C}\mathbf{D}^{\mathrm{T}} \in \mathbb{C}^{N_r \times N_t} \tag{17}$$

end for
Set the expansion coefficient vectors

$$\left[\mathbf{c}_u^1[n], \ldots, \mathbf{c}_u^S[n]\right] = \mathbf{C} \tag{18}$$

4 Leakage-bounded angular beamforming and precoding

In this section, we propose multi-user beamforming and precoding methods that utilize the angular-domain CSI feedback described in the previous section. We consider first multi-user beamforming with a single stream per user in Section 4.1 and extend to multiple streams per user in Section 4.2.

4.1 Leakage-bounded angular beamforming

In this section, we assume that the base station transmits only a single stream $L_u = 1$ to each served user $u \in \mathcal{S}[n]$, i.e., the precoding matrices $\mathbf{F}_u[n]$ reduce to beamforming vectors $\mathbf{f}_u[n]$. From the angular CSI feedback provided by the users, the base station has at time instant n delayed knowledge of the channel expansion vectors $\mathbf{d}_u^s[n-m]$ and the norm of the corresponding expansion coefficient vectors $\|\mathbf{c}_u^s[n-m]\|$, where m represent the delay of the CSI feedback path. To calculate the transmit beamformers, we interpret the expansion vectors $\mathbf{d}_u^s[n-m]$ as specular channel contributions and propose a beamformer optimization that maximizes the expected received signal power of user u over its specular components, while restricting the interference leakage caused to the other users over their respective specular components. To shorten notation, we omit the time indices $[n]$ and $[n-m]$ in the following derivations.

Consider the signal power received over the S specular components corresponding to the available CSIT:

$$\begin{aligned} P_u &= \mathbb{E}\left(\left\|\left(\sum_{s=1}^{S}\mathbf{c}_u^s\,(\mathbf{d}_u^s)^{\mathrm{T}}\right)\mathbf{f}_u s_u\right\|^2\right) \\ &= \mathrm{tr}\left(\sum_{s=1}^{S}\sum_{k=1}^{S}\mathbb{E}((\mathbf{c}_u^k)^{\mathrm{H}}\mathbf{c}_u^s)(\mathbf{d}_u^s)^{\mathrm{T}}\mathbf{f}_u\mathbf{f}_u^{\mathrm{H}}(\mathbf{d}_u^k)^*\right). \end{aligned} \tag{19}$$

The value of the inner product $(\mathbf{c}_u^k)^{\mathrm{H}}\mathbf{c}_u^s = r_u^{k,s}e^{j\varphi_u^{k,s}}$ depends on the relative phase-shift $\varphi_u^{k,s}$ between these expansion coefficient vectors. Since we do not provide feedback information about these relative phase-shifts, we assume $\varphi_u^{k,s} \sim \mathcal{U}(0, 2\pi)$ such that $\mathbb{E}(r_u^{k,s}e^{j\varphi_u^{k,s}}) = 0$. Notice, this assumption also makes sense for our targeted application scenario of high user mobility. In such scenarios, the angular CSI feedback corresponding to the expansion vectors $\mathbf{d}_u^s$ is relatively stable, whereas the phase of the expansion coefficient vectors $\mathbf{c}_u^s$ varies significantly over time. Hence, relying on such phase information at high mobility is not recommended. With this assumption of uncorrelated phases, the signal power is:

$$P_u = \sum_{s=1}^{S}\|\mathbf{c}_u^s\|^2\,(\mathbf{d}_u^s)^{\mathrm{T}}\mathbf{f}_u\mathbf{f}_u^{\mathrm{H}}(\mathbf{d}_u^s)^* \tag{20}$$

and the transmitter has sufficient CSIT to calculate this value. Similarly, we can determine the interference leakage power caused by the transmission to user u and received over specular component k of user j as:

$$L_{j,u}^k = \left\|\mathbf{c}_j^k\right\|^2(\mathbf{d}_j^k)^{\mathrm{T}}\mathbf{f}_u\mathbf{f}_u^{\mathrm{H}}(\mathbf{d}_j^k)^*. \tag{21}$$

With these definitions, we now proceed with the formulation of our optimization problem. We consider independent optimization of the transmit beamformers of the individual users according to the following optimization problem:

$$\begin{aligned} &\max_{\mathbf{f}_u\in\mathbb{C}^{N_t\times 1}} P_u \\ &\text{s. t.: } \|\mathbf{f}_u\|^2 \le 1/|\mathcal{S}|, \\ &\qquad L_{j,u}^k \le L_{\max}, \quad \forall j\in\mathcal{S}, j\neq u, \forall k\in\{1,\ldots,S\}. \end{aligned} \tag{P1}$$

where the parameter $L_{\max}$ denotes the maximum tolerable interference level. Replacing $\max_{\mathbf{f}_u\in\mathbb{C}^{N_t\times 1}} P_u$ equivalently by $\min_{\mathbf{f}_u\in\mathbb{C}^{N_t\times 1}}(-P_u)$, this optimization problem can be formulated as a quadratically constrained quadratic program; however, since the matrix $-\sum_{s=1}^{S}\|\mathbf{c}_u^s\|^2(\mathbf{d}_u^s)^*(\mathbf{d}_u^s)^{\mathrm{T}}$, which determines the quadratic objective function, is negative semidefinite, the problem is in general non-convex and non-deterministic polynomial-time (NP)-hard. An approximate solution is possible by applying a semidefinite programming relaxation (SDR) to the optimization variable $\mathbf{f}_u$ and by recovering a feasible suboptimal rank-one beamforming solution through randomization [56]. The approximation performance of this approach in general deteriorates with increasing dimension N_t of the optimization variable as well as with growing number of constraints in (P1) [57, 58]. Hence, for our envisioned use case of FD-MIMO systems with large N_t and $|\mathcal{S}|$, the achieved approximation performance might not be satisfactory.

We therefore consider a modification of (P1) to avoid the SDR: Instead of maximizing the sum signal power over all S specular components of user u, we rather maximize the power received only over the strongest specular component $s = 1$, while still considering the interference leakage caused over all specular components:

$$\begin{aligned} &\max_{\mathbf{f}_u\in\mathbb{C}^{N_t\times 1}} \|\mathbf{c}_u^1\|^2(\mathbf{d}_u^1)^{\mathrm{T}}\mathbf{f}_u\mathbf{f}_u^{\mathrm{H}}(\mathbf{d}_u^1)^* \\ &\text{s. t.: } \|\mathbf{f}_u\|^2 \le 1/|\mathcal{S}|, \\ &\qquad L_{j,u}^k \le L_{\max}, \quad \forall j\in\mathcal{S}, j\neq u, \forall k\in\{1,\ldots,S\}. \end{aligned} \tag{P2}$$

This problem can be transformed to a convex problem by recognizing that (P2) is invariant w.r.t. the absolute phase of $\mathbf{f}_u$. Hence, we can apply the approach considered, e.g., in [59], and restrict $(\mathbf{d}_u^1)^{\mathrm{T}}\mathbf{f}_u$ to be real valued to obtain the following equivalent convex second-order cone program:

$$\begin{aligned} &\max_{\mathbf{f}_u\in\mathbb{C}^{N_t\times 1}, p_u\in\mathbb{R}} p_u \\ &\text{s. t.: } \|\mathbf{f}_u\|^2 \le 1/|\mathcal{S}|, \\ &\qquad \Re\left((\mathbf{d}_u^1)^{\mathrm{T}}\mathbf{f}_u\right) \ge p_u, \\ &\qquad \Im\left((\mathbf{d}_u^1)^{\mathrm{T}}\mathbf{f}_u\right) = 0, \\ &\qquad L_{j,u}^k \le L_{\max}, \quad \forall j\in\mathcal{S}, j\neq u, \forall k\in\{1,\ldots,S\}. \end{aligned} \tag{P3}$$

We denote the optimal solution of this optimization problem as the *non-robust leakage-bounded angular beamformer*. As this is a second-order cone program, its complexity scales with the square root of the number of cone constraints [60, 61].

The solution to problem (P3) allows to efficiently control the interference leakage caused to other users, provided the angular CSIT is accurate. However, since this CSIT is obtained by limited feedback, angular quantization is unavoidable which impairs the quality of the CSIT. Additionally, in mobile situations, the azimuth and elevation angles representing the specular components change over time, causing a mismatch between the angular CSI feedback and the actual angles, due to the processing delay of the feedback link. These effects increase the inter-user interference and thus deteriorate the performance of the system. To improve the robustness of our beamformer solution w.r.t. angular uncertainty, we consider a robust problem formulation in the following.

Given the angular feedback (ϕ_u^s, θ_u^s), we propose to incorporate angular uncertainty regions into the beamformer optimization problem. We denote the discrete angular uncertainty region corresponding to specular component s of user u as $T_u^s \subseteq \left[-\frac{\pi}{2}, \frac{\pi}{2}\right] \times [0, \pi]$, such that $(\phi_u^s, \theta_u^s) \in T_u^s$. In our investigations, we utilize symmetric rectangular regions around (ϕ_u^s, θ_u^s). The sizes of these regions can, e.g., be set according to the angular variation observed over the previous CSI feedback interval, i.e., $2\left|\phi_u^s[n-m] - \phi_u^s[n-2m]\right|$ and $2\left|\theta_u^s[n-m] - \theta_u^s[n-2m]\right|$, where the factor 2 accounts for the unknown direction of movement. The density of the lattice points considered within these regions, i.e., the angular resolution of T_u^s, must be chosen sufficiently large to ensure that no side-lobes of the radiation pattern are missed. As a guideline, consider a UPA with equal gain beamforming: here, zeros in the radiation pattern in azimuth occur at angles $\sin\phi = \pm\frac{n\lambda}{d_h N_h}$; hence, the angular resolution in azimuth must be chosen to assure that the peaks in-between such zeros are resolved with sufficient accuracy. Notice, though, that the complexity of the proposed optimization problems, more specifically the number of constraints, scales linearly with the size of each set T_u^s; hence, one should avoid unnecessary oversampling of T_u^s to reduce computational complexity.

Utilizing these angular uncertainty regions, we now formulate the optimization problem for the *robust leakage-bounded angular beamformer*:

$$\max_{\mathbf{f}_u \in \mathbb{C}^{N_t \times 1}} \left\|\mathbf{c}_u^1\right\|^2 (\mathbf{d}_u^1)^{\mathrm{T}} \mathbf{f}_u \mathbf{f}_u^{\mathrm{H}} (\mathbf{d}_u^1)^* \tag{P4}$$

$$\text{s. t.: } \|\mathbf{f}_u\|^2 \leq 1/|\mathcal{S}|,$$

$$\left\|\mathbf{c}_j^k\right\|^2 (\mathbf{a}_t(\boldsymbol{\Psi}_{i,j}^k))^{\mathrm{T}} \mathbf{f}_u \mathbf{f}_u^{\mathrm{H}} (\mathbf{a}_t(\boldsymbol{\Psi}_{i,j}^k))^* \leq L_{\max},$$

$$\forall j \in \mathcal{S}, j \neq u, \ \forall k \in \{1, \ldots, S\}, \ \forall \boldsymbol{\Psi}_{i,j}^k \in T_j^k.$$

In this problem, the vector $\mathbf{a}_t(\boldsymbol{\Psi}_{i,j}^k)$ denotes the transmit antenna array response evaluated at angle $\boldsymbol{\Psi}_{i,j}^k \in T_j^k$; here, the same antenna array response as in the CSI feedback dictionary (10) is used. Notice, we account for the angular uncertainty only in the interference terms, but not in the signal term. This is because the system reacts much more sensitive w.r.t uncertainty in the interference directions as compared to the signal direction, since nulls in the beamforming pattern are commonly very narrow whereas peaks are comparatively broad in the angular domain (see also the example in Fig. 2). Another reason for neglecting uncertainty in the angular direction of the intended signal is to keep the problem convex.[3] Problem (P4) can be brought into the form of a second-order cone program, similar to Problem (P3).

4.2 Leakage-bounded angular precoding

In this section, we extend the robust beamformer design to multi-user precoding with multiple streams per user.

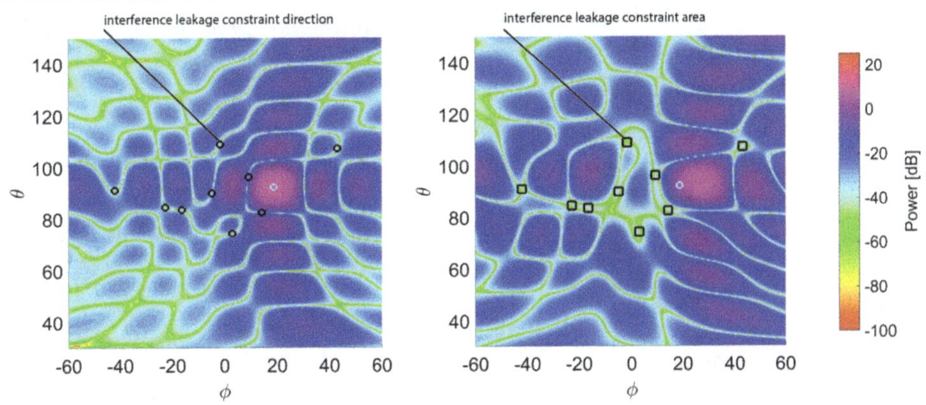

Fig. 2 Radiation beam pattern obtained with non-robust leakage-bounded angular beamforming (left) and with robust leakage-bounded angular beamforming (right). The nine angular uncertainty regions of robust leakage-bounded angular beamforming have a width of $\psi_{\max} = \pm 1.6°$ in this example

When considering multi-stream precoding, it is not sufficient to maximize the sum received power over all L_u streams of user u because this will generally lead to a rank one beamforming solution that steers all signal energy over the largest singular value of the channel. Hence, to obtain a multi-stream precoding solution of rank L_u, we have to consider the achievable rate of a user in terms of the mutual information. With the available CSIT, however, we can only obtain a coarse estimate of the achievable rate as detailed below.

Let us start by considering the achievable rate of user u in terms of the mutual information:

$$R_u = \mathbb{E}\log\left|\mathbf{I}_{N_r} + \left(\sigma_{n,u}^2\mathbf{I}_{N_r} + \sum_{\substack{j\in\mathcal{S}\\ j\neq u}}\mathbf{R}_{u,j}\right)^{-1}\mathbf{R}_u\right| \tag{22}$$

$$\mathbf{R}_u = \mathbf{H}_u\mathbf{F}_u(\mathbf{H}_u\mathbf{F}_u)^{\mathrm{H}},\quad \mathbf{R}_{u,j} = \mathbf{H}_u\mathbf{F}_j(\mathbf{H}_u\mathbf{F}_j)^{\mathrm{H}}, \tag{23}$$

where $\mathbf{R}_u$ and $\mathbf{R}_{u,j}$ denote the signal and interference covariance matrices, respectively. We consider individual optimization of the achievable rate of each user w.r.t. its precoder, putting additional interference constraints on the signal leakage caused to the other users. This means that we do not attempt to jointly optimize the sum rate of all users; notice, though, that for single-stream beamforming, it has been shown in [62, 63] that such an approach can still achieve the maximal sum-rate, provided the leakage constraints are appropriately selected. When optimizing R_u w.r.t. $\mathbf{F}_u$, we can then focus on

$$\tilde{R}_u = \mathbb{E}\log\left|\sigma_{n,u}^2\mathbf{I}_{N_r} + \sum_{\substack{j\in\mathcal{S}\\ j\neq u}}\mathbf{R}_{u,j} + \mathbf{R}_u\right|, \tag{24}$$

since the remaining terms are independent of $\mathbf{F}_u$. Because the CSIT contains no information about the channel estimation error in (8) and the channel decomposition error in (9), the transmitter can only calculate approximations $\hat{\mathbf{R}}_u$ and $\hat{\mathbf{R}}_{u,j}$ of the covariance matrices utilizing the angular-domain CSI feedback; this gives a corresponding approximation $\hat{\tilde{R}}_u$ of $\tilde{R}_u$. In the Appendix, we derive the following upper bound on $\hat{\tilde{R}}_u$:

$$\hat{\tilde{R}}_u \le \log\left|\sigma_{n,u}^2\mathbf{I}_S + (|\mathcal{S}|-1)\frac{SL_{\max}}{\mathrm{tr}\,(\boldsymbol{\Sigma}_u^2)}\boldsymbol{\Sigma}_u^2 + \boldsymbol{\Sigma}_u^2\mathbf{U}_u^{\mathrm{H}}\mathbf{C}_u\mathbf{U}_u\right|, \tag{25}$$

where the matrices $\boldsymbol{\Sigma}_u^2$ and $\mathbf{U}_u$, as defined in (36), can be calculated from the available CSIT and $\mathbf{C}_u = \mathbf{F}_u(\mathbf{F}_u)^{\mathrm{H}}$ denotes the transmit covariance matrix associated to user u.

We now utilize this upper bound to formulate the optimization problem for the *non-robust leakage-bounded angular precoder*:

$$\max_{\mathbf{C}_u\in\mathbb{C}^{N_t\times N_t}}\left|\sigma_{n,u}^2\mathbf{I}_S + (|\mathcal{S}|-1)\frac{SL_{\max}}{\mathrm{tr}\,(\boldsymbol{\Sigma}_u^2)}\boldsymbol{\Sigma}_u^2 + \boldsymbol{\Sigma}_u^2\mathbf{U}_u^{\mathrm{H}}\mathbf{C}_u\mathbf{U}_u\right| \tag{P5}$$

$$\text{s. t.:}\quad \mathrm{tr}\,(\mathbf{C}_u) \le 1/|\mathcal{S}|,\ \mathbf{C}_u \succeq \mathbf{0},$$
$$L_{j,u}^k \le L_{\max},\ \forall j\in\mathcal{S}, j\neq u,\ \forall k\in\{1,\ldots,S\},$$

with $L_{j,u}^k$ as defined in (38). Notice, this actually represents an SDR of the optimization w.r.t. the precoder $\mathbf{F}_u \in \mathbb{C}^{N_t\times L_u}$. In general, the optimal solution $\mathbf{C}_u^{(\mathrm{opt})}$ of (P5) is not guaranteed to be of rank L_u. Due to the definition of $\boldsymbol{\Sigma}_u^2$ and $\mathbf{U}_u$ in (36), rank $\mathbf{C}_u^{(\mathrm{opt})} \le S$. This is because a solution of rank greater than S would steer part of the transmit energy into the null space of $\mathbf{U}_u$, which cannot maximize our objective function. The restriction rank $\mathbf{C}_u^{(\mathrm{opt})} \le S$ also implies that we have to feed back at least as many specular components S as the number of data streams L_u to make sure that the solution $\mathbf{C}_u^{(\mathrm{opt})}$ can potentially support the transmission of L_u streams. Still, even with $S \ge L_u$, it can happen that the optimal solution $\mathbf{C}_u^{(\mathrm{opt})}$ has rank $\mathbf{C}_u^{(\mathrm{opt})} < L_u$; this is likely to be the case at low signal-to-noise ratio SNR where the transmission of less than L_u streams can provide an advantageous beamforming gain. If rank $\mathbf{C}_u^{(\mathrm{opt})} < L_u$, we simply transmit less than L_u streams over the eigenvectors corresponding to the non-zero eigenvalues of $\mathbf{C}_u^{(\mathrm{opt})}$, since this maximizes our objective function. If rank $\mathbf{C}_u^{(\mathrm{opt})} > L_u$, we apply Gaussian randomization to derive a feasible precoder of rank L_u, by appropriately modifying the randomization approaches described in more detail in [56, 64]. Notice, though, if we set $S = L_u$, i.e., we feed back as many specular components as the number of data streams, then the rank of $\mathbf{C}_u^{(\mathrm{opt})}$ is upper bounded by L_u and we therefore do not have to apply randomization at all. Thus, by setting $S = L_u$, we can guarantee that the SDR provides a globally optimal solution for the precoder $\mathbf{F}_u$.

Similar to (P4), we can also implement a *robust leakage-bounded angular precoder* by accounting for the angular uncertainty regions $\mathcal{T}_j^k$:

$$\max_{\mathbf{C}_u\in\mathbb{C}^{N_t\times N_t}}\left|\sigma_{n,u}^2\mathbf{I}_S + (|\mathcal{S}|-1)\frac{SL_{\max}}{\mathrm{tr}\,(\boldsymbol{\Sigma}_u^2)}\boldsymbol{\Sigma}_u^2 + \boldsymbol{\Sigma}_u^2\mathbf{U}_u^{\mathrm{H}}\mathbf{C}_u\mathbf{U}_u\right| \tag{P6}$$

$$\text{s. t.:}\quad \mathrm{tr}\,(\mathbf{C}_u) \le 1/|\mathcal{S}|,\ \mathbf{C}_u \succeq \mathbf{0},$$
$$\left\|\mathbf{c}_j^k\right\|^2 \mathrm{tr}\,(\mathbf{a}_t(\boldsymbol{\Psi}_{i,j}^k))^*(\mathbf{a}_t(\boldsymbol{\Psi}_{i,j}^k))^{\mathrm{T}}\mathbf{C}_u \le L_{\max},$$
$$\forall j\in\mathcal{S}, j\neq u,\ \forall k\in\{1,\ldots,S\},\ \forall\boldsymbol{\Psi}_{i,j}^k\in\mathcal{T}_j^k.$$

Applying a similar step as from (P2) to (P3), both problems (P5) and (P6) can be brought into convex form by adding an extra optimization variable $p_u \in \mathbb{R}$ and maximizing p_u subject to the logarithmic term in (P5), (P6) being greater than or equal to p_u. However, such a determinant constraint is substantially more complex than the corresponding linear constraint in (P3). Hence, we consider in our simulations in Section 6.3 also as an alternative to determinant optimization a simple *per stream optimization*: we apply (P3) or (P4) L_u times to obtain beamformers that maximize the received power over the L_u strongest specular components and we concatenate these beamformers to obtain the L_u dimensional precoding matrix.

In our derivation, we assumed that the precoders of the other $|\mathcal{S}| - 1$ users are unknown when calculating the precoder of user u. Yet, in principle when optimizing the u-th user, the base station could already utilize the knowledge of the precoders calculated for those $u - 1$ users that have been optimized before. Similarly, an alternating optimization of precoders with a proper termination criterion could be applied to obtain an iterative approach that might provide better performance. For complexity reasons, however, we have not implemented such an approach in our simulations.

4.3 Implementation issues

4.3.1 *Computational complexity*

Our derivation in Section 4.1 shows that leakage-bounded transmit optimization as proposed in (P1) is in general an NP-hard quadratically constrained quadratic optimization problem, which cannot be solved efficiently. Approximate solution by means of an SDR is a convex optimization problem, which can be solved efficiently; yet, it is still computationally demanding, since the optimization variable of the SDR is of dimension $N_t \times N_t$ and hence grows quadratically with the size of the FD-MIMO antenna array. This complexity issue is relaxed by problem (P2) and its convex second-order cone programming reformulation (P3), in which the dimension of the optimization variable $\mathbf{f}_u$ grows only linearly with the size of the FD-MIMO antenna array. Since (P3) is a convex problem, it can be solved in polynomial time, e.g., by means of an interior point method; more details on solving second-order cone programs can for example be found in [65]. In problem (P4), we extend the number of constraints of our optimization problem by accounting for the angular uncertainty regions; this directly impacts the computational complexity of the problem. However, since the worst-case complexity of a second-order cone program scales with the square root of the number of constraints [60, 61], the increase in complexity is moderate. Considering the two multi-stream optimization problems (P5) and (P6), they both admit solution by means of an interior point method or a Newton conjugate-gradient approach as proposed in [66]. However, as our computer experiments have shown, this approach can be computationally demanding and slow. Therefore, the per-stream optimization approach proposed in Section 4.2 appears practically more relevant, since its complexity only grows linearly with the number of data streams per user.

4.3.2 *Combination with hybrid architectures*

As mentioned in the introduction of this paper, much research work on FD-MIMO systems is currently devoted to reducing system complexity through so-called hybrid architectures, which divide the precoding operation into an analog RF domain part and a digital base band processing part. The RF domain precoder thereby reduces the dimension of the effective base band channel, i.e., the product of channel matrix and RF domain precoder, which simplifies channel estimation, alleviates the CSI feedback overhead burden, and reduces the number of RF chains required. The RF domain precoder optimization is commonly required to provide a precoder solution with unit modulus matrix entries, to enable implementation with simple phase-shifter elements, and to be constant over the entire system bandwidth, to facilitate application in the analog domain. Both these constraints are not fulfilled by the leakage-bounded precoding optimization problem proposed in this paper. Nevertheless, we consider an extension of our leakage-bounded precoding optimization to hybrid architectures as promising future work.

Specifically, our leakage-bounded angular beamforming approach might be well suited for the design of the RF domain precoder, by applying the per stream optimization mentioned in Section 4.2 for each user more than L_u times in order to obtain a precoder of dimension equal to the number of available RF chains. However, the corresponding optimization problem (P3) or (P4) will in general not output a unit-modulus solution. It is well-known that the unit modulus precoder constraint of RF precoding is a non-convex constraint. A common solution to deal with this issue is to relax the constraint in the precoder optimization and to orthogonally project the resulting solution onto the set of unit modulus matrices [14]; this approach can also be applied to our problem at hand. Notice also, by spending twice the amount of phase-shifters it is actually possible to avoid the unit-modulus constraint all together [67].

A second issue that needs to be addressed is that the channel decomposition (11) is frequency selective due to the multipath delay spread introduced by the channel, which implies that the precoder also needs to be frequency selective and can thus not be implemented in the analog domain. However, in LOS situations, the frequency selectivity is not very distinct and can potentially be neglected for the cost of a slight frequency-dependent mismatch of the decomposition.

In principle, the so obtained analog RF domain precoder can then be combined with any base band precoding approach that is designed for the effective base band channel, similar to the hybrid BD precoding scheme described in Section 5.2 below.

5 Benchmark methods

In this section, we summarize several well-known beamforming and precoding techniques that we employ as benchmarks for our proposed methods. We provide this section as a reference for the interested reader as well as to clarify how these benchmark methods are applied in our considered setup. Additionally, we propose in Section 5.2.4 a modification of the popular signal to leakage and noise ratio SLNR metric to enable its application with the angular-domain CSI feedback proposed in this paper.

5.1 Interference-unaware beamsteering

The following methods do not explicitly account for multi-user interference in the beamformer design. This methods are pure angle-of-departure-based beamsteering techniques, which can be viewed as typical SDMA implementations. As is well known, the performance of SDMA strongly depends on the selection of co-scheduled users [68], i.e., the transmitter should co-schedule users that experience close to orthogonal channels to avoid extensive multi-user interference. In our simulations in Section 6.2, we avoid the impact of imperfect user scheduling by employing an exhaustive search over all possibilities of co-scheduled users.

5.1.1 Max-gain beamsteering

The simplest beamforming/precoding technique is to steer the transmit signal along the L_u strongest angular directions provided by the CSI feedback:

$$\mathbf{F}_u = \frac{1}{\sqrt{|\mathcal{S}|}} \frac{\tilde{\mathbf{F}}_u}{\left\| \tilde{\mathbf{F}}_u \right\|}, \tag{26}$$

$$\tilde{\mathbf{F}}_u = \left[(\mathbf{d}_u^1)^*, \ldots, (\mathbf{d}_u^{L_u})^* \right]. \tag{27}$$

This technique does provide large signal gain; yet, its major weakness is that it does not account for interference among users. Hence, it only works properly if the base station serves users that are a priori well separated in the angular domain.

5.1.2 Chebyshev-tapered beamsteering

Max-gain beamsteering, as considered above, in general, produces large side lobe levels and thus potentially strong inter-user interference. Such large side lobe levels can be reduced by array weight tapering methods [69]. A popular weighting technique is Chebyshev weighting, which achieves the smallest main lobe width for a given side lobe level [70].

In order to steer the main lobe in an intended direction, array tapering is combined with the beamsteering approach of the previous section; i.e., the actual beamformer is obtained as the product of the array taper and the max-gain beamsteering vectors.

5.2 Interference-aware precoding

The following methods take multi-user interference into account and attempt to mitigate it.

5.2.1 Block-diagonalization precoding

BD precoding enables inter-user interference-free transmission of a total number of at most N_t data streams [29]. This is achieved by steering the transmit signal of a user into the null-space of the channel matrices of all other users. In our simulations in Section 6, we do not employ BD precoding in combination with our CSI feedback methods, but we rather assume perfect knowledge of $\mathbf{H}_u$ or its estimate $\hat{\mathbf{H}}_u$. We do this since BD precoding requires phase information which is not provided by our CSI feedback; see, e.g., [33], for an efficient CSI feedback method that can be applied with BD precoding.

5.2.2 Angle of departure block-diagonalization precoding

BD precoding is very sensitive to CSI imperfections since it relies on destructive multipath interference to eliminate multi-user interference [71–73]. It also has heavy CSI requirements as it requires knowledge of the entire channel matrix $\mathbf{H}_u$ of each user. To reduce the CSI requirements, the authors of [74] propose to calculate the precoder not for the entire channel matrix $\mathbf{H}_u$ but rather only for the dominant propagation paths between transmitter and receiver. Specifically, the authors of [74] consider single stream transmission per user and propose to calculate the corresponding ZF beamformer for the single strongest multipath component of each co-scheduled user. Rather than decomposing the frequency response channel matrix as we do in the present paper, the authors of [74] directly consider the time domain channel impulse response (4) and select the strongest multipath component according to the magnitude of α_ℓ; the multi-user ZF beamformer is then calculated from the corresponding transmit antenna array response vectors $\mathbf{a}_t(\boldsymbol{\Psi}_{t,\ell,u})$. This can be extended to multi-stream transmission per user by means of BD precoding in a straightforward way, by calculating the BD precoder from the L_u transmit antenna array response vectors corresponding to the angles of departure of the L_u strongest multipath components.

5.2.3 Hybrid block-diagonalization precoding

The angle of departure BD precoding approach described above does not achieve full multi-user interference cancelation, since it mitigates interference only over the strongest multipath component of each user. The residual multi-user interference leads to a significant performance loss compared to regular BD precoding in case of perfect CSI. To reduce this performance gap, while still keeping CSI requirements at a reasonable level, the authors of [74] propose hybrid BD precoding. Here, the hybrid precoding principle is applied to first reduce the dimension of the effective base band channel matrix of each user, by employing a RF domain precoder. This RF domain precoder $\mathbf{F}_{\mathrm{RF}} \in \mathbb{C}^{N_t \times \sum_{u \in \mathcal{S}[n]} L_u}$ is obtained as the matched filter to the L_u strongest multipath components of each scheduled user. The base band BD precoder is then calculated for the product of channel matrix and RF domain precoder, which is called the effective base band channel. CSI requirements for estimating this effective base band channel are reduced compared to the full channel matrix, since the dimension of the effective channel matrix is smaller.

5.2.4 Incoherent SLNR precoding

SLNR beamforming is a popular coherent beamforming technique for multi-user broadcast and interference channels [75]. If full CSI is available at the base station, that is the full channel matrix $\mathbf{H}_u$ is known for each user u, the SLNR beamformer is obtained by maximizing a corresponding Rayleigh quotient. If we apply the CSI feedback of Section 3, however, the transmitter only has knowledge of the channel decomposition expansion vectors $\mathbf{d}_u^s$ and the norm of their corresponding expansion coefficients $\|\mathbf{c}_u^s\|$. Hence, the transmitter does not know whether the multipath transmissions over the different specular components $\mathbf{d}_u^s$ add up constructively or destructively, since it has no phase information available. We therefore modify the SLNR metric to consider the incoherent ratio of signal and leakage powers, utilizing the delayed angular CSI available at the transmitter:

$$
\begin{aligned}
\mathrm{SL\hat{N}R}_u &= \frac{\mathrm{tr}\,(\mathbf{F}_u^{\mathrm{H}} \hat{\mathbf{P}}_u \mathbf{F}_u)}{\mathrm{tr}\,(\sigma_{n,u}^2 \frac{N_r}{L_u} \mathbf{I}_{L_u} + \mathbf{F}_u^{\mathrm{H}} \hat{\mathbf{L}}_u \mathbf{F}_u)}, \\
\hat{\mathbf{P}}_u &= \sum_{s=1}^{S} \|\mathbf{c}_u^s\|^2 (\mathbf{d}_u^s)^* (\mathbf{d}_u^s)^{\mathrm{T}}, \\
\hat{\mathbf{L}}_u &= \sum_{\substack{j \in \mathcal{S} \\ j \neq u}} \sum_{s=1}^{S} \left\|\mathbf{c}_j^s\right\|^2 (\mathbf{d}_j^s)^* (\mathbf{d}_j^s)^{\mathrm{T}}.
\end{aligned}
\tag{28}
$$

If we restrict to semi-unitary precoding with equal power allocation, such that $\mathbf{F}_u^{\mathrm{H}} \mathbf{F}_u = \frac{1}{L_u |\mathcal{S}|} \mathbf{I}_{L_u}$, we can pull the precoder in the denominator of (28) left and right out of the sum, taking into account that $\mathrm{tr}\,(\mathbf{F}_u^{\mathrm{H}} \mathbf{F}_u) = 1/|\mathcal{S}|$. This leads to a generalized Rayleigh quotient that is maximized by the following semi-unitary precoder:

$$
\mathbf{F}_u = \frac{1}{\sqrt{L_u |\mathcal{S}|}} \tilde{\mathbf{F}}_u, \tag{29}
$$

$$
\tilde{\mathbf{F}}_u = \max_{L_u} \mathrm{eig\,vec} \left(\sigma_{n,u}^2 \frac{N_r}{L_u} |\mathcal{S}| \, \mathbf{I}_{N_t} + \hat{\mathbf{L}}_u \right)^{-1} \hat{\mathbf{P}}_u, \tag{30}
$$

where the operator $\max \mathrm{eig\,vec}_{L_u}$ calculates the L_u orthonormal eigenvectors corresponding to the largest eigenvalues.

6 Simulations

In this section, we evaluate the performance of the proposed beamforming and precoding methods by means of numerical simulations. We first provide in Section 6.1 an example of the radiation beam pattern obtained by leakage-bounded angular beamforming with and without robustness constraints. This facilitates the development of a basic understanding of the behavior of the beamforming solutions produced by the proposed optimization problem, specifically with respect to the influence of the robustness constraints. In Section 6.2, we compare the achievable transmission rate of the proposed leakage-bounded angular beamforming methods to the classical interference-unaware beamsteering methods summarized in Section 5.1. In these simulations we consider a simple Rician-fading channel model, which allows to easily control the compressibility of the CSI in the angular domain via the Rician K-factor. In Section 6.3, we then utilize the more realistic QuaDRiGa channel model [76] to compare the proposed leakage-bounded angular precoding methods to the interference-aware precoding methods summarized in Section 5.2[4]. We first of all evaluate the compressibility of the channel realizations obtained from QuaDRiGa in the angular domain. Based on these results, we then investigate the performance of the different precoding methods in terms of the achievable transmission rate. We furthermore simulate the SER of uncoded transmission as a function of the variance of the channel estimation error σ_e^2 in (8), demonstrating the robustness of the proposed leakage-bounded beamformer. To find solutions to the proposed optimization problems, we apply CVX [77].

6.1 Radiation beam pattern

In our first example, we illustrate the radiation beam pattern obtained by leakage-bounded angular beamforming with and without robustness constraints. We consider single-stream beamforming and assume a UPA of isotropic antenna elements at the transmitter of size $N_t = N_h \cdot N_v = 11 \cdot 11 = 121$. We assume that $U = 10$ users are randomly placed within an angular area stretching in azimuth ϕ from $-60°$ to $60°$ and in elevation θ from $70°$ to $110°$[5]. In this example, we consider only a single specular component $S = 1$ per user; in the robust leakage-bounded optimization problem, we additionally assume an angular uncertainty region of width $\psi_{\max} = \pm 1.6°$ in azimuth and elevation around each specular component. This uncertainty represents, for example, LOS transmission to users that move on a radius of distance $d = 50$ m from the base station with $v = 100$ km/h when high-resolution CSI feedback is provided every $\tau = 50$ ms ($\psi_{\max} = \frac{v\tau}{d}$).

In Fig. 2, we compare the radiation beam patterns of robust and non-robust leakage-bounded angular beamforming for a user that is located at approximately $(\phi, \theta) = (22°, 95°)$ (indicated with the green circle) and nine leakage areas corresponding to the other $U-1$ users (indicated in black). In this example, we set the maximum tolerable interference level to $L_{\max} = -40$ dB relative to the unit transmit power. In Fig. 2 (left), we observe that the radiation pattern of non-robust leakage-bounded angular beamforming shows very narrow valleys wherever interference leakage constraints are active. Therefore, the transmission to the intended user does not cause substantial interference to the other users, provided the angular CSIT is accurate. However, due to the narrow width of the valleys at the interference leakage constraints, the residual inter-user inference grows significantly if the angular CSIT is not accurate. To compensate for such angular uncertainty, we consider interference leakage uncertainty regions of width $\psi_{\max}$ in the robust leakage-bounded angular beamformer. As shown in Fig. 2 (right), the interference leakage is then low over the entire uncertainty regions, thus improving the robustness with respect to imperfect angular CSIT. However, the maximum achievable gain of the intended user is slightly reduced, since it is not possible any more to perfectly steer the signal power towards the intended user while still satisfying all leakage constraints. Therefore, there exists a trade-off between the achievable signal gain of the intended user and the robustness of the interference leakage constraints.

6.2 Comparison to interference-unaware beamsteering

In our second example, we investigate multi-user beamforming with a single stream per user, comparing the interference-unaware angle of departure-based SDMA beamsteering methods of Section 5.1 and our robust/non-robust leakage-bounded angular beamformers. We consider users with a single receive antenna $N_r = 1$ and a transmitter with a UPA of size $N_t = N_h \cdot N_v = 11 \cdot 11$. We employ a Rician-fading channel model to generate the channel vector $\mathbf{h}_u \in \mathbb{C}^{1 \times N_t}$ according to:

$$\mathbf{h}_u = \sqrt{\frac{K}{K+1}} e^{j\beta_u} \mathbf{a}_t^{\mathrm{UPA}}(\phi_u, \theta_u)^{\mathrm{T}} + \sqrt{\frac{1}{K+1}} \mathbf{h}_{u,d}, \quad (31)$$

with $\beta_u \sim \mathcal{U}(-\pi, \pi)$ denoting a random phase-shift, $\phi_u \sim \mathcal{U}(-60°, 60°)$ and $\theta_u \sim \mathcal{U}(70°, 110°)$ representing the LOS directions, and $\mathbf{h}_{u,d} \sim \mathcal{CN}(\mathbf{0}, \mathbf{I})$ being a diffuse background scattering component. The Rician K-factor controls the relative strength of the LOS component. We do not utilize OMP in this example to decompose the channel vector, but we rather assume that the receivers employ a dictionary that is matched to the UPA and that the receivers are able to estimate the LOS angles with an accuracy of $\pm 2°$. Hence, the CSIT utilized for beamforming is $\mathbf{a}_t^{\mathrm{UPA}}(\hat{\phi}_u, \hat{\theta}_u)$ with $\hat{\phi}_u \sim \mathcal{U}(\phi_u - 2°, \phi_u + 2°)$ and $\hat{\theta}_u \sim \mathcal{U}(\theta_u - 2°, \theta_u + 2°)$ corresponding to a dictionary with 4° angular resolution. In our simulations, we consider a total number of $U = 10$ users; out of these ten users we select the best subset $\mathcal{S} \subseteq \{1, \ldots, U\}$ of users through exhaustive search with varying size $|\mathcal{S}| \in [1, U]$ to avoid an impact of imperfect user scheduling. We set the noise variance equal to $\sigma_{n,u}^2 = 10^{-3}$. For Chebyshev-tapered beamsteering, we set the side-lobe level equal to the maximum of noise power and residual interference caused by the diffuse component:

$$\mathrm{SLL}_u = \frac{1}{|\mathcal{S}| - 1} \max\left(\sigma_{n,u}^2, \frac{1}{K+1}\right), \quad (32)$$

such that the multi-user interference caused by the side-lobes is in the order of the maximum of diffuse background interference and noise power.

In Fig. 3, we show the achievable rate performance of the interference-unaware beamsteering methods and our robust/non-robust leakage-bounded angular beamformers for varying K-factors. The performance of the interference-unaware beamsteering methods (max. gain and Cheb. tapered beamsteering) is plotted for the case of perfect angular CSIT; however, their performance with imperfect CSIT of accuracy $\pm 2°$ is virtually the same. The performance of leakage-bounded beamforming is shown for perfect angular CSIT as well as for imperfect angular CSIT with robust/non-robust interference leakage constraints. We observe that our leakage-bounded beamforming method can achieve better performance and can serve more users in parallel than the interference-unaware beamsteering schemes (max. gain and Cheb. tapered beamsteering). With decreasing Rician K-factor (left to right) the performance deteriorates since the residual interference over the diffuse background scattering component is not mitigated by our beamformer optimization. Still, even with the relatively low value of $K = 10$ our proposed method outperforms the interference-unaware beamsteering schemes. We furthermore observe in Fig. 3 that the performance loss due to imperfect angular CSIT can partly be compensated with our robust beamforming method as compared to the non-robust approach.

6.3 Comparison to interference-aware precoding

For the remaining simulations, we utilize the QuaDRiGa channel model to test our proposed methods on more realistic channel realizations. We summarize the most important simulation parameters in Table 1, which have

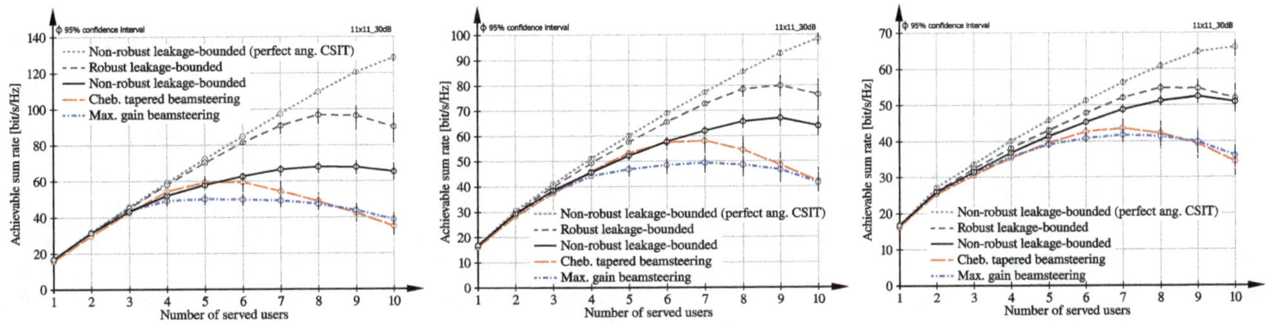

Fig. 3 Comparison of the achievable sum rate of different single-stream beamforming methods with varying K factor. **a** Achievable sum rate with $K = 10^3$. **b** Achievable sum rate with $K = 10^2$. **c** Achievable sum rate with $K = 10$

been selected to be LTE standard compliant. We consider users that are uniformly distributed within a sector of 120° with a maximum distance of 500 m from the transmitter and we assume that the users move perpendicular to the transmit antenna array bore-sight direction with 150 km/h. For such user placement, the variation of the elevation angle is in most cases very small and we therefore consider beamforming only in the azimuth domain utilizing a high-resolution horizontal uniform linear array (ULA) of 50 antenna elements. In our simulations, we do not consider macroscopic pathloss even though it is considered in the output of the QuaDRiGa channel model. We therefore normalize the channel matrices obtained from the QuaDRiGa channel model to satisfy $\mathbb{E}\left(\|\mathbf{H}_u[n]\|^2\right) = N_r N_t$. We apply this normalization since we do not consider multi-user scheduling in this section, which would be necessary in case of strong received signal power disparities of different users. We employ a dictionary/quantization codebook that is matched to the horizontal ULA and we consider an azimuthal angular resolution for CSI quantization of $1^{\circ 6}$. We consider a block-fading channel model with constant channel during each transmission time interval (TTI).

Table 1 Simulation parameters employed to setup the QuaDRiGa channel model

Parameter	Value
System bandwidth	10 MHz (600 OFDM subcarriers)
Carrier frequency	$f_c = 2.1$ GHz
Transmission time interval (TTI)	1 ms
Number of OFDM symbols per TTI	14
Transmitter height	25 m
Receiver height	1.5 m
Scenario	3GPP UMa LOS/NLOS
	3GPP UMi LOS/NLOS
	BERLIN UMa LOS/NLOS
Number of receive antennas	$N_r \in \{1, 2\}$
Receive antenna model	Dipole ($N_r = 1$), x-pol ($N_r = 2$)
Number of transmit antennas	$N_t = N_h \cdot N_v = 50 \cdot 1$
Transmit antenna model	3GPP-3D
Antenna element spacing	Half-wavelength
User placement	Uniform within 120° sector with maximum distance 500 m
User movement	Linear track perpendicular to antenna boresight direction
User speed	150 km/h

In our first simulation of this section, we investigate the compressibility of the per-subcarrier channel matrices $\mathbf{H}_u[n]$ in the angular domain. That is, we apply the OMP algorithm to decompose $\mathbf{H}_u[n]$ in the angular domain using a dictionary that is matched to the employed ULA and we determine the average norm of the expansion coefficient vectors $\left\|\mathbf{c}_u^s[n]\right\|^2$ of the decomposition. The corresponding results are shown in Fig. 4 for different simulation scenarios as specified in the QuaDRiGa channel model. We observe that in the LOS situations, the first expansion coefficient contains the majority of the channel energy, especially in the 3GPP compliant scenarios; this expansion coefficient corresponds to the LOS direction. The second largest coefficient lies already approximately an order of magnitude below the first coefficient. In the non-line of sight (NLOS) scenarios, however, the channel energy distributes more equally over many expansion coefficients. Correspondingly, compressibility with the employed dictionary is worse. In the remaining simulations we therefore consider LOS transmission and employ the 3GPP compliant UMa scenario. We also conducted simulations in NLOS situations, where our proposed method however does not achieve a significant gain over the interference-aware benchmark precoders.

In the next simulation, we compare the achievable transmission rate of the interference-aware precoding schemes summarized in Section 5.2 to our robust leakage-bounded

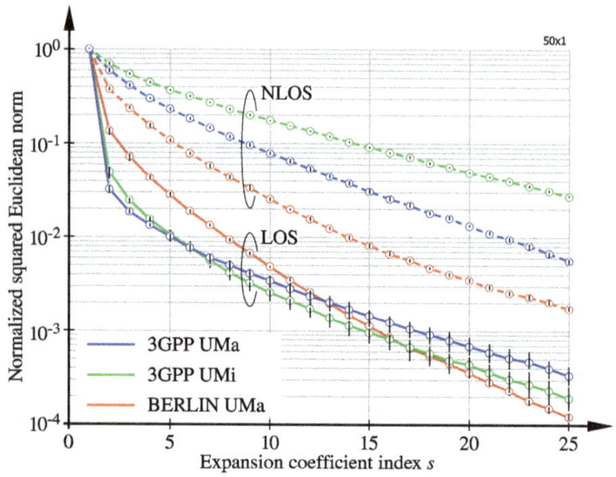

Fig. 4 Angular-domain compressibility of the channel matrices obtained with the QuaDRiGa channel model. Blue: 3GPP UMa; green: 3GPP UMi; red: BERLIN UMa

precoding method. We consider transmission to $U = 5$ users in parallel with $L_u = 2$ streams per user; hence, the total number of spatially multiplexed data streams is ten. We have seen above that in the LOS scenario most channel energy is contained in a single specular component. To efficiently support transmission of two spatial streams per user, we therefore employ a distributed antenna system (DAS) composed of two remote radio heads (RRHs) that are separated by 25 m. Each RRH radiates the same transmit signal; that is, the radio heads are fed with the same RF signal, e.g., via radio-over-fiber connections. Correspondingly, the two RRHs do not appear as independently steerable antenna arrays since they are fed with the same transmit signal; hence, the effective number of actively steerable transmit antennas (and thus the size of the channel matrix that must be estimated) is still $N_t = 50$. In that way, we guarantee that each user observes two strong specular components of approximately equal strength corresponding to the two antenna arrays of the RRHs.

We assume that the receivers estimate and feed back the CSI only at the first TTI, to enable calculation of the precoders; these precoder are then utilized for the entire transmission duration of 150 TTIs. For BD precoding and hybrid BD precoding, we assume perfect unquantized CSIT of the (effective) base band channel at the first TTI. For angle of departure BD precoding, we assume perfect knowledge of the LOS angles w.r.t. the two RRHs and hence perfect knowledge of the corresponding transmit antenna array response vectors. For the other schemes, we utilize our OMP-based channel decomposition to determine the $S = 2$ strongest specular components, in order to facilitate transmission of two spatial streams. We have observed that the performance does not improve if we consider $S > L_u$ expansion coefficients; when S is too large, the performance rather deteriorates since too many interference leakage constraints are active. For BD precoding and its hybrid variant, we assume perfect base band CSIT on each of the 600 subcarriers within the first TTI. For angular BD precoding, only broadband feedback of the angles of departure of the strongest multipath components needs to be provided, since this CSI is obtained from the time domain channel impulse response; notice, though, that this either requires estimation of the time domain channel impulse response or direction estimation of the dominant angles from uplink training signals [74]. For the remaining schemes, we consider a CSI feedback resolution of 60 subcarriers, i.e., the $S = 2$ strongest specular components are provided for every 60th subcarrier. Therefore, to estimate the full channel matrix at the receiver for CSI feedback calculation, we require a total of $600/60 \cdot N_t = 500$ orthogonal pilot-symbols within the first TTI, amounting to an overhead of $500/(14 \cdot 600) \approx 6\,\%$ within the first TTI. However, since we provide CSI feedback only every 150th TTI the pilot overhead for CSI feedback calculation is negligible.

For our robust leakage-bounded precoders, we determine the size of the angular uncertainty regions to match the uncertainty of the LOS direction caused by the user movement. To avoid unnecessarily large uncertainty regions, we recalculate the robust leakage-bounded precoders every 25 TTIs with correspondingly increasing angular uncertainty regions. Notice, in this recalculation, only the size of the uncertainty regions is updated and not the angular CSIT itself; it does therefore not require renewed CSI feedback from the users. In Fig. 5, we show the achievable rate performance of the considered precoding schemes. We observe that regular BD precoding as well as hybrid BD precoding achieve very high data

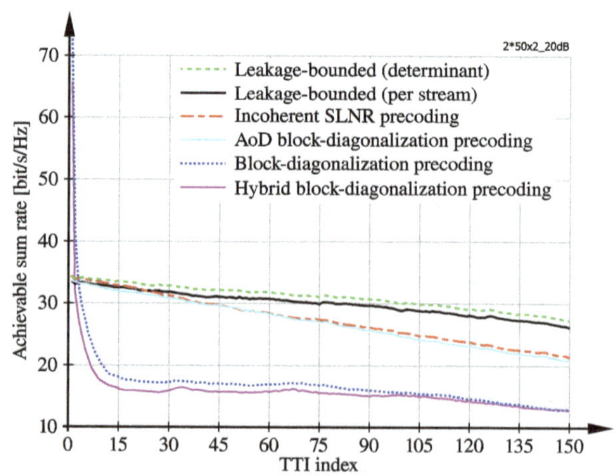

Fig. 5 Time-evolution of the achievable rate of different precoding schemes assuming constant precoders over time and time-variant channels

rate in the first TTI since at this time the CSIT is perfect and thus interference-free transmission is achieved; notice, we assume a noise variance of $\sigma_{n,u}^2 = 10^{-2}$ in this simulation. However, the performance of BD precoding and its hybrid variant deteriorates very quickly over time due to increasing residual interference caused by growing precoder mismatch, since the base band channel varies quickly over time. Our proposed leakage-bounded precoding scheme cannot compete with BD precoding in the first TTI. This is because it does not achieve interference-free transmission, since there is always residual interference caused by the channel decomposition error. Even with S very large and therefore a small channel decomposition error, our method does not achieve the multiplexing capabilities of BD precoding because it is an incoherent scheme that does not make use of coherent multipath interference, as BD precoding does to eliminate multi-user interference. Yet, our method is much less sensitive to the temporal variation of the channel matrices, since the employed angular channel decomposition is comparatively stable over time. This means that the angular directions of the largest channel expansion coefficients change only slowly over time and therefore the leakage-based precoding solution performs well even after many TTIs. Notice, the incoherent SLNR precoder as well as the angle of departure BD precoder achieve virtually the same performance as our scheme at the beginning of the transmission; however, their performance deteriorates faster because they do not consider angular uncertainty regions. Nevertheless, both methods provide a reasonably good trade-off between robustness and computational complexity. We did attempt to incorporate angular uncertainty regions in the SLNR metric, but we were not successful. In Fig. 5, we also compare the performance of leakage-bounded precoding utilizing the determinant optimization approach and the per stream optimization approach described in Section 4.2. We observe that per stream optimization exhibits only a small rate degradation compared to determinant optimization and therefore might be practically more relevant due to lower computational complexity.

In our final example, we investigate the sensitivity of leakage-bounded beamforming and BD precoding with respect to channel estimation errors. For this purpose, we distort the output $\mathbf{H}_u[n]$ of the QUADRIGA channel model to emulate a channel estimation error:

$$\hat{\mathbf{H}}_u[n] = \sqrt{1-\sigma_e^2}\,\mathbf{H}_u[n] + \sigma_e^2\,\mathbf{E}_u[n], \tag{33}$$

with $[\mathbf{E}_u[n]]_{i,j} \sim \mathcal{CN}(0,1)$ denoting the normalized channel estimation error and σ_e^2 controlling its relative strength, such that $\mathbb{E}\left(\left\|\hat{\mathbf{H}}_u[n]\right\|^2\right) = N_r N_t$[7]. We consider uncoded 16 QAM transmission to a varying number of users $U \in \{3,5,7\}$ with $L_u = 1$ stream per user. We use high-resolution angular quantization of a single $S = 1$ specular component, such that the CSI uncertainty of this single expansion vector is only due to the channel estimation error.

In Fig. 6, we show the obtained SER as a function of the channel estimation error variance σ_e^2. We observe that leakage-bounded beamforming exhibits an SER floor due to residual interference caused by the channel decomposition error, which grows with the number of users served in parallel. Yet, the performance of the method is hardly impacted by the channel estimation error; this robustness is achieved by the signal denoising properties of the employed angular basis decomposition. Only for very large channel estimation errors the decomposition fails due to the growing impact of $\mathbf{E}_u[n]$. BD precoding, on the other hand, is much more sensitive w.r.t. channel estimation errors, since such errors directly cause residual inter-user interference; yet, for small error variance it outperforms our proposed method. We want to stress that the performance shown in (8) is for uncoded transmission; with channel coding, the SER can of course be substantially reduced.

7 Conclusions

Multi-user beamforming and precoding approaches are in general very sensitive to CSI imperfections at the transmitter. Most existing approaches are not efficiently applicable in high-mobility situations, since channel matrices change so quickly over time that already small processing delays can lead to substantial inter-user interference. In this paper, we provide a remedy to this problem by considering non-coherent beamforming/precoding in the

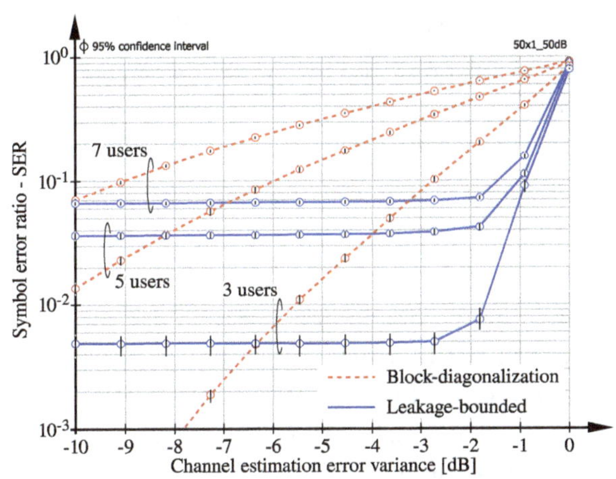

Fig. 6 Sensitivity of the SER achieved with BD precoding and leakage-bounded beamforming with respect to the channel estimation error for uncoded 16 QAM transmission. Red: Block diagonalization; blue: leakage-bounded

angular domain with robustness w.r.t. angular uncertainty. In addition to improving the robustness of the beamformer/precoder, this has the further advantage that only a minimal amount of CSIT is required, namely the angles of departure of the most significant specular components together with their respective gains as obtained from a low-rank basis decomposition. We show that the relevant CSI can efficiently be quantized by applying a variant of the OMP algorithm. Naturally, under ideal circumstances of perfect CSIT, such a non-coherent approach cannot compete with coherent methods in terms of the achievable multiplexing gain. Also, it is not well suited for strong scattering environments with a large number of significant specular components because the low-rank channel decomposition has a large residual error in such cases. Nevertheless, the proposed technique can provide substantial improvements in situations with few dominant specular components, as for example expected in the millimeter wave regime, and it enables robust multi-user beamforming/precoding with low requirements on CSIT accuracy.

Endnotes

[1] A basis of a vector space is said to be over-complete if it is complete even after the removal of a vector from the basis.

[2] A related low-rank basis expansion model has recently also been investigated in [78]; however, in [78] the authors consider an orthogonal basis expansion model, whereas we consider an over-complete basis set containing far more entries $|\mathcal{D}|$ than the dimension N_t of the vector space.

[3] We investigate a related max-min problem that accounts for uncertainty in the signal direction in [39]; it can only be approximately solved via SDR.

[4] The simulation results were generated using QuaDRiGa Version 1.4.8-571.

[5] Notice, the antenna boresight direction is $(\phi, \theta) = (0°, 90°)$.

[6] Notice, for multi-user beamforming with a single stream per user we investigate the impact of the angular CSI resolution in [40] and demonstrate that an angular resolution of 4° achieves close to optimal performance for a UPA of size 10 · 10. Since the spatial resolution of the 50 element ULA considered here is better, we increase the angular CSI resolution to 1°.

[7] The definition of σ_e^2 here is slightly different as compared to (8), in order to keep the energy of the estimated channel constant irrespective of σ_e^2.

Appendix

Upper bound on the approximation $\hat{\tilde{R}}_u$

In this appendix we derive an upper bound on the approximation $\hat{\tilde{R}}_u$ of (24) that can be calculated from the provided CSI feedback. We start by applying Sylvester's determinant theorem and Jensen's inequality to (24):

$$\tilde{R}_u \leq \log \left| \sigma_{n,u}^2 \mathbf{I}_{N_t} + \mathbb{E}\left(\mathbf{H}_u^{\mathrm{H}} \mathbf{H}_u \sum_{j \in \mathcal{S}} \mathbf{F}_j \mathbf{F}_j^{\mathrm{H}} \right) \right| \tag{34}$$

Let us next consider the approximation of a single interference term in detail:

$$\begin{aligned}\mathbb{E}(\mathbf{H}_u^{\mathrm{H}} \mathbf{H}_u \mathbf{F}_j \mathbf{F}_j^{\mathrm{H}}) &\approx \mathbb{E}\left(\sum_{s=1}^{S} \sum_{k=1}^{S} \left(\mathbf{d}_u^k\right)^* \left(\mathbf{c}_u^k\right)^{\mathrm{H}} \mathbf{c}_u^s \left(\mathbf{d}_u^s\right)^{\mathrm{T}} \mathbf{C}_j \right) \\ &= \mathbb{E}\left(\sum_{s=1}^{S} \left\|\mathbf{c}_u^s\right\|^2 \left(\mathbf{d}_u^s\right)^* \left(\mathbf{d}_u^s\right)^{\mathrm{T}} \mathbf{C}_j \right),\end{aligned} \tag{35}$$

with $\mathbf{C}_j = \mathbf{F}_j \mathbf{F}_j^{\mathrm{H}}$ being the transmit covariance matrix. In this equation, we again utilized the phase independence of different specular components as in the steps from (19) to (20). We next apply an eigen-decomposition to the sum of the specular components:

$$\sum_{s=1}^{S} \left\|\mathbf{c}_u^s\right\|^2 (\mathbf{d}_u^s)^* (\mathbf{d}_u^s)^{\mathrm{T}} = \mathbf{U}_u \mathbf{\Sigma}_u^2 \mathbf{U}_u^{\mathrm{H}}, \tag{36}$$

$$\mathbf{U}_u \in \mathbb{C}^{N_t \times S}, \mathbf{U}_u^{\mathrm{H}} \mathbf{U}_u = \mathbf{I}_S, \ \mathbf{\Sigma}_u^2 \in \mathbb{R}^{S \times S}, \tag{37}$$

where we assume that $S \leq N_t$ as our focus is on large-scale antenna arrays and low-rank channel decomposition.

Let us next consider the interference leakage power caused by the transmission to user j w.r.t. all specular components of user u:

$$\begin{aligned}L_{u,j} &= \sum_{s=1}^{S} L_{u,j}^s = \sum_{s=1}^{S} \operatorname{tr}\left(\mathbf{c}_u^s (\mathbf{d}_u^s)^{\mathrm{T}} \mathbf{C}_j (\mathbf{d}_u^s)^* (\mathbf{c}_u^s)^{\mathrm{H}} \right) \\ &= \sum_{s=1}^{S} \left\|\mathbf{c}_u^s\right\|^2 (\mathbf{d}_u^s)^{\mathrm{T}} \mathbf{C}_j (\mathbf{d}_u^s)^* = \operatorname{tr}\left(\mathbf{U}_u \mathbf{\Sigma}_u^2 \mathbf{U}_u^{\mathrm{H}} \mathbf{C}_j \right).\end{aligned} \tag{38}$$

In the precoder optimization problem (P5), we restrict this interference leakage to satisfy $L_{u,j}^s \leq L_{\max}$ and hence $L_{u,j} \leq S L_{\max}$. From this upper bound on the interference leakage power, we can determine an upper bound on the interference leakage covariance (35). For that purpose, we decompose the precoding matrix in terms of the S dimensional basis $\mathbf{U}_u$ and a basis $\mathbf{U}_u^{\perp}$ of its $N_t - S$ dimensional orthogonal complement:

$$\mathbf{F}_j = \mathbf{U}_u \mathbf{F}_j^{\parallel} + \mathbf{U}_u^{\perp} \mathbf{F}_j^{\perp}, \quad (39)$$

$$\mathbf{F}_j^{\parallel} = \mathbf{U}_u^{\mathrm{H}} \mathbf{F}_j,\ \mathbf{F}_j^{\perp} = \left(\mathbf{U}_u^{\perp}\right)^{\mathrm{H}} \mathbf{F}_j. \quad (40)$$

With this, we can reformulate the leakage power as:

$$L_{u,j} = \mathrm{tr}\left(\boldsymbol{\Sigma}_u^2 \mathbf{F}_j^{\parallel} \left(\mathbf{F}_j^{\parallel}\right)^{\mathrm{H}}\right) \leq SL_{\max}. \quad (41)$$

We next assume $\mathbf{F}_j^{\parallel} \in \mathbb{C}^{S \times L_j}$ to be isotropically distributed, such that in combination with (41) we get

$$\mathbb{E}\left(\mathbf{F}_j^{\parallel} \left(\mathbf{F}_j^{\parallel}\right)^{\mathrm{H}}\right) = k_j \mathbf{I}_S, \quad (42)$$

$$k_j \leq \frac{SL_{\max}}{\mathrm{tr}\left(\boldsymbol{\Sigma}_u^2\right)}. \quad (43)$$

Utilizing these results in (35), we obtain:

$$\mathbb{E}\left(\mathbf{U}_u \boldsymbol{\Sigma}_u^2 \mathbf{U}_u^{\mathrm{H}} \mathbf{C}_j\right) \preceq \frac{SL_{\max}}{tr(\boldsymbol{\Sigma}_u^2)} \mathbf{U}_u \boldsymbol{\Sigma}_u^2 \mathbf{U}_u^{\mathrm{H}}, \quad (44)$$

where $\preceq$ means that the difference between the right-hand side and left-hand side is positive semidefinite. This brings us to the desired upper bound on the approximation $\hat{\hat{R}}_u$:

$$\tilde{R}_u \approx \hat{\hat{R}}_u \leq \log\left|\sigma_{n,u}^2 \mathbf{I}_S + (|\mathcal{S}| - 1)\frac{SL_{\max}}{\mathrm{tr}\left(\boldsymbol{\Sigma}_u^2\right)} \boldsymbol{\Sigma}_u^2 + \boldsymbol{\Sigma}_u^2 \mathbf{U}_u^{\mathrm{H}} \mathbf{C}_u \mathbf{U}_u\right|, \quad (45)$$

where we again applied Sylvester's determinant theorem.

Abbreviations

3GPP: Third generation partnership project; AD: Analog to digital; BD: Block diagonalization; BS: Base station; CoMP: Coordinated multipoint transmission; CSI: Channel state information; CSIT: Channel state information at the transmitter; DAS: Distributed antenna system; DA: Digital to analog; DoF: Degrees of freedom; FD: Full-dimension; FDD: Frequency division duplex; FD-MIMO: Full-dimension multiple-input multiple-output; FRT: Fluctuating two-ray; GRT: Generalized two-ray; IA: Interference alignment; i.i.d.: Independent and identically distributed; ITS: Intelligent transport systems; LOS: Line of sight; LTE: Long term evolution; MBMS: Multimedia broadcast multicast service; eMBMS: Evolved MBMS; MC-IC: Multicast interference channel; MIMO: Multiple-input multiple-output; MISO: Multiple-input single-output; NLOS: Non-line of sight; NP: Non-deterministic polynomial-time; NR: New radio; OFDM: Orthogonal frequency division multiplexing; OMP: Orthogonal matching pursuit; QAM: Quadrature amplitude modulation; RF: Radio frequency; RRH: Remote radio head; SDMA: Space-division multiple access; SDR: Semidefinite programming relaxation; SER: Symbol error ratio; SINR: Signal to interference and noise ratio; SLNR: Signal to leakage and noise ratio; SLR: Signal to leakage ratio; SNR: Signal to noise ratio; SVD: Singular value decomposition; TDD: Time division duplex; TTI: Transmission time interval; TWDP: Two-wave diffuse power; UC-IC: Unicast interference channel; UE: User equipment; UMTS: Universal mobile telecommunications system; ULA: Uniform linear array; UPA: Uniform planar array; ZF: Zero forcing

Acknowledgements

The financial support by the Austrian Federal Ministry of Science, Research and Economy and the National Foundation for Research, Technology and Development is gratefully acknowledged. The author acknowledges the TU Wien University Library for the financial support through its Open Access Funding Program.

Funding

Not applicable.

Authors' contributions

All contributions are from the sole author SS. The author read and approved the final manuscript.

Authors' information

Not applicable.

Competing interests

The author declares that he has no competing interests.

References

1. H Yang, T Marzetta, Performance of conjugate and zero-forcing beamforming in large-scale antenna systems. IEEE J. Sel. Areas Commun. **31**(2), 172–179 (2013)
2. Y Kim, H Ji, J Lee, YH Nam, BL Ng, I Tzanidis, Y Li, J Zhang, Full dimension MIMO (FD-MIMO): The next evolution of MIMO in LTE systems. IEEE Wirel. Commun. **21**(3), 92–100 (2014)
3. YH Nam, MS Rahman, Y Li, G Xu, E Onggosanusi, J Zhang, JY Seol, in *Information Theory and Applications Workshop (ITA)*. Full dimension MIMO for LTE-Advanced and 5G (IEEE, San Diego, 2015), pp. 143–148
4. M Kim, YH Lee, MSE-based hybrid RF/baseband processing for millimeter-wave communication systems in MIMO interference channels. IEEE Trans. Veh. Technol. **64**(6), 2714–2720 (2015)
5. F Sohrabi, W Yu, Hybrid digital and analog beamforming design for large-scale antenna arrays. IEEE J. Sel. Topics Signal Process. **10**(3), 501–513 (2016)
6. R Rajashekar, L Hanzo, Hybrid beamforming in mm-Wave MIMO systems having a finite input alphabet. IEEE Trans. Commun. **64**(8), 3337–3349 (2016)
7. A Liu, VKN Lau, Impact of CSI knowledge on the codebook-based hybrid beamforming in massive MIMO. IEEE Trans. Signal Process. **64**(24), 6545–6556 (2016)
8. H Wang, W Wang, VKN Lau, Z Zhang, Hybrid limited feedback in 5G cellular systems with massive MIMO. IEEE Syst. J. **11**(1), 50–61 (2017)
9. C Rusu, R Méndez-Rial, N González-Prelcic, RW Heath, Low complexity hybrid precoding strategies for millimeter wave communication systems. IEEE Trans. Wirel. Commun. **15**(12), 8380–8393 (2016)
10. QUA Nadeem, A Kammoun, M Debbah, MS Alouini, Design of 5G full dimension massive MIMO systems. IEEE Trans. Commun. **PP**(99), 1–1 (2017)
11. N Li, Z Wei, H Yang, X Zhang, D Yang, Hybrid precoding for mmWave massive MIMO systems with partially connected structure. IEEE Access. **5**, 2–15151 (1514)
12. O Orhan, E Erkip, S Rangan, in *2015 Information Theory and Applications Workshop (ITA)*. Low power analog-to-digital conversion in millimeter wave systems: impact of resolution and bandwidth on performance (IEEE, San Diego, 2015), pp. 191–198
13. S Jacobsson, G Durisi, M Coldrey, U Gustavsson, C Studer, Throughput analysis of massive MIMO uplink with low-resolution ADCs. IEEE Trans. Wirel. Commun. **16**(6), 4038–4051 (2017)
14. WB Abbas, F Gomez-Cuba, M Zorzi, Millimeter wave receiver efficiency: a comprehensive comparison of beamforming schemes with low resolution ADCs. IEEE Trans. Wirel. Commun. **16**(12), 8131–8146 (2017)
15. K Roth, JA Nossek, Achievable rate and energy efficiency of hybrid and digital beamforming receivers with low resolution ADC. IEEE J. Sel. Areas Commun. **35**(9), 2056–2068 (2017)
16. TE Bogale, LB Le, Massive MIMO and mmWave for 5G wireless HetNet: potential benefits and challenges. IEEE Veh. Technol. Mag. **11**(1), 64–75 (2016)
17. LN Ribeiro, S Schwarz, M Rupp, ALF de Almeida, Energy efficiency of mmWave massive MIMO precoding with low-resolution DACs. CoRR. **abs/1709.05139** (2017)
18. J Zuo, J Zhang, C Yuen, W Jiang, W Luo, Multicell multiuser massive MIMO transmission with downlink training and pilot contamination precoding. IEEE Trans. Veh. Technol. **65**(8), 6301–6314 (2016)
19. Y Han, J Lee, Uplink pilot design for multi-cell massive MIMO networks. IEEE Commun. Lett. **20**(8), 1619–1622 (2016)
20. S Schwarz, R Heath, M Jr. Rupp, Single-user MIMO versus multi-user MIMO in distributed antenna systems with limited feedback. EURASIP J. Adv. Sig. Proc. **2013**(1), 54 (2013)

21. W Shen, L Dai, Y Shi, B Shim, Z Wang, Joint channel training and feedback for FDD massive MIMO systems. IEEE Trans. Veh. Technol. **65**(10), 8762–8767 (2016)
22. MS Sim, J Park, CB Chae, RW Heath, Compressed channel feedback for correlated massive MIMO systems. J. Commun. Netw. **18**(1), 95–104 (2016)
23. S Schwarz, T Philosof, M Rupp, Signal processing challenges in cellular assisted vehicular communications. IEEE Signal Process. Mag. **34**(2), 47–59 (2017)
24. MA Maddah-Ali, D Tse, Completely stale transmitter channel state information is still very useful. IEEE Trans. Inf. Theory. **58**(7), 4418–4431 (2012)
25. S Yang, M Kobayashi, D Gesbert, X Yi, Degrees of freedom of time correlated MISO broadcast channel with delayed CSIT. IEEE Trans. Inf. Theory. **59**(1), 315–328 (2013)
26. J Chen, P Elia, Toward the performance versus feedback tradeoff for the two-user MISO broadcast channel. IEEE Trans. Inf. Theory. **59**(12), 8336–8356 (2013)
27. H Joudeh, B Clerckx, Sum-rate maximization for linearly precoded downlink multiuser MISO systems with partial CSIT: A rate-splitting approach. IEEE Trans. Commun. **64**(11), 4847–4861 (2016)
28. C Peel, B Hochwald, A Swindlehurst, A vector-perturbation technique for near-capacity multiantenna multiuser communication-part I: channel inversion and regularization. IEEE Trans. Commun. **53**(1), 195–202 (2005)
29. Q Spencer, A Swindlehurst, M Haardt, Zero-forcing methods for downlink spatial multiplexing in multiuser MIMO channels. IEEE Trans. Signal Process. **52**(2), 461–471 (2004)
30. V Cadambe, S Jafar, Interference alignment and degrees of freedom of the K-user interference channel. IEEE Trans. Inf. Theory. **54**(8), 3425–3441 (2008)
31. M Maddah-Ali, A Motahari, A Khandani, Communication over MIMO X channels: interference alignment, decomposition, and performance analysis. IEEE Trans. Inf. Theory. **54**(8), 3457–3470 (2008)
32. J Zhang, RW Heath Kountouris, JG Andrews, Mode switching for the multi-antenna broadcast channel based on delay and channel quantization. EURASIP J. Adv. Signal Process. **2009**(1), 802548 (2009)
33. S Schwarz, R Heath, Rupp Jr M, in *38th International Conference on Acoustics, Speech and Signal Processing*. Adaptive quantization on the Grassmann-manifold for limited feedback multi-user MIMO systems (IEEE, Vancouver, 2013), pp. 5021–5025
34. SA Vorobyov, Principles of minimum variance robust adaptive beamforming design.Signal Process. **93**(12), 3264–3277 (2013). special Issue on Advances in Sensor Array Processing in Memory of Alex B. Gershman
35. WS Chua, C Yuen, YL Guan, F Chin, Robust multi-antenna multi-user precoding based on generalized multi-unitary decomposition with partial CSI feedback. IEEE Trans. Veh. Technol. **62**(2), 596–605 (2013)
36. S Schwarz, M Rupp, Evaluation of distributed multi-user MIMO-OFDM with limited feedback. IEEE Trans. Wirel. Commun. **13**(11), 6081–6094 (2014)
37. J Wang, M Bengtsson, B Ottersten, DP Palomar, Robust MIMO precoding for several classes of channel uncertainty. IEEE Trans. Signal Process. **61**(12), 3056–3070 (2013)
38. F Khalid Speidel, Robust hybrid precoding for multiuser MIMO wireless communication systems. IEEE Trans. Wirel. Commun. **13**(6), 3353–3363 (2014)
39. S Schwarz, T Philosof, M Rupp, in *50th Asilomar Conference on Signals Systems and Computers*. Convex-optimization based geometric beamforming for FD-MIMO arrays (IEEE, Asilomar, 2016)
40. S Schwarz, M Rupp, in *50th Asilomar Conference on Signals Systems and Computers*. Limited feedback based double-sided full-dimension MIMO for mobile backhauling (IEEE, Asilomar, 2016)
41. P Almers, E Bonek, A Burr, N Czink, M Debbah, V Degli-Esposti, H Hofstetter, P Kyösti, D Laurenson, G Matz, A Molisch, C Oestges, H Özcelik, Survey of channel and radio propagation models for wireless MIMO systems. EURASIP J. Wirel. Commun. Netw. **2007**(019070), 19 (2007)
42. H Özcelik, *Indoor MIMO channel models, 2004.* (Wien, Technische Universität, Dissertation, 2005). http://resolver.obvsg.at/urn:nbn:at:at-ubtuw:1-17999
43. CA Balanis, *Antenna theory: analysis and design*, 3rd. (Wiley, New Jersey, 2005)
44. H Ji, Y Kim, J Lee, E Onggosanusi, Y Nam, J Zhang, B Lee, B Shim, Overview of full-dimension MIMO in LTE-advanced pro. IEEE Commun. Mag. **55**(2), 176–184 (2017)
45. J Choi, V Va, N Gonzalez-Prelcic, R Daniels, CR Bhat, RW Heath, Millimeter-wave vehicular communication to support massive automotive sensing. IEEE Commun. Mag. **54**(12), 160–167 (2016)
46. GD Durgin, TS Rappaport, DA de Wolf, New analytical models and probability density functions for fading in wireless communications. IEEE Trans. Commun. **50**(6), 1005–1015 (2002)
47. M Rao, FJ Lopez-Martinez, A Goldsmith, in *2014 48th Annual Conference on Information Sciences and Systems*. Statistics and system performance metrics for the two wave with diffuse power fading model (IEEE, Princeton, 2014), pp. 1–6
48. M Rao, FJ Lopez-Martinez, MS Alouini, A Goldsmith, MGF approach to the analysis of generalized two-ray fading models. IEEE Trans. Wirel. Commun. **14**(5), 2548–2561 (2015)
49. JM Romero-Jerez, FJ Lopez-Martinez, JF Paris, A Goldsmith, in *2016 IEEE Globecom Workshops*. The fluctuating two-ray fading model for mmWave communications (IEEE, Washington, 2016), pp. 1–6
50. S Schwarz, Outage investigation of beamforming over random-phase finite-scatterer MISO channels. IEEE Signal Process. **24**(7), 1029–1033 (2017)
51. J Zhang, W Zeng, X Li, Q Sun, KP Peppas, New results on the fluctuating two-ray model with arbitrary fading parameters and its applications. IEEE Trans. Veh. Technol. **PP**(99), 1–1 (2017)
52. E Dahlman, S Parkvall, J Sköld, *4G LTE/LTE-advanced for mobile broadband.* (Elsevier Academic Press, Oxford, 2011)
53. O El Ayach, R Heath Jr., Grassmannian differential limited feedback for interference alignment. IEEE Trans. Signal Process. **60**(12), 6481–6494 (2012)
54. YC Eldar, G Kutyniok, *Compressed sensing: theory and applications.* (Cambridge Univ. Press, Cambridge, 2012)
55. YC Pati, R Rezaiifar, PS Krishnaprasad, in *Proceedings of 27th Asilomar Conference on Signals, Systems and Computers*. Orthogonal matching pursuit: recursive function approximation with applications to wavelet decomposition, (1993), pp. 40–44
56. Z-Q Luo, W-K Ma, A-C So, Y Ye, S Zhang, Semidefinite relaxation of quadratic optimization problems. IEEE Signal Process. Mag. **27**(3), 20–34 (2010)
57. Z Luo, ND Sidiropoulos, P Tseng, S Zhang, Approximation bounds for quadratic optimization with homogeneous quadratic constraints. SIAM J. Optim. **18**(1), 1–28 (2007)
58. T-H Chang, Z-Q Luo, C-Y Chi, Approximation bounds for semidefinite relaxation of max-min-fair multicast transmit beamforming problem. IEEE Trans. Signal Process. **56**(8), 3932–3943 (2008)
59. R Zhang, YC Liang, Exploiting multi-antennas for opportunistic spectrum sharing in cognitive radio networks. IEEE J. Sel. Topics Signal Process. **2**(1), 88–102 (2008)
60. Y Nesterov, A Nemirovski, *Interior-point polynomial algorithms in convex programming: theory and applications.* (Philadelphia: Society for Industrial and Applied Mathematics, SIAM, 1994)
61. MS Lobo, L Vandenberghe, S Boyd, H Lebret, Applications of second-order cone programming. Linear Algebra Appl. **284**(1), 193–228 (1998). international Linear Algebra Society (ILAS) Symposium on Fast Algorithms for Control, Signals and Image Processing
62. R Zhang, S Cui, Cooperative interference management with MISO beamforming. IEEE Trans. Signal Process. **58**(10), 5450–5458 (2010)
63. S Schwarz, M Rupp, Transmit optimization for the MISO multicast interference channel. IEEE Trans. Commun. **63**(12), 4936–4949 (2015)
64. Z-Q Luo, T-H Chang, in *Convex Optimization in Signal Processing and Communications*, ed. by D Palomar, Y Eldar. SDP relaxation of homogeneous quadratic optimization: approximation and applications (Cambridge University Press, 2010)
65. F Alizadeh, D Goldfarb, Second-order cone programming. Math Program. **95**, 3–51 (2001)
66. C Wang, D Sun, K-C Toh, Solving log-determinant optimization problems by a newton-cg primal proximal point algorithm. SIAM J. Optim. **20**(6), 2994–3013 (2010)
67. YP Lin, SH Tsai, in *2017 IEEE 18th International Workshop on Signal Processing Advances in Wireless Communications (SPAWC)* (Thic structures for rf beamforming, 2017), pp. 1–5
68. Y Lee, Y Nam, J So, in *2012 18th Asia-Pacific Conference on Communications (APCC)*. Optimal user selection algorithm for opportunistic space division multiple access systems (IEEE, Jeju Island, 2012), pp. 922–923

69. HL Van Trees, *Detection, estimation, and modulation theory. Part IV., Optimum array processing*. (Wiley-Interscience, New York, 2002)
70. F-I Tseng, DK Cheng, Optimum scannable planar arrays with an invariant sidelobe level. Proc IEEE. **56**(11), 1771–1778 (1968)
71. N Jindal, MIMO broadcast channels with finite-rate feedback. IEEE Trans. Inf. Theory. **52**(11), 5 (2006)
72. N Ravindran, N Jindal, Limited feedback-based block diagonalization for the MIMO broadcast channel. IEEE J. Sel. Areas Commun. **26**(8), 1473–1482 (2008)
73. S Schwarz, M Rupp, Subspace quantization based combining for limited feedback block-diagonalization. IEEE Trans. Wirel. Commun. **12**(11), 5868–5879 (2013)
74. X Zhang, J Tadrous, E Everett, F Xue, A Sabharwal, in *2015 49th Asilomar Conference on Signals, Systems and Computers*. Angle-of-arrival based beamforming for FDD massive MIMO, (2015), pp. 704–708
75. M Sadek, A Tarighat, A Sayed, A leakage-based precoding scheme for downlink multi-user MIMO channels. IEEE Trans. Wirel. Commun. **6**(5), 1711–1721 (2007)
76. S Jaeckel, L Raschkowski, K Börner, L Thiele, Quadriga: A 3-D multi-cell channel model with time evolution for enabling virtual field trials. IEEE Trans. Antennas Propag. **62**(6), 3242–3256 (2014)
77. M Grant, S Boyd, CVX: Matlab software for disciplined convex programming, version 2.1. (2014). http://cvxr.com/cvx
78. H Xie, F Gao, S Zhang, S Jin, A unified transmission strategy for TDD/FDD massive MIMO systems with spatial basis expansion model. IEEE Trans. Veh. Technol. **66**(4), 3170–3184 (2017)

12

Channel estimation via gradient pursuit for mmWave massive MIMO systems with one-bit ADCs

In-soo Kim and Junil Choi*

Abstract

In millimeter wave (mmWave) massive multiple-input multiple-output (MIMO) systems, 1 bit analog-to-digital converters (ADCs) are employed to reduce the impractically high power consumption, which is incurred by the wide bandwidth and large arrays. In practice, the mmWave band consists of a small number of paths, thereby rendering sparse virtual channels. Then, the resulting maximum a posteriori (MAP) channel estimation problem is a sparsity-constrained optimization problem, which is NP-hard to solve. In this paper, iterative approximate MAP channel estimators for mmWave massive MIMO systems with 1 bit ADCs are proposed, which are based on the gradient support pursuit (GraSP) and gradient hard thresholding pursuit (GraHTP) algorithms. The GraSP and GraHTP algorithms iteratively pursue the gradient of the objective function to approximately optimize convex objective functions with sparsity constraints, which are the generalizations of the compressive sampling matching pursuit (CoSaMP) and hard thresholding pursuit (HTP) algorithms, respectively, in compressive sensing (CS). However, the performance of the GraSP and GraHTP algorithms is not guaranteed when the objective function is ill-conditioned, which may be incurred by the highly coherent sensing matrix. In this paper, the band maximum selecting (BMS) hard thresholding technique is proposed to modify the GraSP and GraHTP algorithms, namely, the BMSGraSP and BMSGraHTP algorithms, respectively. The BMSGraSP and BMSGraHTP algorithms pursue the gradient of the objective function based on the band maximum criterion instead of the naive hard thresholding. In addition, a fast Fourier transform-based (FFT-based) fast implementation is developed to reduce the complexity. The BMSGraSP and BMSGraHTP algorithms are shown to be both accurate and efficient, whose performance is verified through extensive simulations.

Keywords: MmWave, Massive MIMO, 1 bit ADC, MAP channel estimation, GraSP, GraHTP, CoSaMP, HTP, CS, BMS hard thresholding technique, FFT

1 Introduction

In millimeter wave (mmWave) massive multiple-input multiple-output (MIMO) systems, the wide bandwidth of the mmWave band in the range of 30–300 GHz offers a high data rate, which guarantees a significant performance gain [1–4]. However, the power consumption of analog-to-digital converters (ADCs) is scaled quadratically with the sampling rate and exponentially with the ADC resolution, which renders high-resolution ADCs impractical for mmWave systems [5]. To reduce the power consumption, low-resolution ADCs were suggested as a possible solution, which recently gained popularity [6–9]. Coarsely quantizing the received signal using low-resolution ADCs results in an irreversible loss of information, which might cause a significant performance degradation. In this paper, we consider the extreme scenario of using 1 bit ADCs for mmWave systems.

In practice, the mmWave band consists of a small number of propagation paths, which results in sparse virtual channels. In the channel estimation point of view, sparse channels are favorable because the required complexity and measurements can be reduced. Sparsity-constrained channel distributions, however, cannot be described in closed forms, which makes it difficult to exploit Bayesian channel estimation. In [10, 11], channel estimators for massive MIMO systems with 1 bit ADCs were proposed, which account for the effect of the coarse quantization.

*Correspondence: junil@kaist.ac.kr
School of Electrical Engineering, KAIST, Daejeon, South Korea

The near maximum likelihood (nML) channel estimator [10] selects the maximizer of the likelihood function subject to the L^2-norm constraint as the estimate of the channel, which is solved using the projected gradient descent method [12]. However, the channel sparsity was not considered in [10]. In [11], the Bussgang linear minimum mean squared error (BLMMSE) channel estimator was derived by linearizing 1 bit ADCs based on the Bussgang decomposition [13]. The BLMMSE channel estimator is an LMMSE channel estimator for massive MIMO systems with 1 bit ADCs, whose assumption is that the channel is Gaussian. Therefore, the sparsity of the channel is not taken into account in [11] either.

To take the channel sparsity into account, iterative approximate MMSE estimators for mmWave massive MIMO systems with 1 bit ADCs were proposed in [14, 15]. The generalized expectation consistent signal recovery (GEC-SR) algorithm in [14] is an iterative approximate MMSE estimator based on the turbo principle [16], which can be applied to any nonlinear function of the linearly mapped signal to be estimated. Furthermore, the only constraint on the distribution of the signal to be estimated is that its elements must be independent and identically distributed (i.i.d.) random variables. Therefore, the GEC-SR algorithm can be used as an approximate MMSE channel estimator for any ADC resolutions ranging from 1 bit to high-resolution ADCs. However, the inverse of the sensing matrix is required at each iteration, which is impractical in massive MIMO systems in the complexity point of view.

The generalized approximate message passing-based (GAMP-based) channel estimator for mmWave massive MIMO systems with low-resolution ADCs was proposed in [15], which is another iterative approximate MMSE channel estimator. In contrast to the GEC-SR algorithm, only matrix-vector multiplications are required at each iteration, which is favorable in the complexity point of view. As in the GEC-SR algorithm, the GAMP-based algorithm can be applied to any ADC resolutions and any channel distributions as long as the elements of channel are i.i.d. random variable. The performance of the GEC-SR and GAMP algorithms, however, cannot be guaranteed when the sensing matrix is ill-conditioned, which frequently occurs in the mmWave band. To prevent the sensing matrix from becoming ill-conditioned, the GAMP-based channel estimator constructs the virtual channel representation using discrete Fourier transform (DFT) matrices, whose columns are orthogonal. However, such virtual channel representation results in a large gridding error, which leads to performance degradation.

Our goal is to propose an iterative approximate maximum a posteriori (MAP) channel estimator for mmWave massive MIMO systems with 1 bit ADCs. Due to the sparse nature, the MAP channel estimation problem is converted to a sparsity-constrained optimization problem, which is NP-hard to solve *[17]*. To approximately solve such problems iteratively, the gradient support pursuit (GraSP) and gradient hard thresholding pursuit (GraHTP) algorithms were proposed in [17] and [18], respectively. The GraSP and GraHTP algorithms pursue the gradient of the objective function at each iteration by hard thresholding. These algorithms are the generalizations of the compressive sampling matching pursuit (CoSaMP) [19] and hard thresholding pursuit (HTP) [20] algorithms, respectively, in compressive sensing (CS).

With highly coherent sensing matrix, however, the GraSP and GraHTP algorithms do not perform appropriately since the objective function becomes ill-conditioned. To remedy such break down, we exploit the band maximum selecting (BMS) hard thresholding technique, which is then applied to the GraSP and GraHTP algorithms to propose the BMSGraSP and BMSGraHTP algorithms, respectively. The proposed BMS-based algorithms perform hard thresholding for the gradient of the objective function based on the proposed band maximum criterion, which tests whether an index is the ground truth index or the by-product of another index. To reduce the complexity of the BMS-based algorithms, a fast Fourier transform-based (FFT-based) fast implementation of the objective function and gradient is proposed. The BMS-based algorithms are shown to be both accurate and efficient, which is verified through extensive simulations.

The rest of this paper is organized as follows. In Section 2, mmWave massive MIMO systems with 1 bit ADCs are described. In Section 3, the MAP channel estimation framework is formulated. In Section 4, the BMS hard thresholding technique is proposed, which is applied to the GraSP and GraHTP algorithms. In addition, an FFT-based fast implementation is proposed. In Section 5, the results and discussion are presented, and the conclusions are followed in Section 6.

Notation: a, $\mathbf{a}$, and $\mathbf{A}$ denote a scalar, vector, and matrix, respectively. $\|\mathbf{a}\|_0$, $\|\mathbf{a}\|_1$, and $\|\mathbf{a}\|$ represent the L^0-, L^1-, and L^2-norm of $\mathbf{a}$, respectively. $\|\mathbf{A}\|_F$ is the Frobenius norm of $\mathbf{A}$. The transpose, conjugate transpose, and conjugate of $\mathbf{A}$ are denoted as $\mathbf{A}^T$, $\mathbf{A}^H$, and $\overline{\mathbf{A}}$, respectively. The element-wise vector multiplication and division of $\mathbf{a}$ and $\mathbf{b}$ are denoted as $\mathbf{a} \odot \mathbf{b}$ and $\mathbf{a} \oslash \mathbf{b}$, respectively. The sum of all of the elements of $\mathbf{a}$ is denoted as sum($\mathbf{a}$). The vectorization of $\mathbf{A}$ is denoted as vec($\mathbf{A}$), which is formed by stacking all of the columns of $\mathbf{A}$. The unvectorization of $\mathbf{a}$ is denoted as unvec($\mathbf{a}$), which is the inverse of vec($\mathbf{A}$). The Kronecker product of $\mathbf{A}$ and $\mathbf{B}$ is denoted as $\mathbf{A} \otimes \mathbf{B}$. The support of $\mathbf{a}$ is denoted as supp($\mathbf{a}$), which is the set of indices formed by collecting all of the indices of the non-zero elements of $\mathbf{a}$. The best s-term approximation of $\mathbf{a}$ is denoted as $\mathbf{a}|_s$, which is formed by leaving only the s largest (in absolute value) elements of $\mathbf{a}$ and replacing the other elements with 0. Similarly, the vector obtained by leaving only the elements of $\mathbf{a}$ indexed by the set $\mathcal{A}$ and replacing the other elements with 0 is denoted as $\mathbf{a}|_{\mathcal{A}}$. The absolute value of a scalar a and cardinality of a set $\mathcal{A}$ are denoted as $|a|$ and $|\mathcal{A}|$, respectively. The set difference between the sets $\mathcal{A}$ and $\mathcal{B}$, namely, $\mathcal{A}\cap\mathcal{B}^{c}$, is denoted as $\mathcal{A}\setminus\mathcal{B}$. $\phi(\mathbf{a})$ and $\Phi(\mathbf{a})$ are element-wise standard normal PDF and CDF functions of $\mathbf{a}$, whose ith elements are $\frac{1}{\sqrt{2\pi}}e^{-\frac{a_i^2}{2}}$ and $\int_{-\infty}^{a_i}\frac{1}{\sqrt{2\pi}}e^{-\frac{x^2}{2}}dx$, respectively. The $m\times 1$ zero vector and $m\times m$ identity matrix are denoted as $\mathbf{0}_m$ and $\mathbf{I}_m$, respectively.

2 mmWave massive MIMO systems with 1 bit ADCs

2.1 System model

As shown in Fig. 1, consider a mmWave massive MIMO system with uniform linear arrays (ULAs) at the transmitter and receiver. The N-antenna transmitter transmits a training sequence of length T to the M-antenna receiver. Therefore, the received signal $\mathbf{Y}=[\mathbf{y}[1]\ \ \mathbf{y}[2]\ \ \cdots\ \ \mathbf{y}[T]]\in\mathbb{C}^{M\times T}$ is

$$\mathbf{Y}=\sqrt{\rho}\mathbf{H}\mathbf{S}+\mathbf{N}, \tag{1}$$

which is the collection of the t-th received signal $\mathbf{y}[t]\in\mathbb{C}^{M}$ over $t\in\{1,\ldots,T\}$. In the mmWave band, the channel $\mathbf{H}\in\mathbb{C}^{M\times N}$ consists of a small number of paths, whose parameters are the path gains, angle-of-arrivals (AoAs), and angle-of-departures (AoDs) [21]. Therefore, $\mathbf{H}$ is

$$\mathbf{H}=\sum_{\ell=1}^{L}\alpha_\ell\mathbf{a}_{\text{RX}}(\theta_{\text{RX},\ell})\mathbf{a}_{\text{TX}}(\theta_{\text{TX},\ell})^{\text{H}} \tag{2}$$

where L is the number of paths, $\alpha_\ell\in\mathbb{C}$ is the path gain of the ℓ-th path, and $\theta_{\text{RX},\ell}\in[-\pi/2,\pi/2]$ and $\theta_{\text{TX},\ell}\in[-\pi/2,\pi/2]$ are the AoA and AoD of the ℓth path, respectively. The steering vectors $\mathbf{a}_{\text{RX}}(\theta_{\text{RX},\ell})\in\mathbb{C}^{M}$ and $\mathbf{a}_{\text{TX}}(\theta_{\text{TX},\ell})\in\mathbb{C}^{N}$ are

$$\mathbf{a}_{\text{RX}}(\theta_{\text{RX},\ell})=\frac{1}{\sqrt{M}}\left[1\ \cdots\ e^{-j\pi(M-1)\sin(\theta_{\text{RX},\ell})}\right]^{\text{T}}, \tag{3}$$

$$\mathbf{a}_{\text{TX}}(\theta_{\text{TX},\ell})=\frac{1}{\sqrt{N}}\left[1\ \cdots\ e^{-j\pi(N-1)\sin(\theta_{\text{TX},\ell})}\right]^{\text{T}} \tag{4}$$

where the inter-element spacing is half-wavelength. The training sequence $\mathbf{S}=[\mathbf{s}[1]\ \ \mathbf{s}[2]\ \ \cdots\ \ \mathbf{s}[T]]\in\mathbb{C}^{N\times T}$ is the collection of the tth training sequence $\mathbf{s}[t]\in\mathbb{C}^{N}$ over $t\in\{1,\ldots,T\}$, whose power constraint is $\|\mathbf{s}[t]\|^2=N$. The additive white Gaussian noise (AWGN) $\mathbf{N}=[\mathbf{n}[1]\ \ \mathbf{n}[2]\ \ \cdots\ \ \mathbf{n}[T]]\in\mathbb{C}^{M\times T}$ is the collection of the tth AWGN $\mathbf{n}[t]\in\mathbb{C}^{M}$ over $t\in\{1,\ldots,T\}$, which is distributed as $\text{vec}(\mathbf{N})\sim\mathcal{CN}(\mathbf{0}_{MT},\mathbf{I}_{MT})$. The signal-to-noise ratio (SNR) is defined as ρ.

At the receiver, the real and imaginary parts of the received signal are quantized by 1 bit ADCs. The quantized received signal $\hat{\mathbf{Y}}$ is

$$\begin{aligned}\hat{\mathbf{Y}}&=\text{Q}(\mathbf{Y})\\&=\text{Q}(\sqrt{\rho}\mathbf{H}\mathbf{S}+\mathbf{N})\end{aligned} \tag{5}$$

where $\text{Q}(\cdot)$ is the 1 bit quantization function, whose threshold is 0. Therefore, $\text{Q}(\mathbf{Y})$ is

$$\text{Q}(\mathbf{Y})=\text{sign}(\text{Re}(\mathbf{Y}))+j\text{sign}(\text{Im}(\mathbf{Y})) \tag{6}$$

where $\text{sign}(\cdot)$ is the element-wise sign function. The goal is to estimate $\mathbf{H}$ by estimating $\{\alpha_\ell\}_{\ell=1}^{L}$, $\{\theta_{\text{RX},\ell}\}_{\ell=1}^{L}$, and $\{\theta_{\text{TX},\ell}\}_{\ell=1}^{L}$ from $\hat{\mathbf{Y}}$.

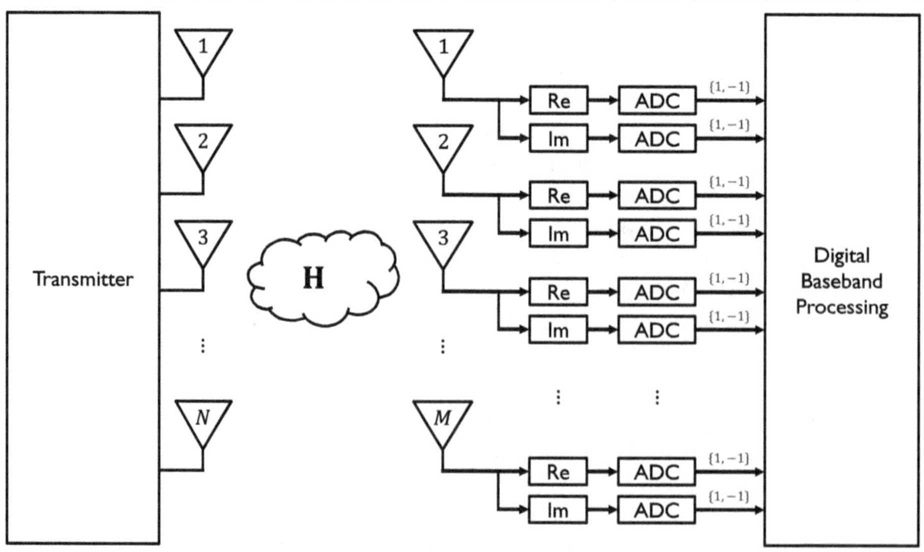

Fig. 1 A mmWave massive MIMO system with an N-antenna transmitter and M-antenna receiver. The real and imaginary parts of the received signal are quantized by 1 bit ADCs

2.2 Virtual channel representation

In the mmWave channel model in (2), $\{\theta_{\text{RX},\ell}\}_{\ell=1}^{L}$ and $\{\theta_{\text{TX},\ell}\}_{\ell=1}^{L}$ are hidden in $\{\mathbf{a}_{\text{RX}}(\theta_{\text{RX},\ell})\}_{\ell=1}^{L}$ and $\{\mathbf{a}_{\text{TX}}(\theta_{\text{TX},\ell})\}_{\ell=1}^{L}$, respectively. The non-linear mapping of $\{\theta_{\text{RX},\ell}\}_{\ell=1}^{L}$ and $\{\theta_{\text{TX},\ell}\}_{\ell=1}^{L}$ to $\mathbf{Y}$ renders a non-linear channel estimation problem. To convert the non-linear channel estimation problem to a linear channel estimation problem, we adopt the virtual channel representation [22].

The virtual channel representation of $\mathbf{H}$ is

$$\mathbf{H} \approx \mathbf{A}_{\text{RX}}\mathbf{X}^*\mathbf{A}_{\text{TX}}^{\text{H}} \tag{7}$$

where the dictionary pair $\mathbf{A}_{\text{RX}} \in \mathbb{C}^{M\times B_{\text{RX}}}$ and $\mathbf{A}_{\text{TX}} \in \mathbb{C}^{N\times B_{\text{TX}}}$ is the collection of $B_{\text{RX}} \geq M$ steering vectors and $B_{\text{TX}} \geq N$ steering vectors, respectively. Therefore, $\mathbf{A}_{\text{RX}}$ and $\mathbf{A}_{\text{TX}}$ are

$$\mathbf{A}_{\text{RX}} = \begin{bmatrix} \mathbf{a}_{\text{RX}}(\omega_{\text{RX},1}) & \cdots & \mathbf{a}_{\text{RX}}(\omega_{\text{RX},B_{\text{RX}}}) \end{bmatrix}, \tag{8}$$

$$\mathbf{A}_{\text{TX}} = \begin{bmatrix} \mathbf{a}_{\text{TX}}(\omega_{\text{TX},1}) & \cdots & \mathbf{a}_{\text{TX}}(\omega_{\text{TX},B_{\text{TX}}}) \end{bmatrix}, \tag{9}$$

whose gridding AoAs $\{\omega_{\text{RX},i}\}_{i=1}^{B_{\text{RX}}}$ and AoDs $\{\omega_{\text{TX},j}\}_{j=1}^{B_{\text{TX}}}$ are selected so as to form overcomplete DFT matrices. The gridding AoAs and AoDs are the B_{RX} and B_{TX} points from $[-\pi/2, \pi/2]$, respectively, to discretize the AoAs and AoDs because the ground truth AoAs and AoDs are unknown. To make a dictionary pair of the overcomplete DFT matrix form, the gridding AoAs and AoDs are given as $\omega_{\text{RX},i} = \sin^{-1}(-1 + 2/B_{\text{RX}} \cdot (i-1))$ and $\omega_{\text{RX},j} = \sin^{-1}(-1 + 2/B_{\text{TX}} \cdot (j-1))$, respectively. We prefer overcomplete DFT matrices because they are relatively well-conditioned, and DFT matrices are friendly to the FFT-based implementation, which will be discussed in Section 4. The virtual channel $\mathbf{X}^* \in \mathbb{C}^{B_{\text{RX}}\times B_{\text{TX}}}$ is the collection of $\{\alpha_\ell\}_{\ell=1}^{L}$, whose (i,j)th element is α_ℓ whenever $(\omega_{\text{RX},i}, \omega_{\text{TX},j})$ is the nearest to $(\theta_{\text{RX},\ell}, \theta_{\text{TX},\ell})$ but 0 otherwise. In general, the error between $\mathbf{H}$ and $\mathbf{A}_{\text{RX}}\mathbf{X}^*\mathbf{A}_{\text{TX}}^{\text{H}}$ is inversely proportional to B_{RX} and B_{TX}. To approximate $\mathbf{H}$ using (7) with negligible error, the dictionary pair must be dense, namely, $B_{\text{RX}} \gg M$ and $B_{\text{TX}} \gg N$.

Before we proceed, we provide a supplementary explanation on the approximation in (7). In this work, we attempt to estimate the L-sparse $\mathbf{X}^*$ in (7) because the sparse assumption on $\mathbf{X}^*$ is favorable when the goal is to formulate the channel estimation problem as a sparsity-constrained problem. The cost of assuming that $\mathbf{X}^*$ is L-sparse is, as evident, the approximation error shown in (7). Alternatively, the approximation error can be perfectly removed by considering $\mathbf{X}^*$ satisfying $\mathbf{H} = \mathbf{A}_{\text{RX}}\mathbf{X}^*\mathbf{A}_{\text{TX}}^{\text{H}}$, i.e., equality instead of approximation. One well-known $\mathbf{X}^*$ satisfying the equality is the minimum Frobenius norm solution, i.e., $\mathbf{X}^* = \mathbf{A}_{\text{RX}}^{\text{H}}(\mathbf{A}_{\text{RX}}\mathbf{A}_{\text{RX}}^{\text{H}})^{-1}\mathbf{H}(\mathbf{A}_{\text{TX}}\mathbf{A}_{\text{TX}}^{\text{H}})^{-1}\mathbf{A}_{\text{TX}}$. Such $\mathbf{X}^*$, however, has no evident structure to exploit in channel estimation, which is the reason why we assume that $\mathbf{X}^*$ is L-sparse at the cost of the approximation error in (7).

In practice, the arrays at the transmitter and receiver are typically large to compensate the path loss in the mmWave band, whereas the number of line-of-sight (LOS) and near LOS paths is small [23]. Therefore, $\mathbf{X}^*$ is sparse when the dictionary pair is dense because only L elements among $B_{\text{RX}}B_{\text{TX}}$ elements are non-zero where $L \ll MN \ll B_{\text{RX}}B_{\text{TX}}$. In the sequel, we use the shorthand notation $B = B_{\text{RX}}B_{\text{TX}}$.

To facilitate the channel estimation framework, we vectorize (1) and (5) in conjunction with (7). First, note that

$$\mathbf{Y} = \sqrt{\rho}\mathbf{A}_{\text{RX}}\mathbf{X}^*\mathbf{A}_{\text{TX}}^{\text{H}}\mathbf{S} + \mathbf{N} + \mathbf{E} \tag{10}$$

where the gridding error $\mathbf{E} \in \mathbb{C}^{M\times T}$ represents the mismatch in (7).[1] Then, the vectorized received signal $\mathbf{y} = \text{vec}(\mathbf{Y}) \in \mathbb{C}^{MT}$ is

$$\mathbf{y} = \sqrt{\rho}\mathbf{A}\mathbf{x}^* + \mathbf{n} + \mathbf{e} \tag{11}$$

where

$$\begin{aligned} \mathbf{A} &= \mathbf{S}^{\text{T}}\overline{\mathbf{A}}_{\text{TX}} \otimes \mathbf{A}_{\text{RX}} \\ &= \begin{bmatrix} \mathbf{a}_1 & \mathbf{a}_2 & \cdots & \mathbf{a}_B \end{bmatrix}, \end{aligned} \tag{12}$$

$$\begin{aligned} \mathbf{x}^* &= \text{vec}(\mathbf{X}^*) \\ &= \begin{bmatrix} x_1^* & x_2^* & \cdots & x_B^* \end{bmatrix}^{\text{T}}, \end{aligned} \tag{13}$$

$$\mathbf{n} = \text{vec}(\mathbf{N}), \tag{14}$$

$$\mathbf{e} = \text{vec}(\mathbf{E}). \tag{15}$$

The vectorized quantized received signal $\hat{\mathbf{y}} = \text{vec}(\hat{\mathbf{Y}}) \in \mathbb{C}^{MT}$ is

$$\begin{aligned} \hat{\mathbf{y}} &= \text{Q}(\mathbf{y}) \\ &= \text{Q}(\sqrt{\rho}\mathbf{A}\mathbf{x}^* + \mathbf{n} + \mathbf{e}). \end{aligned} \tag{16}$$

The goal is to estimate L-sparse $\mathbf{x}^*$ from $\hat{\mathbf{y}}$.

3 Problem formulation

In this section, we formulate the channel estimation problem using the MAP criterion. To facilitate the MAP channel estimation framework, the real counterparts of the complex forms in (16) are introduced. Then, the likelihood function of $\mathbf{x}^*$ is derived.

The real counterparts are the collections of the real and imaginary parts of the complex forms. Therefore, the real counterparts $\hat{\mathbf{y}}_{\text{R}} \in \mathbb{R}^{2MT}$, $\mathbf{A}_{\text{R}} \in \mathbb{R}^{2MT\times 2B}$, and $\mathbf{x}_{\text{R}}^* \in \mathbb{R}^{2B}$ are

[1] In practice, $\mathbf{X}^*$ may be either approximately sparse or exactly sparse to formulate (10). If $\mathbf{X}^*$ is approximately sparse, the leakage effect is taken into account so the mismatch in (7) becomes 0, namely, $\text{vec}(\mathbf{E}) = \mathbf{0}_{MT}$. In contrast, the mismatch in (7) must be taken into account with a non-zero $\mathbf{E}$ when $\mathbf{X}^*$ is exactly sparse. Fortunately, $\mathbf{E}$ is inversely proportional to B_{RX} and B_{TX}. Therefore, we adopt the latter definition of $\mathbf{X}^*$ and propose our algorithm ignoring $\mathbf{E}$ assuming that $B_{\text{RX}} \gg M$ and $B_{\text{TX}} \gg N$. The performance degradation from $\mathbf{E}$ will become less as B_{RX} and B_{TX} become sufficiently large.

$$\begin{aligned}\hat{\mathbf{y}}_{\mathrm{R}} &= \left[\,\mathrm{Re}(\hat{\mathbf{y}})^{\mathrm{T}}\ \mathrm{Im}(\hat{\mathbf{y}})^{\mathrm{T}}\,\right]^{\mathrm{T}}\\ &= \left[\,\hat{y}_{\mathrm{R},1}\ \hat{y}_{\mathrm{R},2}\ \cdots\ \hat{y}_{\mathrm{R},2MT}\,\right]^{\mathrm{T}}, \end{aligned} \tag{17}$$

$$\begin{aligned}\mathbf{A}_{\mathrm{R}} &= \begin{bmatrix}\mathrm{Re}(\mathbf{A}) & -\mathrm{Im}(\mathbf{A})\\ \mathrm{Im}(\mathbf{A}) & \mathrm{Re}(\mathbf{A})\end{bmatrix}\\ &= \left[\,\mathbf{a}_{\mathrm{R},1}\ \mathbf{a}_{\mathrm{R},2}\ \cdots\ \mathbf{a}_{\mathrm{R},2MT}\,\right]^{\mathrm{T}}, \end{aligned} \tag{18}$$

$$\begin{aligned}\mathbf{x}_{\mathrm{R}}^{*} &= \left[\,\mathrm{Re}(\mathbf{x}^{*})^{\mathrm{T}}\ \mathrm{Im}(\mathbf{x}^{*})^{\mathrm{T}}\,\right]^{\mathrm{T}}\\ &= \left[\,x_{\mathrm{R},1}^{*}\ x_{\mathrm{R},2}^{*}\ \cdots\ x_{\mathrm{R},2B}^{*}\,\right]^{\mathrm{T}}, \end{aligned} \tag{19}$$

which are the collections of the real and imaginary parts of $\hat{\mathbf{y}}$, $\mathbf{A}$, and $\mathbf{x}^*$, respectively. In the sequel, we use the complex forms and the real counterparts interchangeably. For example, $\mathbf{x}^*$ and $\mathbf{x}_{\mathrm{R}}^*$ refer to the same entity.

Before we formulate the likelihood function of $\mathbf{x}^*$, note that $\mathbf{e}$ is hard to analyze. However, $\mathbf{e}$ is negligible when the dictionary pair is dense. Therefore, we formulate the likelihood function of $\mathbf{x}^*$ without $\mathbf{e}$. The price of such over-simplification is negligible when $B_{\mathrm{RX}} \gg M$ and $B_{\mathrm{TX}} \gg N$, which is to be shown in Section 5 where $\mathbf{e} \neq \mathbf{0}_{MT}$. To derive the likelihood function of $\mathbf{x}^*$, note that

$$\sqrt{\rho}\mathbf{A}\mathbf{x}^* + \mathbf{n} \sim \mathcal{CN}(\sqrt{\rho}\mathbf{A}\mathbf{x}^*, \mathbf{I}_{MT}) \tag{20}$$

given $\mathbf{x}^*$. Then, from (20) in conjunction with (16), the log-likelihood function $f(\mathbf{x})$ is [10]

$$\begin{aligned}f(\mathbf{x}) &= \log \Pr\left[\,\hat{\mathbf{y}} = \mathrm{Q}(\sqrt{\rho}\mathbf{A}\mathbf{x} + \mathbf{n}) \mid \mathbf{x}\,\right]\\ &= \sum_{i=1}^{2MT} \log \Phi\left(\sqrt{2\rho}\hat{y}_{\mathrm{R},i}\mathbf{a}_{\mathrm{R},i}^{\mathrm{T}}\mathbf{x}_{\mathrm{R}}\right). \end{aligned} \tag{21}$$

If the distribution of $\mathbf{x}^*$ is known, the MAP estimate of $\mathbf{x}^*$ is

$$\underset{\mathbf{x}\in\mathbb{C}^B}{\operatorname{argmax}}\ (f(\mathbf{x}) + g_{\mathrm{MAP}}(\mathbf{x})) \tag{22}$$

where $g_{\mathrm{MAP}}(\mathbf{x})$ is the logarithm of the PDF of $\mathbf{x}^*$. In practice, however, $g_{\mathrm{MAP}}(\mathbf{x})$ is unknown. Therefore, we formulate the MAP channel estimation framework based on $\{\alpha_\ell\}_{\ell=1}^L$, $\{\theta_{\mathrm{RX},\ell}\}_{\ell=1}^L$, and $\{\theta_{\mathrm{TX},\ell}\}_{\ell=1}^L$ where we assume the followings:

1. $\alpha_\ell \sim \mathcal{CN}(0,1)$ for all ℓ
2. $\theta_{\mathrm{RX},\ell} \sim \mathrm{unif}([-\pi/2, \pi/2])$ for all ℓ
3. $\theta_{\mathrm{TX},\ell} \sim \mathrm{unif}([-\pi/2, \pi/2])$ for all ℓ
4. $\{\alpha_\ell\}_{\ell=1}^L$, $\{\theta_{\mathrm{RX},\ell}\}_{\ell=1}^L$, and $\{\theta_{\mathrm{TX},\ell}\}_{\ell=1}^L$ are independent.

Then, the MAP estimate of $\mathbf{x}^*$ considering the channel sparsity is

$$\underset{\mathbf{x}\in\mathbb{C}^B}{\operatorname{argmax}}\ (f(\mathbf{x}) + g(\mathbf{x}))\ \text{ s.t. } \|\mathbf{x}\|_0 \leq L \tag{23}$$

where $g(\mathbf{x}) = -\|\mathbf{x}_{\mathrm{R}}\|^2$ is the logarithm of the PDF of $\mathcal{CN}(\mathbf{0}_B, \mathbf{I}_B)$ ignoring the constant factor. However, note that only the optimization problems (22) and (23) are equivalent in the sense that their solutions are the same, not $g_{\mathrm{MAP}}(\mathbf{x})$ and $g(\mathbf{x})$. In the ML channel estimation framework, the ML estimate of $\mathbf{x}^*$ is

$$\underset{\mathbf{x}\in\mathbb{C}^B}{\operatorname{argmax}} f(\mathbf{x})\ \text{ s.t. } \|\mathbf{x}\|_0 \leq L. \tag{24}$$

In the sequel, we focus on solving (23) because (23) reduces to (24) when $g(\mathbf{x}) = 0$. In addition, we denote the objective function and the gradient in (23) as $h(\mathbf{x})$ and $\nabla h(\mathbf{x})$, respectively. Therefore,

$$h(\mathbf{x}) = f(\mathbf{x}) + g(\mathbf{x}), \tag{25}$$

$$\begin{aligned}\nabla h(\mathbf{x}) &= \nabla f(\mathbf{x}) + \nabla g(\mathbf{x})\\ &= \left[\,\nabla h(x_1)\ \nabla h(x_2)\ \cdots\ \nabla h(x_B)\,\right]^{\mathrm{T}} \end{aligned} \tag{26}$$

where the differentiation is with respect to $\mathbf{x}$.

4 Channel estimation via gradient pursuit

In this section, we propose the BMSGraSP and BMSGraHTP algorithms to solve (23), which are the variants of the GraSP [17] and GraHTP [18] algorithms, respectively. Then, an FFT-based fast implementation is proposed. In addition, we investigate the limit of the BMSGraSP and BMSGraHTP algorithms in the high SNR regime in 1 bit ADCs.

4.1 Proposed BMSGraSP and BMSGraHTP algorithms

Note that $h(\mathbf{x})$ in (23) is concave because $f(\mathbf{x})$ and $g(\mathbf{x})$ are the sums of the logarithms of $\Phi(\cdot)$ and $\phi(\cdot)$, respectively, which are log-concave [24]. However, (23) is not a convex optimization problem because the sparsity constraint is not convex. Furthermore, solving (23) is NP-hard because of its combinatorial complexity. To approximately optimize convex objective functions with sparsity constraints iteratively by pursuing the gradient of the objective function, the GraSP and GraHTP algorithms were proposed in [17] and [18], respectively.

To solve (23), the GraSP and GraHTP algorithms roughly proceed as follows at each iteration when $\mathbf{x}$ is the current estimate of $\mathbf{x}^*$ where the iteration index is omitted for simplicity. First, the best L-term approximation of $\nabla h(\mathbf{x})$ is computed, which is

$$T_L(\nabla h(\mathbf{x})) = \nabla h(\mathbf{x})|_L \tag{27}$$

where $T_L(\cdot)$ is the L-term hard thresholding function. Here, $T_L(\cdot)$ leaves only the L largest elements (in absolute value) of $\nabla h(\mathbf{x})$, and sets all the other remaining elements to 0. Then, after the estimate of $\mathrm{supp}(\mathbf{x}^*)$ is updated by selecting

$$\mathcal{I} = \mathrm{supp}(T_L(\nabla h(\mathbf{x}))), \tag{28}$$

i.e., $\mathcal{I}$ is the set of indices formed by collecting the L indices of $\nabla h(\mathbf{x})$ corresponding to its L largest elements (in absolute value), the estimate of $\mathbf{x}^*$ is updated by solving the following optimization problem

$$\underset{\mathbf{x}\in\mathbb{C}^B}{\operatorname{argmax}}\, h(\mathbf{x}) \text{ s.t. } \operatorname{supp}(\mathbf{x}) \subseteq \mathcal{I}, \tag{29}$$

which can be solved using convex optimization because the support constraint is convex [24]. The GraSP and GraHTP algorithms are the generalizations of the CoSaMP [19] and HTP [20] algorithms, respectively. This follows because the gradient of the squared error is the scaled proxy of the residual.

To solve (23) using the GraSP and GraHTP algorithms, $h(\mathbf{x})$ is required either to have a stable restricted Hessian [17] or to be strongly convex and smooth [18]. These conditions are the generalizations of the restricted isometry property (RIP) in CS [25], which means that $h(\mathbf{x})$ is likely to satisfy these conditions when $\mathbf{A}$ is either a restricted isometry, well-conditioned, or incoherent. In practice, however, $\mathbf{A}$ is highly coherent because the dictionary pair is typically dense to reduce the mismatch in (7).

To illustrate how the GraSP and GraHTP algorithms fail to solve (23) when $\mathbf{A}$ is highly coherent, consider the real counterpart of $\nabla h(\mathbf{x})$. The real counterpart $\nabla h(\mathbf{x}_\mathrm{R}) \in \mathbb{R}^{2B}$ is

$$\begin{aligned}\nabla h(\mathbf{x}_\mathrm{R}) &= \left[\, \operatorname{Re}(\nabla h(\mathbf{x}))^\mathrm{T} \;\; \operatorname{Im}(\nabla h(\mathbf{x}))^\mathrm{T} \,\right]^\mathrm{T} \\ &= \sum_{i=1}^{2MT} \lambda\left(\sqrt{2\rho}\hat{y}_{\mathrm{R},i}\mathbf{a}_{\mathrm{R},i}^\mathrm{T}\mathbf{x}_\mathrm{R}\right)\sqrt{2\rho}\hat{y}_{\mathrm{R},i}\mathbf{a}_{\mathrm{R},i} - 2\mathbf{x}_\mathrm{R},\end{aligned} \tag{30}$$

which follows from $\nabla \log \Phi(\mathbf{a}_\mathrm{R}^\mathrm{T}\mathbf{x}_\mathrm{R}) = \lambda(\mathbf{a}_\mathrm{R}^\mathrm{T}\mathbf{x}_\mathrm{R})\mathbf{a}_\mathrm{R}$ and $\nabla\|\mathbf{x}_\mathrm{R}\|^2 = 2\mathbf{x}_\mathrm{R}$ where $\lambda(\cdot) = \phi(\cdot) \oslash \Phi(\cdot)$ is the inverse Mills ratio function[2]. Then, the following observation holds from directly computing $\nabla h(x_i)$, whose real and imaginary parts are the i-th and $(i+B)$-th elements of $\nabla h(\mathbf{x}_\mathrm{R})$, respectively.

Observation 1 $\nabla h(x_i) = \nabla h(x_j)$ *if* $\mathbf{a}_i = \mathbf{a}_j$ *and* $x_i = x_j$.

However, Observation 1 is meaningless because $\mathbf{a}_i \neq \mathbf{a}_j$ unless $i = j$. To establish a meaningful observation, consider the coherence between $\mathbf{a}_i$ and $\mathbf{a}_j$, which reflects the proximity between $\mathbf{a}_i$ and $\mathbf{a}_j$ according to [26, 27]

$$\mu(i,j) = \frac{|\mathbf{a}_i^\mathrm{H}\mathbf{a}_j|}{\|\mathbf{a}_i\|\,\|\mathbf{a}_j\|}. \tag{31}$$

Then, using the η-coherence band, which is [26]

$$B_\eta(i) = \{j \mid \mu(i,j) \geq \eta\} \tag{32}$$

where $\eta \in (0,1)$, we establish the following conjecture when η is sufficiently large.

Conjecture 1 $\nabla h(x_i) \approx \nabla h(x_j)$ *if* $j \in B_\eta(i)$ *and* $x_i = x_j$.

At this point, we use Conjecture 1 to illustrate how the GraSP and GraHTP algorithms fail to estimate $\operatorname{supp}(\mathbf{x}^*)$ from (28) by naive hard thresholding when $\mathbf{A}$ is highly coherent. To proceed, consider the following example, which assumes that $\mathbf{x}^*$ and $\hat{\mathbf{y}}$ are realized with $\mathbf{x}$ representing the current estimate of $\mathbf{x}^*$ so as to satisfy

1) $i = \underset{k\in\{1,2,\dots,B\}}{\operatorname{argmax}} |x_k^*|$
2) $i = \underset{k\in\{1,2,\dots,B\}}{\operatorname{argmax}} |\nabla h(x_k)|$
3) $\mathcal{J} \cap \operatorname{supp}(\mathbf{x}^*) = \emptyset$

where i is the index corresponding to the largest element of the ground truth[3] virtual channel $\mathbf{x}^*$, and

$$\mathcal{J} = \{j \mid j \in B_\eta(i), x_i = x_j\} \setminus \{i\} \tag{33}$$

is the by-product of i. Here, $\mathcal{J}$ is called the by-product of i because

$$\begin{aligned}|\nabla h(x_j)| &\approx |\nabla h(x_i)| \\ &= \max_{k\in\{1,2,\dots,B\}} |\nabla h(x_k)|,\end{aligned} \tag{34}$$

which follows from Conjecture 1, holds despite $x_j^* = 0$ for all $j \in \mathcal{J}$. In other words, the by-product of i refers to the fact that $\nabla h(x_i)$ and $\nabla h(x_j)$ are indistinguishable for all $j \in \mathcal{J}$ according to (34), but the elements of $\mathbf{x}^*$ indexed by $\mathcal{J}$ are 0 according to 3). Therefore, when we attempt to estimate $\operatorname{supp}(\mathbf{x}^*)$ by hard thresholding $\nabla h(\mathbf{x})$, the indices in $\mathcal{J}$ will likely be erroneously selected as the estimate of $\operatorname{supp}(\mathbf{x}^*)$ because $\nabla h(x_j)$ and the maximum element of $\nabla h(\mathbf{x})$, which is $\nabla h(x_i)$ according to 2), are indistinguishable for all $j \in \mathcal{J}$.

To illustrate how (28) cannot estimate $\operatorname{supp}(\mathbf{x}^*)$ when $\mathbf{A}$ is highly coherent, consider another example where $\nabla h(\mathbf{x})$ and $T_L(\nabla h(\mathbf{x}))$ are shown in Figs. 2 and 3, respectively. In this example, $\operatorname{supp}(\mathbf{x}^*)$ is widely spread, whereas most of $\operatorname{supp}(T_L(\nabla h(\mathbf{x})))$ are in the coherence band of the index of the maximum element of $\nabla h(\mathbf{x})$. This shows that hard thresholding $\nabla h(\mathbf{x})$ is not sufficient to distinguish whether an index is the ground truth index or the by-product of another index. To solve this problem, we propose the BMS hard thresholding technique.

The BMS hard thresholding function $T_{\mathrm{BMS},L}(\cdot)$ is an L-term hard thresholding function, which is proposed based on Conjecture 1. The BMS hard thresholding technique is presented in Algorithm 1. Line 3 selects the index of the

[2]The element-wise vector division in the inverse Mills ratio function is meaningless because the arguments of the inverse Mills ratio function are scalars in (30). The reason we use the element-wise vector division in the inverse Mills ratio function will become clear in (37), whose arguments are vectors.

[3]We use the term "ground truth" to emphasize that the ground truth $\mathbf{x}^*$ is the true virtual channel which actually gives the quantized received signal $\hat{\mathbf{y}}$ from (16), whereas $\mathbf{x}$ merely represents the point where $\nabla h(\mathbf{x})$ is computed to estimate $\operatorname{supp}(\mathbf{x}^*)$ via hard thresholding.

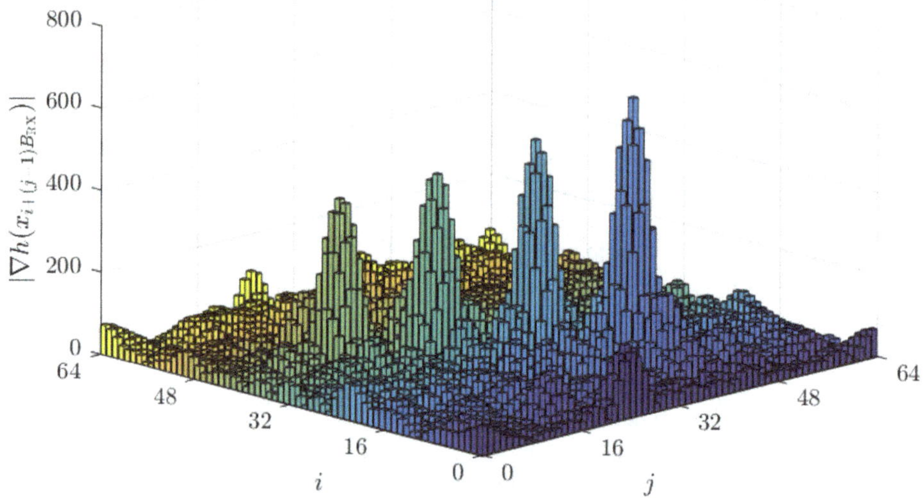

Fig. 2 The magnitude of unvec($\nabla h(\mathbf{x})$) $\in \mathbb{C}^{B_{\text{RX}} \times B_{\text{TX}}}$ at $\mathbf{x} = \mathbf{0}_B$, namely, before hard thresholding

maximum element of $\nabla h(\mathbf{x})$ from the unchecked index set as the current index. Line 4 constructs the by-product testing set. Line 5 checks whether the current index is greater than the by-product testing set. In this paper, Line 5 is referred to as the band maximum criterion. If the band maximum criterion is satisfied, the current index is selected as the estimate of supp($\mathbf{x}^*$) in Line 6. Otherwise, the current index is not selected as the estimate of supp($\mathbf{x}^*$) because the current index is likely to be the by-product of another index rather than the ground truth index. Line 8 updates the unchecked index set.

Note that Algorithm 1 is a hard thresholding technique applied to $\nabla h(\mathbf{x})$. If the BMS hard thresholding technique is applied to $\mathbf{x} + \kappa \nabla h(\mathbf{x})$ where κ is the step size, $\nabla h(\mathbf{x})$ is replaced by $\mathbf{x} + \kappa \nabla h(\mathbf{x})$ in the input, output, and Lines 3, 5, and 10 of Algorithm 1. This can be derived using the

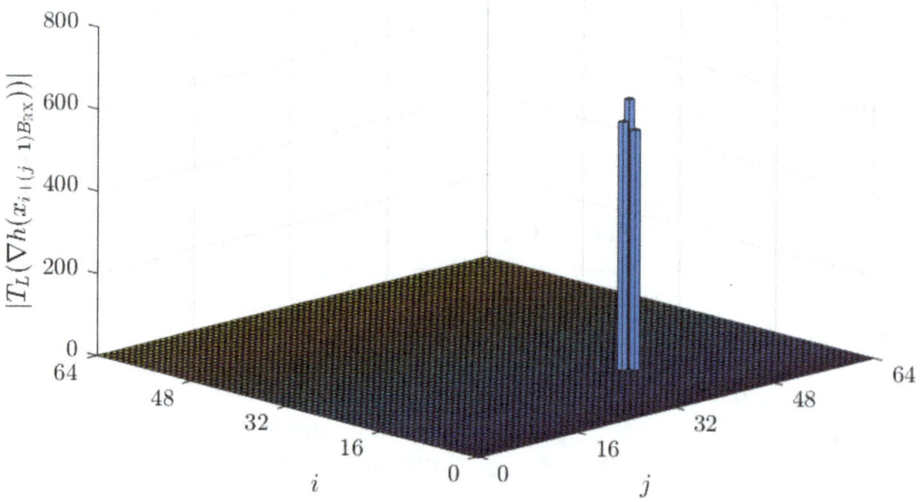

Fig. 3 The magnitude of unvec($T_L(\nabla h(\mathbf{x}))$) $\in \mathbb{C}^{B_{\text{RX}} \times B_{\text{TX}}}$ at $\mathbf{x} = \mathbf{0}_B$, namely, after hard thresholding. This shows how hard thresholding on $\nabla h(\mathbf{x})$ results in an incorrect estimate of supp($\mathbf{x}^*$) when $\mathbf{A}$ is highly coherent. In this example, $M = N = 16$, $B_{\text{RX}} = B_{\text{TX}} = 64$, $T = 20$, $L = 4$, SNR $= 20$ dB, and supp($\mathbf{x}^*$) is widely spread

Algorithm 1 BMS hard thresholding technique

Input: $\mathbf{x}$, $\nabla h(\mathbf{x}), L$
Output: $T_{\text{BMS},L}(\nabla h(\mathbf{x}))$

$\mathcal{S} = \emptyset, \mathcal{I} = \{1, 2, \ldots, B\}$
while $|\mathcal{S}| < L$ **do**
  $i = \underset{j \in \mathcal{I}}{\text{argmax}}\ |\nabla h(x_j)|$
  $\mathcal{J} = \{j \mid j \in B_\eta(i), x_i = x_j\} \setminus \{i\}$
  if $|\nabla h(x_i)| > \max_{j \in \mathcal{J}} |\nabla h(x_j)|$ **then**
    $\mathcal{S} = \mathcal{S} \cup \{i\}$
  end if
  $\mathcal{I} = \mathcal{I} \setminus \{i\}$
end while
$T_{\text{BMS},L}(\nabla h(\mathbf{x})) = \nabla h(\mathbf{x})|_{\mathcal{S}}$

same logic based on Conjecture 1. Now, we propose the BMSGraSP and BMSGraHTP algorithms to solve (23).

The BMSGraSP and BMSGraHTP algorithms are the variants of the GraSP and GraHTP algorithms, respectively. The difference between the BMS-based and non-BMS-based algorithms is that the hard thresholding function is $T_{\text{BMS},L}(\cdot)$ instead of $T_L(\cdot)$. The BMSGraSP and BMSGraHTP algorithms are presented in Algorithms 2 and 3, respectively. Lines 3, 4, and 5 of Algorithms 2 and 3 roughly proceed based on the same logic. Line 3 computes the gradient of the objective function. Line 4 selects $\mathcal{I}$ from the support of the hard thresholded gradient of the objective function. Line 5 maximizes the objective function subject to the support constraint. This can be solved using convex optimization because the objective function and support constraint are concave and convex, respectively. In addition, $\mathbf{b}$ is hard thresholded in Line 6 of Algorithm 2 because $\mathbf{b}$ is at most $3L$-sparse. A natural halting condition of Algorithms 2 and 3 is to halt when the current and previous $\text{supp}(\tilde{\mathbf{x}})$ are the same. The readers who are interested in a more in-depth analyses of the GraSP and GraHTP algorithms are referred to [17] and [18], respectively.

Algorithm 2 BMSGraSP algorithm

Input: $h(\cdot), L$
Output: $\tilde{\mathbf{x}}$

1: $\tilde{\mathbf{x}} = \mathbf{0}_B$
2: **while** halting condition **do**
3:   $\mathbf{z} = \nabla h(\tilde{\mathbf{x}})$
4:   $\mathcal{I} = \text{supp}(T_{\text{BMS},2L}(\mathbf{z})) \cup \text{supp}(\tilde{\mathbf{x}})$
5:   $\mathbf{b} = \underset{\mathbf{x} \in \mathbb{C}^B}{\text{argmax}}\ h(\mathbf{x})$ s.t. $\text{supp}(\mathbf{x}) \subseteq \mathcal{I}$
6:   $\tilde{\mathbf{x}} = T_L(\mathbf{b})$
7: **end while**

Algorithm 3 BMSGraHTP algorithm

Input: $h(\cdot), L$
Output: $\tilde{\mathbf{x}}$

1: $\tilde{\mathbf{x}} = \mathbf{0}_B$
2: **while** halting condition **do**
3:   $\mathbf{z} = \tilde{\mathbf{x}} + \kappa \nabla h(\tilde{\mathbf{x}})$
4:   $\mathcal{I} = \text{supp}(T_{\text{BMS},L}(\mathbf{z}))$
5:   $\tilde{\mathbf{x}} = \underset{\mathbf{x} \in \mathbb{C}^B}{\text{argmax}}\ h(\mathbf{x})$ s.t. $\text{supp}(\mathbf{x}) \subseteq \mathcal{I}$
6: **end while**

Remark 1 *Instead of hard thresholding* $\mathbf{b}$, *we can solve*

$$\tilde{\mathbf{x}} = \underset{\mathbf{x} \in \mathbb{C}^B}{\text{argmax}}\ h(\mathbf{x}) \ \text{s.t.}\ \text{supp}(\mathbf{x}) \subseteq \text{supp}(T_L(\mathbf{b})), \tag{35}$$

which is a convex optimization problem, to obtain $\tilde{\mathbf{x}}$ *in Line 6 of Algorithm 2. This is the debiasing variant of Algorithm 2* [17]. *The advantage of the debiasing variant of Algorithm 2 is a more accurate estimate of* $\mathbf{x}^*$. *However, the complexity is increased, which is incurred by solving* (35).

Remark 2 *In this paper, we assume that only* $h(\mathbf{x})$ *and* $\nabla h(\mathbf{x})$ *are required at each iteration to solve* (23) *using Algorithms 2 and 3, which can be accomplished when the first order method is used to solve convex optimization problems in Line 5 of Algorithms 2 and 3. An example of such first order method is the gradient descent method with the backtracking line search* [24].

4.2 Fast implementation via FFT

In practice, the complexity of Algorithms 2 and 3 is demanding because $h(\mathbf{x})$ and $\nabla h(\mathbf{x})$ are required at each iteration, which are high-dimensional functions defined on $\mathbb{C}^B$ where $B \gg MN$. In recent works on channel estimation and data detection in the mmWave band [14, 15, 28], the FFT-based implementation is widely used because $\mathbf{H}$ can be approximated by (7) using overcomplete DFT matrices. In this paper, an FFT-based fast implementation of $h(\mathbf{x})$ and $\nabla h(\mathbf{x})$ is proposed, which is motivated by [14, 15, 28].

To facilitate the analysis, we convert the summations in $h(\mathbf{x})$ and $\nabla h(\mathbf{x}_{\text{R}})$ to matrix-vector multiplications by algebraically manipulating (21) and (30). Then, we obtain

$$\begin{aligned} &h(\mathbf{x}) \\ &= \text{sum}(\log \Phi(\sqrt{2\rho}\hat{\mathbf{y}}_{\text{R}} \odot \mathbf{A}_{\text{R}}\mathbf{x}_{\text{R}})) - \|\mathbf{x}_{\text{R}}\|^2, \end{aligned} \tag{36}$$

$$\begin{aligned} &\nabla h(\mathbf{x}_{\text{R}}) \\ &= \mathbf{A}_{\text{R}}^{\text{T}}(\lambda(\sqrt{2\rho}\hat{\mathbf{y}}_{\text{R}} \odot \mathbf{A}_{\text{R}}\mathbf{x}_{\text{R}}) \odot \sqrt{2\rho}\hat{\mathbf{y}}_{\text{R}}) - 2\mathbf{x}_{\text{R}} \end{aligned} \tag{37}$$

where we see that the bottlenecks of $h(\mathbf{x})$ and $\nabla h(\mathbf{x})$ come from the matrix-vector multiplications involving $\mathbf{A}_{\text{R}}$ and $\mathbf{A}_{\text{R}}^{\text{T}}$ resulting from the large size of $\mathbf{A}$. For example, the size of $\mathbf{A}$ is 5120×65536 in Section 5 where $M = N = 64$, $B_{\text{RX}} = B_{\text{TX}} = 256$, and $T = 80$.

To develop an FFT-based fast implementation of the matrix-vector multiplications involving $\mathbf{A}_{\mathrm{R}}$ and $\mathbf{A}_{\mathrm{R}}^{\mathrm{T}}$, define $\mathbf{c}_{\mathrm{R}} \in \mathbb{R}^{2MT}$ as $\mathbf{c}_{\mathrm{R}} = \lambda(\sqrt{2\rho}\hat{\mathbf{y}}_{\mathrm{R}} \odot \mathbf{A}_{\mathrm{R}}\mathbf{x}_{\mathrm{R}}) \odot \sqrt{2\rho}\hat{\mathbf{y}}_{\mathrm{R}}$ from (37) with $\mathbf{c} \in \mathbb{C}^{MT}$ being the complex form of $\mathbf{c}_{\mathrm{R}}$. From the fact that

$$\mathbf{A}_{\mathrm{R}}\mathbf{x}_{\mathrm{R}} = \left[\, \mathrm{Re}(\mathbf{A}\mathbf{x})^{\mathrm{T}} \;\; \mathrm{Im}(\mathbf{A}\mathbf{x})^{\mathrm{T}} \,\right]^{\mathrm{T}}, \tag{38}$$

$$\mathbf{A}_{\mathrm{R}}^{\mathrm{T}}\mathbf{c}_{\mathrm{R}} = \left[\, \mathrm{Re}(\mathbf{A}^{\mathrm{H}}\mathbf{c})^{\mathrm{T}} \;\; \mathrm{Im}(\mathbf{A}^{\mathrm{H}}\mathbf{c})^{\mathrm{T}} \,\right]^{\mathrm{T}}, \tag{39}$$

we now attempt to compute $\mathbf{A}\mathbf{x}$ and $\mathbf{A}^{\mathrm{H}}\mathbf{c}$ via the FFT. Then, $\mathbf{A}\mathbf{x}$ and $\mathbf{A}^{\mathrm{H}}\mathbf{c}$ are unvectorized according to

$$\begin{aligned}\mathrm{unvec}(\mathbf{A}\mathbf{x}) &= \mathbf{A}_{\mathrm{RX}}\mathbf{X}\mathbf{A}_{\mathrm{TX}}^{\mathrm{H}}\mathbf{S} \\ &= \underbrace{\mathbf{A}_{\mathrm{RX}}(\underbrace{\mathbf{S}^{\mathrm{H}}(\underbrace{\mathbf{A}_{\mathrm{TX}}\mathbf{X}^{\mathrm{H}}}_{\text{FFT}})}_{\text{IFFT}})^{\mathrm{H}}}_{\text{FFT}},\end{aligned} \tag{40}$$

$$\begin{aligned}\mathrm{unvec}(\mathbf{A}^{\mathrm{H}}\mathbf{c}) &= \mathbf{A}_{\mathrm{RX}}^{\mathrm{H}}\mathbf{C}\mathbf{S}^{\mathrm{H}}\mathbf{A}_{\mathrm{TX}} \\ &= \underbrace{\mathbf{A}_{\mathrm{RX}}^{\mathrm{H}}(\underbrace{\mathbf{A}_{\mathrm{TX}}^{\mathrm{H}}(\underbrace{\mathbf{S}\mathbf{C}^{\mathrm{H}}}_{\text{FFT}})}_{\text{IFFT}})^{\mathrm{H}}}_{\text{IFFT}}\end{aligned} \tag{41}$$

where $\mathbf{X} = \mathrm{unvec}(\mathbf{x}) \in \mathbb{C}^{B_{\mathrm{RX}} \times B_{\mathrm{TX}}}$ and $\mathbf{C} = \mathrm{unvec}(\mathbf{c}) \in \mathbb{C}^{M \times T}$. If the matrix multiplication involving $\mathbf{S}$ can be implemented using the FFT, e.g., Zadoff-Chu (ZC) [29] or DFT [11] training sequence, (40) and (41) can be implemented using the FFT because $\mathbf{A}_{\mathrm{RX}}$ and $\mathbf{A}_{\mathrm{TX}}$ are overcomplete DFT matrices. For example, each column of $\mathbf{A}_{\mathrm{TX}}\mathbf{X}^{\mathrm{H}}$ in (40) can be computed using the B_{TX}-point FFT with pruned outputs, i.e., retaining only N outputs, because we constructed $\mathbf{A}_{\mathrm{TX}}$ as an overcomplete DFT matrix.

In particular, the matrix multiplications involving $\mathbf{A}_{\mathrm{TX}}$, $\mathbf{S}^{\mathrm{H}}$, and $\mathbf{A}_{\mathrm{RX}}$ in (40) can be implemented with B_{TX}-point FFT with pruned outputs repeated B_{RX} times, T-point IFFT with pruned inputs repeated B_{RX} times, and B_{RX}-point FFT with pruned outputs repeated T times, respectively.[4] Using the same logic, the matrix multiplications involving $\mathbf{S}$, $\mathbf{A}_{\mathrm{TX}}^{\mathrm{H}}$, and $\mathbf{A}_{\mathrm{RX}}^{\mathrm{H}}$ in (41) can be implemented using T-point FFT with pruned outputs repeated M times, B_{TX}-point IFFT with pruned inputs repeated M times, and B_{RX}-point IFFT with pruned inputs repeated B_{TX} times, respectively. Therefore, the complexity of the FFT-based implementation of (40) and (41) is $O(B_{\mathrm{RX}}B_{\mathrm{TX}}\log B_{\mathrm{TX}} + B_{\mathrm{RX}}T\log T + TB_{\mathrm{RX}}\log B_{\mathrm{RX}})$ and $O(MT\log T + MB_{\mathrm{TX}}\log B_{\mathrm{TX}} + B_{\mathrm{TX}}B_{\mathrm{RX}}\log B_{\mathrm{RX}})$, respectively.

To illustrate the efficiency of the FFT-based implementation of (40) and (41), M/N, M/B_{RX}, M/B_{TX}, and M/T are assumed to be fixed. Then, the complexity of the FFT-based implementation of $\mathbf{A}\mathbf{x}$ and $\mathbf{A}^{\mathrm{H}}\mathbf{c}$ is $O(M^2\log M)$, whereas the complexity of directly computing $\mathbf{A}\mathbf{x}$ and $\mathbf{A}^{\mathrm{H}}\mathbf{c}$ is $O(M^4)$. Therefore, the complexity of Algorithms 2 and 3 is reduced when $h(\mathbf{x})$ and $\nabla h(\mathbf{x})$ are implemented using the FFT operations.

[4]The inputs and outputs are pruned because $\mathbf{A}_{\mathrm{RX}}$, $\mathbf{A}_{\mathrm{TX}}$, and $\mathbf{S}$ are rectangular, not square. The details of the pruned FFT are presented in [30–32].

Remark 3 *Line 5 of Algorithms 2 and 3 is equivalent to solving*

$$\underset{\mathbf{x}_{\mathcal{I}} \in \mathbb{C}^{|\mathcal{I}|}}{\mathit{argmax}}\; h_{\mathcal{I}}(\mathbf{x}_{\mathcal{I}}) = \underset{\mathbf{x}_{\mathcal{I}} \in \mathbb{C}^{|\mathcal{I}|}}{\mathit{argmax}}\; (f_{\mathcal{I}}(\mathbf{x}_{\mathcal{I}}) + g_{\mathcal{I}}(\mathbf{x}_{\mathcal{I}})) \tag{42}$$

where

$$f_{\mathcal{I}}(\mathbf{x}_{\mathcal{I}}) = \log Pr\left[\, \hat{\mathbf{y}} = \mathrm{Q}(\sqrt{\rho}\mathbf{A}_{\mathcal{I}}\mathbf{x}_{\mathcal{I}} + \mathbf{n}) \mid \mathbf{x}_{\mathcal{I}} \,\right], \tag{43}$$

$$g_{\mathcal{I}}(\mathbf{x}_{\mathcal{I}}) = -\|\mathbf{x}_{\mathcal{I}}\|^2, \tag{44}$$

and $\mathbf{A}_{\mathcal{I}} \in \mathbb{C}^{MT \times |\mathcal{I}|}$ is the collection of $\mathbf{a}_i$ with $i \in \mathcal{I}$. Therefore, only $h_{\mathcal{I}}(\mathbf{x}_{\mathcal{I}})$ and $\nabla h_{\mathcal{I}}(\mathbf{x}_{\mathcal{I}})$ are required in Line 5 of Algorithms 2 and 3, which are low-dimensional functions defined on $\mathbb{C}^{|\mathcal{I}|}$ where $|\mathcal{I}| = O(L)$. If $h_{\mathcal{I}}(\mathbf{x}_{\mathcal{I}})$ and $\nabla h_{\mathcal{I}}(\mathbf{x}_{\mathcal{I}})$ are computed based on the same logic in (40) and (41) but $\mathbf{A}$ replaced by $\mathbf{A}_{\mathcal{I}}$, the complexity of Algorithms 2 and 3 is reduced further because the size of the FFT is reduced in Line 5.

5 Results and discussion

In this section, we evaluate the performance of Algorithms 2 and 3 from different aspects in terms of the accuracy, achievable rate, and complexity. Throughout this section, we consider a mmWave massive MIMO system with 1 bit ADCs, whose parameters are $M = N = 64$ and $T = 80$. The rest vary from simulation to simulation, which consist of B_{RX}, B_{TX}, and L. In addition, we consider $\mathbf{S}$, whose rows are the circular shifts of the ZC training sequence of length T as in [15, 33]. Furthermore, $\mathbf{H}$ is either random or deterministic. If $\mathbf{H}$ is random, $\alpha_\ell \sim \mathcal{CN}(0,1)$, $\theta_{\mathrm{RX},\ell} \sim \mathrm{unif}([-\pi/2, \pi/2])$, and $\theta_{\mathrm{TX},\ell} \sim \mathrm{unif}([-\pi/2, \pi/2])$ are independent. Otherwise, we consider different $\mathbf{H}$ from simulation to simulation.

The MSEs of $\{\alpha_\ell\}_{\ell=1}^L$, $\{\theta_{\mathrm{RX},\ell}\}_{\ell=1}^L$, and $\{\theta_{\mathrm{TX},\ell}\}_{\ell=1}^L$ are

$$\mathrm{MSE}(\{\alpha_\ell\}_{\ell=1}^L) = \mathbb{E}\left\{\frac{1}{L}\sum_{\ell=1}^{L}|\tilde{\alpha}_\ell - \alpha_\ell|^2\right\}, \tag{45}$$

$$\mathrm{MSE}(\{\theta_{\mathrm{RX},\ell}\}_{\ell=1}^L) = \mathbb{E}\left\{\frac{1}{L}\sum_{\ell=1}^{L}(\tilde{\theta}_{\mathrm{RX},\ell} - \theta_{\mathrm{RX},\ell})^2\right\}, \tag{46}$$

$$\mathrm{MSE}(\{\theta_{\mathrm{TX},\ell}\}_{\ell=1}^L) = \mathbb{E}\left\{\frac{1}{L}\sum_{\ell=1}^{L}(\tilde{\theta}_{\mathrm{TX},\ell} - \theta_{\mathrm{TX},\ell})^2\right\} \tag{47}$$

where $(\tilde{\alpha}_\ell, \tilde{\theta}_{\mathrm{RX},\ell}, \tilde{\theta}_{\mathrm{TX},\ell})$ corresponds to some non-zero element of $\tilde{\mathbf{X}} = \mathrm{unvec}(\tilde{\mathbf{x}}) \in \mathbb{C}^{B_{\mathrm{RX}} \times B_{\mathrm{TX}}}$. The normalized MSE (NMSE) of $\mathbf{H}$ is

$$\mathrm{NMSE}(\mathbf{H}) = \mathbb{E}\left\{\frac{\|\tilde{\mathbf{H}} - \mathbf{H}\|_{\mathrm{F}}^2}{\|\mathbf{H}\|_{\mathrm{F}}^2}\right\} \tag{48}$$

where $\tilde{\mathbf{H}} = \mathbf{A}_{\text{RX}}\tilde{\mathbf{X}}\mathbf{A}_{\text{TX}}$. In (45)–(48), the symbol $\tilde{}$ is used to emphasize that the quantity is an estimate.

Throughout this section, we consider the debiasing variant of Algorithm 2. The halting condition of Algorithms 2 and 3 is to halt when the current and previous $\text{supp}(\tilde{\mathbf{x}})$ are the same. The gradient descent method is used to solve convex optimization problems, which consist of (35) and Line 5 of Algorithms 2 and 3. The backtracking line search is used to compute the step size in the gradient descent method and κ in Line 3 of Algorithm 3. In addition, η is selected so that Conjecture 1 is satisfied. In this paper, we select the maximum η satisfying

$$\min_{i\in\{1,2,\dots,B\}} |B_\eta(i)| > 1. \tag{49}$$

For example, the maximum η satisfying (49) is $\eta = 0.6367$ when $B_{\text{RX}} = 2M$ and $B_{\text{TX}} = 2N$. The channel estimation criterion of Algorithms 2 and 3 is either MAP or ML, which depends on whether $\mathbf{H}$ is random or deterministic. To compare the BMS-based and non-BMS-based algorithms, the performance of the GraSP and GraHTP algorithms is shown as a reference in Figs. 4, 5, 6, and 7. The GraSP and GraHTP algorithms forbid $B_{\text{RX}} \gg M$ and $B_{\text{TX}} \gg N$ because the GraSP and GraHTP algorithms diverge when $\mathbf{A}$ is highly coherent. Therefore, the parameters are selected as $B_{\text{RX}} = M$ and $B_{\text{TX}} = N$ when the GraSP and GraHTP algorithms are implemented. Such B_{RX} and B_{TX}, however, are problematic because the mismatch in (7) is inversely proportional to B_{RX} and B_{TX}.

In Figs. 4 and 5, we compare the accuracy of the BMS-based and band excluding-based (BE-based) algorithms at different SNRs using $\text{MSE}(\{\alpha_\ell\}_{\ell=1}^{L})$, $\text{MSE}(\{\theta_{\text{RX},\ell}\}_{\ell=1}^{L})$, $\text{MSE}(\{\theta_{\text{TX},\ell}\}_{\ell=1}^{L})$, and $\text{NMSE}(\mathbf{H})$. The BE hard thresholding technique was proposed in [26], which was applied to the orthogonal matching pursuit (OMP) algorithm [34]. In this paper, we apply the BE hard thresholding technique to the GraSP algorithm, which results in the BEGraSP algorithm. In this example, $B_{\text{RX}} = B_{\text{TX}} = 256$ for the BMS-based and BE-based algorithms. We assume that

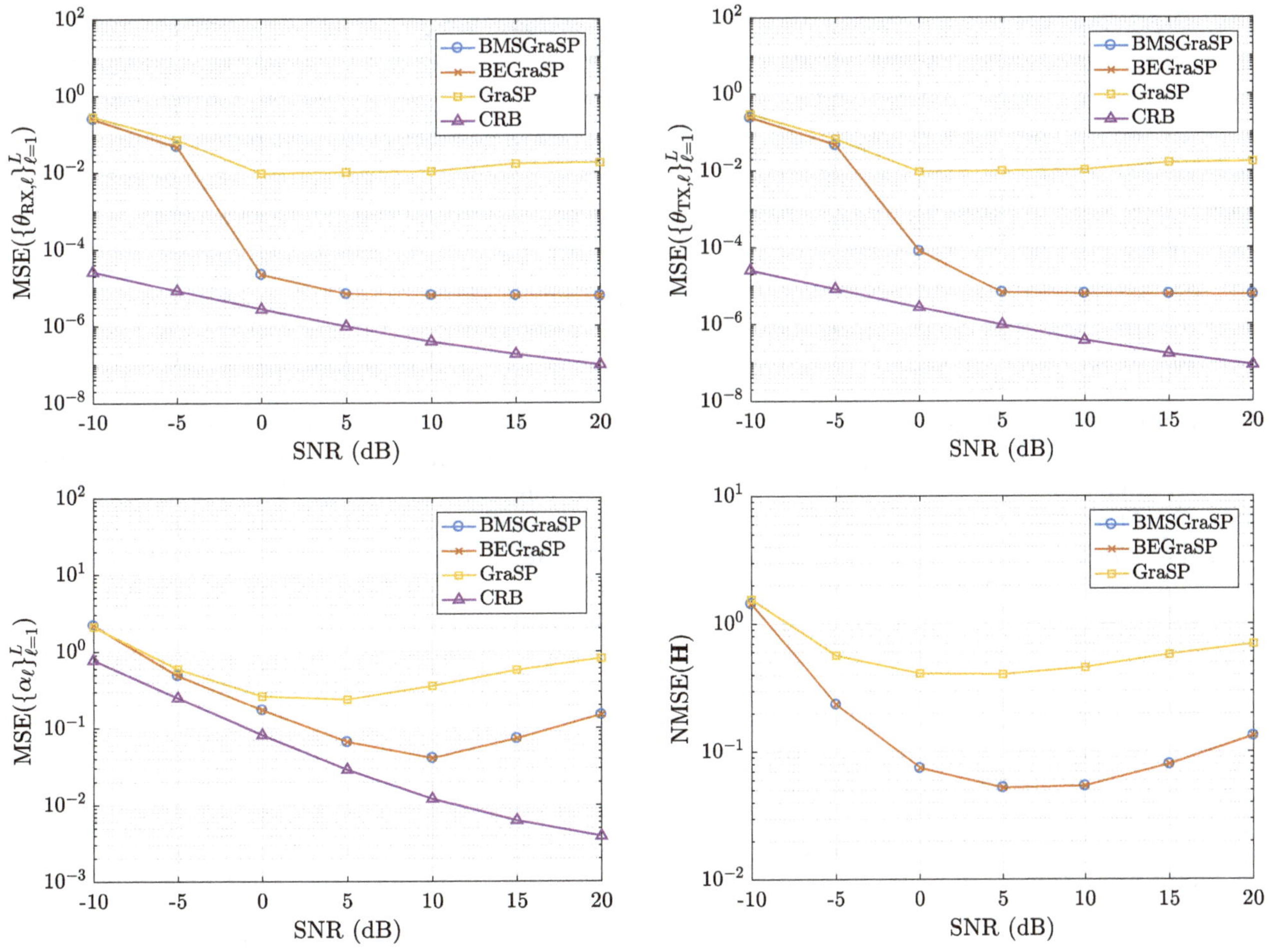

Fig. 4 MSEs of the BMS-based and BE-based GraSP algorithms for widely spread $\theta_{\text{RX},\ell} = \theta_{\text{TX},\ell} = \frac{\pi}{18}(\ell - 1)$ with $M = N = 64$, $T = 80$, $L = 8$, and $\alpha_\ell = (0.8 + 0.1(\ell - 1))e^{j\frac{\pi}{4}(\ell-1)}$. The CRB is provided as a reference, which was derived in [36]

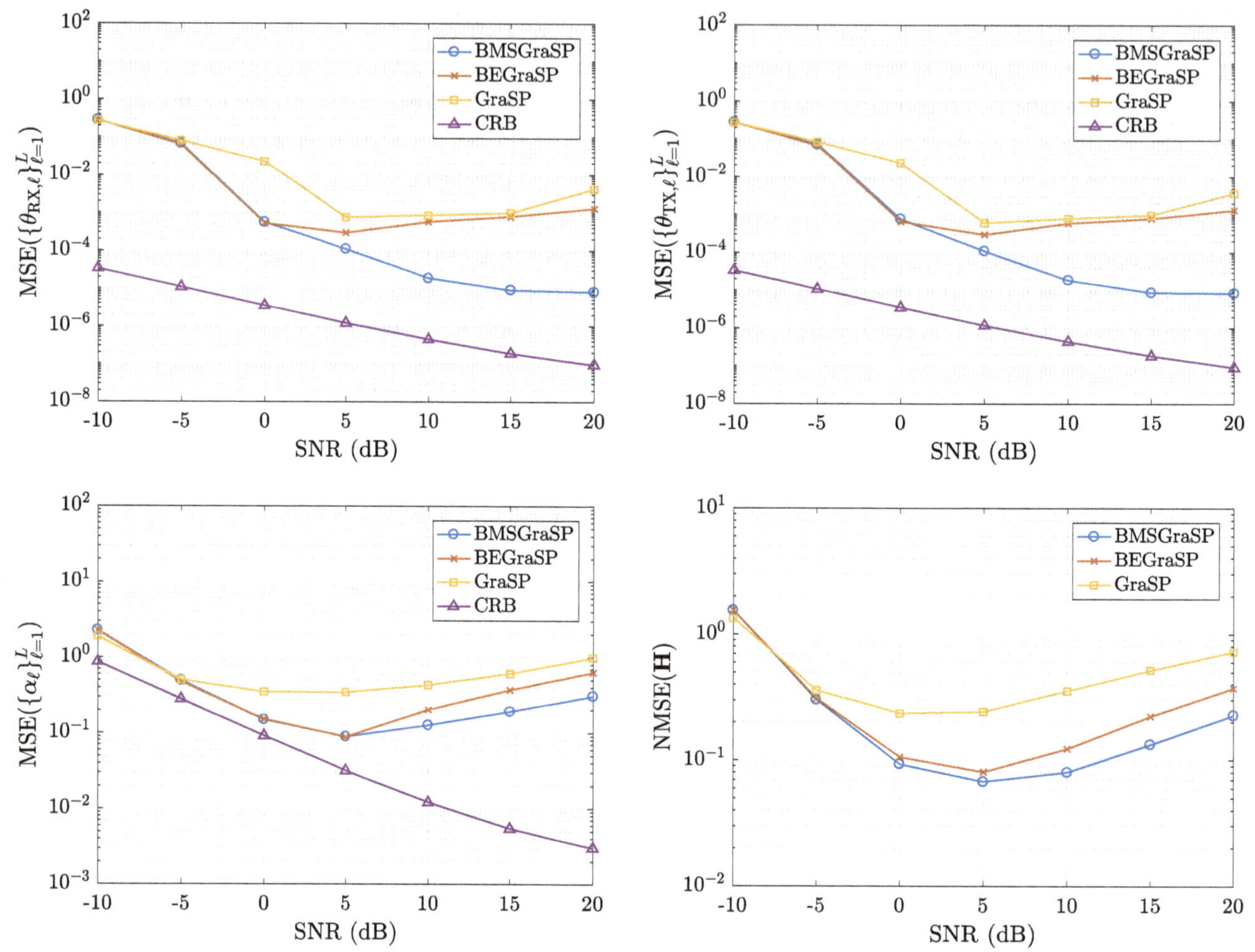

Fig. 5 MSEs of the BMS-based and BE-based GraSP algorithms for closely spread $\theta_{\mathrm{RX},\ell} = \theta_{\mathrm{TX},\ell} = \frac{\pi}{36}(\ell - 1)$ with $M = N = 64, T = 80, L = 8$, and $\alpha_\ell = (0.8 + 0.1(\ell - 1))e^{j\frac{\pi}{4}(\ell-1)}$. The CRB is provided as a reference, which was derived in [36]

$L = 8$ and $\mathbf{H}$ is deterministic where $\alpha_\ell = (0.8 + 0.1(\ell - 1))e^{j\frac{\pi}{4}(\ell-1)}$. However, $\{\theta_{\mathrm{RX},\ell}\}_{\ell=1}^{L}$ and $\{\theta_{\mathrm{TX},\ell}\}_{\ell=1}^{L}$ vary from simulation to simulation, which are either widely spread (Fig. 4) or closely spread (Fig. 5). In Figs. 4 and 5, the notion of widely and closely spread paths refer to the fact that the minimum 2-norm distance between the paths are either relatively far or close, i.e., $\min_{i \neq j} \|(\theta_{\mathrm{RX},i} - \theta_{\mathrm{RX},j}, \theta_{\mathrm{TX},i} - \theta_{\mathrm{TX},j})\|_2$ of Fig. 4, which is $\sqrt{(\pi/18)^2 + (\pi/18)^2}$, is greater than that of Fig. 5, which is $\sqrt{(\pi/36)^2 + (\pi/36)^2}$. The path gains, AoAs, and AoDs are assumed to be deterministic because the CRB is defined for deterministic parameters only [35]. A variant of the CRB for random parameters is known as the Bayesian CRB, but adding the Bayesian CRB to our work is left as a future work because applying the Bayesian CRB to non-linear measurements, e.g., 1 bit ADCs, is not as straightforward.

According to Figs. 4 and 5, the BMS-based algorithms succeed to estimate both widely spread and closely spread paths, whereas the BE-based algorithms fail to estimate closely spread paths. This follows because the BE hard thresholding technique was derived based on the assumption that $\mathrm{supp}(\mathbf{x}^*)$ is widely spread. In contrast, the BMS hard thresholding technique is proposed based on Conjecture 1 without any assumption on $\mathrm{supp}(\mathbf{x}^*)$. This means that when $\mathrm{supp}(\mathbf{x}^*)$ is closely spread, the BE hard thresholding technique cannot properly estimate $\mathrm{supp}(\mathbf{x}^*)$ because the BE hard thresholding technique, by its nature, excludes the elements near the maximum element of $\mathbf{x}^*$ from its potential candidate. The BMS hard thresholding technique, in contrast, uses the elements near the maximum element of $\mathbf{x}^*$ to construct the by-product testing set only, i.e., Line 4 of Algorithm 1. Therefore, the BMS-based algorithms are superior to the BE-based algorithms when the paths are closely spread. The Cramér-Rao bounds (CRBs) of $\mathrm{MSE}(\{\alpha_\ell\}_{\ell=1}^{L})$, $\mathrm{MSE}(\{\theta_{\mathrm{RX},\ell}\}_{\ell=1}^{L})$, and $\mathrm{MSE}(\{\theta_{\mathrm{TX},\ell}\}_{\ell=1}^{L})$ are provided, which were derived in [36]. The gaps between

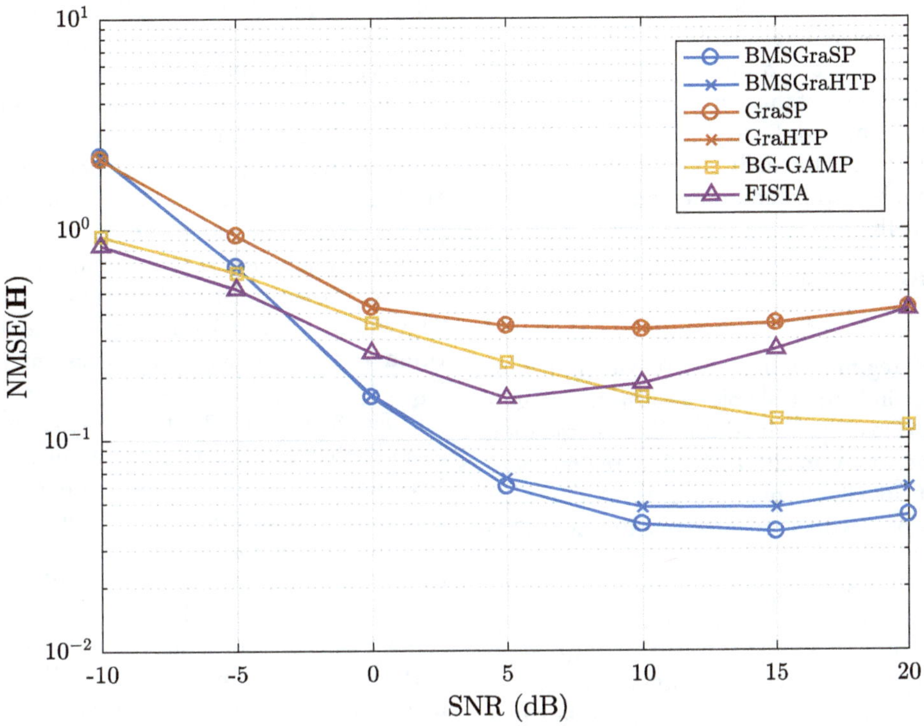

Fig. 6 NMSE vs. SNR where $M = N = 64$, $T = 80$, and $L = 4$ with varying B_{RX} and B_{TX} from algorithm to algorithm

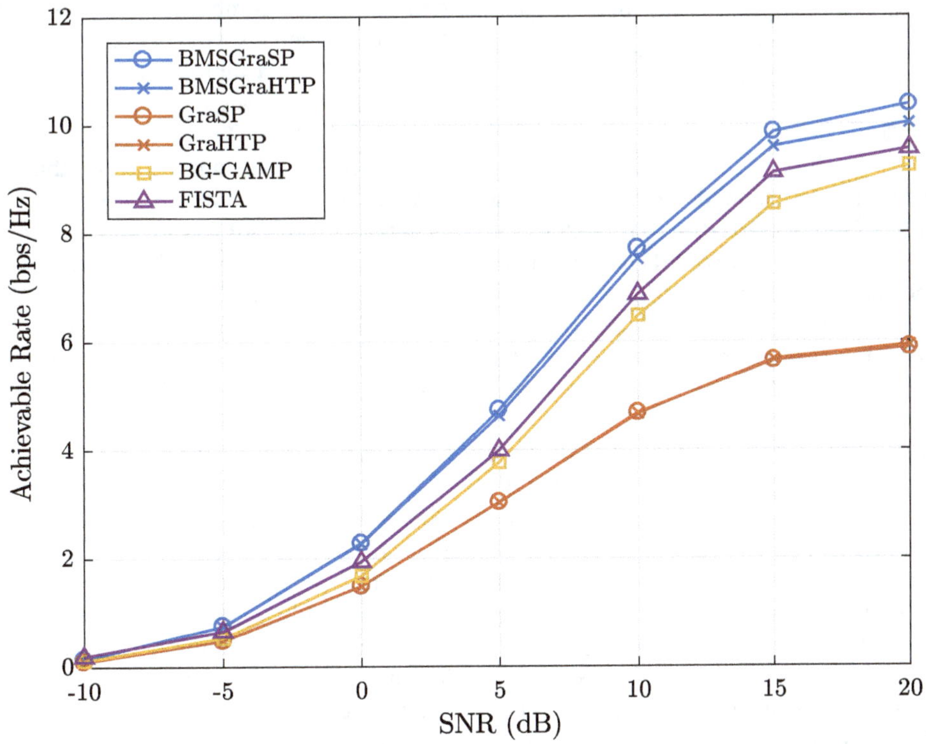

Fig. 7 Achievable rate lower bound [15] vs. SNR where $M = N = 64$, $T = 80$, and $L = 4$ with varying B_{RX} and B_{TX} from algorithm to algorithm

the MSEs and their corresponding CRBs can be interpreted as a performance limit incurred by the discretized AoAs and AoDs. To overcome such limit, the AoAs and AoDs must be estimated based on the off-grid method, which is beyond the scope of this paper.

In addition, note that $\mathrm{MSE}(\{\alpha_\ell\}_{\ell=1}^{L})$ and $\mathrm{NMSE}(\mathbf{H})$ worsen as the SNR enters the high SNR regime. To illustrate why $\mathbf{x}^*$ cannot be estimated in the high SNR regime in 1 bit ADCs, note that

$$\begin{aligned}\mathrm{Q}(\sqrt{\rho}\mathbf{A}\mathbf{x}^* + \mathbf{n}) &\approx \mathrm{Q}(\sqrt{\rho}\mathbf{A}\mathbf{x}^*)\\ &= \mathrm{Q}(\sqrt{\rho}\mathbf{A}c\mathbf{x}^*)\end{aligned} \tag{50}$$

in the high SNR regime with $c > 0$, which means that $\mathbf{x}^*$ and $c\mathbf{x}^*$ are indistinguishable because the magnitude information is lost by 1 bit ADCs. The degradation of the recovery accuracy in the high SNR regime with 1 bit ADCs is an inevitable phenomenon, as observed from other previous works on low-resolution ADCs [11, 14, 15, 33, 37].

In Figs. 6 and 7, we compare the performance of Algorithms 2 and 3, and other estimators when $\mathbf{H}$ is random. The Bernoulli Gaussian-GAMP (BG-GAMP) algorithm [15] is an iterative approximate MMSE estimator of $\mathbf{x}^*$, which was derived based on the assumption that x_i^* is distributed as $\mathcal{CN}(0, 1)$ with probability L/B but 0 otherwise, namely, the BG distribution. The fast iterative shrinkage-thresholding algorithm (FISTA) [38] is an iterative MAP estimator of $\mathbf{x}^*$, which was derived based on the assumption that the logarithm of the PDF of $\mathbf{x}^*$ is $g_{\mathrm{FISTA}}(\mathbf{x}) = -\gamma\|\mathbf{x}\|_1$ ignoring the constant factor, namely, the Laplace distribution. Therefore, the estimate of $\mathbf{x}^*$ is

$$\underset{\mathbf{x}\in\mathbb{C}^B}{\operatorname{argmax}}\,(f(\mathbf{x}) + g_{\mathrm{FISTA}}(\mathbf{x})), \tag{51}$$

which is solved using the accelerated proximal gradient descent method [38]. The regularization parameter γ is selected so that the expected sparsity of (51) is $3L$ for a fair comparison, which was suggested in [17]. In this example, $L = 4$, whereas B_{RX} and B_{TX} vary from algorithm to algorithm. In particular, we select $B_{\mathrm{RX}} = B_{\mathrm{TX}} = 256$ for Algorithms 2, 3, and the FISTA, whereas $B_{\mathrm{RX}} = M$ and $B_{\mathrm{TX}} = N$ for the BG-GAMP algorithm.

In Fig. 6, we compare the accuracy of Algorithms 2, 3, and other estimators at different SNRs using $\mathrm{NMSE}(\mathbf{H})$. According to Fig. 6, Algorithms 2 and 3 outperform the BG-GAMP algorithm and FISTA as the SNR enters the medium SNR regime. The accuracy of the BG-GAMP algorithm is disappointing because the mismatch in (7) is inversely proportional to B_{RX} and B_{TX}. However, increasing B_{RX} and B_{TX} is forbidden because the BG-GAMP algorithm diverges when $\mathbf{A}$ is highly coherent. The accuracy of the FISTA is disappointing because the Laplace distribution does not match the distribution of $\mathbf{x}^*$. Note that (23), which is the basis of Algorithms 2 and 3, is indeed the MAP estimate of $\mathbf{x}^*$, which is in contrast to the FISTA. According to Fig. 6, $\mathrm{NMSE}(\mathbf{H})$ worsens as the SNR enters the high SNR regime, which follows from the same reason as in Figs. 4 and 5.

In Fig. 7, we compare the achievable rate lower bound of Algorithms 2, 3, and other estimators at different SNRs when the precoders and combiners are selected based on $\tilde{\mathbf{H}}$. The achievable rate lower bound shown in Fig. 7 is presented in [15], which was derived based on the Bussgang decomposition [13] in conjunction with the fact that the worst-case noise is Gaussian. According to Fig. 7, Algorithms 2 and 3 outperform the BG-GAMP algorithm and FISTA, which is consistent with the result in Fig. 6.

In Fig. 8, we compare the complexity of Algorithms 2, 3, and other estimators at different B_{RX} and B_{TX} when $\mathbf{H}$ is random. To analyze the complexity, note that Algorithms 2, 3, and the FISTA require $h(\mathbf{x})$ and $\nabla h(\mathbf{x})$ at each iteration, whose bottlenecks are $\mathbf{A}\mathbf{x}$ and $\mathbf{A}^{\mathrm{H}}\mathbf{c}$, respectively, while the BG-GAMP algorithm requires $\mathbf{A}\mathbf{x}$ and $\mathbf{A}^{\mathrm{H}}\mathbf{c}$ at each iteration. Therefore, the complexity is measured based on the number of complex multiplications performed to compute $\mathbf{A}\mathbf{x}$ and $\mathbf{A}^{\mathrm{H}}\mathbf{c}$, which are implemented based on the FFT. In this example, $L = 4$, whereas SNR is either 0 dB or 10 dB.

In this paper, the complexity of the BG-GAMP algorithm is used as a baseline because the BG-GAMP algorithm is widely used. The normalized complexity is defined as the number of complex multiplications performed divided by the per-iteration complexity of the BG-GAMP. For example, the normalized complexity of the FISTA with $B_{\mathrm{RX}} = B_{\mathrm{TX}} = 256$ is 160 when the complexity of the FISTA with $B_{\mathrm{RX}} = B_{\mathrm{TX}} = 256$ is equivalent to the complexity of the 160-iteration BG-GAMP algorithm with $B_{\mathrm{RX}} = B_{\mathrm{TX}} = 256$. In practice, the BG-GAMP algorithm converges in 15 iterations when $\mathbf{A}$ is incoherent [39]. In this paper, an algorithm is said to be as efficient as the BG-GAMP algorithm when the normalized complexity is below the target threshold, which is 15. As a sidenote, our algorithms, namely, the BMSGraSP and BMSGraHTP, requires 2.1710 and 2.0043 iterations in average, respectively, across the entire SNR range.

According to Fig. 8, the complexity of the FISTA is impractical because the objective function of (51) is a high-dimensional function defined on $\mathbb{C}^B$ where $B \gg MN$. In contrast, the complexity of Algorithms 2 and 3 is dominated by (42), whose objective function is a low-dimensional function defined on $\mathbb{C}^{|\mathcal{I}|}$ where $|\mathcal{I}| = O(L)$. The normalized complexity of Algorithms 2 and 3 is below the target threshold when $B_{\mathrm{RX}} \geq 192$ and $B_{\mathrm{TX}} \geq 192$. Therefore, we conclude that Algorithms 2 and 3 are as efficient as the BG-GAMP algorithm when $B_{\mathrm{RX}} \gg M$ and $B_{\mathrm{TX}} \gg N$.

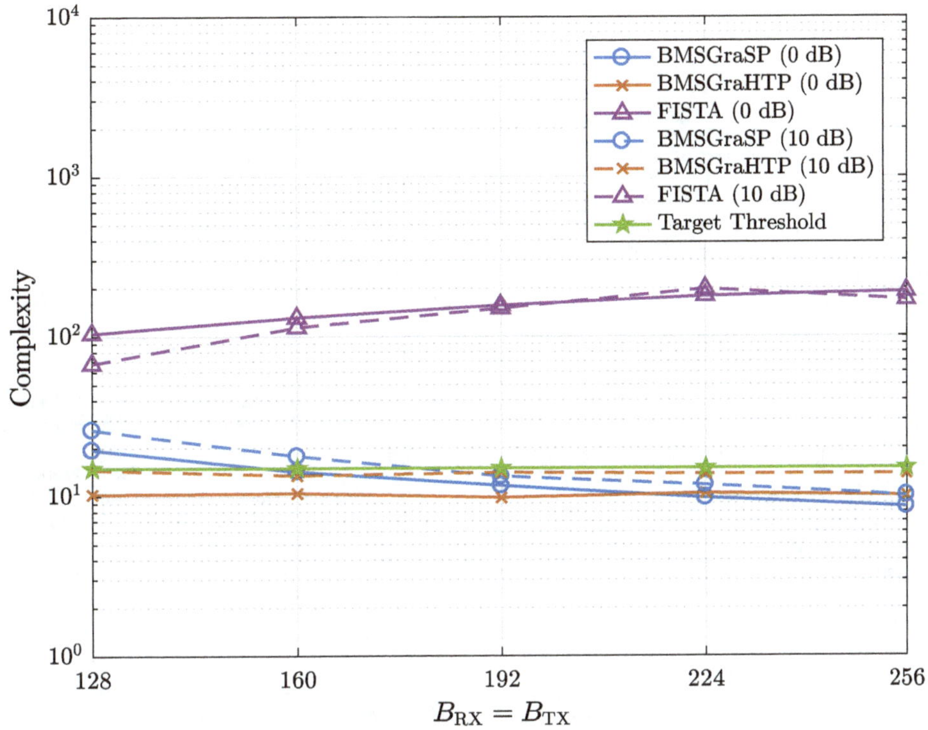

Fig. 8 Normalized complexity vs. $B_{RX} = B_{TX}$ where $M = N = 64$, $T = 80$, and $L = 4$ at SNR = 0 dB and SNR = 10 dB

6 Conclusions

In the mmWave band, the channel estimation problem is converted to a sparsity-constrained optimization problem, which is NP-hard to solve. To approximately solve sparsity-constrained optimization problems, the GraSP and GraHTP algorithms were proposed in CS, which pursue the gradient of the objective function. The GraSP and GraHTP algorithms, however, break down when the objective function is ill-conditioned, which is incurred by the highly coherent sensing matrix. To remedy such break down, we proposed the BMS hard thresholding technique, which is applied to the GraSP and GraHTP algorithms, namely, the BMSGraSP and BMSGraHTP algorithms, respectively. Instead of directly hard thresholding the gradient of the objective function, the BMS-based algorithms test whether an index is the ground truth index or the by-product of another index. We also proposed an FFT-based fast implementation of the BMS-based algorithms, whose complexity is reduced from $O(M^4)$ to $O(M^2 \log M)$. In the simulation, we compared the performance of the BMS-based, BE-based, BG-GAMP, and FISTA algorithms from different aspects in terms of the accuracy, achievable rate, and complexity. The BMS-based algorithms were shown to outperform other estimators, which proved to be both accurate and efficient. Our algorithms, however, addressed only the flat fading scenario, so an interesting future work would be to propose a low-complexity channel estimator capable of dealing with the wideband scenario.

7 Methods/experimental

The aim of this study is to propose an accurate yet efficient channel estimator for mmWave massive MIMO systems with 1 bit ADCs. Our channel estimator was proposed based on theoretical analysis. To be specific, we adopted and modified CS algorithms to exploit the sparse nature of the mmWave virtual channels. In addition, we carefully analyzed the proposed channel estimator to reduce the complexity. To verify the accuracy and complexity of the proposed channel estimator, we conducted extensive (Monte-Carlo) simulations.

Abbreviations

ADC: Analog-to-digital converter; AoA: Angle-of-arrival; AoD: Angle-of-departure; AWGN: Additive white Gaussian noise; BE: Band excluding; BG-GAMP: Bernoulli Gaussian-generalized approximate message passing; BLMMSE: Bussgang linear minimum mean squared error; BMS: Band maximum selecting; CoSaMP: Compressive sampling matching pursuit; CRB: Cramér-Rao bound; CS: Compressive sensing; DFT: Discrete fourier transform; FFT: Fast Fourier transform; FISTA: Fast iterative shrinkage-thresholding algorithm; GAMP: Generalized approximate message passing; GEC-SR: Generalized expectation consistent signal recovery; GraHTP: Gradient hard thresholding pursuit; GraSP: Gradient support pursuit; HTP: Hard thresholding pursuit; i.i.d.: Independent and identically distributed; LOS: Line-of-sight; MAP: Maximum a posteriori; MIMO: Multiple-input multiple-output; mmWave: Millimeter wave; nML: Near maximum likelihood; NMSE: Normalized mean squared error; OMP: Orthogonal matching pursuit; RIP: Restricted isometry property; SNR: Signal-to-noise ratio; ULA: Uniform linear array; ZF: Zadoff-Chu

Authors' contributions
IK and JC led the research and wrote the manuscript. Both authors read and approved the final manuscript.

Funding
This work was partly supported by Institute for Information & communications Technology Promotion(IITP) grant funded by the Korea government(MSIT) (No. 2016-0-00123, Development of Integer-Forcing MIMO Transceivers for 5G & Beyond Mobile Communication Systems) and by the National Research Foundation (NRF) grant funded by the MSIT of the Korea government (2019R1C1C1003638).

Competing interests
The authors declare that they have no competing interests.

References

1. A. L. Swindlehurst, E. Ayanoglu, P. Heydari, F. Capolino, Millimeter-wave massive MIMO: the next wireless revolution? IEEE Commun. Mag. **52**(9), 56–62 (2014). https://doi.org/10.1109/MCOM.2014.6894453
2. Z. Gao, L. Dai, D. Mi, Z. Wang, M. A. Imran, M. Z. Shakir, mmWave massive-MIMO-based wireless backhaul for the 5G Ultra-dense network. IEEE Wirel. Commun. **22**(5), 13–21 (2015). https://doi.org/10.1109/MWC.2015.7306533
3. T. E. Bogale, L. B. Le, Massive MIMO and mmWave for 5G Wireless HetNet: potential benefits and challenges. IEEE Veh. Technol. Mag. **11**(1), 64–75 (2016). https://doi.org/10.1109/MVT.2015.2496240
4. F. Boccardi, R. W. Heath, A. Lozano, T. L. Marzetta, P. Popovski, Five disruptive technology directions for 5G. IEEE Commun. Mag. **52**(2), 74–80 (2014). https://doi.org/10.1109/MCOM.2014.6736746
5. Bin Le, T. W. Rondeau, J. H. Reed, C. W. Bostian, Analog-to-digital converters. IEEE Sig. Proc. Mag. **22**(6), 69–77 (2005). https://doi.org/10.1109/MSP.2005.1550190
6. C. Mollén, J. Choi, E. G. Larsson, R. W. Heath, Uplink performance of wideband massive MIMO with one-bit ADCs. IEEE Trans. Wirel. Commun. **16**(1), 87–100 (2017). https://doi.org/10.1109/TWC.2016.2619343
7. L. Fan, S. Jin, C. Wen, H. Zhang, Uplink achievable rate for massive MIMO systems with low-resolution ADC. IEEE Commun. Lett. **19**(12), 2186–2189 (2015). https://doi.org/10.1109/LCOMM.2015.2494600
8. J. Zhang, L. Dai, S. Sun, Z. Wang, On the spectral efficiency of massive MIMO systems with low-resolution ADCs. IEEE Commun. Lett. **20**(5), 842–845 (2016). https://doi.org/10.1109/LCOMM.2016.2535132
9. S. Jacobsson, G. Durisi, M. Coldrey, U. Gustavsson, C. Studer, Throughput analysis of massive MIMO uplink with low-resolution ADCs. IEEE Trans. Wirel. Commun. **16**(6), 4038–4051 (2017). https://doi.org/10.1109/TWC.2017.2691318
10. J. Choi, J. Mo, R. W. Heath, Near maximum-likelihood detector and channel estimator for uplink multiuser massive MIMO systems with one-bit ADCs. IEEE Trans. Commun. **64**(5), 2005–2018 (2016). https://doi.org/10.1109/TCOMM.2016.2545666
11. Y. Li, C. Tao, G. Seco-Granados, A. Mezghani, A. L. Swindlehurst, L. Liu, Channel estimation and performance analysis of one-bit massive MIMO systems. IEEE Trans. Sign. Proc. **65**(15), 4075–4089 (2017). https://doi.org/10.1109/TSP.2017.2706179
12. D. P. Bertsekas, Nonlinear Programming. Journal of the Operational Research Society. **48**(3), 334–334 (1997). https://doi.org/10.1057/palgrave.jors.2600425
13. J. J. Bussgang, Research Laboratory of Electronics, Massachusetts Institute of Technology. Technical report. **216**, 1–14 (1952)
    Article type: Technical report Institution: Research Laboratory of Electronics, Massachusetts Institute of Technology Volume: 216 Page: 1-14 Year: 1952
14. H. He, C. Wen, S. Jin, Bayesian optimal data detector for hybrid mmWave MIMO-OFDM systems with low-resolution ADCs. IEEE J. Sel. Top. Sig. Proc. **12**(3), 469–483 (2018). https://doi.org/10.1109/JSTSP.2018.2818063
15. J. Mo, P. Schniter, R. W. Heath, Channel estimation in broadband millimeter wave MIMO systems with Few-Bit ADCs. IEEE Trans. Sign. Proc. **66**(5), 1141–1154 (2018). https://doi.org/10.1109/TSP.2017.2781644
16. T. Liu, C. Wen, S. Jin, X. You, in *2016 IEEE International Symposium on Information Theory (ISIT)*. Generalized turbo signal recovery for nonlinear measurements and orthogonal sensing matrices, (2016), pp. 2883–2887. https://doi.org/10.1109/ISIT.2016.7541826
17. S. Bahmani, B. Raj, P. T. Boufounos, Greedy Sparsity-Constrained Optimization. J. Mach. Learn. Res. **14**(Mar), 807–841 (2013)
18. X.-T. Yuan, P. Li, T. Zhang, Gradient Hard Thresholding Pursuit. J. Mach. Learn. Res. **18**(166), 1–43 (2018)
19. D. Needell, J. A. Tropp, CoSaMP: Iterative signal recovery from incomplete and inaccurate samples. Appl. Comput. Harmon. Anal. **26**(3), 301–321 (2009). https://doi.org/10.1016/j.acha.2008.07.002
20. S. Foucart, Hard Thresholding Pursuit: An Algorithm for Compressive Sensing. SIAM J. Numer. Anal. **49**(6), 2543–2563 (2011). https://doi.org/10.1137/100806278
21. M. R. Akdeniz, Y. Liu, M. K. Samimi, S. Sun, S. Rangan, T. S. Rappaport, E. Erkip, Millimeter wave channel modeling and cellular capacity evaluation. IEEE J. Sel. Areas Commun. **32**(6), 1164–1179 (2014). https://doi.org/10.1109/JSAC.2014.2328154
22. A. M. Sayeed, Deconstructing Multiantenna Fading Channels. IEEE Trans. Sign. Proc. **50**(10), 2563–2579 (2002). https://doi.org/10.1109/TSP.2002.803324
23. W. Hong, K. Baek, Y. Lee, Y. Kim, S. Ko, Study and Prototyping of Practically Large-Scale mmWave Antenna Systems for 5G Cellular Devices. IEEE Commun. Mag. **52**(9), 63–69 (2014). https://doi.org/10.1109/MCOM.2014.6894454
24. S. Boyd, L. Vandenberghe, *Convex Optimization*. (Cambridge University Press, Cambridge, 2004)
25. Y. C. Eldar, G. Kutyniok, *Compressed Sensing: Theory and Applications*. (Cambridge University Press, Cambridge, 2012)
26. A. Fannjiang, W. Liao, Coherence Pattern–Guided Compressive Sensing with Unresolved Grids. SIAM J. Imaging Sci. **5**(1), 179–202 (2012). https://doi.org/10.1137/110838509
27. N. Jindal, MIMO broadcast channels with finite-rate feedback. IEEE Trans. Inf. Theory. **52**(11), 5045–5060 (2006). https://doi.org/10.1109/TIT.2006.883550
28. Z. Marzi, D. Ramasamy, U. Madhow, Compressive channel estimation and tracking for large arrays in mm-wave Picocells. IEEE J. Sel. Top. Sig. Proc. **10**(3), 514–527 (2016). https://doi.org/10.1109/JSTSP.2016.2520899
29. D. Chu, Polyphase Codes with Good Periodic Correlation Properties (Corresp.) IEEE Trans. Inf. Theory. **18**(4), 531–532 (1972). https://doi.org/10.1109/TIT.1972.1054840
30. J. Markel, FFT Pruning. IEEE Trans. Audio Electroacoustics. **19**(4), 305–311 (1971). https://doi.org/10.1109/TAU.1971.1162205
31. D. Skinner, Pruning the Decimation In-Time FFT Algorithm. IEEE Trans. Acoust. Speech Sig. Proc. **24**(2), 193–194 (1976). https://doi.org/10.1109/TASSP.1976.1162782
32. T. Sreenivas, P. Rao, FFT Algorithm for both input and output pruning. IEEE Trans. Acoust. Speech Sig. Proc. **27**(3), 291–292 (1979). https://doi.org/10.1109/TASSP.1979.1163246
33. Y. Ding, S. Chiu, B. D. Rao, Bayesian Channel estimation algorithms for massive MIMO systems with hybrid analog-digital processing and low-resolution ADCs. IEEE J. Sel. Top. Sig. Proc. **12**(3), 499–513 (2018). https://doi.org/10.1109/JSTSP.2018.2814008
34. J. A. Tropp, A. C. Gilbert, Signal Recovery from Random Measurements via Orthogonal Matching Pursuit. IEEE Trans. Inf. Theory. **53**(12), 4655–4666 (2007). https://doi.org/10.1109/TIT.2007.909108
35. H. V. Poor, *An Introduction to Signal Detection and Estimation*. (Springer, Berlin, 2013)
36. P. Wang, J. Li, M. Pajovic, P. T. Boufounos, P. V. Orlik, in *2017 51st Asilomar Conference on Signals, Systems, and Computers*. On Angular-Domain Channel Estimation for One-Bit Massive MIMO Systems with Fixed and Time-Varying Thresholds, (2017), pp. 1056–1060. https://doi.org/10.1109/ACSSC.2017.8335511
37. R. P. David, J. Cal-Braz, in *ICASSP 2019 - 2019 IEEE International Conference on Acoustics, Speech and Signal Processing (ICASSP)*. Feedback-Controlled Channel Estimation with Low-Resolution ADCs in Multiuser MIMO Systems, (2019), pp. 4674–4678. https://doi.org/10.1109/ICASSP.2019.8683652
38. A. Beck, M. Teboulle, A Fast Iterative Shrinkage-Thresholding Algorithm for Linear Inverse Problems. SIAM J. Imaging Sci. **2**(1), 183–202 (2009). https://doi.org/10.1137/080716542
39. J. P. Vila, P. Schniter, Expectation-maximization Gaussian-mixture approximate message passing. IEEE Trans. Sign. Proc. **61**(19), 4658–4672 (2013). https://doi.org/10.1109/TSP.2013.2272287

13

Robust beamforming and spatial precoding for quasi-OSTBC massive MIMO communications

Ju-Hong Lee[1*] and Wei-En Sun[2]

Abstract

This paper considers the problem of massive multiple-input-multiple-output (MIMO) wireless communication systems with quasi-orthogonal space-time block code (QOSTBC) transmission in the presence of spatial correlation effect (SCE) and mutual coupling effect (MCE). Conventional MIMO channels with QOSTBC transmission suffer from several drawbacks. The number of transmit antennas is restricted to a few in order to achieve full transmission rate. Therefore, it is difficult to make a quasi-orthogonal space time block coded massive MIMO system to achieve full transmission rate. Moreover, MIMO channel with QOSTBC transmission is usually considered for the case where the number of user equipments (UEs) is one. We present a joint beamforming and spatial precoding method to deal with the above drawbacks. The proposed method incorporates a beamforming scheme and spatial precoding to formulate an appropriate optimization process to effectively alleviate the considered problems. The resulting optimization problem will be solved by using a cooperative coevolutionary particle swarm optimization algorithm under two proposed fitness functions. During the optimization process, the proposed method finds the optimal beamforming coefficients of the precoding matrix, the optimal normalized positions of transmit and receive antenna elements for the case of using linear antenna arrays, and the optimal angle differences of transmit and receive antenna elements for the case of using circular arrays. Based on the proposed method, we are able to cure the performance degradation of a quasi-orthogonal space time block coded massive MIMO system due to the SCE and MCE. Moreover, the proposed method makes QOSTBC MIMO communications with full transmission rate for any number of transmit antennas achievable. Several simulation examples are presented to show the superior bit error rate (BER) performances of QOSTBC wireless MIMO scenarios with linear as well as circular antenna arrays by using the proposed method as compared to the existing methods.

Keywords: Massive MIMO, Wireless communication, Spatial precoding, Spatial correlation, Mutual coupling, Beamforming, Quasi-orthogonal space-time block codes, Population-based stochastic optimization, Cooperative coevolutionary particle swarm optimization, Bit error rate

1 Introduction

Achieving higher transmission rates and increasing number of users have become the essential requirements of modern wireless MIMO communications. However, the performance and capacity of MIMO wireless communication systems are degraded due to the interference problem at the mobile users. Nowadays, it is well known that employing array beamforming techniques, we can make the transmitted information stream weighted with appropriate phases and amplitudes at each antenna sensor to create a directed beam form. As a result, the interference problem due to multiple UEs can be effectively mitigated. However, downlink beamformnig for alleviating the interference problem at the mobile users is not an easy task for the frequency-division duplex (FDD) mode of transmission in MIMO systems. This is due to only limited knowledge of the downlink channel available for the base station (BS). In the literature, the existing beamforming

*Correspondence: jhlee@ntu.edu.tw
[1]Department of Electrical Engineering, Graduate Institute of Communication Engineering, and Graduate Institute of Biomedical Electronics and Bioinformatics, National Taiwan University, No. 1, Sec. 4, Roosevelt Rd., Taipei 10617, Taiwan
Full list of author information is available at the end of the article

and precoding methods, such as [1–3], employ a set of pre-defined fixed beamforming coefficients to incorporate the necessary precoding design. The key idea of using the beamforming-based spatial precoding (BBSP) scheme of [3] is to gather all the antennas deployed at BS into several beam groups and employ a set of pre-defined fixed beamforming coefficients to design the necessary precoding matrix. As a result, each antenna in one beam group can transmit the downlink training signal to UEs simultaneously, and hence, the user equipments (UEs) only need to feed the selected beam indices back to the BS; the feedback overhead is drastically reduced. It has been shown [3] that the BBSP scheme preserves an acceptable downlink transmission performance with reduced downlink training overhead and has a significant reduction of the total feedback bits from each UE. Nevertheless, these methods cause severe interference problem between UEs. Consequently, the bit error rate (BER) performance of the whole system becomes poor. Recently, a joint beamforming and spatial precoding technique was presented by [4] to optimally design the beamforming coefficients to effectively cure the interference problem between UEs. It has been shown in [4] that this technique not only can cure the problem of the BBSP scheme, but also preserves the advantages of requiring lower overheads of the downlink training and the channel state information (CSI) feedback for massive wireless MIMO systems.

For developing the next generation wireless mobile communications, it has been realized that employing wireless MIMO systems with massive antenna arrays deployed at transmitter and receiver is the key manner of achieving the necessary requirements of channel capacity, energy efficiency, and spectral efficiency [5, 6]. Unfortunately, massive MIMO wireless systems suffer from the degradation of diversity gain due to the spatial correlation effect (SCE) of the fading signals between the array elements with limited spacing. Moreover, a phenomenon called mutual coupling effect (MCE) in addition to the SCE inevitably occurs in a massive antenna array. Due to antenna elements in close proximity, the MCE is caused by a part of data in one antenna element outflow to other antennas when the antennas operate simultaneously. The performance degradations of a massive MIMO wireless system due to the SCE and MCE have been reported in [7–10], respectively. Recently, optimally finding the antenna deployment and antenna location in massive MIMO systems has been considered in the literature. In [11], the uplink transmission of a single-cell multiuser distributed massive MIMO system with a large number of distributed antennas deployed at BS which receives information from multiple single-antenna users is considered. The radius of the distributed circular BS antenna array can be optimally designed for maximizing the average achievable rate of the distributed massive MIMO system. Later on, Koyuncu [12] considers the single-cell multiuser MIMO uplink transmission systems with several single antenna transmitters/users and one BS with massive antennas. For a massive MIMO uplink where the BS antennas are evenly distributed to n different locations of the cell. It has been shown that a per-user rate is achievable by optimizing the n different antenna locations of the cell. However, there are practically no papers concerning the design problem of beamforming and spatial precoding for the massive MIMO systems with QOSTBC transmission and proposing effective approaches to mitigate the SCE and MCE simultaneously.

In this paper, we consider the robust beamforming and spatial precoding to enhance the BER performance of massive MIMO systems with QOSTBC transmission in the presence of MCE as well as SCE. A robust method is developed to incorporate the cooperative coevolutionary particle swarm optimization (CCPSO) [13] with the beamforming-selection spatial precoding (BSSP) scheme [4]. Recently, there are few works concerning quantized feedback-based precoding for multiple-input single-output (MISO) systems [14–16]. Quantized feedback method is developed to achieve performance gain in MIMO system by transmitting a few bits feedback about CSI from receiver to transmitter for reducing feedback overhead. Quantized feedback-based precoding is a closed-loop system proposed to alleviate the performance degradation due to erroneous CSI feedback. It exploits the channel information available by transmitting few bits feedback about the CSI from the receiver to transmitter. The research works [14–16] focus on the imperfect quantized feedback-based diagonal precoding for orthogonal/non-orthogonal space-time block codes in MISO systems with only one receive antenna. The spatial precoding matrices used are all set to diagonal matrices. However, the robust method proposed in this paper explores the joint design of beamforming and precoding for QOSTBC massive MIMO systems with any number of transmit antennas and receive antennas to deal with the SCE and MCE. Moreover, the spatial precoding matrix designed by the proposed method is not restricted to be a diagonal matrix.

The main contributions of the robust method over the work of [4] are briefly described as follows. An efficient scheme is developed to incorporate the joint beamforming and spatial precoding with QOSTBC transmission. To achieve the optimal power distribution to the beam groups partitioned by the BSSP scheme, we introduce an appropriate manner to decide the optimal power parameters during the CCPSO optimization process. Moreover, to mitigate the SCE and MCE simultaneously, we present an efficient approach to optimally adjust the positions of the antenna array elements by maximizing the average mutual information (AMI) associated with the downlink channel. Based on an approximated average BER (AA-BER) of wireless MIMO systems proposed for the fitness function of the CCPSO, the robust method finds the optimal beamforming and precoding coefficients by minimizing the AA-BER of wireless MIMO systems. The robust method not only can provide better average BER performance than the BSSP method, but also preserves the advantages of the BSSP method having lower overheads of the downlink training and the CSI feedback in massive wireless MIMO systems. Moreover, the robust method makes QOSTBC MIMO communications with full transmission rate for any number of transmit antennas achievable. Simulation examples confirm the effectiveness of the robust method in the QOSTBC wireless MIMO channel scenarios with SCE as well as MCE. The main contributions of this paper are summarized as follows:

(I) We develop a robust joint beamforming and spatial precoding method to effectively deal with QOSTBC MIMO communication systems in the presence of mutual coupling and spatial correlation effects due to using massive antennas. Through the proposed beam selection spatial precoding scheme in conjunction with the appropriate optimization formulation, the considered problems can be effectively alleviated as shown by simulation examples.

(II) The proposed joint beamforming and spatial precoding method can preserve the advantage of reducing the overheads for downlink training and feedback of channel state information between the base station (BS) and user equipments (UEs) for the frequency-division duplex (FDD) mode of downlink transmission in MIMO systems.

(III) In the literature, it is well known that the QOSTBC can achieve 1/2 of the full transmission rate for any number of transmission antennas as well as the codes with 3/4 of the full transmission rate for the specific cases of three and four transmit antennas. There are practically no papers developing the spatial precoding for QOSTBC MIMO communications deployed massive antennas with full transmission rate. Through the proposed beam selection spatial precoding scheme in conjunction with the QOSTBC strategy, the proposed joint beamforming and spatial precoding method provides a novel manner to make QOSTBC MIMO communications with full transmission rate for any number of transmit antennas achievable.

This paper is organized as follows. In Section 2, we briefly describe the downlink transmission model in wireless MIMO systems with/without spatial correlation as well as mutual coupling effects. The principle of wireless MIMO communication systems using QOSTBC transmission is also summarized. In Section 3, we propose a robust method based on the CCPSO algorithm for QOSTBC MIMO communication systems. Appropriate fitness functions for implementing the CCPSO algorithm to tackle the considered problem are also presented. Section 4 provides the simulation results for illustration and comparison. Finally, we conclude this paper in Section 5.

We use the following notation in the paper: Superscript $.^H$ denotes the transpose-conjugate operation. Superscript $.^T$ denotes the transpose operation. $\mathbf{S}(:,j)$ stands for the jth column vector of matrix $\mathbf{S}$. The Frobenius norm of a matrix $\mathbf{S}$ is denoted by $||\mathbf{S}||$.

2 System model and fundamentals

2.1 Downlink transmission model

For the downlink transmission of a wireless MIMO channel, Fig. 1 depicts the basic system model of MIMO channels for the BS with N_B antennas and each UE with N_U antennas. Consider that the BS is equipped with a linear array of N_B antennas and K UEs with a linear array of N_U antennas deployed at each UE, we assume that the linear arrays of BS and each UE have the reference point located at the first antenna element as shown in Fig. 2. A typical $N_U \times N_B$ MIMO channel matrix $\mathbf{H}$ between the BS

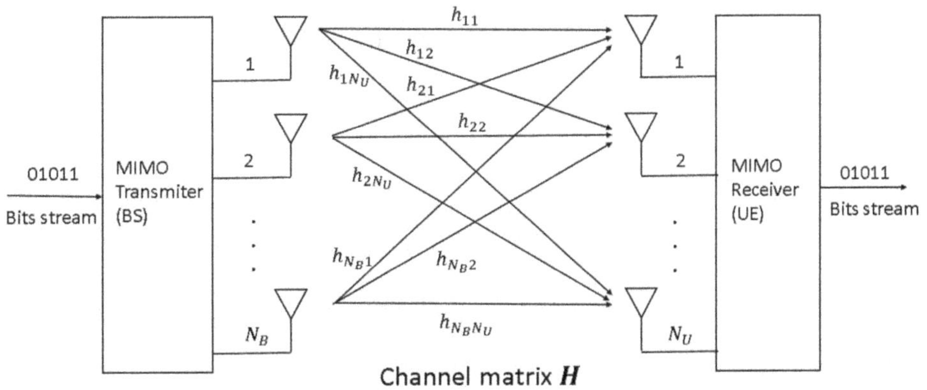

Fig. 1 Downlink block diagram of MIMO channels for the BS with N_B antennas and each UE with N_U antennas

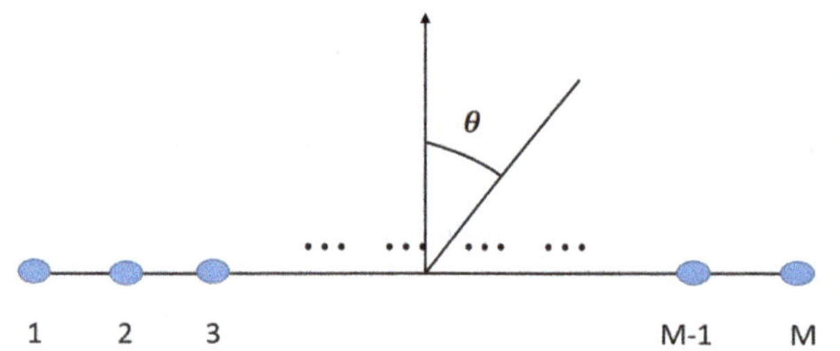

Fig. 2 A uniform linear array with signal direction angle θ

and each UE for the corresponding MIMO channel can be written as follows [17]:

$$\mathbf{H}=\begin{bmatrix} h_{1,1} & \cdots & h_{1,N_B} \\ \vdots & \ddots & \vdots \\ h_{N_U,1} & \cdots & h_{N_U,N_B} \end{bmatrix} =\sum_{l=1}^{L} a_l\sqrt{N_U N_B/L}\,\mathbf{e}_r(\theta_{r,l})\mathbf{e}_t(\theta_{t,l})^H, \tag{1}$$

where $\theta_{r,l}$ and $\theta_{t,l}$ denote the corresponding direction angles of arrival and departure, respectively. Moreover, the vectors $\mathbf{e}_r(\theta_{r,l})$ and $\mathbf{e}_t(\theta_{t,l})$ represent the receive and transmit array response vectors associated with the angles of $\theta_{r,l}$ and $\theta_{t,l}$, respectively. a_l is the complex gain of the lth downlink transmission path. L is the total number of transmission paths. Moreover, the unit-norm array response vector of an M-element linear array with direction angle = θ off array broadside, irrespective of transmit and receive antennas, can be expressed as follows:

$$\mathbf{e}(\theta) = \sqrt{1/M}\begin{pmatrix} 1 \\ e^{j\pi(A_1 sin\theta)} \\ \vdots \\ e^{j\pi(A_M sin\theta)} \end{pmatrix}, \tag{2}$$

where $A_i, i = 1, 2, ..., M$ represents the location normalized by the half signal wavelength of the ith antenna element. On the other hand, if the BS is equipped with a circular array of N_B antennas and K UEs with a circular array of N_U antennas deployed at each UE. The corresponding $N_U \times N_B$ MIMO channel matrix $\mathbf{H}$ between the BS and each UE is given by [17]:

$$\mathbf{H}=\begin{bmatrix} h_{1,1} & \cdots & h_{1,N_B} \\ \vdots & \ddots & \vdots \\ h_{N_U,1} & \cdots & h_{N_U,N_B} \end{bmatrix} =\sum_{l=1}^{L} a_l\sqrt{N_U N_B/L}\,\mathbf{e}_r\left(\phi_{r,l},\theta_{r,l}\right)\mathbf{e}_t\left(\phi_{t,l},\theta_{t,l}\right)^H, \tag{3}$$

where $(\phi_{r,l}, \theta_{r,l})$ and $(\phi_{t,l}, \theta_{t,l})$ denote the corresponding (azimuth, elevation) angles of arrival and departure, respectively. Moreover, the vectors $\mathbf{e}_r(\phi_{r,l}, \theta_{r,l})$ and $\mathbf{e}_t(\phi_{t,l}, \theta_{t,l})$ represent the receive and transmit array response vectors associated with the angles of $(\phi_{r,l}, \theta_{r,l})$ and $(\phi_{t,l}, \theta_{t,l})$, respectively. Finally, a_l is the complex gain of the lth downlink transmission path. Assume that the circular arrays of BS and each UE have the reference point located at the center of the circular array and radius equal to R as shown in Fig. 3, then the unit-norm array response vector of an M-element circular array with azimuth angle = ϕ and elevation angle = θ, irrespective of transmit and receive antennas, can be expressed as follows:

$$\mathbf{e}(\phi,\theta) = \sqrt{1/M}\begin{pmatrix} e^{j\pi\rho(sin(\theta)cos(\phi-\varphi_1))} \\ \vdots \\ e^{j\pi\rho(sin(\theta)cos(\phi-\varphi_M))} \end{pmatrix}, \tag{4}$$

where φ_k represents the angle difference between the first and the kth antenna elements of the circular array. For a uniform circular array, $\varphi_k = \frac{2\pi(k-1)}{M}$. $\rho = \frac{1}{\sqrt{2(1-cos(2\pi/M))}}$ represents the radius normalized by the half of the signal wavelength.

For the downlink of a multiuser MIMO wireless system with K UEs , the received downlink training signal matrix $\mathbf{Y}_k^{DT}$ at the kth UE can be written by [4]:

$$\mathbf{Y}_k^{DT} = \mathbf{H}_k\mathbf{S}_k + \mathbf{N}_k, \tag{5}$$

where $\mathbf{S}_k$ denotes the $N_B \times N_B$ training signal matrix and $\mathbf{N}_k$ the $N_U \times N_B$ background noise matrix received by the kth UE. Moreover, the $N_U \times 1$ received signal vector $\mathbf{y}_k^{DL}$ at the kth UE can be written as [17]:

$$\mathbf{y}_k^{DL} = \mathbf{H}_k\mathbf{W}_k\mathbf{x}_k + \sum_{i\neq k}^{K}\mathbf{H}_k\mathbf{W}_i\mathbf{x}_i + \mathbf{n}_k, \tag{6}$$

where $\mathbf{x}_k$ denotes the data symbol vector and $\mathbf{n}_k$ the $N_U \times 1$ background noise vector. $\mathbf{W}_k$ denotes the corresponding precoding matrix with appropriate size and $\mathbf{H}_k$ with $N_U \times N_B$ the downlink channel matrix associated with the kth UE.

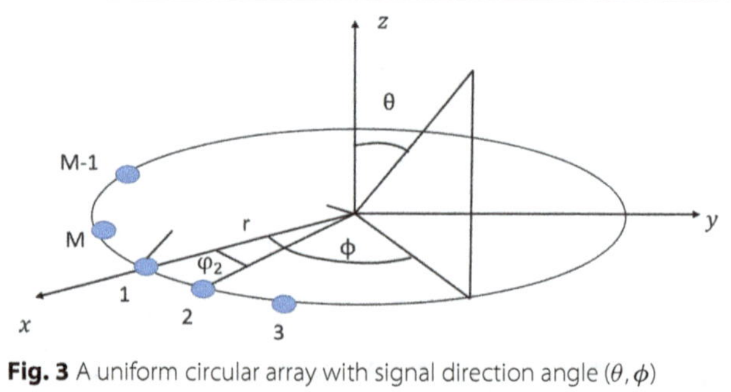

Fig. 3 A uniform circular array with signal direction angle (θ, ϕ)

In the literature, a point-to-point wireless MIMO system with spatial correlation effect only was considered in [18]. In the presence of the spatial correlation and mutual coupling effects for a downlink communication, according to the analysis presented in [19, 20], the channel matrix **H** for the non line of sight (NLOS) propagation condition (i.e., $L > 1$) can be simplified as the following expression:

$$\mathbf{H} = \begin{bmatrix} h_{1,1} & \dots & h_{1,N_B} \\ \vdots & \ddots & \vdots \\ h_{N_U,1} & \dots & h_{N_U,N_B} \end{bmatrix}$$

$$= \sum_{l=1}^{L} a_l \sqrt{N_U N_B / L} \left(\mathbf{R}_U^{MC}\right)^{1/2} \mathbf{H}_{w,l} \left(\mathbf{R}_B^{MC}\right)^{1/2}, \tag{7}$$

where $\mathbf{H}_{w,l}$ denotes the matrix with i.i.d. Gaussian entries for the lth downlink transmission path. $\mathbf{R}_B^{MC}$ and $\mathbf{R}_U^{MC}$ denote the matrices containing the normalized spatial correlation coefficients in the presence of mutual coupling between the antenna elements deployed at BS and UEs, respectively. Moreover, the (i,j)th elements of $\mathbf{R}_U^{MC}$ and $\mathbf{R}_B^{MC}$ are respectively given as follows [20]:

$$R_U^{MC}(i,j) = \frac{1}{\sqrt{P_{Ui} P_{Uj}}} \sum_{m=1}^{N_U} \sum_{n=1}^{N_U} C_U(i,m) C_U(j,n)^* R_U(m,n), \tag{8}$$

and

$$R_B^{MC}(i,j) = \frac{1}{\sqrt{P_{Bi} P_{Bj}}} \sum_{m=1}^{N_B} \sum_{n=1}^{N_B} C_B(i,m) C_B(j,n)^* R_B(m,n), \tag{9}$$

where $C_U(i,m)$ and $C_B(j,n)$ represent the (i,m)th and (j,n)th elements of the mutual coupling matrices $\mathbf{C}_U$ and $\mathbf{C}_B$ associated with the receiver antenna array and the transmitter antenna array, respectively. They can be calculated by [21]:

$$\mathbf{C} = (Z_A + Z_T) + (\mathbf{Z} + Z_T \mathbf{I})^{-1}, \tag{10}$$

where

$$\mathbf{Z} = \begin{bmatrix} Z_{1,1} & \dots & Z_{1,M} \\ \vdots & \ddots & \vdots \\ Z_{M,1} & \dots & Z_{M,M} \end{bmatrix} \tag{11}$$

and M is the number of antennas. Consider the dipole length is half wavelength ($l = \frac{\lambda}{2}$). The (a,b)th element of **Z** is generated as follows:

$$Z_{a,b} = \begin{cases} \frac{\eta}{4\pi}(\zeta + \ln(2kl) - C_i(2kl)) + j\frac{\eta}{4\pi}S_i(2kl) = Z_A \\ , a = b \\ \frac{\eta}{4\pi}\left\{(2C_i(kd_{a,b})) - C_i\left(k\left(\sqrt{d_{a,b}^2 + l^2} + l\right)\right)\right. \\ \left. -C_i\left(k\left(\sqrt{d_{a,b}^2 + l^2} - l\right)\right)\right\} \\ -j\frac{\eta}{4\pi}\left\{(2S_i(kd_{a,b})) - S_i\left(k\left(\sqrt{d_{a,b}^2 + l^2} + l\right)\right)\right. \\ \left. -S_i\left(k\left(\sqrt{d_{a,b}^2 + l^2} - l\right)\right)\right\}, a \neq b \end{cases}, \tag{12}$$

where

$$Z_T = Z_A^*,$$

$$S_i(x) = \int_0^x \frac{\sin(\tau)}{\tau} d\tau,$$

$$C_i(x) = -\int_x^\infty \frac{\cos(\tau)}{\tau} d\tau,$$

η denotes intrinsic impedance $\simeq 120\pi(ohms)$, $\zeta = \ln(1.781) \simeq 0.5772$, $k = \frac{2\pi}{\lambda}$, and $d_{a,b}$ is the distance between the ath and the bth antennas (in λ), a and b $= 1, 2, ..., M$. Moreover, P_{Ui} and P_{Bi} represent the average powers of the ith antennas at UEs and BS, respectively. According to [20], they can be computed as follows:

$$P_{Ui} = \Re\left[\sum_{m=1}^{N_U} \sum_{n=1}^{N_U} C_U(i,m) C_U(i,n)^* R_U(m,n)\right] \tag{13}$$
$$\sum_{n=1}^{N_U} |C_U(i,n)|^2 R_U(n,n),$$

and

$$P_{Bi} = \Re\left[\sum_{m=1}^{N_B} \sum_{n=1}^{N_B} C_B(i,m) C_B(i,n)^* R_B(m,n)\right] \tag{14}$$
$$\sum_{n=1}^{N_B} |C_B(i,n)|^2 R_B(n,n),$$

respectively, where $\Re[x]$ denotes the real part of x. $R_U(m,n)$ and $R_B(m,n)$ represent the (m,n)th elements of the $\mathbf{R}_U$ and $\mathbf{R}_B$ matrices containing the normalized spatial correlation coefficients associated with the antennas deployed at UEs and BS, respectively. Moreover, the formulas for computing $R_U(m,n)$ and $R_B(m,n)$ regarding circular or linear antennas are provided by [20].

2.2 Downlink communication using QOSTBC transmission

Space-time block coding (STBC) was presented by [22] to combine coding, modulation, and signal processing for achieving transmit diversity of a wireless communication. The space-time block encoder generates the simplest type of spatial temporal codes to exploit the transmission diversity provided by multiple transmit antennas in wireless communications. For downlink transmission, the STBC at BS with M antennas encodes the data to be transmitted into M transmitted code symbols. At each UE, the signal received by each receive antenna is the M transmitted signals plus noise. It was shown in [22] that full diversity can be achieved and a very simple maximum-likelihood decoding algorithm can be employed at the decoder with reduced data rates. However, it was shown in [23] that the data rate of a full-diversity code is less than or equal to one. To achieve the goal of a full-rate full-diversity STBC, a novel transmit diversity scheme was proposed by [24]. This so-called two-branch transmit diversity scheme is developed based on a 2×2 block code carried out in sets of two modulated symbols. As a result, the transmission rate can be one. Nevertheless, this STBC scheme is only suitable for two transmit antennas.

In order to search for similar schemes available for MIMO systems with more than two transmit antennas to achieve diversity level higher than two, orthogonal space-time block code (OSTBC) was then introduced by [25]. OSTBC is a generalization of the scheme of [24] to arbitrary number of transmit antennas. Moreover, OSTBC preserves the advantage of having linear maximum-likelihood decoding with full transmit diversity. Unfortunately, it is also shown [25] that a complex orthogonal design and the corresponding space-time block code which provides full diversity and full transmission rate is not possible for more than two transmit antennas. Moreover, a large MIMO system using OSTBC transmission usually requires an OSTBC of very large dimension in order to take the advantages of OSTBC for transmission. Nevertheless, employing a large dimension OSTBC is not possible for complex constellations, and the data transmission rate of the resulting OSTBC MIMO system will be significantly reduced even if it can be achieved. Recently, an OSTBC transmission scheme embedding the OSTBCs of small dimensions based on the concept of null space to achieve good rate for large MIMO systems was proposed in [26]. It is shown by simulation of [26] that the code rate can be improved at the cost of degraded performance. For practical applications, as mentioned by [5], when large-scale or massive MIMO systems are considered, the BS is deployed with massive antennas to simultaneously serve a large number of UEs within the same frequency band and achieve higher spectral and energy efficiency. However, it has been shown in [26] that the proposed scheme of [26] is not work if the number of transmit antennas is larger than the number of receive antennas. Based on the similar concept of embedding the small OSTBCs, a code transmission scheme for large MIMO system was presented in [27]. Pilot matrices are employed in the proposed scheme to increase the data rate. Another work [28] considered the drawbacks caused by using STBCs and OSTBCs. A space-time transmission scheme (STTS) is then proposed in [28] for large MIMO communication systems. In this work, the proposed scheme compromises diversity order to obtain good data rate and decouple decoding. It was also mentioned that the proposed scheme may be useful for base station (BS) to BS communication. In the literature, some more works employ differential/non-differential orthogonal/non-orthogonal space-time block code (STBC) on the subject of fastly varying channels for arbitrarily correlated Rayleigh/Ricean channels have been presented in [29, 30]. On the other hand, a precoder design for orthogonal/non-orthogonal STBC over correlated Ricean MIMO channels with invertible correlation matrices has been developed in [31].

This drawback of the OSTBC motivates the research work [32] on developing the so-called quasi-orthogonal space-time block code (QOSTBC). It relaxes the orthogonality constraints due to OSTBC by constraining subsets of data symbols orthogonal to each other instead of constraining every single symbol orthogonal to any other. It has been shown in [32] that the QOSTBC can achieve 1/2 of the full transmission rate for any number of transmission antennas as well as the codes with 3/4 of the full transmission rate for the specific cases of three and four transmit antennas. A typical example of a full rate quasi-orthogonal code that achieves full transmission rate for four transmit antennas is as follows:

$$\mathbf{S} = \begin{bmatrix} S_{1,2} & S_{3,4} \\ S_{3,4} & S_{1,2} \end{bmatrix} = \begin{bmatrix} s_1 & s_2 & s_3 & s_4 \\ -s_2^* & s_1^* & -s_4^* & s_3^* \\ -s_3^* & -s_4^* & s_1^* & s_2^* \\ s_4 & -s_3 & -s_2 & s_1 \end{bmatrix} \tag{15}$$

where s_i, $i = 1, 2, 3, 4$, are the symbols selected based on the input bits. Consider the simplest mobile communication system where BS is equipped with four transmit antennas and the UE is equipped with one receive antenna. Assume that the system has a slow fading channel. The signal vector received at the UE can be modeled as:

$$\mathbf{y} = \begin{bmatrix} y(1) \\ y(2) \\ y(3) \\ y(4) \end{bmatrix} = \left(\begin{bmatrix} h_{1,1} & h_{1,2} & h_{1,3} & h_{1,4} \end{bmatrix} \mathbf{W} \begin{bmatrix} s_1 & s_2 & s_3 & s_4 \\ -s_2^* & s_1^* & -s_4^* & s_3^* \\ -s_3^* & -s_4^* & s_1^* & s_2^* \\ s_4 & -s_3 & -s_2 & s_1 \end{bmatrix} \right)^T + \begin{bmatrix} n(1) \\ n(2) \\ n(3) \\ n(4) \end{bmatrix} \tag{16}$$

where $y(i)$ and $n(i)$ represent the signal and the noise received by the receive antenna at the ith time slot. $\mathbf{W}$ denotes the corresponding 4×4 spatial precoding matrix.

3 The robust beamforming and spatial precoding method

In this section, we present a robust method to design beamforming and spatial precoding for dealing with the mutual coupling effect (MCE) and the spatial correlation effect (SCE) for QOSTBC massive wireless MIMO channels. To construct the spatial precoding matrix, the optimization of beamforming coefficients taking into account the spatial correlation as well as the mutual coupling effects between the antennas is established. To tackle the SCE and MCE which are due to antenna spacing and array geometry, we construct an appropriate optimization problem which can be solved by using the CCPSO algorithm to find the optimal beamforming coefficients for the spatial precoding matrix and the optimal locations for the antenna elements at the BS and UEs simultaneously. As shown below, the proposed method makes a QOSTBC massive MIMO wireless channel that achieves full-rate transmission for any number of transmit antennas achievable.

Instead of using a preset fixed beamforming coefficients as shown in [3] for the B beam groups of the N_B antennas at the BS, we set a unit-norm steering vector of the bth beam group according to the BSSP scheme [4] as follows:

$$\mathbf{v}_b = \frac{1}{\sqrt{a_{b,1} + a_{b,2} + ... + a_{b,N_B}}} \begin{pmatrix} a_{b,1}e^{j\theta_{b,1}} \\ a_{b,2}e^{j\theta_{b,2}} \\ \vdots \\ a_{b,N_B}e^{j\theta_{b,N_B}} \end{pmatrix}, \quad (17)$$

where $b = 1, 2, \cdots, B$, $a_{b,k}$ and $\theta_{b,k}$, $k = 1, 2, \cdots, N_B$, are the variables to be adjusted by the CCPSO. In order to provide an appropriate precoding and reduce the unacceptable feedback overhead in the downlink transmission of massive MIMO channels, each UE selects N beams from the precoding matrix $\mathbf{P} = [\mathbf{v}_1, \mathbf{v}_2..., \mathbf{v}_B]$ as its precoder and feeds the selected beam indices back to the BS. Therefore, the proposed method employs beamforming to estimate channel state information (CSI) by using a small number of downlink training symbols for FDD massive MIMO systems. The is due to that partitioning all the antennas of the BS into several beam groups enables each antenna in one beam group to transmit the downlink training signal to UEs simultaneously. Since UEs only need to feed the selected beam indices back to the BS, the CSI feedback overhead in bits required by the proposed method is almost the same as that of the original BBSP method of [3]. It can be written as $N_f = K \times N \times log_2 B$. Moreover, the UE selects a beam according to the channel gain of each beam. For simplicity, assume that the kth UE selects the first N beams from the B beam groups as its serving beams. Accordingly, the signal vector received at the kth UE in the downlink transmission with power constraints on the selected beams becomes:

$$\mathbf{y}_k^{DL} = \mathbf{H}_k\mathbf{W}_{pk}\mathbf{T}_k\mathbf{x}_k + \sum_{i\neq k}^{K} \mathbf{H}_k\mathbf{W}_{pi}\mathbf{T}_i\mathbf{x}_i + \mathbf{n}_k, \quad (18)$$

where $\mathbf{x}_k$ denotes an $N \times 1$ data symbol vector and K the number of UEs. The precoder $\mathbf{W}_{pk}$ is given by:

$$\mathbf{W}_{pk} = [\mathbf{v}_1\mathbf{v}_2...\mathbf{v}_N]. \quad (19)$$

Moreover, $\mathbf{T}_k$ represents the power constraint matrix associated with the selected N beams and is given by:

$$\mathbf{T}_k = \begin{bmatrix} \sqrt{t_{k1}} & \cdots & 0 \\ \vdots & \ddots & \vdots \\ 0 & \cdots & \sqrt{t_{kN}} \end{bmatrix}, \quad (20)$$

where t_{ki} denotes the power distributed to the ith beam selected by the kth UE and the total power transmitted to the N beams is constrained by:

$$\sum_{i}^{N} t_{ki} = P_t. \quad (21)$$

To incorporate the advantages of full transmission rate and low ML decoding complexity over the conventional multiuser beamforming methods (e.g., zero-forcing beamforming and MMSE beamforming) by using a simple 4×4 QOSTBC with the proposed joint beamforming and spatial precoding method, we can easily set the number of beam groups of the proposed beamforming scheme to 4, i.e., $B = 4$. As a result, we can see that the proposed method preserves the advantages of computational complexity reduction (low ML decoding complexity) and full-rate transmission as compared to using some arbitrary linear dispersion code for massive MIMO communication systems.

Next, consider a QOSTBC that achieves full transmission rate for four-transmit antennas as described in (16) for downlink transmission, we note from (18) and (16) that the signal matrix received at each UE can be reformulated as follows:

$$\mathbf{Y}_k^{DL} = \mathbf{H}_k\mathbf{W}\mathbf{T}\mathbf{S} + \mathbf{n}_k, \quad (22)$$

where $\mathbf{Y}_k^{DL}$ represents the $N_U \times 4$ data matrix. The ith row of $\mathbf{Y}_k^{DL}$ is the 1×4 data vector received at the ith antenna of the kth UE. $\mathbf{H}_k$ is still the $N_U \times N_B$ channel matrix associated with the transmission. $\mathbf{W}$ represents the $N_B \times 4$ spatial precoding matrix and given by $\mathbf{W} = [\mathbf{w}_1, \mathbf{w}_2, \mathbf{w}_3, \mathbf{w}_4]$, where each $\mathbf{w}_i, i = 1, 2, 3, 4$, represents the unit-norm steering vector of the ith beam group under the QOSTBC transmission. $\mathbf{T}$ denotes the corresponding 4×4 power constraint matrix given by:

$$\mathbf{T} = \begin{bmatrix} \sqrt{t_1} & 0 & 0 & 0 \\ 0 & \sqrt{t_2} & 0 & 0 \\ 0 & 0 & \sqrt{t_3} & 0 \\ 0 & 0 & 0 & \sqrt{t_4} \end{bmatrix} \quad (23)$$

and **S** the 4 × 4 QOSTBC symbol matrix as shown by (15). This formulation reveals that the resulting QOSTBC transmission system can provide a significant advantage of a full transmission rate for massive wireless communication systems using $N_B >> 4$ transmit antennas.

3.1 The formulation of the optimization problem

In this section, we present the problem formulation for the robust method based on the results developed above. The resulting optimization problem will be solved by using the CCPSO algorithm under appropriate fitness functions. The CCPSO algorithm demonstrates its potential for effectively solving multimodal and nonconvex highly nonlinear optimization problems with significantly higher dimensions [13]. As shown by (17), there are $2N_B$ parameters, namely, $a_{b,1}, a_{b,2}, ..., a_{b,N_B}, \theta_{b,1}, \theta_{b,2}, ..., \theta_{b,N_B}$ which can be adjusted during the optimization process. To mitigate the SCE and MCE, we aim at adjusting the normalized locations of the antenna elements deployed at BS and UEs. For the case of using linear arrays with M antennas, $A_i, i = 1, 2, ..., M$ are to be optimally decided. In contrast, the parameters $\varphi_k, k = 1, 2, ..., M$ are to be varied optimally when using a circular arrays with M antennas. Moreover, the elements $t_i, i = 1, 2, ..., 4$ of the power constraint matrix **T** are to be optimally adjusted. Accordingly, we define a parameter vector which is to be found to achieve the robust beamforming and spatial precoding when N_B transmit antennas and N_U receive antennas are deployed at BS and each UE, respectively as follows:

$$\mathbf{F} = \left[a_{1,1}, a_{1,2}, ..., a_{B,N_B}, \theta_{1,2}, ..., \theta_{B,N_B}, A_1^t, ..., A_{N_B}^t, A_1^r, ..., A_{N_U}^r, t_1, ..., t_4\right], \tag{24}$$

for the case of using linear antenna arrays, where the sets of $A_1^t, ..., A_{N_B}^t$ and $A_1^r, ..., A_{N_B}^r$ represent the normalized positions of transmit and receive antenna elements, respectively. In contrast,

$$\mathbf{F} = \left[a_{1,1}, a_{1,2}, ..., a_{B,N_B}, \theta_{1,2}, ..., \theta_{B,N_B}, \varphi_1^t, ..., \varphi_{N_B}^t, \varphi_1^r, ..., \varphi_{N_U}^r, t_1, ..., t_4\right], \tag{25}$$

for the case of using circular arrays, where the sets of $\varphi_1^t, ..., \varphi_{N_B}^t$ and $\varphi_1^r, ..., \varphi_{N_U}^r$ represent the angle differences of transmit and receive antenna elements, respectively. From (24) and (25), we note that the number of entries in the parameter vector **F** is linearly increased by N_B and N_U. Hence, the computational complexity required for finding the optimal entries of **F** becomes huge during the optimization process for massive wireless communication systems. To alleviate this difficulty, a two-step optimization procedure is presented as follows. In the first step, we employ the CCPSO algorithm to find the optimal parameter set $\left(A_1^t, ..., A_{N_B}^t, A_1^r, ..., A_{N_U}^r\right)$ related to the positions of the linear transmit and receive antenna elements, or the optimal parameter set $\left(\varphi_1^t, ..., \varphi_{N_B}^t, \varphi_1^r, ..., \varphi_{N_U}^r\right)$ related to the positions of the circular transmit and receive antenna elements. After obtaining the optimal locations for the antenna elements, the channel matrix $\mathbf{H}_K$ as shown by (7) can be determined. In the second step, based on the channel matrix $\mathbf{H}_K$ obtained in the first step, the CCPSO algorithm is utilized to find the optimal parameter set $(a_{1,1}, a_{1,2}, ..., a_{4,N_B}, \theta_{1,2}, ..., \theta_{4,N_B}, t_1, ..., t_4)$ related to beamforming coefficients and the power distribution. According to our experience, the two-step optimization procedure reduces the computation time significantly when performing simulation examples for illustration.

During the the first-step optimization process by using the CCPSO algorithm, each particle is treated as a point in the $D(= N_U + N_B)$-dimensional optimization problem space and a swarm consists of D random particles. Then, the CCPSO searches for the best position (solution or optimum) by updating generations until obtaining a relatively steady position or reaching the limit of a preset iteration number. At the tth moment or iteration, the position vector of the ith particle is given by $\mathbf{x_i}(t) = \left[A_1^t, ..., A_{N_B}^t, A_1^r, ..., A_{N_U}^r\right]$ for the linear transmit and receive antenna elements or $\mathbf{x_i}(t) = \left[\varphi_1^t, ..., \varphi_{N_B}^t, \varphi_1^r, ..., \varphi_{N_U}^r\right]$ for the circular transmit and receive antenna elements. The personal best position vector $\mathbf{p_i}(t)$ and the global best position vector $\mathbf{g}(t)$ are obtained by evaluating the performances in terms of the fitness values associated with the current population of particles. Moreover, we update the position and velocity characterizing a particle status on the search space according to the formulas given by [13] and convert a particle position vector into a candidate solution vector by employing an appropriate mapping. The performance of the mapped solution vector evaluated by a massive MIMO channel is viewed as the fitness of the corresponding particle. At the end of iteration, the global best position vector $\mathbf{g}(t_{max})$ is regarded as the optimal position vector for the linear or circular transmit and receive antenna elements. In general, the optimization process will be terminated if the specified maximum iteration number t_{max} is reached or the best particle position of the whole swarm remains almost static for a significantly large number of successive iterations.

During the second-step optimization process, each particle is treated as a point in the $D(= 8N_B + 4)$-dimensional optimization problem space and a swarm consists of D random particles. Then, the CCPSO searches for the best position (solution or optimum) by updating generations until obtaining a relatively steady position or reaching the limit of a preset iteration number. At the tth moment or iteration, the position vector of the ith particle is given by $\mathbf{x_i}(t) =$

$\left[a_{1,1}, a_{2,1}, \cdots, a_{4,1}, a_{1,2}, \cdots, a_{4,N_B}, \theta_{1,2}, ..., \theta_{4,N_B}, t_1, ..., t_4\right]$. The personal best position vector $\mathbf{p_i}(t)$ and the global best position vector $\mathbf{g}(t)$ are obtained by evaluating the performances in terms of the fitness values associated with the current population of particles. Moreover, we update the position and velocity characterizing a particle status on the search space according to the formulas given by [13] and convert a particle position vector into a candidate solution vector by employing an appropriate mapping. The performance of the mapped solution vector evaluated by a massive MIMO channel is viewed as the fitness of the corresponding particle. At the end of iteration, the global best position vector $\mathbf{g}(t_{max})$ is regarded as the optimal beamforming coefficient vector. Again, the optimization process will be terminated if the specified maximum iteration number t_{max} is reached or the best particle position of the whole swarm remains almost static for a significantly large number of successive iterations.

3.2 Fitness functions for CCPSO

3.2.1 *Fitness function for the first-step CCPSO*

In the first-step optimization process, we aim at finding the optimal positions for antenna elements. As shown by (7), the positions of antenna elements determine the entries of $\mathbf{R}_U^{MC}$ and $\mathbf{R}_B^{MC}$. Hence, the characteristics of the downlink channel matrix is significantly affected by the positions of antenna elements. According to information theory, the mutual information is used to measure how much information the output of a channel contains about the input. It has been shown in the literature like [33] that the average mutual information (AMI) can be viewed as an index for evaluating the characteristics of a channel matrix. Accordingly, it is appropriate to find the optimal positions of transmit and receive antenna elements by maximizing the AMI associated with the downlink transmission channel matrix $\mathbf{H}$. According to [33], the AMI of a channel matrix $\mathbf{H}$ with size $N_U \times N_B$ is given by:

$$\mathbf{AMI} = E\left[\log_2\left(\det\left[\mathbf{I} + \frac{SNR}{N_B}\mathbf{H}\mathbf{H}^H\right]\right)\right], \tag{26}$$

where $\mathbf{I}$ denotes the $N_U \times N_U$ identity matrix and $\det[\mathbf{X}]$ the determinant of the matrix $\mathbf{X}$. SNR represents the signal-to-noise power ratio at each receive antenna. Therefore, we define an appropriate fitness for implementing the first-step optimization process by using the CCPSO algorithm as follows:

$$\mathbf{Fitness1} = \frac{1}{Q}\sum_{q=1}^{Q}\log_2\left(\det\left[\mathbf{I} + \frac{SNR}{N_B}\mathbf{H}_q\mathbf{H}_q^H\right]\right), \tag{27}$$

where Q denotes the number of the Monte Carlo random channel realizations for the MIMO channel matrices and $\mathbf{H}_q$ the qth channel realization.

3.2.2 *Fitness function for the second-step CCPSO*

For the second-step optimization process, we develop an appropriate fitness function based on the bit error rate (BER) performance for implementing the CCPSO to search the optimal spatial precoding matrix and the power constraints for downlink communication in massive MIMO channels. For practical applications, massive MIMO channels with linear receivers, such as the minimum mean square error (MMSE) receiver have been considered in several emerging standards, e.g., IEEE 802.11n and 802.16e. We consider the MMSE receiver $\mathbf{r}$ at each UE for the case of 4 UEs, i.e., $K = 4$, for simplicity. Based on the downlink transmission model for the received signal shown by (22) and the MMSE receiver, we can express the signal received at the kth UE as follows:

$$\mathbf{y}_k^{DL} = \mathbf{r}\mathbf{Y}_k^{DL} = \mathbf{r}(\mathbf{H}_k\mathbf{W}\mathbf{T}\mathbf{S} + \mathbf{n}_k), \tag{28}$$

where $\mathbf{r}$ is the MMSE receiver. Following the derivation presented in [4], the $\mathbf{r}$ with size $1 \times N_U$ is given by:

$$\mathbf{r} = \mathbf{w}_k^H\mathbf{H}_k^H\left(\sum_{i=1}^{4}\mathbf{H}_k\mathbf{w}_i\mathbf{w}_i^H\mathbf{H}_k^H + (1/\Gamma)\mathbf{I}_{N_U}\right)^{-1}, \tag{29}$$

where Γ denotes the signal-to-noise power ratio (SNR) and $\mathbf{w}_k$ the unit-norm steering vector of the kth beam group selected by the kth UE. Accordingly, the signal received at the kth UE during the four time slots can be obtained from (16) and (28) as follows:

$$\mathbf{y}_k^{DL} = \begin{bmatrix} y_k(1) \\ y_k(2) \\ y_k(3) \\ y_k(4) \end{bmatrix} = \left(\begin{bmatrix} \hat{h}_1 & \hat{h}_2 & \hat{h}_3 & \hat{h}_4 \end{bmatrix}\begin{bmatrix} s_1 & s_2 & s_3 & s_4 \\ -s_2^* & s_1^* & -s_4^* & s_3^* \\ -s_3^* & -s_4^* & s_1^* & s_2^* \\ s_4 & -s_3 & -s_2 & s_1 \end{bmatrix}\right)^T + \begin{bmatrix} n(1) \\ n(2) \\ n(3) \\ n(4) \end{bmatrix}, \tag{30}$$

where

$$\begin{bmatrix} \hat{h}_1 & \hat{h}_2 & \hat{h}_3 & \hat{h}_4 \end{bmatrix} = \mathbf{r}\mathbf{H}_k\mathbf{W}\mathbf{T} \tag{31}$$

Moreover, it is easy to show that (30) can be rewritten as:

$$\mathbf{y}_k^{DL} = \begin{bmatrix} y_k(1) \\ y_k{}^*(2) \\ y_k{}^*(3) \\ y_k(4) \end{bmatrix} = \left(\begin{bmatrix} \hat{h}_1 & -\hat{h}_2 & -\hat{h}_3 & \hat{h}_4 \\ \hat{h}_2^* & \hat{h}_1^* & -\hat{h}_4^* & -\hat{h}_3^* \\ \hat{h}_3^* & -\hat{h}_4^* & \hat{h}_1^* & -\hat{h}_2^* \\ \hat{h}_4 & \hat{h}_3 & \hat{h}_2 & \hat{h}_1 \end{bmatrix}\begin{bmatrix} s_1 \\ s_2^* \\ s_3^* \\ s_4 \end{bmatrix}\right)^T + \begin{bmatrix} n(1) \\ n^*(2) \\ n^*(3) \\ n(4) \end{bmatrix} \tag{32}$$

Let

$$\hat{\mathbf{H}}_k = \begin{bmatrix} \hat{h}_1 & -\hat{h}_2 & -\hat{h}_3 & \hat{h}_4 \\ \hat{h}_2^* & \hat{h}_1^* & -\hat{h}_4^* & -\hat{h}_3^* \\ \hat{h}_3^* & -\hat{h}_4^* & \hat{h}_1^* & -\hat{h}_2^* \\ \hat{h}_4 & \hat{h}_3 & \hat{h}_2 & \hat{h}_1 \end{bmatrix} \quad (33)$$

and

$$\hat{\mathbf{n}}_k = \begin{bmatrix} n(1) \\ n^*(2) \\ n^*(3) \\ n(4) \end{bmatrix}, \quad (34)$$

(30) becomes

$$\mathbf{y}_k^{DL} = \begin{bmatrix} y_k(1) \\ y_k{}^*(2) \\ y_k{}^*(3) \\ y_k(4) \end{bmatrix} = \hat{\mathbf{H}}_k(:,i)s_i + \sum_{j=1,j\neq k}^{4} \mathbf{H}_k(:,j)s_j + \hat{\mathbf{n}}_k, \quad (35)$$

where $\hat{\mathbf{H}}_k(:,i)$ represents the ith column of $\hat{\mathbf{H}}_k$. After performing MMSE receiver operation on $\mathbf{y}_k^{DL}$, the kth UE obtains:

$$\hat{s}_i = \hat{\mathbf{R}}_k \begin{bmatrix} y_k(1) \\ y_k{}^*(2) \\ y_k{}^*(3) \\ y_k(4) \end{bmatrix} = \hat{\mathbf{R}}_k\hat{\mathbf{H}}_k \begin{bmatrix} s_1 \\ s_2^* \\ s_3^* \\ s_4 \end{bmatrix} + \hat{\mathbf{R}}_k\hat{\mathbf{n}}_k, \quad (36)$$

where $\hat{\mathbf{R}}_k$ with size 1×4 represents the corresponding MMSE receiver which can be derived as follows. According to the orthogonality principle, we have:

$$E\left[\epsilon\left(\mathbf{y}_k^{DL}\right)^H\right] = \mathbf{0}, \quad (37)$$

where

$$\epsilon = \hat{\mathbf{R}}_k\left(\hat{\mathbf{H}}_k(:,i)s_i + \sum_{j=1,j\neq k}^{4} \hat{\mathbf{H}}_k(:,j)s_j + \hat{\mathbf{n}}_k\right) - s_i. \quad (38)$$

Moreover, assume that s_k, s_i, and $\mathbf{n}_k$ are independent of each other and their expectations are 0. From (37) and (38), we can get:

$$\hat{\mathbf{R}}_k \sum_{j=1}^{4} \hat{\mathbf{H}}_k(:,j)\hat{\mathbf{H}}_k^H(:,j) - \hat{\mathbf{H}}_k^H(:,i) + \hat{\mathbf{R}}_k\mathbf{r}\mathbf{r}^H(1/\Gamma\mathbf{I}_4) = \mathbf{0}. \quad (39)$$

where $\mathbf{I}_4$ denotes the 4×4 identity matrix. Consequently, the MMSE receiver $\hat{\mathbf{R}}_k$ can be obtained from (39) as follows:

$$\hat{\mathbf{R}}_k = \hat{\mathbf{H}}_k^H(:,i)\left(\sum_{j=1}^{4} \hat{\mathbf{H}}_k(:,j)\hat{\mathbf{H}}_k^H(:,j) + \mathbf{r}\mathbf{r}^H\frac{\mathbf{I}_4}{\Gamma}\right)^{-1}, \quad (40)$$

After the demodulation of $\mathbf{y}_k^{DL}$, we then compare the demodulated bit and the source bit to decide the BER as follows. Consider the binary phase-shift keying (BPSK) for modulation and the maximum likelihood (ML) criterion for data decision after demodulation. Assume that the bit value of each UE is 0 or 1 with equal probability and the noise has complex Gaussian distribution. Let the 1×4 matrix $\mathbf{A}_q = \hat{\mathbf{R}}_q\hat{\mathbf{H}}_q$ represent the gain vector of the transmission from BS to each UE in the qth Monte Carlo random channel realization for the MIMO channel matrices by using the channel model (7) for simulation. Moreover, assume that $A_q(i)$ is the ith entry of $\mathbf{A}_q$. It represents the gain for the signal transmitted to each UE. The other three entries $A_q(j)$ of $\mathbf{A}_q$, $j = 1, 2, 3, 4$ and $j \neq i$, represent the gains of the interference received at each UE. According to the BER calculation developed in [4], the total estimated BER for each UE is written as:

$$P_e(q) = \frac{1}{2^{4-1}} \sum_{a_1=0}^{1} \cdots \sum_{a_{4-1}=0}^{1} Q\left(\Re(A_q(i) + \sum_{c=1,c\neq i}^{4} (-1)^{a_c} \times A_q(c))\sqrt{\frac{2\Gamma}{||\mathbf{r}_q||^2||\hat{\mathbf{R}}_q||^2}}\right), \quad (41)$$

where the $1 \times N_U$ vector $\mathbf{r}_q$ denotes the qth Monte Carlo random channel realization of the corresponding MMSE receiver $\mathbf{r}$ at the UEs.

Finally, we compute the average of the approximated BERs (AA-BER) as follows:

$$\textbf{Fitness2} = \frac{1}{Q}\sum_{q=1}^{Q} P_e(q), \quad (42)$$

where Q denotes the total number of the Monte Carlo random channel realizations for the MIMO channel matrices by using the channel model (7) for simulation. The **Fitness2** of (42) is set to be the fitness function to be minimized by the second-step CCPSO.

3.3 An algorithm form of the proposed method

To make the implementation of the proposed method more comprehensive, we summarize the procedure of implementing the proposed method step by step as follows.

3.3.1 *The first-step optimization*

Step 1. Determine the numbers of transmit and receive antennas, N_B and N_U, respectively. Set the number t of iterations to zero.

Step 2. At the tth iteration, set the parameter vectors as shown by (24) and (25), respectively to be the position vector of the ith particle of the CCPSO given by $\mathbf{x}_\mathbf{i}(t) = \left[A_1^t, \ldots, A_{N_B}^t, A_1^r, \ldots, A_{N_U}^r\right]$ for the linear transmit and receive antenna elements or $\mathbf{x}_\mathbf{i}(t) = \left[\varphi_1^t, \ldots, \varphi_{N_B}^t, \varphi_1^r, \ldots, \varphi_{N_U}^r\right]$ for the circular transmit and receive antenna elements.

Step 3. Generate the channel matrix $\mathbf{H}$ as shown by (7) for downlink transmission.
Step 4. Compute the average mutual information (AMI) associated with the downlink transmission channel matrix $\mathbf{H}$ according to (26).
Step 5. Compute the fitness function **Fitness1** for implementing the first-step optimization process by using the CCPSO algorithm according to (27).
Step 6. Perform the CCPSO algorithm of [13] for minimizing the fitness function **Fitness1** to obtain the most appropriate parameter set $\mathbf{x}_i(t) = \left[A_1^t, ..., A_{N_B}^t, A_1^r, ..., A_{N_U}^r\right]$ for the positions of the linear transmit and receive antenna elements or $\mathbf{x}_i(t) = \left[\varphi_1^t, ..., \varphi_{N_B}^t, \varphi_1^r, ..., \varphi_{N_U}^r\right]$ for the positions the circular transmit and receive antenna elements.
Step 7. Form the channel matrix $\mathbf{H}$ as shown by (7) by utilizing the positions for the antenna elements obtained from *Step* 6.
Step 8. Terminate the process if the specified maximum iteration number t_{max} is reached or the best particle position of the whole swarm remains almost static for a significantly large number of successive iterations. Otherwise, update the positions of antenna elements based on the results obtained from *Step* 6 and set the iteration number t to $t + 1$. Then, go to *Step* 2.

3.3.2 The second-step optimization

After obtaining the optimal locations for the antenna elements, the channel matrix $\mathbf{H}_K$ as shown by (7) can be determined. In the second step, based on the channel matrix $\mathbf{H}_K$ obtained in the first step, the CCPSO algorithm is utilized to find the optimal parameter set $(a_{1,1}, a_{1,2}, ..., a_{4,N_B}, \theta_{1,2}, ..., \theta_{4,N_B}, t_1, ..., t_4)$ related to the beamforming coefficients and the power distribution.
Step 1. Determine the channel matrix $\mathbf{H}_K$ as shown by (7) by using the optimal locations for the antenna elements, respectively obtained in the first-step optimization. Set the number t of iterations to zero.
Step 2. At the tth iteration, set the position vector of the ith particle of the CCPSO to $\mathbf{x}_i(t) = \left[a_{1,1}, a_{2,1}, \cdots, a_{4,1}, a_{1,2}, \cdots, a_{4,N_B}, \theta_{1,2}, ..., \theta_{4,N_B}, t_1, ..., t_4\right]$.
Step 3. Generate the channel matrix $\mathbf{H}$ as shown by (7).
Step 4. Compute the matrix $\hat{\mathbf{H}}_k$ associated with the downlink transmission channel matrix $\mathbf{H}$ according to (33).
Step 5. Compute the MMSE receiver $\hat{\mathbf{R}}_k$ associated with the $\hat{\mathbf{H}}_k$ according to (40).
Step 6. Calculate the total estimated BER for each UE according to (41).
Step 7. Compute the fitness function **Fitness2** for implementing the second-step optimization process by using the CCPSO algorithm according to (42).
Step 8. Perform the CCPSO algorithm of [13] for minimizing the fitness function **Fitness2** to obtain the most appropriate parameter set $\mathbf{x}_i(t) = \left[a_{1,1}, a_{2,1}, \cdots, a_{4,1}, a_{1,2}, \cdots, a_{4,N_B}, \theta_{1,2}, ..., \theta_{4,N_B}, t_1, ..., t_4\right]$ for the beamforming coefficients and the power distribution.
Step 9. Terminate the process if the specified maximum iteration number t_{max} is reached or the best particle position of the whole swarm remains almost static for a significantly large number of successive iterations. Otherwise, update the beamforming coefficients and the power distribution based on the results obtained from *Step* 8 and set the iteration number t to $t + 1$. Then, go to *Step* 2.

3.4 Computational complexity of the proposed method

From the procedure described above for implementing the proposed method, we note that the proposed method requires the additional computational complexity due to using the CCPSO for finding the optimal precoding coefficients without costing additional feedback overhead as compared with the original BBSP method and the method of [4] . According to the CCPSO [13] employed by the proposed method, its computational complexity is dominated by the number t_{max} of iterations, the number s of the group sizes, the number M of particles, the computational complexity of the fitness functions, and the computational complexity of the particles' position updating rule. Therefore, the computational complexity of the CCPSO can be expressed as follows:

$$O_{CCPSO} = O\left(M \times s \times t_{max} \times O_{CCPSO}^{U}\right), \tag{43}$$

where O_{CCPSO}^{U} denotes the total computational complexity of the particles' position updating rule given by [13] and the fitness functions given by (27) and (42), respectively. As a result, the proposed method requires the additional computational complexity given by (43) as compared with the method of [4]. To further evaluate the computational complexity of the proposed method for some setup like 100×100 system as an illustration, we present the more detail descriptions regarding O_{CCPSO}^{U}. The required computational complexity for calculating the fitness function of the first-step optimization is dominated by the calculations of the AMI of (26) and Fitness1 of (27). Their computational complexities are about $O(QN_BN_U)$ and $O\left(Q\left(N_U^2 + B^2\right)\right)$, respectively, where Q denotes the total number of the Monte Carlo random channel realizations for the MIMO channel matrices. In addition, the required computational complexity for calculating the fitness function of the second-step optimization is dominated by calculating Fitness2 of (42). To calculate Fitness2 of (42), we have to compute the squared norms of the $1 \times N_U$ vector $\mathbf{r}q$ and the 1×4 vector $\hat{\mathbf{R}}_q$, respectively. Hence, the required computational complexity for calculating Fitness2 of (42) is about $O(Q(N_U^2 + 16))$, where Q denotes the total number of the Monte Carlo random channel realizations for the MIMO channel matrices. As a result, we have O_{CCPSO}^{U} approximately given by

$O^{U}_{CCPSO} \approx$ the total computational complexity of the particles' position updating rule given by [13] $+O(QN_BN_U)+O\left(Q\left(N_U^2+B^2\right)\right)+O\left(Q\left(Q(N_U^2+16\right)\right)$. Consider the case of 100×100 ($N_B \times N_U$) system. The resulting computational complexity is approximately given by $O_{CCPSO} = O(M \times s \times t_{max} \times O^{U}_{CCPSO}) \approx O(M \times s \times t_{max}\times$ [the total computational complexity of the particles' position updating rule given by [13] $+O(QN_BN_U) + O\left(Q\left(N_U^2+B^2\right)\right) + O\left(\left(N_U^2+16\right)\right)\right]) \approx O(M \times s \times t_{max}\times$ [the total computational complexity of the particles' position updating rule given by [13] $+3O(10000Q)]$).

4 Simulation results and discussion

This section presents simulation examples for illustration and comparison. The following 2 examples are presented to compare the average BER versus SNR in the scenarios considered above under different numbers of antennas used at BS. Example 1 (a large MIMO case) is presented for the case of $N_B = 8$ and $N_U = 2$. In contrast, the simulation results for Example 2 (a massive MIMO case) is presented for the case of $N_B = 32$ and $N_U = 2$. The purpose of presenting Example 1 is to make comparisons between the performance differences of using large MIMO and massive MIMO wireless communications. To simplify the illustration and comparison for the proposed method, we consider the simple case of linear arrays as shown in Fig. 2 and the simple case of circular arrays as shown in Fig. 3 instead of using the complicated antenna configurations for 3D channel model like those employed by the authors [19, 20].

4.1 Example 1: The BER performance of wireless MIMO channel with $N_B = 8$ and $N_U = 2$

In this example, we utilize the spatial channel model similar to (7) for simulation. We employ this channel model to generate 40 links with each link sampled 10 times for obtaining a set of $Q = 400$ channel matrices for illustration and comparison. The parameters used are as follows. $B = K = 4$. For the first-step CCPSO procedure, we combine all of the $L = 10$ multipath components produced by sampling each link to get the qth channel matrix $\mathbf{H}_q = \sum_{l=1}^{L} \mathbf{H}_l$ for simulation, where $\mathbf{H}_l$ denote the lth multipath component. The dimension $s \in S = \{1, 2, 4, 8\}$, swarm number $V = \frac{8}{s}$, $t_{max} = 100$, the particle number of each swarm is 450, and the fitness function is given by (27) with $Q = 400$. The value for the parameter Z is set to 0.6. $\Gamma = 12$, $N_B = 8$, $N_U = 2$. For the second-step CCPSO procedure, the dimension $s \in S = \{2, 4, 17, 34\}$, swarm number $V = \frac{68}{s}$, $t_{max} = 100$, the particle number of each swarm is 450, and the fitness function is given by (42) with $Q = 400$. The value for the parameter Z is set to 0.6. $\Gamma = 12$, $N_B = 8$, $N_U = 2$. The constraint for the total power P_t is set to 4 watts. For the case of using linear antennas, the signal direction angle θ is assumed to be a truncated Gaussian random variable with mean MOA = 0° and standard deviation $\sigma = 30^o$. For the case of using circular antennas, we set the elevation angle $\theta = 90^o$ and the azimuth angle ϕ to be a truncated Gaussian random variable with mean MOA = 0° and standard deviation $\sigma = 30^o$. For simplicity, we assume that each UE selects one beam group as its serving beam, i.e., $N = 1$. Figure 4 plots the AMI versus SNR of using the existing method [4] and the proposed method. We note from the figure that the AMI are significantly improved after adjusting the positions of array elements for both cases of using linear and circular arrays. Moreover, the results reveal that using circular antennas provides better AMI than using linear antennas under the same number of antenna elements. Figure 5 plots the average BER versus SNR of using the existing method [4] and the proposed method, respectively. As we can observe from the simulation results, the proposed method outperforms the existing method [4]. Moreover, the results reveal that using circular antennas provides better BER performance than using linear antennas under the same number of antenna elements.

4.2 Example 2: The BER performance of wireless MIMO channel with $N_B = 32$ and $N_U = 2$

Here, the specifications and parameters used are the same as those used by *Example* 1 except that the following parameters are reset. The number of array elements deployed at BS is set to $N_B = 32$ under the consideration of massive wireless MIMO scenarios. For the first-step CCPSO procedure, the dimension $s \in S = \{2, 4, 8, 16\}$ and swarm number $V = \frac{32}{s}$. For the second-step CCPSO procedure, the dimension $s \in S = \{2, 4, 5, 13\}$ and swarm number $V = \frac{260}{s}$. Figure 6 plots the AMI versus SNR of using the existing method [4] and the proposed method. Again, we note from the figure that the AMI are significantly improved after adjusting the positions of array elements for both cases of using linear and circular arrays. Moreover, the results reveal that using circular antennas provides better AMI than using linear antennas under the same number of antenna elements. Figure 7 plots the average BER versus SNR of using the existing method [4] and the proposed method, respectively. Again, as we can observe from the simulation results, the proposed method outperforms the existing method [4]. The results also reveal that using circular antennas provides better BER performance than using linear antennas under the same number of antenna elements. From the simulation results showing that using circular antennas provides better BER performance than using linear antennas under the same number of antenna elements, we reason that this is due to the following phenomena: (a) Because only small amount of power transmitted by utilizing circular antenna array would be reflected back, circular antenna array provides better return loss as compared to linear antenna array. (b)

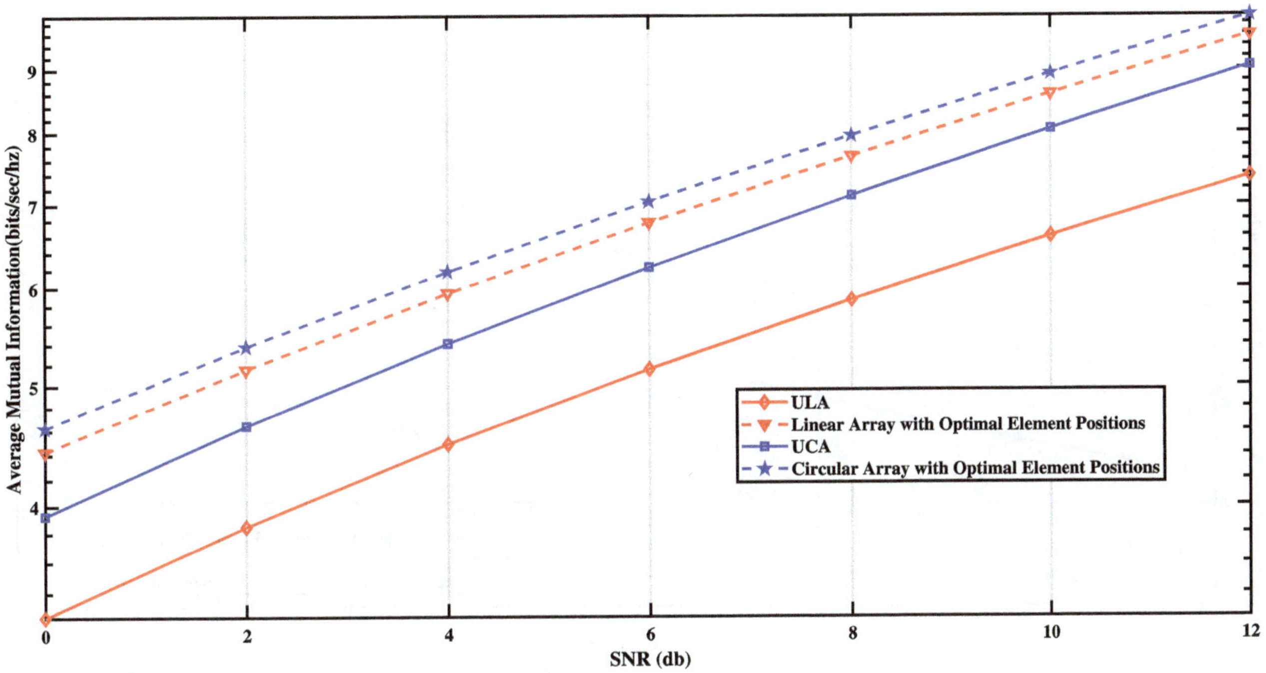

Fig. 4 The average mutual information for **Example 1** with $N_B = 8$ and $N_U = 2$

Because the percentage of bandwidth and the half-power beamwidth for circular antenna array are wider than those for linear antenna array, circular antenna array provides higher gain than linear antenna array. (c) To mitigate MCE, circular antenna array is a more suitable geometrical arrangement than linear antenna array. The more array elements are being used, the better and higher gain could be achieved. Finally, comparing the results of Examples 1 and 2, we see that the system performance can be considerably enhanced by using massive antennas at BS. From the simulation results, we note that the proposed method is very effective in simultaneously dealing with the mutual

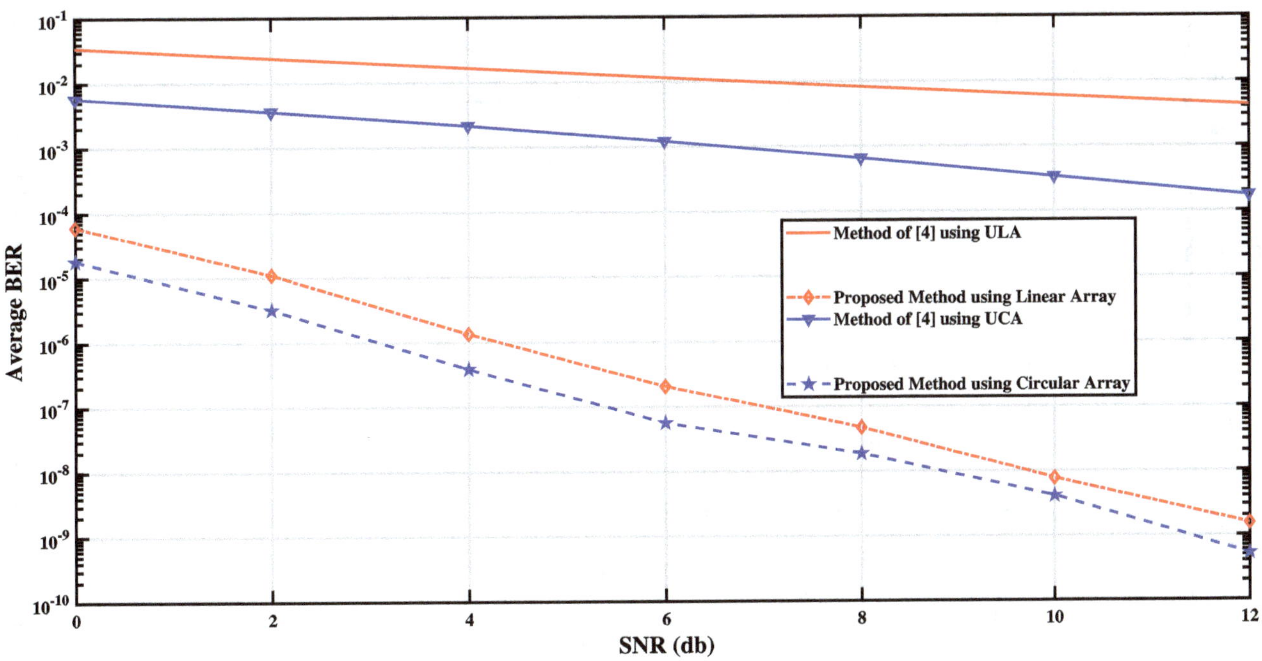

Fig. 5 Average BER performance versus SNR for **Example 1** with $N_B = 8$ and $N_U = 2$

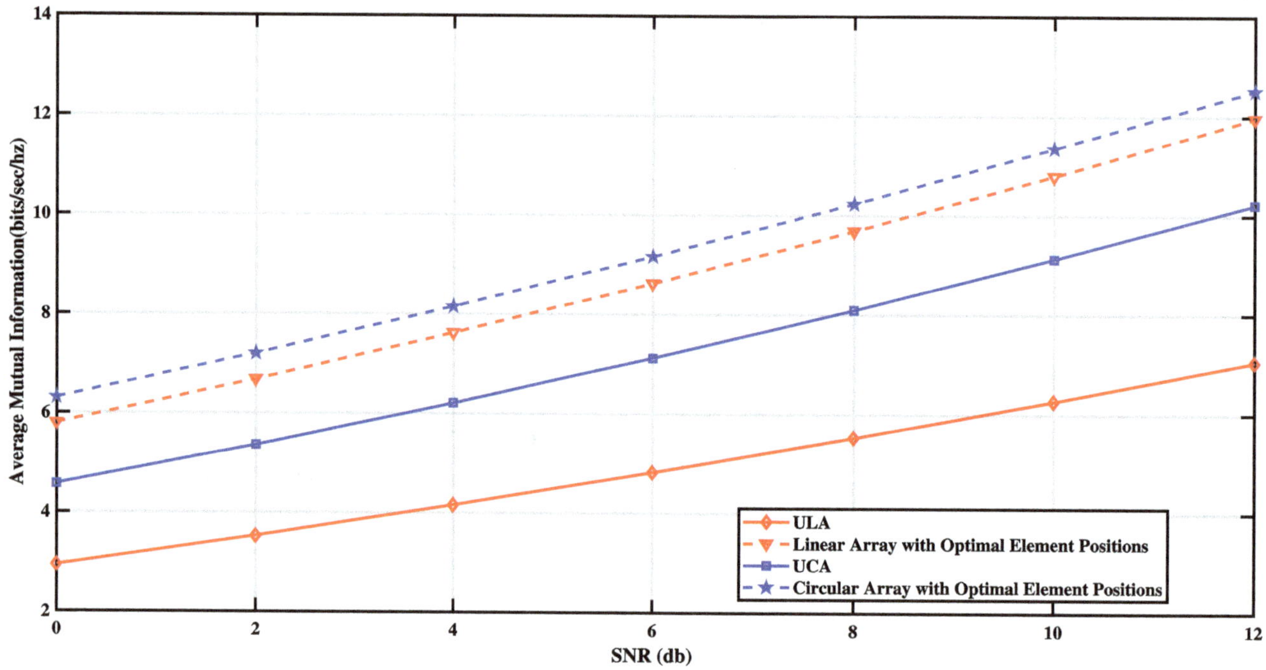

Fig. 6 The average mutual information for **Example 2** with $N_B = 32$ and $N_U = 2$

coupling and spatial correlation problems in QOSTBC massive wireless MIMO communications.

5 Conclusion

In this paper, we have presented an optimal joint beamforming and spatial precoding method based on a population-based stochastic optimization for QOSTBC massive wireless MIMO communication systems. Based on the concept of beamforming selection, the proposed method can preserve the advantage of reducing the overheads for downlink training and feedback of channel state information between the base station (BS) and user

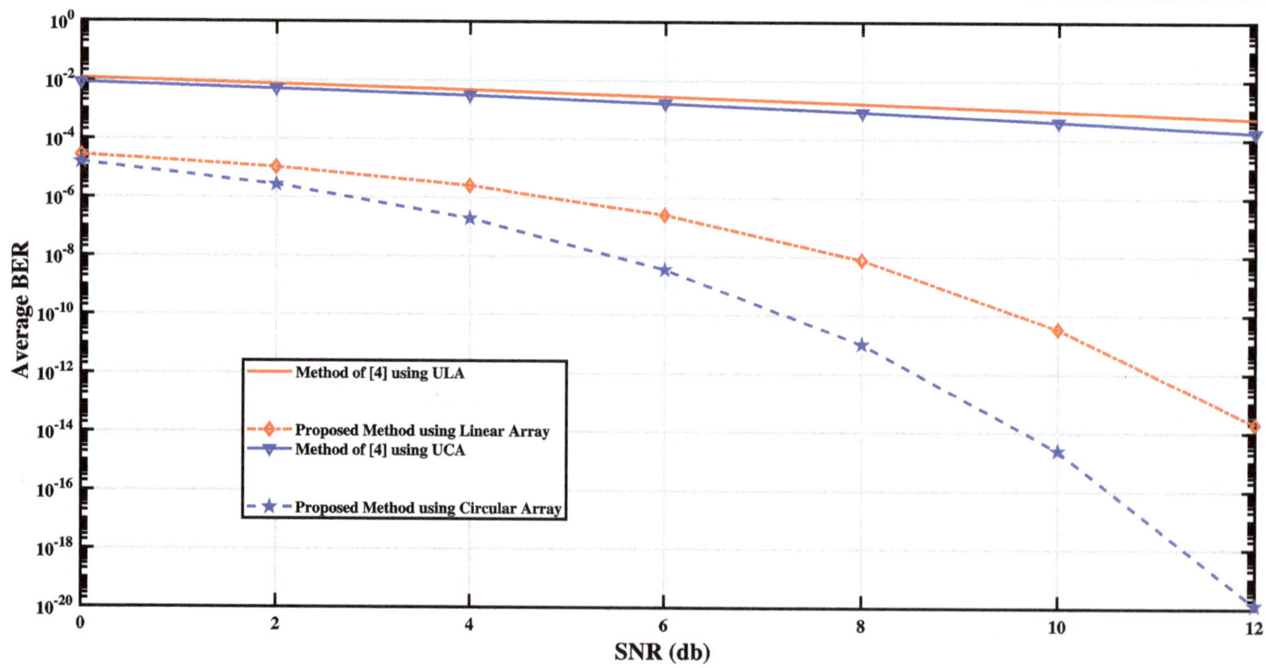

Fig. 7 Average BER performance versus SNR for **Example 2** with $N_B = 32$ and $N_U = 2$

equipments (UEs). The proposed method utilizes a two-step cooperative coevolutionary particle swarm optimization (CCPSO) to find the optimal beamforming coefficients and the optimal positions of array elements. As a result, the proposed method can significantly mitigate the mutual coupling and spatial correlation problems for downlink transmission, particularly when massive antennas are deployed at the BS. Moreover, appropriate fitness functions based on the average mutual information and an estimated average bit error rate (BER) are proposed for implementing the two-step CCPSO scheme. Finally, the proposed method makes QOSTBC MIMO communications with full transmission rate for any number of transmit antennas achievable. Simulation results show that the proposed method can indeed achieve much better BER performance than the existing methods under the scenarios of massive antennas with antenna mutual coupling and spatial correlation. We reason that the proposed method offers a more appropriate beamforming and spatial precoding scheme for future QOSTBC massive wireless MIMO communications. As to analytical results regarding the proposed method, the exact performance analysis of large or massive MIMO system is not an easy task. In the literature, most of the papers present the bit error rate (BER) performance of QOSTBC-based massive MIMO systems through the numerical results without theoretical analysis. To obtain the complete mathematical analysis regarding the proposed method, we are currently conducting the further research on the exact analytical results on the achievable performance gains, mathematical analysis of the proposed schemes/antenna placements in the presence of mutual coupling and spatial correlation effects.

Abbreviations

AA-BER: Approximated average bit error rate; AMI: Average mutual information; BER: Bit error rate; BS: Base station; BSSP: Beam selection spatial precoding; CCPSO: Cooperative coevolutionary particle swarm optimization; CSI: Channel state information; FDD: Frequency-division duplex; MCE: Mutual coupling effect; MIMO: Multiple-input multiple-output; MMSE: Minimum mean square error; OSTBC: Orthogonal space-time block code; PSO: Particle swarm optimization; QOSTBC: Quasi-orthogonal space-time block code; STBC: Space-time block code; SCE: Spatial correlation effect; SCF: Spatial correlation function; UCA: Uniform circular array; UEs: User equipments; ULA: Uniform linear array

Acknowledgements

Not applicable.

Funding

This work was supported by the Ministry of Science and Technology of Taiwan under Grants MOST 106-2221-E-002-062 and MOST 107-2221-E-002-121-MY3. The Ministry of Science and Technology of Taiwan supports the financial grant required for conducting the research. The authors complete all of the design of the study and collection, analysis, and interpretation of data and writing the manuscript.

Authors' contributions

JHL completed the main work and updated the manuscript, as the first author and the correspondence author of this paper. WES performed all of MATLAB programming and prepared the simulation data and figures for presenting in the manuscript. All authors have read and approved the final manuscript for submission.

Authors' information

Ju-Hong Lee was born in I-Lan, Taiwan. He received the B.S. degree from National Cheng Kung University, Tainan, Taiwan, in 1975; the M.S. degree from National Taiwan University (NTU), Taipei, Taiwan, in 1977; and the Ph.D. degree from Rensselaer Polytechnic Institute, Troy, NY, USA, in 1984, all in electrical engineering. From September 1980 to July 1984, he was a Research Assistant and was involved in research on multidimensional recursive digital filtering with the Department of Electrical, Computer, and Systems Engineering, Rensselaer Polytechnic Institute. From August 1984 to July 1986, he was a Visiting Associate Professor and, in August 1986, became an Associate Professor with the Department of Electrical Engineering, NTU. Since August 1989, he has been a Professor with NTU. He was appointed as a Visiting Professor with the Department of Computer Science and Electrical Engineering, University ofMaryland, Baltimore, MD, USA, during a sabbatical leave in 1996. His current research interests include multidimensional digital signal processing, multirate signal and image processing, detection and estimation theory, analysis and processing of joint vibration signals for the diagnosis of cartilage pathology, statistical signal processing, and adaptive signal processing for smart antennas with applications in mobile wireless communication systems. Dr. Lee received the Excellence Research Awards from the National Science Council (NSC) of Taiwan in the academic years 1988, 1989, and 1991-1994; the Outstanding Research Awards from the NSC in the academic years 1998-2004; and the NSC Research Fellowships for the academic years 2005-2008 and 2011-2014. In 2015, he received the Merit MOST Research Fellow Award from the Ministry of Science and Technology of Taiwan. He has been appointed as NTU's Tenured Distinguished Professor since August 2006. Wei-En Sun was born in Taiwan. He received the B.S. degree in electrical engineering from the National Chung Hsing University, Taichung city, Taiwan, in 2016 and the M.S. degree in communication engineering from the National Taiwan University, Taipei, Taiwan, in 2018. His current research interests are in the areas of multiple-antenna multiple-input-multiple-output systems in communications.

Competing interests

The authors declare that they have no competing interests.

Author details

[1]Department of Electrical Engineering, Graduate Institute of Communication Engineering, and Graduate Institute of Biomedical Electronics and Bioinformatics, National Taiwan University, No. 1, Sec. 4, Roosevelt Rd., Taipei 10617, Taiwan. [2]Graduate Institute of Communication Engineering, National Taiwan University, No. 1, Sec. 4, Roosevelt Rd., Taipei 10617, Taiwan.

References

1. C. Peel, Q. Spencer, A. Swindlehurst, B. Hochwald, in *Proceedings of 2004 IEEE Sensor Array and Multichannel Signal Processing Workshop*. Downlink transmit beamforming in multiuser MIMO systems (IEEE Sensor Array and Multichannel Signal Processing Workshop, 18-21 July, 2004, Barcelona, Spain, 2004), pp. 43–51
2. R. Doostnejad, T. J. Lim, E. Sousa, Joint precoding and beamforming design for the downlink in a multiuser MIMO system. IEEE Int. Conf. Wirel. Mob. Comput. Netw. Commun. **1**, 153–159 (2005)
3. M. F. Tang, M. Y. Lee, B. Su, C. P. Yen, Beamforming-based spatial precoding in FDD massive MIMO systems. IEEE 48th Asilomar Conf. Syst. Comput. **1**, 2073–2077 (2014)
4. J.-H. Lee, J. Y. Lee, Optimal beamforming-selection spatial precoding using population-based stochastic optimization for massive wireless MIMO communication systems. J. Frankl. Inst. **354**, 4247–4272 (2017)
5. T. Marzetta, Noncooperative cellular wireless with unlimited numbers of base station antennas. IEEE Trans. Wirel. Commun. **9**, 3590–3600 (2010)
6. E. Larsson, O. Edfors, F. Tufvesson, T. Marzetta, Massive MIMO for next generation wireless systems. IEEE Commun. Mag. **52**, 186–195 (2014)
7. D. Shiu, G. J. Foschini, M. J. Gans, J. M. Kahn, Fading correlation and its effect on the capacity of multi-element antenna systems. IEEE Trans. Commun. **48**, 502–513 (2000)
8. J. A. Tsai, M. Buehrer, B. D. Woerner, BER performance of a uniform circular array versus a uniform linear array in a mobile radio environment. IEEE Trans. Wirel. Commun. **3**, 695–700 (2004)

9. Z. Li, Z. Du, M. Takahashi, K. Saito, K. Ito, Reducing mutual coupling of MIMO antennas with parasitic elements for mobile terminals. IEEE Trans. Antennas Propag. **60**, 473–481 (2012)
10. L. Sun, P. Li, M. R. McKay, R. D. Murch, Capacity of MIMO systems with mutual coupling: transmitter optimization with dual power constraints. IEEE Trans. Signal Process. **60**, 848–861 (2012)
11. A. Yang, Y. Jing, C. Xing, Z. Fei, J. Kuang, Performance analysis and location optimization for massive MIMO systems with circularly distributed antennas. IEEE Trans. Wirel. Commun. **14**, 5659–5671 (2015)
12. E. Koyuncu, Performance gains of optimal antenna deployment in massive MIMO systems. IEEE Trans. Wirel. Commun. **17**, 2633–2644 (2018)
13. X. Li, Y. Yao, Cooperatively coevolving particle swarms for large scale optimization. IEEE Trans. Evol. Comput. **16**, 210–224 (2012)
14. A. Garg, M. R. Bhatnagar, O. Berder, B. Vrigneau, Quantized feedback-based diagonal precoding for $N \times 1$ MISO system with generalized orthogonal space-time block codes. IEEE Trans. Commun. **66**, 91–104 (2018)
15. A. Garg, V. R. Venu, M. R. Bhatnagar, One bit feedback based diagonal precoding for non-orthogonal space-time block codes. IEEE Trans. Wirel. Commun. Lett. **4**, 573–576 (2015)
16. A. Garg, M. R. Bhatnagar, in *Proceedings of 2016 IEEE 83rd Vehicular Technology Conference (VTC Spring), 15–18 May 2016, Nanjing, China.* Performance analysis of diagonal precoding for Alamouti STBC over Nakagami-m fading channels, (2016), pp. 1–5
17. O. E. Ayach, S. Rajagopal, S. Abu-Surra, Z. Pi, J. R. W. Heath, Spatially sparse precoding in millimeter wave MIMO systems. IEEE Trans. Wirel. Commun. **13**, 1499–1513 (2014)
18. A. Forenza, D. J. Love, R. W. Heath, Simplified spatial correlation models for clustered MIMO channels with different array configurations. IEEE Trans. Vechicular Technol. **56**, 1924–1934 (2007)
19. Q. U. A. E. Nadeem, A. Kammoun, M. Debbah, M. S. Alouini, A generalized spatial correlation model for 3D MIMO channels based on the Fourier coefficients of power spectrum. IEEE Trans. Signal Process. **63**, 3671–3686 (2015)
20. Q. U. A. E. Nadeem, A. Kammoun, M. Debbah, M. S. Alouini, Performance alalysis of compacted FD-MIMO antenna arrays in a correlated environment. IEEE Access. **5**, 4163–4178 (2017)
21. C. A. Balanis, *Antenna Theory Analysis and Design*. (John Wiley and Sons, New York, 1997)
22. V. Tarokh, H. Jafarkhani, A. R. Calderbank, Space-time block coding for wireless communications: performance results. IEEE J. Sel. Areas Commun. **17**, 451–460 (1999)
23. V. Tarokh, H. Jafarkhani, A. R. Calderbank, Space-time codes for high data rate wireless communication: Performance analysis and code construction. IEEE Trans. Inf. Theory. **44**, 744–765 (1998)
24. S. M. Alamouti, A simple transmitter diversity scheme for wireless communications. IEEE J. Sel. Areas Commun. **16**, 1451–1458 (1998)
25. V. Tarokh, H. Jafarkhani, A. R. Calderbank, Space-time block codes from orthogonal designs. IEEE Trans. Inf. Theory. **45**, 1456–1467 (1999)
26. M. K. Arti, OSTBC Transmission in Large MIMO Systems. IEEE Commun. Lett. **29**, 2308–2311 (2016)
27. M. K. Arti, S. Seema, in *2017 International Conference on Computer, Communications and Electronics (Comptelix), Manipal University Jaipur, Malaviya National Institute of Technology Jaipur & IRISWORLD, July 01-02.* High-rate code design and transmission in large MIMO systems, (2017), pp. 373–378
28. M. K. Arti, A Space-Time Transmission Scheme for Large MIMO Systems. IEEE Wirel. Commun. Lett. **7**, 62–65 (2018)
29. M. R. Bhatnagar, A. Hjorungnes, L. Song, Procoded differential orthogonal space-time modulation over correlated ricean MIMO channels. IEEE J. Sel. Top. Sign. Process. **2**, 124–134 (2008)
30. M. R. Bhatnagar, A. Hjorungnes, L. Song, Differential coding for non-orthogonal space-time block codes with non-unitary constellations over arbitrarily correlated Rayleigh channels. IEEE Trans. Wirel. Commun. **8**, 3985–3995 (2009)
31. M. R. Bhatnagar, A. Hjorungnes, Linear precoding of STBC over correlated ricean MIMO channels. IEEE Wirel. Commun. **9**, 1832–1836 (2010)
32. H. Jafarkhani, A quasi-orthogonal space-time block code. IEEE Trans. Commun. **49**, 1–4 (2001)
33. D. Tse, P. Viswanath, *Fundamentals of Wireless Communication*. (Cambridge University Press, Cambridge, 2005)

14

Joint interference alignment and power allocation for NOMA-based multi-user MIMO systems

Mohamed Rihan[1,2], Lei Huang[1] and Peichang Zhang[1*]

Abstract

Interference alignment (IA) and non-orthogonal multiple access (NOMA) are key technologies for achieving the capacity scaling required by next generation networks to overcome the unprecedented growth of data network traffic. Each of these technologies was proved to present excellent performance for MIMO systems. In this article, we propose a joint IA and power allocation (PA) framework for NOMA-based multi-user MIMO (MU-MIMO) systems. Different approaches for applying IA in downlink NOMA-based MU-MIMO systems will be addressed while implementing a PA technique that fully exploits the characteristics of NOMA-based systems. The proposed framework aims to maximize the sum-rate of the NOMA-based MU-MIMO system through combining IA with PA. The process begins by initially grouping the system users into clusters for optimum implementation of NOMA. The sum-rate maximization is carried out under cluster power budget, user quality-of-service (QoS), and robust successive interference cancellation (SIC) constraints. Meanwhile, it uses the power domain multiplexing strategy to allow the users within each cluster to share the data streams without exerting interference to one another. Three iterative joint IA and PA algorithms are proposed for NOMA-based MU-MIMO systems. Moreover, these algorithms are compared with orthogonal multiple access (OMA)-based MU-MIMO counterpart as well as the state-of-the-art techniques presented for NOMA-based MU-MIMO systems. Numerical simulations verify that the proposed framework can greatly improve the performance of NOMA-based MU-MIMO systems in terms of the achievable sum-rate when compared with OMA-based MU-MIMO and the state-of-the-art NOMA-based MU-MIMO systems.

Keywords: Non-orthogonal multiple access (NOMA), Interference alignment (IA), Power allocation, Joint optimization, Grassmann manifold

1 Introduction

One of the key challenges facing the fifth generation (5G) mobile networks is the overwhelming growth of data network traffic. Accordingly, non-orthogonal multiple access (NOMA) has recently attracted much attention as a promising radio access technology in 5G mobile networks due to its superior spectral efficiency [1]. The concept behind NOMA is the exploitation of the power domain for implementing a multiple access mechanism in mobile networks. Specifically, the signals of NOMA users are assigned with different power allocation (PA) coefficients according to their channel conditions. Users with poor channel conditions are assigned more power level, and users with better channel conditions are assigned lower power level [2, 3]. One of the major advantages of the NOMA technique is its excellent ability to balance between sum-rate and fairness, and accordingly achieves an optimized spectral efficiency for all the served users [2, 4].

The NOMA technique was the core of many research studies in the last few years [1–7]. In [5], a comparison between NOMA and its orthogonal counterparts, in terms of the achievable sum-rate, has been accomplished and the results demonstrated the superiority of NOMA as a radio access technology for future 5G cellular networks. In [6], the NOMA technique is used to implement a cooperative transmission strategy for spectrum-sharing

*Correspondence: pzhang@szu.edu.cn
[1]Multi-dimensional Signal Processing Lab., College of Information Engineering, Shenzhen University, Shenzhen, China
Full list of author information is available at the end of the article

cognitive radio networks. The user fairness for NOMA-based cellular systems has been addressed in [4]. NOMA technique is also used in cognitive radio networks in order to maintain some pre-defined quality-of-service (QoS) conditions [7]. The application of MIMO-NOMA to the downlink mobile communication networks is addressed in [2]. Specifically, they implement MIMO-NOMA to both cellular and cognitive inspired wireless networks. Additionally, they explored the outage probability for MIMO-NOMA systems and study the sum-rate gap between NOMA-based networks and their orthogonal multiple access (OMA) counterparts.

During the last decade, interference alignment (IA) technique is also proposed as an excellent solution to the interference problem arising in wireless multi-user (MU) communication networks that significantly improve its sum-rate [8]. Specifically, implementing IA technique in wireless networks results in sum-rate that can scale linearly and without bound with the number of users in the network at high signal-to-noise power ratio (SNR) [8, 9]. The key idea behind the IA technique is to align interference signals into a reduced dimensional subspace leaving the remaining subspace for the transmission of useful signal without any interference. Accordingly, maximum degrees of freedom (DoFs) for the whole network can be achieved. The IA scheme is studied for many different networks, e.g., X channel [10], K-user interference channel (IC) [8, 11], heterogeneous networks [12–14], and cognitive radio networks [15, 16]. Moreover, the importance of the channel state information (CSI) for successful IA implementation is addressed in many works [17, 18]. Additionally, the feasibility conditions for IA implementation were the core of careful research studies [8–10].

Evaluating the capacity of a general IC is still a difficult goal for researchers in wireless communications and information theory [19] communities. However, IA technique is introduced as a DoFs optimal approach to interference management [10, 11]. This means that it can achieve the capacity of the IC at high SNR value. Note that IA approach can be achieved in time, frequency, and space dimensions. However, applying IA approach in the space dimension is the most popular due to the widespread use of MIMO technology. In MIMO networks, IA technique is applied using transmit beamforming matrices that help keeping all undesired received signals at each receiver within the same minimum dimensional sub-space, leaving the desired signal sub-spaces interference-free. Then, a receive beamforming matrix orthogonal to the interference sub-spaces at each receiver is used to completely eliminate the undesired interference signals [18, 20].

Recently, maximizing the capacity and accordingly the sum-rate for NOMA-based MU-MIMO communication networks becomes a target for many research works [21–24]. In [21], the problem of maximizing the sum-rate for NOMA-based MIMO communication systems is studied under both total transmit power and minimum rate per user constraints. However, this study gives no attention to the applicability of successive interference cancellation (SIC) technique for networks with large number of users. In this work, they proposed PA scheme based on the CSI corresponding to full-rate transmission condition. The concept of signal alignment for both uplink and downlink transmissions in NOMA-based MIMO systems is addressed in [22]. Specifically, the authors used stochastic geometry to evaluate the performance of the proposed transmission framework with both randomly deployed users and interferers. The authors in [23] proposed a user-clustering algorithm for conventional NOMA-based MU-MIMO systems. They also investigated the performance of NOMA-based MIMO systems compared to OMA-based MIMO systems and concludes that NOMA-based MIMO systems are offering better capacity than the conventional OMA-based MIMO counterparts. Unlike our work, the method presented in [23] does not consider working with IA-based networks. In [24], the authors proposed a resource allocation scheme based on IA for NOMA-based networks. Specifically, they proposed a PA algorithm for 2-users NOMA network that implements the singular value decomposition (SVD)-based IA scheme which is not scalable to networks with large number of users. Additionally, they targeted optimizing the sum-rate under total power constraint. However, the generalization of the PA to the case where there are $K > 2$ users in the network is done in heuristically non-optimal manner based on 2-users pairing. Moreover, the solutions presented in [24] totally ignores the practical feasibility of SIC technique as well as the QoS requirements at each user, which are all considered in our proposed joint optimization algorithms.

In this article, we propose a joint IA and PA framework for optimizing the sum-rate of the NOMA-based MU-MIMO systems. The main contributions in our article can be summarized as follows:

- We propose a system and signal model for NOMA-based MU-MIMO systems that implements both the IA and PA techniques.
- We formulate the IA problem for NOMA-based MU-MIMO systems as an optimization problem and then find the optimum solutions according to the approach of IA employed for maximizing the system sum-rate.
- A PA technique for NOMA-based MU-MIMO system that employs IA transceivers is introduced, aiming to maximize the sum-rate under total power, robust SIC feasibility, and user QoS constraints.
- We devise three iterative algorithms for solving the optimization problem in the previous item, through

jointly optimizing the IA transceivers and PA coefficients of the system users.

- Compare the performance of the joint optimization-based iterative algorithms in both OMA and NOMA-based MU-MIMO systems.

The remaining sections are organized as follow: Section 2 introduces the signal and system model for the considered downlink NOMA-based MU-MIMO system. Then, different approaches for realizing the IA conditions for the downlink NOMA-based MU-MIMO systems with full CSI available at the base-station (BS) are addressed in Section 3 together with a simple solution for the case of no CSI available at the BS. This is followed by exploiting the role of both the optimum PA and clustering in maximizing the achievable sum-rate, in Section 4. Simulation results with their discussion are presented in Section 5, and Section 6 concludes our work.

Notation: Vectors and matrices are written in boldface lower-case and upper-case letters, respectively. The $\mathbf{A}^\dagger$ and $\mathbf{A}^*$ are referred to as the complex Hermitian transpose, and the complex conjugate of matrix $\mathbf{A}$, respectively. The symbols $\mathrm{tr}(\mathbf{A})$ and $\|\mathbf{A}\|_2$ represent the trace and 2−norm of matrix $\mathbf{A}$, respectively. Moreover, $\|\mathbf{a}\|$ represents the absolute value for the vector $\mathbf{a}$. The matrix $\mathbf{I}_n$ stands for the identity matrix of size $n \times n$. The $\mathbf{x} \sim \mathcal{CN}(\mu, \Sigma)$ means that $\mathbf{x}$ is complex Gaussian distributed with mean μ and covariance matrix Σ. The expression $\sigma^2_{max}(\mathbf{H})$ refers to the maximum eigenvalue of the matrix $\mathbf{H}$. The null($\mathbf{A}$) refers to the null space of the matrix $\mathbf{A}$.

2 System and signal models

Consider a downlink MU-MIMO communication scenario where a BS with M transmit antennas is communicating with multiple UEs, each equipped with N receiving antennas. The served UEs are grouped into M clusters with K UEs in each cluster (Fig. 1). In this work, we are considering scenarios in which the number of antennas at each user, namely N, is greater than the number of antennas equipped at the BS, namely M, that is to say N ≥ M. This assumption is popular in some 5G scenarios such as ultra-dense small cells and cloud-radio access networks (C-RANs) [2]. Through this assumption we are trying to consider some of the realistic scenarios that all 5G communities and mobile communication companies agreed that it will be challenging in the near future. One of the main pillars upon which the next generation mobile networks will based for achieving the 1000 times capacity scaling is the ultra-dense small-cell networks. In such network design, low-power low-cost small-cell BSs will be employed for mobile data offloading. So, it is very likely that it owns the same number of antennas as the UE or even less, given the rapid progress in increasing the capabilities of such UEs. Another network design that support our assumption is the C-RANs, in which UEs are served by a small number of low-cost remote radio heads (RRHs) to reduce the fronthaul overhead [2].

The power-domain multiplexed signal of the users' signals in cluster m is expressed as:

$$\tilde{s}_m = \alpha_{m,1}s_{m,1} + \alpha_{m,2}s_{m,2} + \cdots + \alpha_{m,K}s_{m,K}, \tag{1}$$

where $s_{m,k}$ is the useful information signal to be transmitted to the kth user in the mth cluster, and $\alpha_{m,k}$ is its corresponding NOMA PA coefficient. The signals to be transmitted by the BS in the downlink direction is firstly precoded using the beamforming filter $\mathbf{V} \in \mathbb{C}^{\mathrm{M}\times\mathrm{M}}$. Accordingly, the BS downlink transmitted signal can be written as:

$$\mathbf{x} = \mathbf{V}\tilde{\mathbf{s}}, \tag{2}$$

where $\mathbf{x} = [x_1 x_2 \cdots x_M]^T \in \mathbb{C}^{M\times 1}$ is the combined signal transmitted from the BS to all users in different clusters, with $x_m = \sum_{k=1}^{K} \alpha_{m,k}s_{m,k}$ is the data to be transmitted from the BS to the mth cluster (Fig. 1). Let the radio channel over which the BS transmits its signals is denoted as $\mathbf{H} = \left[\mathbf{H}_1{}^T\mathbf{H}_2{}^T\mathbf{H}_3{}^T\cdots\mathbf{H}_M{}^T\right]^T \in \mathbb{C}^{MKN\times M}$, where $\mathbf{H}_m \in \mathbb{C}^{KN\times M}$ are the channels between the BS and users in cluster m which are all Rayleigh fading channels, and the channel between the BS and the kth user in the mth cluster is denoted as $\mathbf{H}_{m,k}$. The vector $\tilde{\mathbf{s}}$ represents the power-domain multiplexed signals for all the M clusters, which can be expressed in matrix form as:

$$\tilde{\mathbf{s}} = \begin{bmatrix} \tilde{s}_1 \\ \vdots \\ \tilde{s}_M \end{bmatrix} = \begin{bmatrix} \alpha_{1,1}s_{1,1} + \alpha_{1,2}s_{1,2} + \cdots + \alpha_{1,K}s_{1,K} \\ \vdots \\ \alpha_{M,1}s_{M,1} + \alpha_{M,2}s_{M,2} + \cdots + \alpha_{M,K}s_{M,K} \end{bmatrix}. \tag{3}$$

Accordingly, the signal received at the kth user in the mth cluster is decoded using $\mathbf{u}_{m,k}$ to give the detected signal as

$$\mathbf{u}^H_{m,k}\mathbf{y}_{m,k} = \mathbf{u}^H_{m,k}\mathbf{H}_{m,k}\mathbf{V}\tilde{\mathbf{s}} + \mathbf{u}^H_{m,k}\mathbf{n}_{m,k}. \tag{4}$$

Assuming that $\mathbf{V} = [\mathbf{v}_1\mathbf{v}_2\cdots\mathbf{v}_M]$ and $\mathbf{U}_i = \left[\mathbf{u}_{i,1}\mathbf{u}_{i,2}\cdots\mathbf{u}_{i,K}\right] \in \mathbb{C}^{N\times K}, \forall i \in \{1,2,\cdots,M\}$. The signal model in (4) can be rewritten as:

$$\mathbf{u}^H_{m,k}\mathbf{y}_{m,k} = \mathbf{u}^H_{m,k}\mathbf{H}_{m,k}\mathbf{v}_m\left(\alpha_{m,1}s_{m,1} + \cdots + \alpha_{m,K}s_{m,K}\right) + \mathbf{u}^H_{m,k}H_{m,k}\sum_{\substack{i=1\\i\neq m}}^{M}\mathbf{v}_i\tilde{\mathbf{s}}_i + \mathbf{u}^H_{m,k}\mathbf{n}_{m,k}. \tag{5}$$

The interference signals generated in the assumed scenario can be divided into two parts, namely, the intra-cluster interference and the inter-cluster interference. The intra-cluster interference results from the intentional overlapping/superimposing of signals that are to be

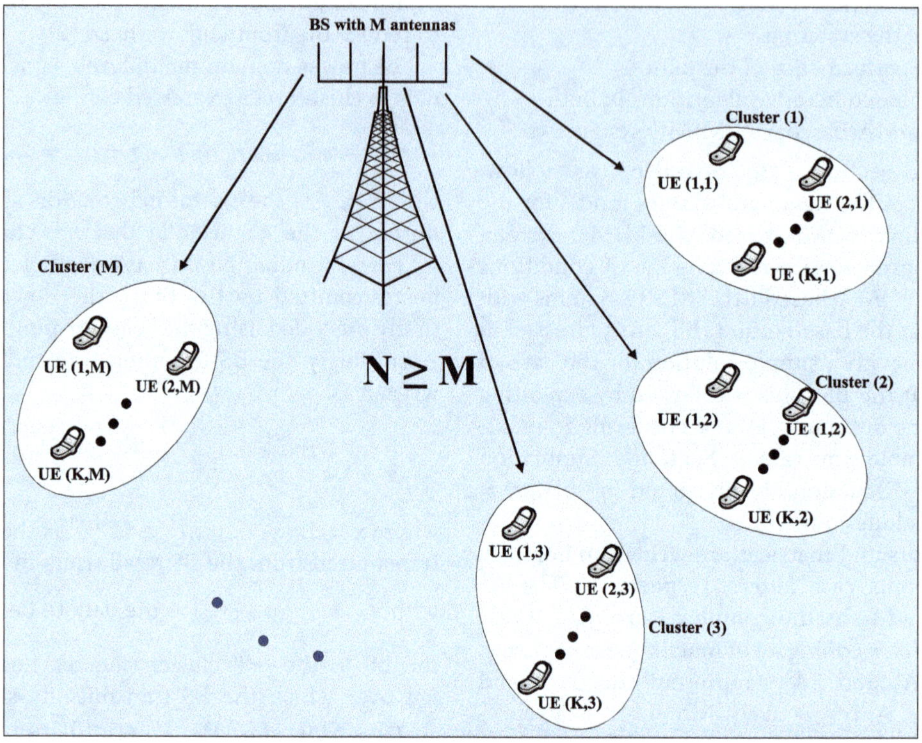

Fig. 1 One-cell downlink multi-user scenario composed of one BS equipped with M transmit antennas. The BS serves some users each equipped with N receive antennas and the users are grouped to M clusters with N ≥ M

transmitted to users belonging to the cluster of the user of interest, namely, the self-interference generated due to the implementation of the NOMA technique. On the other hand, the inter-user interference originated due to the transmission of signals to users who are not belonging to the same cluster of the considered user. Using the notation of intra-cluster and inter-cluster interference, the signal model in (5) can be detailed as:

$$\mathbf{u}_{m,k}^{H}\mathbf{y}_{m,k} = \underbrace{\mathbf{u}_{m,k}^{H}\mathbf{H}_{m,k}\mathbf{v}_{m}\alpha_{m,k}s_{m,k}}_{\text{Desired Signal}} + \underbrace{\mathbf{u}_{m,k}^{H}\mathbf{H}_{m,k}\sum_{\substack{j=1\\ j\neq k}}^{K}\mathbf{v}_{m}\alpha_{m,j}s_{m,j}}_{\text{Intra-cluster Interference}}$$
$$+ \underbrace{\mathbf{u}_{m,k}^{H}\mathbf{H}_{m,k}\sum_{\substack{i=1\\ i\neq m}}^{M}\mathbf{v}_{i}\tilde{\mathbf{s}}_{i}}_{\text{Inter-cluster Interference}} + \underbrace{\mathbf{u}_{m,k}^{H}\mathbf{n}_{m,k}}_{\text{Noise}}. \tag{6}$$

where $i, m \in \{1, 2, \cdots, M\}$, and $j, k \in \{1, 2, \cdots, K\}$. Equation (6) shows how IA and NOMA schemes are integrated together to optimize the sum-rate of MU-MIMO network. The IA technique is applied through implementing the transmit and receive beamformers, $\mathbf{u}_{m,k}$ and $\mathbf{v}_m$. On the other hand, NOMA is applied through superimposing the signals of all users in the cluster together using power domain multiplexing, and this is achieved through careful evaluation of the PA coefficients. Our proposed algorithms will jointly optimize the beamforming vectors and the PA coefficients based on different objectives that are all related to the system sum-rate. The knowledge of the channel conditions is very critical for the implementation of NOMA systems. Accordingly and without loss of generality, we will assume the channels such that the effective channel gains are ordered as follows:

$$\left|\mathbf{u}_{m,1}^{H}\mathbf{H}_{m,1}\mathbf{v}_{m}\right|^{2} \geq \cdots\cdots\cdots \geq \left|\mathbf{u}_{m,K}^{H}\mathbf{H}_{m,K}\mathbf{v}_{m}\right|^{2}, \tag{7}$$

and according to the principles of NOMA technique, the PA coefficients of the users with in the mth cluster are ordered as follow:

$$\alpha_{m,1}^{2} \leq \cdots\cdots\cdots \leq \alpha_{m,K}^{2}. \tag{8}$$

Based on (6), the signal-to-interference-plus-noise power (SINR) ratio for the Kth user, the user with smallest effective channel gain in the mth cluster, is given by

$$\text{SINR}_{m,K} = \frac{\left|\mathbf{u}_{m,K}^{H}\mathbf{H}_{m,K}\mathbf{v}_{m}\right|^{2}\alpha_{m,K}^{2}}{\sum_{j=1}^{K-1}\left|\mathbf{u}_{m,K}^{H}\mathbf{H}_{m,K}\mathbf{v}_{m}\right|^{2}\alpha_{m,j}^{2} + \sum_{\substack{i=1\\ i\neq m}}^{M}\left|\mathbf{u}_{m,K}^{H}\mathbf{H}_{m,K}\mathbf{v}_{i}\right|^{2} + \left|\mathbf{u}_{m,K}\right|^{2}\frac{1}{\rho}}, \tag{9}$$

where ρ refers to the transmit SNR. According to the principles of the NOMA technique, the k^{th}user, $\forall\, 1 < k < K$ in the mth cluster, needs to decode the messages sent to other users with poorer channel conditions first before decoding its own message. Accordingly, the message $s_{m,j}$, $K \geq j \geq (k+1)$, can be detected at k^{th} user in the mth cluster with SINR expressed as:

$$\mathrm{SINR}^{j}_{m,k} = \frac{\left|\mathbf{u}^{H}_{m,k}\mathbf{H}_{m,k}\mathbf{v}_{m}\right|^{2}\alpha^{2}_{m,j}}{\sum_{l=1}^{j-1}\left|\mathbf{u}^{H}_{m,k}\mathbf{H}_{m,k}\mathbf{v}_{m}\right|^{2}\alpha^{2}_{m,l}+\sum_{\substack{i=1\\i\neq m}}^{M}\left|\mathbf{u}^{H}_{m,k}\mathbf{H}_{m,k}\mathbf{v}_{i}\right|^{2}+\left|\mathbf{u}_{m,k}\right|^{2}\frac{1}{\rho}}. \tag{10}$$

It is the role of the SIC to remove the message $s_{m,j}$ from the observation of the kth user. This message can be decoded successfully only when meeting the condition $\log(1+\mathrm{SINR}^{j}_{m,k}) > R_{m,j}$, with $R_{m,j}$ denoting the jth user target data rate. The operation of the SIC technique will continue until the kth user decodes its own message with SINR equals $\mathrm{SINR}^{k}_{m,k}$. The first user in the mth cluster, which is the user with largest effective channel gain, is responsible for decoding all the messages of other users in the cluster . If it is successful, it can decode its own message with SINR equals

$$\mathrm{SINR}^{1}_{m,1} = \frac{\left|\mathbf{u}^{H}_{m,1}\mathbf{H}_{m,1}\mathbf{v}_{m}\right|^{2}\alpha^{2}_{m,1}}{\sum_{\substack{i=1\\i\neq m}}^{M}\left|\mathbf{u}^{H}_{m,1}\mathbf{H}_{m,1}\mathbf{v}_{i}\right|^{2}+\left|\mathbf{u}_{m,1}\right|^{2}\frac{1}{\rho}}. \tag{11}$$

IA and PA techniques are both used to improve the sum-rate for many different wireless communication scenarios [25]. However, up to the best of our knowledge, the study of combining IA approach with PA technique in NOMA-based MU-MIMO environment is not sufficiently conducted. In our proposed framework, the design of the PA scheme depends mainly on the IA strategy to be employed. So, the design of the precoding and decoding filters based on the principles of IA is accomplished first and consequently the PA strategies will be achieved. In the following section, the problem formulation for designing the IA-based precoding and decoding matrices for different IA approaches will be manipulated. This will be followed with the PA design problem which will be addressed in Section 4.

3 IA solutions for multi-user MIMO-NOMA

The design of the precoding and decoding filters for all the network nodes depends on the objective of the IA design process and the availability of the CSI. In this section, we consider the case where no global CSI available at the BS followed by the case with full global CSI available at the BS and IA transceivers design in each case. In our designs for NOMA-based MU-MIMO system with full global CSI at the BS, the considered objectives of IA technique are SINR maximization approach (max-SINR), interference leakage minimization approach (MIL), and sum-rate maximization approach (Max-SR). The derivations and details of each approach are discussed in the below subsections.

3.1 The case with no CSI available at the BS

In the proposed framework, the IA technique is responsible for removing the inter-cluster interference leaving the intra-cluster interference to the SIC technique implicitly implemented in the NOMA design. Accordingly, the IA conditions that guarantee the removal of inter-cluster interference are expressed as follows:

$$\mathbf{u}^{H}_{m,k}\mathbf{H}_{m,k}\mathbf{v}_{i} = 0, \qquad \forall\, i \neq m \tag{12}$$

The availability of CSI at the BS can be considered as a great system overhead. Accordingly, designing IA precoding and decoding filters with no CSI available at the BS, however, it is non-optimal, but it is considered in many practical scenarios because of its reduced system overhead of acquiring the global CSI from all the system nodes. One of the possible solutions to (12) is to choose $\mathbf{V}$, as $\mathbf{V} = \mathbf{I}_{M}$. Choosing $\mathbf{V}$ in this form means that the BS broadcasts user messages without any processing, which reduces the overhead due to handshaking messages in acquiring and forwarding CSI in the network. Accordingly, the decoding filters can be evaluated by substituting in (12) as:

$$\mathbf{u}^{H}_{m,k}\mathbf{h}_{i,mk} = 0, \qquad \forall\, i \neq m \tag{13}$$

where $\mathbf{h}_{i,mk}$ is the ith column of $\mathbf{H}_{m,k}$. As a result of that, the IA constraints at the kth user in the mth cluster can be written as:

$$\mathbf{u}^{H}_{m,k}\underbrace{\left[\mathbf{h}_{1,mk}\cdots\mathbf{h}_{m-1,mk}\mathbf{h}_{m+1,mk}\cdots\mathbf{h}_{M,mk}\right]}_{\tilde{\mathbf{H}}_{m,k}\in\mathbb{C}^{N\times M}} = 0, \tag{14}$$

The above equation can be solved for $\mathbf{u}_{m,k}$ as:

$$\mathbf{u}_{m,k} = \mathrm{null}\left(\tilde{\mathbf{H}}_{m,k}\right), \quad \forall k \in \{1,2,\cdots,K\}, \text{and } m \in \{1,2,\cdots,M\} \tag{15}$$

By using the precoding and decoding matrices derived in this section, the inter-cluster interference will be eliminated, and the SINR for K^{th} user in the mth cluster will be given by:

$$\mathrm{SINR}^{j}_{m,K} = \frac{\left|\mathbf{u}^{H}_{m,K}\mathbf{h}_{m,mK}\right|^{2}\alpha^{2}_{m,K}}{\sum_{j=1}^{K-1}\left|\mathbf{u}^{H}_{m,K}\mathbf{h}_{m,mK}\right|^{2}\alpha^{2}_{m,j}+\left|\mathbf{u}_{m,K}\right|^{2}\frac{1}{\rho}}, \tag{16}$$

similarly, messages of users j, $K \geq j \geq k+1 \geq 1$ will be successfully detected at the kth user, $K > k > 1$ with SINR:

$$\mathrm{SINR}^{j}_{m,k} = \frac{\left|\mathbf{u}^{H}_{m,k}\mathbf{h}_{m,mk}\right|^{2}\alpha^{2}_{m,j}}{\sum_{l=1}^{j-1}\left|\mathbf{u}^{H}_{m,k}\mathbf{h}_{m,mk}\right|^{2}\alpha^{2}_{m,l}+\left|\mathbf{u}_{m,k}\right|^{2}\frac{1}{\rho}}, \tag{17}$$

The SIC scheme implemented with NOMA technique will take care of the remaining intra-cluster interference. Specifically, for the users k, and j, with $K \geq j \geq k+1 \geq 1$, when the message $s_{j,1}$ is successfully detected at the kth user, it will be removed from the kth user's superimposed received signal, and SIC scheme will continue working until its own message is received with SINR equals $\mathrm{SINR}^{k}_{m,k}$. The evaluation of the optimum values of the PA coefficients will be discussed in Section 4.

3.2 The case with full global CSI available at the BS

In this section, we discuss different approaches for evaluating IA-based precoding and decoding filters according to the objective of the IA optimization problem. For each approach, we will formulate the optimization problem which is used at the BS to determine the optimal IA transceiver design for all users and the BS. Specifically, we explain how to design the optimal IA precoder and decoder for each user-BS pair using the MIL, Max-SINR, and Max-SR approaches through solving their respective optimization problems. The general idea of using IA in our scenario is to align the unwanted received signals at the user of interest (the power multiplexed signals assigned to other clusters), into an interference subspace and reducing its projection within the desired signal subspace (the power multiplexed signals of users in its cluster). For example, for the kth user in the mth cluster, the IA constraints are:

$$\operatorname{rank}\left(\mathbf{u}_{m,k}^{H}\mathbf{H}_{m,k}\mathbf{v}_{m}\right) = \mathrm{M}, \tag{18}$$

$$\mathbf{u}_{m,k}^{H}\mathbf{H}_{m,k}\mathbf{v}_{i} = 0, \qquad \forall\, i \neq m \tag{19}$$

$$\forall m \in \{1,2,\cdots,M\}\,\&\, k \in \{1,2,\cdots,K\}$$

IA is used to align the effect of all other clusters by considering them as interference and directing their effects into the interference subspace. This will leave each user only with the effect of its cluster-partners which can be dealt with the SIC technique which is already the core of the NOMA radio access technology.

3.2.1 Interference leakage minimization approach

As the name implies, this approach targets minimizing the other clusters-interference signals deliberated to the desired signal subspace at the user, and the process can be accomplished for the whole cluster at one shot by solving the corresponding optimization problem. The optimization problem corresponding to the mth cluster is formulated as:

$$\begin{aligned} &\underset{\{\mathbf{U}_m,\mathbf{V}\}}{\text{minimize}}\; l_m \\ &\text{s.t. } \mathbf{U}_m\mathbf{U}_m^H = \mathbf{I}_N, \\ &\mathbf{V}\mathbf{V}^H = \mathbf{I}_M, \end{aligned} \tag{20}$$

where $l_m = \operatorname{tr}\left[\mathbf{U}_m^H\mathbf{Q}_m\mathbf{U}_m\right]$ is the total interference leakage deliberated to the useful subspace of cluster m, and $\mathbf{Q}_m$ is the interference covariance matrix for the mth cluster, $\mathbf{Q}_m = \sum_{i=1;\, i\neq m}^{M} \sum_{k=1}^{K} \alpha_{m,k}\mathbf{H}_{m,k}\mathbf{V}\mathbf{V}^H\mathbf{H}_{m,k}^H$ [8]. This optimization problem can be solved by fixing a subset of variables (either $\mathbf{U}_m$ or $\mathbf{V}$), and then optimize for the others, then alternate the roles between the fixed constant variables and the optimization variables. This technique tries to minimize interference leakage by alternatively optimizing the IA beamforming filters. Thus, at the BS, suppose the transmission is carried out in a specific communication direction, the optimization problem (20) is subject to $\mathbf{U}_m\mathbf{U}_m^H = \mathbf{I}_N$, where we optimize for the decoding filters $\mathbf{U}_m$. On the other hand, when the communication direction is reversed, the precoding and decoding filters are interchanged, and the optimization problem is now constrained to $\mathbf{V}\mathbf{V}^H = \mathbf{I}_M$ instead. If we denote d_m as $d_m = \min(\mathrm{M},\mathrm{N})$, the resulted optimization problem in each direction can be solved iteratively using alternative minimization by finding the d_m eigenvectors corresponding to smallest d_m number of eigenvalues of the interference covariance matrix $\mathbf{Q}_m$ at each iteration [8]. Therefore, the d_m columns of $\mathbf{U}_m$ are given by:

$$\mathbf{U}_m^{[d]} = \nu_d\left[\mathbf{Q}_m\right], \qquad \forall d = \{1,\cdots,d_m\} \tag{21}$$

where $\nu_d\left[\mathbf{A}\right]$ refers to the eigenvectors corresponding to the d smallest eigenvalues of $\mathbf{A}$.

3.2.2 Maximum-SINR approach

Another criteria that typically used as an objective to the IA design is the SINR-maximization approach. Specifically, the IA transceiver filters can be designed to maximize the SINR instead of only minimizing the interference leakage, where the MIL approach gives no attempt to maximize the desired signal power within the desired signal subspace. In other words, the MIL approach does not depend at all on the channels through which the desired signal arrives at the intended receiver. According to [8], the IA filters are obtained by:

$$\mathbf{u}_{m,k} = \frac{\left(\mathbf{B}_{m,k}\right)^{-1}\mathbf{H}_{m,k}\mathbf{v}_m}{\left\|\left(\mathbf{B}_{m,k}\right)^{-1}\mathbf{H}_{m,k}\mathbf{v}_m\right\|}, \tag{22}$$

where

$$\mathbf{B}_{m,k} = \sum_{i=1}^{M}\sum_{j=1}^{K}\alpha_{i,j}^2\mathbf{H}_{i,j}\mathbf{v}_m\mathbf{v}_m^H\mathbf{H}_{i,j}^H - \alpha_{m,k}^2\mathbf{H}_{m,k}\mathbf{v}_m\mathbf{v}_m^H\mathbf{H}_{m,k}^H - \mathbf{I}. \tag{23}$$

The last criteria and optimization problem that is used to design the IA transceiver filters at the BS is the sum-rate maximization. The goal is to design the optimum transceivers that maximize the sum-rate.

3.2.3 Sum-rate maximization approach

With this approach, we want to obtain the optimum transceiver filters that maximize the total system sum-rate. Accordingly, the optimization problem that will be implemented to evaluate the IA transceivers have the following form:

$$\begin{aligned} &\underset{\{\mathbf{u}_{m,k},\mathbf{v}_m\}}{\text{minimize}} && R_{sum} = \sum_{m=1}^{M}\sum_{k=1}^{K} R_{m,k} \\ &\text{s.t.} && \mathbf{U}_m\mathbf{U}_m^H = \mathbf{I}_N, \\ &&& \mathbf{V}\mathbf{V}^H = \mathbf{I}_M, \end{aligned} \tag{24}$$

where $R_{m,k}$ represents the mutual information rate between the BS and the kth user in the mth cluster, and it is given by:

$$R_{m,k} = \log_2 \frac{\left|\mathbf{I} + \sum_{i=1}^{M}\sum_{k=1}^{K}\alpha_{i,k}^2\mathbf{u}_{m,k}^H\mathbf{H}_{i,k}\mathbf{v}_i\mathbf{v}_i^H\mathbf{H}_{i,k}^H\mathbf{u}_{m,k}\right|}{\left|\mathbf{I} + \sum_{i=1,\, i\neq m}^{M}\sum_{k=1}^{K}\alpha_{i,k}^2\mathbf{u}_{m,k}^H\mathbf{H}_{i,k}\mathbf{v}_i\mathbf{v}_i^H\mathbf{H}_{i,k}^H\mathbf{u}_{m,k}\right|}, \tag{25}$$

For solving the optimization problem in (24), we will use an iterative algorithm based on Riemannian optimization method [26], which can be considered as a generalization of the standard euclidean optimization by formulating the optimization problem over smooth manifolds instead of the standard euclidean space [27–29]. The update rule for the IA iterative algorithm that based on maximizing the sum-rate is given by the Riemannian optimization over the Grassmann manifold and based on geodesics of a straight line in the euclidean space to the manifold. The update rule of the decoding matrix $\mathbf{U}_m$, $m \in \{1, 2, \cdots, M\}$ over the Grassmann manifold Gr with gradient descent method is expressed as [27]:

$$\mathbf{U}_m^{n+1} = \phi_{Gr}\left(\mathbf{U}_m^n, \text{grad}_{\mathbf{U}_m} \text{R}_{\text{sum}}, \mu\right), \tag{26}$$

$$= \exp\left(\mu\left(\nabla \text{R}_{\text{sum}}(\mathbf{U}_m)\mathbf{U}_m^H - \mathbf{U}_m \nabla \text{R}_{\text{sum}}(\mathbf{U}_m)^H\right)\right)\mathbf{U}_m, \tag{27}$$

where $\phi_{Gr}(.)$ is the geodesic emanating from $\mathbf{U}_m$ on the Grassmann manifolds Gr, $\mathbf{U}_m \in Gr$, in the direction of the gradient of the function R_{sum}, expressed as "$\text{grad}_{\mathbf{U}_m} \text{R}_{\text{sum}}$", and μ is the step size. The gradient, $\text{grad}_{\mathbf{U}_m} \text{R}_{\text{sum}}$, on the Grassmann manifold Gr is computed as:

$$\text{grad}_{\mathbf{U}_m}^{Gr} \text{R}_{\text{sum}} = \nabla \text{R}_{\text{sum}}(\mathbf{U}_m) - \mathbf{U}_m \mathbf{U}_m^H \nabla \text{R}_{\text{sum}}(\mathbf{U}_m). \tag{28}$$

The natural gradient of $\text{R}_{\text{sum}}(\mathbf{U}_m)$, expressed as $\nabla \text{R}_{\text{sum}}(\mathbf{U}_m)$ is a real valued function. However, $\mathbf{U}_m$ is a matrix whose components are complex, so according to [30], the gradient can be evaluated as

$$\nabla \text{R}_{\text{sum}}(\mathbf{U}_m) = 2\frac{\delta\left(\text{R}_{\text{sum}}(\mathbf{U}_m)\right)}{\delta \mathbf{U}_m^*} = 2\left(\frac{\delta\left(\text{R}_{\text{sum}}(\mathbf{U}_m)\right)}{\delta \mathbf{U}_m}\right)^* \tag{29}$$

$$= \frac{2}{\ln 2}\sum_{i=1}^{M}\sum_{k=1}^{K} \alpha_{i,k} \mathbf{U}_m \mathbf{H}_{i,k}^H \mathbf{X}_i^{-1} \mathbf{H}_{i,k} - \frac{2}{\ln 2}\sum_{\substack{i=1,\\ i\neq m}}^{M}\sum_{k=1}^{K} \alpha_{i,k} \mathbf{U}_m \mathbf{H}_{i,k}^H \mathbf{Y}_i^{-1} \mathbf{H}_{i,k}, \tag{30}$$

where $\frac{\delta(f(x))}{\delta x}$ refers to the partial derivative of the function $f(x)$ with respect to x. The matrices $\mathbf{X}$ and $\mathbf{Y}$ are expressed as:

$$\mathbf{X}_m = \mathbf{I} + \sum_{i=1}^{M}\sum_{k=1}^{K} \alpha_{i,k} \mathbf{U}_m^H \mathbf{H}_{i,k} \mathbf{H}_{i,k}^H \mathbf{U}_m.$$

$$\mathbf{Y}_m = \mathbf{I} + \sum_{\substack{i=1,\\ i\neq m}}^{M}\sum_{k=1}^{K} \alpha_{i,k} \mathbf{U}_m^H \mathbf{H}_{i,k} \mathbf{H}_{i,k}^H \mathbf{U}_m.$$

The update rule for the iterative algorithm that computes the decoding matrices $\mathbf{U}_m$ is derived by substituting (28) and (30) in (27).

4 The proposed joint PA and IA methods

In this section, the applied user clustering models are first explained; then, the concepts and details of the proposed PA algorithms are manipulated. Finally, the proposed joint PA and IA algorithms is proposed based on both the clustering and PA concepts.

4.1 User clustering models

User pairing is demonstrated to be very beneficial for the implementation of the NOMA technique in downlink MU scenarios [2]. In a similar fashion, we will illustrate the gain of grouping the users into clusters in the case of downlink NOMA-based MU-MIMO system with the implementation of joint IA and PA optimization. The clustering process and its optimization is beyond the scope of this article. However, we will implement two extreme models for user clustering to prove the importance of using it, in terms of sum-rate, when dealing with IA and PA for downlink NOMA-based MU-MIMO systems. The first model depends on grouping the K users with the best channels together in the first cluster, then the following K best channels' users are grouped within the second cluster, and so on. On the other hand, the second model depends on distributing the M users with best channel gains, one in each of the M clusters, then the following M best channels is distributed in the same fashion, and so on. Figure 2 illustrates the concepts of the two clustering models. In the following discussion, we will refer to the first and second models as the best-with-best and the best-with-poor models, respectively. In both clustering models, the user with the best channel gain can be considered as the cluster head.

4.2 Power allocation approach

We consider that the downlink NOMA-based MU-MIMO system is divided into clusters and the beamforming process is designed such that a single beam used to send all the data messages to their respective users within a specific cluster. Since each cluster contains the same number of users and their channel gains follow the same random distribution, we assume the power budget of the BS will be divided equally between all the clusters. If we assume that the power budget of the BS is denoted as P_{BS}, subsequently, this power is allocated equally between the M clusters with each cluster allocated an amount equal to $(\text{P}_{\text{BS}}/\text{M})$. The power budget for each cluster will be allocated among the scheduled users within the cluster according to the principles of the NOMA technique. For the PA among the users within the cluster, the sum-rate is maximized under cluster power budget constraints, minimum user sum-rate (as quality of service metric) constraints, and constraints related to the implementation of SIC technique, i.e. minimum power differences among NOMA received signals as illustrated in [31]. The PA strategy will be applied separately with each cluster. Without loss of generality, we assume that the effective channel gains of the users within the mth cluster satisfying $|\mathbf{u}_{m,1}\mathbf{H}_{m,1}\mathbf{V}_m| > |\mathbf{u}_{m,2}\mathbf{H}_{m,2}\mathbf{V}_m| > \cdots > |\mathbf{u}_{m,k}\mathbf{H}_{m,k}\mathbf{V}_m| > \cdots > |\mathbf{u}_{m,K}\mathbf{H}_{m,K}\mathbf{V}_m|$. Additionally, we refer to the minimum sum-rate values that must be guaranteed by all users within the cluster as $R_{m,1}, R_{m,2}, \cdots, R_{m,K}$ where $R_{m,k} > 0$, $\forall m$ and $\forall k$. Since we are choosing the number of clusters equal to the number of transmitting antennas at the BS, each cluster will

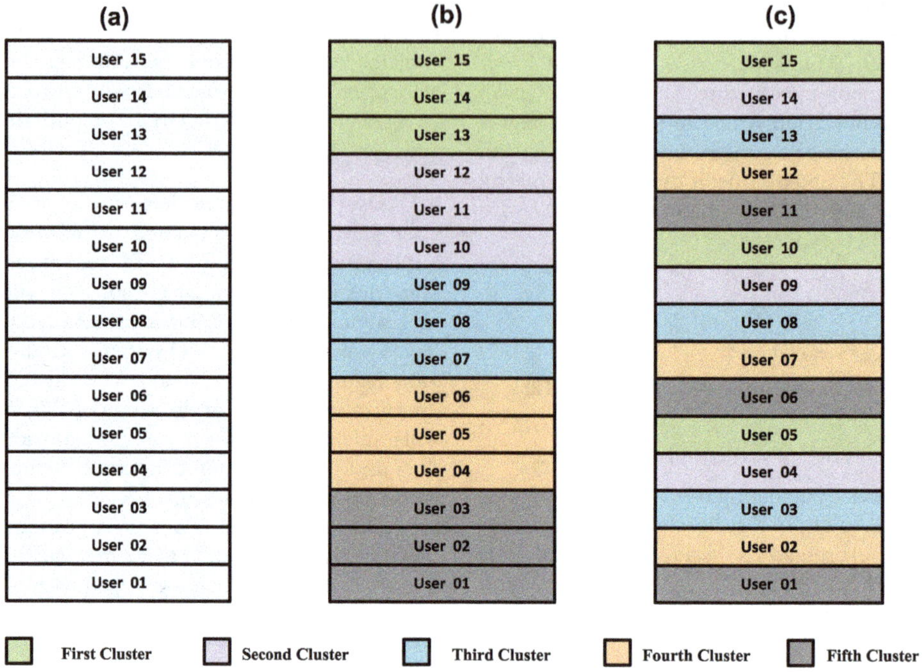

Fig. 2 The clustering models for downlink multi-user MIMO-NOMA with joint IA and PA optimization: **a** The users' effective channel gains are ordered in descending fashion. **b** The users are clustered according to the best-with-best model. **c** The users are clustered according to the best-with-poor model

use the whole system bandwidth BW to serve its users. The PA optimization problem for the users within the mth cluster can be reformulated as:

$$\begin{aligned}
&\underset{\{\alpha_{m,k}^2\}_{k=1}^K}{\text{maximize}} \sum_{k=1}^{K} \text{BW} \log_2\left(1+\frac{\alpha_{m,k}^2 \left|\mathbf{u}_{m,k}\mathbf{H}_{m,k}\mathbf{V}_m\right|^2}{\left|\mathbf{u}_{m,k}\mathbf{H}_{m,k}\mathbf{V}_m\right|^2 \sum_{j=1}^{k-1}\alpha_{m,j}^2 + 1/\rho}\right)\\
&\text{s.t.}: \quad \mathbf{C}_1: \sum_{k=1}^{K} \alpha_{m,k}^2 \leq 1,\\
&\qquad \mathbf{C}_2: \text{BW} \log_2\left(1+\frac{\alpha_{m,k}^2 \left|\mathbf{u}_{m,k}\mathbf{H}_{m,k}\mathbf{V}_m\right|^2}{\left|\mathbf{u}_{m,k}\mathbf{H}_{m,k}\mathbf{V}_m\right|^2 \sum_{j=1}^{k-1}\alpha_{m,j}^2 + 1/\rho}\right) \geq R_{m,k}, \forall k\\
&\qquad \mathbf{C}_3: \left(\alpha_{m,k}^2 - \sum_{j=1}^{k-1}\alpha_{m,j}^2\right)\left|\mathbf{u}_{m,k-1}\mathbf{H}_{m,k-1}\mathbf{V}_m\right|^2 \geq P_{th}, \forall k \neq 1
\end{aligned} \tag{31}$$

where P_{th} denotes the threshold minimum received power difference between users' signals required for carrying out the SIC technique. The previous optimization problem is similar in notation to those mentioned in [31, 32], and accordingly the closed form solution presented in [32] can be applied directly to solve the optimization problem in (31). Let B and C denote the complementary set of users in the mth cluster that meet the minimum sum-rate for the users and SIC visibility constraints, respectively. The optimal PA for the first user within the mth cluster is given by:

$$\begin{aligned}
\alpha_{m,1}^2 =& \frac{1}{\prod\limits_{\substack{j=2\\ j\notin B}}^{K} 2^{\left[\frac{R_{m,j}}{BW}\right]} \prod\limits_{\substack{j=2\\ j\in B}}^{K} 2} - \sum_{\substack{j=2\\ j\notin B}}^{K} \frac{\left(2^{\left[\frac{R_{m,j}}{BW}\right]}-1\right)}{\left(\frac{P_{BS}}{M}\right)\left|\mathbf{u}_{m,j}\mathbf{H}_{m,j}\mathbf{V}_m\right|^2 \prod\limits_{\substack{i=2\\ i\notin B}}^{j} 2^{\left[\frac{R_{m,j}}{BW}\right]} \prod\limits_{\substack{i=2\\ i\in B}}^{j} 2}\\
&- \sum_{\substack{j=2\\ j\notin C}}^{K} \frac{P_{th}}{\left(\frac{2P_{BS}}{M}\right)\left|\mathbf{u}_{m,j-1}\mathbf{H}_{m,j-1}\mathbf{V}_m\right|^2 \prod\limits_{\substack{i=2\\ i\notin B}}^{j-1} 2^{\left[\frac{R_{m,j}}{BW}\right]} \prod\limits_{\substack{i=2\\ i\in B}}^{j-1} 2}.
\end{aligned} \tag{32}$$

Additionally, the PA coefficient for the kth user, with $k \neq 1$, within the mth cluster can be expressed as:

- If $k \notin B$

$$\begin{aligned}
\alpha_{m,k}^2 =& \left[\frac{1}{\prod\limits_{\substack{j=k\\ j\notin B}}^{K} 2^{\left[\frac{R_{m,j}}{BW}\right]} \prod\limits_{\substack{j=k\\ j\in B}}^{K} 2} - \sum_{\substack{j=k\\ j\notin B}}^{K} \frac{\left(2^{\left[\frac{R_{m,j}}{BW}\right]}-1\right)}{\left(\frac{P_{BS}}{M}\right)\left|\mathbf{u}_{m,j}\mathbf{H}_{m,j}\mathbf{V}_m\right|^2 \prod\limits_{\substack{i=k\\ i\notin B}}^{j} 2^{\left[\frac{R_{m,i}}{BW}\right]} \prod\limits_{\substack{i=k\\ i\in B}}^{j} 2}\right.\\
&- \sum_{\substack{j=k\\ j\notin C}}^{K} \frac{P_{th}}{\left(\frac{2P_{BS}}{M}\right)\left|\mathbf{u}_{m,j-1}\mathbf{H}_{m,j-1}\mathbf{V}_m\right|^2 \prod\limits_{\substack{i=k\\ i\notin B}}^{j-1} 2^{\left[\frac{R_{m,i}}{BW}\right]} \prod\limits_{\substack{i=k\\ i\in B}}^{j-1} 2}\\
&\left. + \frac{1}{\left(\frac{P_{BS}}{M}\right)\left|\mathbf{u}_{m,k}\mathbf{H}_{m,k}\mathbf{V}_m\right|^2}\right] \times \left(2^{\left[\frac{R_{m,k}}{BW}\right]}-1\right).
\end{aligned} \tag{33}$$

- If $k \in B$

$$\alpha_{m,k}^2 = \frac{1}{\prod_{\substack{j=k \\ j \notin B}}^{K} 2^{\left[\frac{R_{m,j}}{BW}\right]} \prod_{\substack{j=k \\ j \in B}}^{K} 2} - \sum_{\substack{j=k \\ j \notin B}}^{K} \frac{\left(2^{\left[\frac{R_{m,j}}{BW}\right]} - 1\right)}{\left(\frac{\mathrm{P_{BS}}}{M}\right) \left|\mathbf{u}_{m,j}\mathbf{H}_{m,j}\mathbf{V}_m\right|^2 \prod_{\substack{i=k \\ i \notin B}}^{j} 2^{\left[\frac{R_{m,i}}{BW}\right]} \prod_{\substack{i=k \\ i \in B}}^{j} 2}$$
$$- \sum_{\substack{j=k \\ j \notin C}}^{K} \frac{P_{th}}{\left(\frac{2\mathrm{P_{BS}}}{M}\right) \left|\mathbf{u}_{m,j-1}\mathbf{H}_{m,j-1}\mathbf{V}_m\right|^2 \prod_{\substack{i=k \\ i \notin B}}^{j-1} 2^{\left[\frac{R_{m,i}}{BW}\right]} \prod_{\substack{i=k \\ i \in B}}^{j-1} 2}$$
$$+ \frac{P_{th}}{\left(\frac{\mathrm{P_{BS}}}{M}\right) \left|\mathbf{u}_{m,k-1}\mathbf{H}_{m,k-1}\mathbf{V}_m\right|^2}. \tag{34}$$

The detailed steps for the proposed joint IA and PA framework based on different IA approaches are given in Algorithms 1, 2, and 3.

Algorithm 1 The MIL based IA approach for NOMA-based MU-MIMO systems

---

1: Initialize the beamforming filters, $\mathbf{V}_m$, $\forall\ m \in \{1, 2, \cdots, M\}$, for the clusters with unitary matrices, and initialize the PA coefficients $\alpha_{m,k}$, $\forall\ m \in \{1, 2, \cdots, M\}$, and $k \in \{1, 2, \cdots, K\}$ while keeping the NOMA PA constraints.

2: Calculate the interference covariance matrix for each cluster, m, as:

$$\mathbf{Q}_m = \sum_{i=1;\ i \neq m}^{M} \sum_{k=1}^{K} \alpha_{m,k} \mathbf{H}_{m,k} \mathbf{V} \mathbf{V}^H \mathbf{H}_{m,k}^H,$$

3: Compute the decoding matrices that minimize the interference leakage due to undesired messages, $\mathbf{U}_m$, according to (21).

4: Compute the beamforming matrices using Steps 2 and 3, where we assume a reverse communication direction and initializing the decoding matrices by the values obtained in the previous step and replace the decoding matrices with the beamforming matrices and vice-versa.

5: Update the PA coefficients for all users using (32), (33), and (34).

6: Repeat the Steps from 2 to 5 until the algorithm convergences.

7: Distribute the calculated transceiver filters, and the PA coefficients to the corresponding users.

Algorithm 2 The Max-SINR based IA approach for NOMA-based MU-MIMO systems

---

1: Initialize the beamforming filters, $\mathbf{V}_m$, $\forall\ m \in \{1, 2, \cdots, M\}$, for the clusters with unitary matrices, and initialize the PA coefficients $\alpha_{m,k}$, $\forall\ m \in \{1, 2, \cdots, M\}$, and$k \in \{1, 2, \cdots, K\}$ while keeping the NOMA PA constraints.

2: Calculate the matrix $\mathbf{B}_{m,k}$ for each user using (23).

3: Compute the decoding matrices that maximize the SINR, $\mathbf{U}_m$, according to (22).

4: Compute the decoding matrices using Steps 2 and 3, where we assume a reverse communication direction and initializing the decoding matrices by the values obtained in the previous step and replace the decoding matrices with the beamforming matrices and vice-versa.

5: Update the PA coefficients for all users using (32), (33), and (34).

6: Repeat the Steps from 2 to 5 until the algorithm convergences.

7: Distribute the calculated transceiver filters, and the PA coefficients to the corresponding users.

Algorithm 3 The Max-SR based IA approach for NOMA-based MU-MIMO systems

---

1: Initialize the beamforming filters, $\mathbf{V}_m$, $\forall\ m \in \{1, 2, \cdots, M\}$, for the clusters with unitary matrices, the step size μ with 0.1, and initialize the PA coefficients $\alpha_{m,k}$, $\forall\ m \in \{1, 2, \cdots, M\}$, and$k \in \{1, 2, \cdots, K\}$ while keeping the NOMA PA constraints.

2: Calculate the matrix $\mathbf{B}_{m,k}$ for each user using (23).

3: Compute the decoding matrices that maximize the SINR, $\mathbf{U}_m$, according to (22).

4: Update the beamforming matrices, $\mathbf{V}_m$ along the geodesic over the Grassmann manifold in the direction given by the gradient of $\mathrm{R_{sum}}$ using (27) and (30).

5: Update the PA coefficients for all users using (32), (33), and (34).

6: Update the optimization step size $\mu = \mu \times 0.95$.

7: Repeat the Steps from 2 to 6 in the reverse communication direction until the algorithm convergences.

8: Distribute the calculated transceiver filters, and the PA coefficients to the corresponding users.

Note that the PA coefficients are initialized in all the three algorithms as $\alpha_{m,k} = 1/K$. Each of the algorithms is dependent on a specific criteria for achieving the IA conditions as explained in Subsection 3.2. All these algorithms rely on the availability of CSI, and all apply the same power allocation algorithm introduced in Section 4. Since the evaluation of the matrices $\mathbf{Q}_m$, and $\mathbf{B}_{m,k}$ depends on the PA coefficients, the PA coefficients accordingly affect the design of the transmit and decode beamforming matrices. As a result, the proposed algorithms are jointly optimizing both the PA coefficients and the transceiver filters aiming to eventually optimize the system sum-rate. The joint optimization is solved under total power budget, user QoS, and robust SIC constraints as explained in details in Section 4. It is worth noting that the proposed iterative IA techniques are carried out offline as we are assuming stationary block fading channels, which remains constant during the transmission process. Accordingly, the complexity analysis of the algorithms is not as important as the gain in the system sum-rate.

5 Simulation results and discussion

In this section, we introduce simulation results to illustrate the sum-rate improvement resulted due to joint optimization of IA transceivers and PA coefficients for downlink NOMA-based MU-MIMO systems. Additionally, the simulation results for the proposed NOMA-based MU-MIMO system with joint IA and PA optimization is compared with those of the conventional NOMA-based MU-MIMO as well as the OMA-based MU-MIMO systems. The channels between the BS and the users are all assumed to be Gaussian distributed Rayleigh fading with zero mean and unit variance in addition to the shadowing and the pathloss effects with parameters as mentioned in [32]. Additionally, we assume that perfect full global CSI is available at the BS. In other words, the BS owns a copy of the channel between it and each user in the system [8–18, 32]. The simulation results are obtained through averaging the measurements over 5000 channel realizations. For the proposed NOMA-based MU-MIMO network with clustering model employed, we have assumed that the number of clusters in the system equals the number of antennas at the BS, and all clusters have the same size (number of users in the cluster). A list of all the simulation parameters used in evaluating our results are in Table 1. The algorithms that will be involved in the comparison are MIL-IA implemented in OMA-based MU-MIMO system (MIL-IA-MIMO-OMA), MIL-IA implemented in conventional NOMA-based MU-MIMO system (MIL-IA-Conv-MIMO-NOMA), MIL-IA implemented in the proposed NOMA-based MU-MIMO system (MIL-IA-Proposed-MIMO-NOMA), Max-SINR-IA implemented in OMA-based MU-MIMO system (Max-SINR-IA-MIMO-OMA), Max-SINR-IA implemented in conventional NOMA-based MU-MIMO system (Max-SINR-IA-Conv-MIMO-NOMA), Max-SINR-IA implemented in the proposed NOMA-based MU-MIMO system (Max-SINR-IA-Proposed-MIMO-NOMA), Max-SR-IA implemented in OMA-based MU-MIMO system (Max-SR-IA-MIMO-OMA), Max-SR-IA implemented in conventional NOMA-based MU-MIMO system (Max-SR-IA-Conv-MIMO-NOMA), Max-SR-IA implemented in the proposed NOMA-based MU-MIMO (Max-SR-IA-Proposed-MIMO-NOMA), SVD-based IA implemented in OMA-based MU-MIMO (SVD-IA-MIMO-OMA), SVD-based IA implemented in conventional NOMA-based MU-MIMO system introduced in [24] (SVD-IA-Conv-MIMO-NOMA in [24]), SVD-based IA implemented in the proposed NOMA-based MU-MIMO system (SVD-IA-Proposed-MIMO-NOMA), and the conventional NOMA-based MU-MIMO system without IA (Conv-NOMA without IA [23]).

Figure 3 shows the variation of the sum-rate versus the transmitted power, P_{tr}, with different IA approaches for the proposed NOMA-based MU-MIMO as well as the conventional

Table 1 Simulation parameters

Parameter name	Value
System effective bandwidth (BW)	20 MHz
Bandwidth of a resource block	200 KHz
Number of available resource blocks	100
Transmit power budget	46 dBm
Detection threshold at SIC receiver (P_{th})	10 dBm
Cell diameter	600 m
Path-loss Exponent	4
Noise power density (N_0)	− 174 dBm/Hz
No. of users per cluster (K)	4
No. of transmitting antennas (M)	2
No. of receiving antennas (N)	4
Shadowing standard deviation	8 dB
Path-loss equation at 2 GHz band	$15.3 + 37.6log_{10}(d_0)$
Initial power allocation coefficients $\left(\alpha_{m,k}^2\right)$	$1/K$

MIMO-NOMA and MIMO-OMA systems. In these simulation results, we have assumed that users having the same order within the clusters will be assigned the same minimum sum-rate value, which is inserted in (31) as the minimum sum-rate constraint, mathematically speaking, we assume that $R_{m-1,k} = R_{m,k} = R_{m+1,k}, \forall m \in \{1, 2, \cdots, M\}$. It is obvious from the results that, the IA approach that depends on maximizing the system sum-rate over the Grassmann manifold outperforms both the MIL and Max-SINR approaches with both the proposed and the conventional systems. The cause behind that fact is that the sum-rate maximization approach considers optimizing all the Shanonn's capacity equation's parameters, namely the spatial DoFs, desired signal power, and the undesired interference power, while other approaches consider optimizing only one or two of these parameters. Additionally, the proposed NOMA-based MU-MIMO system that employs clustering, IA, and optimum PA obtained by solving (31) provides the most higher sum-rate performance, followed by the conventional NOMA-based MU-MIMO system, and the OMA-based MU-MIMO system provides the worst performance in the comparison. Moreover, it is obvious that all the proposed algorithms outperform the state-of-the-art algorithms [23, 24].

Figure 4 shows the effect of choosing the clustering model within the proposed NOMA-based MU-MIMO system. In our work, we employed the two clustering models shown in Fig. 2. It is very clear from the results that the proposed NOMA-based MU-MIMO system performs better with the best-with-poor clustering model than with the case of best-with-best clustering model. This is due to the effectiveness of the SIC technique with the best-with-poor model than with the best-with-best model. In other words, the clustering model somehow governs and keeps the minimum received power differences among the signals of different users with the NOMA-based MU-MIMO system improving the performance of the SIC technique and accordingly provides optimum interference cancellation within

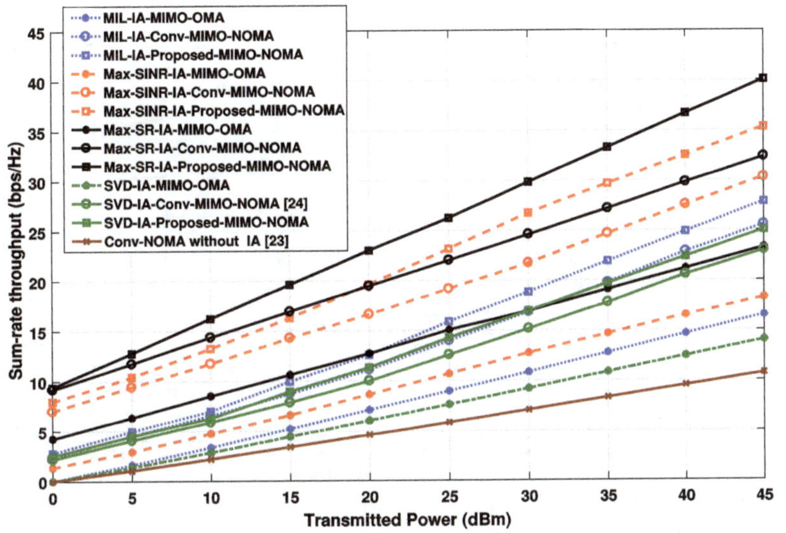

Fig. 3 The transmitted power P_{tr} in (dBm) versus the sum-rate in (bits/sec/Hz) for the proposed MIMO-NOMA approach (With joint IA and PA), in comparison with the conventional MIMO-NOMA approach (no clustering and the power distributed equally between users), and the case of IA with MIMO-OMA

the cluster. Another important observation is that the superiority of the best-with-poor clustering model than the best-with-best clustering model is guaranteed with any of the IA approaches. At a transmitted power of 35 dB, the proposed MIMO-NOMA with Max-SR IA approach achieves around 11 (bits\sec\Hz) more sum-rate with best-with-poor clustering model than with best-with-best model. Similar conclusions can be reported with other IA approaches at different transmitted power levels.

Figure 5 shows the effect of the cluster size, defined as the number of users per cluster, on the system sum-rate. The sum-rate grows almost linearly with the cluster size for all IA approaches until reaching a specific cluster size, 10 and 12 users at transmission power levels 15dBm and 35dBm, respectively. After that, the sum-rate begins to decay due to the lower efficiency of the SIC technique and accordingly the growth of intra-cluster interference. In other words, the sum-rate for the NOMA-based MU-MIMO system grows with the cluster size as long as the system meets the constraint of minimum received power differences among the users. Once the system violates this constraint, the SIC technique provides lower efficiency in canceling the intra-cluster interference and the sum-rate begins to decrease. This behavior is common among all IA approaches with different transmission powers.

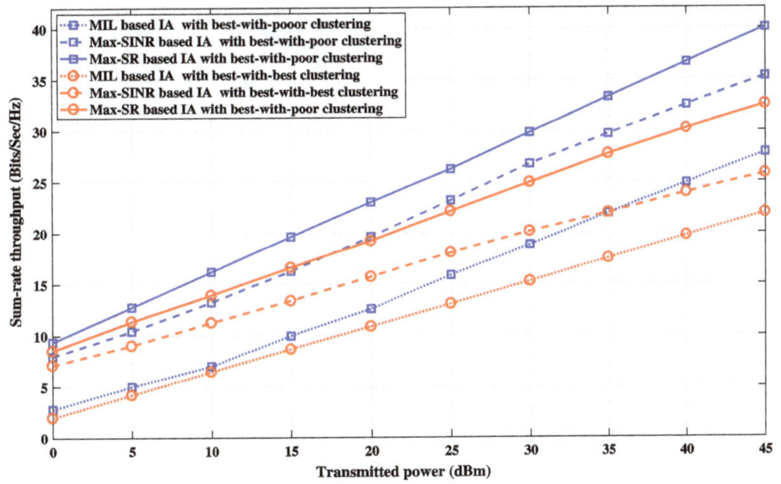

Fig. 4 The transmitted power in (dBm) versus the sum-rate in (Bits/Sec/Hz) for the proposed MIMO-NOMA approach (with joint IA and PA) with different IA approaches and different clustering models

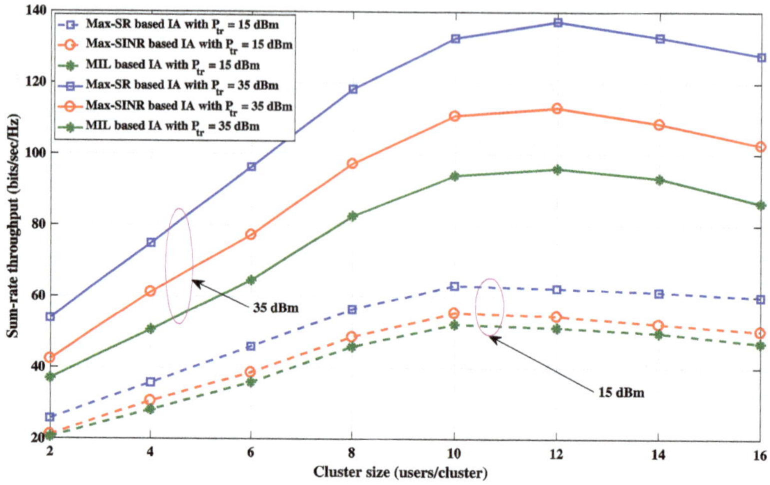

Fig. 5 The cluster size in (users/cluster) versus the sum-rate in (Bits/sec/Hz) for different IA approaches with the proposed MIMO-NOMA with transmitted power of 15dBm, and 35dBm

6 Conclusions

In this article, we have studied the application of different IA approaches to downlink NOMA-based MU-MIMO systems. Specifically, we have proposed a joint IA and PA framework for maximizing the sum-rate of the NOMA-based MU-MIMO system under different approaches of IA. It turns out from the simulation results that IA is still an excellent scheme for accessing the maximum DoFs of the next generation NOMA-based networks. Additionally, user clustering is proven to be a critical step for the joint optimization of the PA coefficients and IA transceivers in NOMA-based MU-MIMO systems. Finally, we concluded that accompanying the NOMA access technology with IA combined with optimum PA algorithm can be considered as a key solution for achieving the capacity scaling targeted by next generation 5G networks. As a future work, the performance of the proposed algorithms would be studied under both instantaneous and statistical CSI. Moreover, the performance of the IA-based transceivers in NOMA-based MU-MIMO system will be investigated with partial CSI.

Abbreviations

BS: Base-station; CSI: Channel state information; C-RAN: Cloud-radio access networks; DoFs: Degrees of freedom; IA: Interference alignment; IC: Interference channel; MU-MIMO: Multi-user multiple-input multiple-output; NOMA: Non-orthogonal multiple access; OMA: Orthogonal multiple access; PA: Power allocation; QoS: Quality of service; RRH: Remote radio heads; SIC: Successive interference cancellation; SINR: Signal to interference plus noise ratio; SNR: Signal to noise ratio; SVD: Singular value decomposition; UE: User equipment; 5G: Fifth generation

Acknowledgements

The authors would like to thank the anonymous reviewers for their insightful comments.

Funding

The EURASIP Journal on Wireless Communications and Networking supports the publication of this article.

Authors' contributions

All authors discussed the experiments; MR performed the experiments and wrote the paper. LH and PZ have made some useful comments on the paper. All authors have read and approved the final manuscript.

Competing interests

The authors declare that they have no competing interests.

Author details

[1]Multi-dimensional Signal Processing Lab., College of Information Engineering, Shenzhen University, Shenzhen, China. [2]Department of Electronics and Communication Engineering, Faculty of Electronic Engineering, Menoufia University, Shibin Al-Kawm 21974, Egypt.

References

1. L. Dai, B. Wang, Z. Ding, Z. Wang, S. Chen, L. Hanzo, A Survey of Non-Orthogonal Multiple Access for 5G. IEEE Surv. Tutor. Accessed on 8 July 2018. [Online]. available https://doi.org/10.1109/COMST.2018.2835558
2. Z. Ding, F. Adachi, V. Poor, The application of MIMO to non-orthogonal multiple access. IEEE Trans. Wirel. Commun. **15**(1), 537–552 (2016)
3. Z. Ding, X. Lei, G.K. Karagiannidis, R. Schober, J. Yuan, V.K. Bhargava, A Survey on Non-Orthogonal Multiple Access for 5G Networks: Research Challenges and Future Trends. IEEE J. Sel. Areas Commun. **35**(10), 2181–2195 (2017)
4. S. Timotheou, I. Krikidis, Fairness for non-orthogonal multiple access in 5G systems. IEEE Signal Sig. Process. **22**(10), 1647–1651 (2015)
5. L. Dai, B. Wang, Y. Yuan, S. Han, I. Chih-lin, Z Wang, Non-orthogonal multiple access for 5G: solutions, challenges, opportunities, and future research trends. IEEE Commun. Mag. **53**(9), 74–81 (2015)
6. L. Lv, Q. Ni, Z. Ding, J. Chen, Application of non-orthogonal multiple access in coopertaive spectrum-sharing networks over Nakagami−m fading channels. IEEE Trans. Veh. Technol. **66**(06), 5506–5511 (2016)
7. S. Islam, M. Zeng, O. Dobre, K. Kwak, Resource Allocation for Downlink NOMA Systems: Key Techniques and Open Issues. IEEE Wirel. Commun. **25**(2), 40–47 (2018)
8. K. Gomadam, V.R. Cadambe, S.A. Jafar, A distributed numerical approach to interference alignment and applications to wireless interference networks. IEEE Trans. Inf. Theory. **57**(06), 3309–3322 (2011)
9. O. El Ayach, S.W. Peters, R.W. Heath Jr., The feasibility of interference alignment over measured MIMO-OFDM channels. IEEE Trans. Veh. Technol. **59**(09), 4309–4321 (2010)
10. T. Ying, W. Feng, G. Liu, Space-Time Interference Alignment: DoF of Two-User MIMO X-Channel with Alternating CSIT. IEEE Commun. Lett. **21**(5), 1167–1170 (2017)
11. X. Jing, L. Mo, H. Liu, C. Zhang, Linear Space-Time Interference Alignment for K-User MIMO Interference Channel. IEEE Access. **6**, 3085–3095 (2017)

12. M. Rihan, M. Elsabrouty, O. Muta, H. Furukawa, in *IEEE vehicular technology conference (VTC Spring)*. Interference alignment with limited feedback for macrocell-femtocell heterogeneous networks (IEEE, Glasgow, 2015), pp. 1–5
13. M. Rihan, M. Elsabrouty, O. Muta, H. Furukawa, Interference mitigation framework based on interference alignment for femtocell-macrocell two tier cellular systems. IEICE Trans. Commun. **E98-B**(03), 467–476 (2015)
14. T.T. Vu, H.H. Kha, O. Muta, M. Rihan, Energy-efficient interference mitigation with hierarchical partial coordination for MIMO heterogeneous networks. IEICE Trans. Commun. **E100-B**(06), 3247–3255 (2016)
15. H. Men, N. Zhao, M. Jin, J. M. Kim, Optimal transceiver design for interference alignment based cognitive radio networks. IEEE Commun. Lett. **19**(08), 1442–1445 (2015)
16. B. Guler, A. Yener, Selective interference alignment for MIMO cognitive femtocell networks. IEEE J. Sel. Areas Commun. **32**(03), 439–450 (2014)
17. S. Cho, K. Huang, D.K. Kim, V.K.N. Lau, H. Chae, H. Seo, B. Kim, Feedback-topology designs for interference alignment in MIMO interference channels. IEEE Trans. Sig. Process. **60**(12), 6561–6575 (2012)
18. O. El Ayach, S.W. Peters, R.W. Heath Jr., The practical challenges of interference alignment. IEEE Wirel. Commun. **20**(01), 35–42 (2013)
19. N. Zhao, F. Richard Yu, M. Jin, Q. Yan, V.C.M. Leung, Interference Alignment and Its Applications: A Survey, Research Issues and Challenges. IEEE Commun. Surv. Tutor. **18**(3), 1779–1803 (2016)
20. Y. Fadlallah, K. Amis, A. Aissa-El-Bey, R. Pyndiah, Interference Alignment for a multi-user SISO interference channel. EURASIP J. Wirel. Commun. Netw. https://doi.org/10.1186/1687-1499-2014-79
21. Q. Sun, S. Han, Z. Xu, S. Wang, I. Chih-Lin, Z. Pan, in *IEEE wireless communications and networking conference (WCNC)*. Sum rate optimization for MIMO non-orthogonal multiple access systems (IEEE, Istanbul, 2015), pp. 747–752
22. Z. Ding, R. Schober, H.V. Poor, A general MIMO framework for NOMA downlink and uplink transmission based on signal alignment. IEEE Trans. Wirel. Commun. **15**(06), 4438–4454 (2016)
23. Y. Liu, G. Pan, H. Zhang, M. Song, On the capacity comparison between MIMO-NOMA and MIMO-OMA. IEEE Access. **04**, 2123–2129 (2016)
24. Z.Q. Al-Abbasi, D.K.C. So, J. Tang, in *IEEE international communications conference (ICC)- wireless communications symposium*. Resource Allocation for MU-MIMO non-orthogonal multiple access (NOMA) system with interference alignment (IEEE, Paris, 2017), pp. 1–6
25. H. Sung, S.H. Park, K.J. Lee, I. Lee, Linear precoder designs for K user interference channels. IEEE Trans. Wirel. Commun. **9**(01), 291–301 (2010)
26. I. Santamaria, O. Gonzalez, R. Heath, S. Peters, in *IEEE GLOBECOM*. Maximum sum rate interference alignment algorithms for MIMO channels (IEEE, Miami, 2010), pp. 1–6
27. T.E. Brudan, J. Eeiksson, V. Koivunen, Steepest descent algorithms for optimization under unitary matrix constraint. IEEE Trans. Sig. Process. **56**(03), 1134–1147 (2006)
28. Y. Nishimori, S. Akaho, S. Abdallah, M. Plumbley, in *IEEE conference on acoustics, speech, and signal processing*. Flag Manifolds for Subspace ICA Problems (IEEE, Honolulu-HI-USA, 2017)
29. A. Edelman, T.A. Arias, S.T. Smith, The Geometry of algorithms with orthogonality constraints. SIAM J. Matrix Anal. Appl. **20**(2), 303–353 (1998)
30. S. Ye, R.S. Blum, Optimized signaling for MIMO interference systems with feedback. IEEE Trans. Sig. Process. **51**(11), 2839–2848 (2003)
31. M.S. Ali, E. Hossain, D.I. Kim, Non-orthogonal multiple access (NOMA) for downlink multiuser MIMO systems: User clustering, beamforming, and power allocation. IEEE Access J. Phys. Mediu. Access Control. Layer Adv. 5G Wirel. Netw. **05**, 565–577 (2016)
32. M.S. Ali, H. Tabassum, E. Hossain, Dynamic user clustering and power allocation in non-orthogonal multiple access (NOMA) systems. IEEE Access. **04**, 6325–6343 (2016)

Permissions

All chapters in this book were first published in EURASIP JWCN, by Springer; hereby published with permission under the Creative Commons Attribution License or equivalent. Every chapter published in this book has been scrutinized by our experts. Their significance has been extensively debated. The topics covered herein carry significant findings which will fuel the growth of the discipline. They may even be implemented as practical applications or may be referred to as a beginning point for another development.

The contributors of this book come from diverse backgrounds, making this book a truly international effort. This book will bring forth new frontiers with its revolutionizing research information and detailed analysis of the nascent developments around the world.

We would like to thank all the contributing authors for lending their expertise to make the book truly unique. They have played a crucial role in the development of this book. Without their invaluable contributions this book wouldn't have been possible. They have made vital efforts to compile up to date information on the varied aspects of this subject to make this book a valuable addition to the collection of many professionals and students.

This book was conceptualized with the vision of imparting up-to-date information and advanced data in this field. To ensure the same, a matchless editorial board was set up. Every individual on the board went through rigorous rounds of assessment to prove their worth. After which they invested a large part of their time researching and compiling the most relevant data for our readers.

The editorial board has been involved in producing this book since its inception. They have spent rigorous hours researching and exploring the diverse topics which have resulted in the successful publishing of this book. They have passed on their knowledge of decades through this book. To expedite this challenging task, the publisher supported the team at every step. A small team of assistant editors was also appointed to further simplify the editing procedure and attain best results for the readers.

Apart from the editorial board, the designing team has also invested a significant amount of their time in understanding the subject and creating the most relevant covers. They scrutinized every image to scout for the most suitable representation of the subject and create an appropriate cover for the book.

The publishing team has been an ardent support to the editorial, designing and production team. Their endless efforts to recruit the best for this project, has resulted in the accomplishment of this book. They are a veteran in the field of academics and their pool of knowledge is as vast as their experience in printing. Their expertise and guidance has proved useful at every step. Their uncompromising quality standards have made this book an exceptional effort. Their encouragement from time to time has been an inspiration for everyone.

The publisher and the editorial board hope that this book will prove to be a valuable piece of knowledge for researchers, students, practitioners and scholars across the globe.

List of Contributors

Kui Xu, Zhexian Shen, Yurong Wang and Xiaochen Xia
Institute of Communication Engineering, Army Engineering University of PLA, No. 2, Biaoying Street, Nanjing, China

Javier Rubio and Antonio Pascual-Iserte
Department of Signal Theory and Communications, Universitat Politècnica de Catalunya (UPC), Jordi Girona 1-3, Building D5, 08034 Barcelona, Spain

Dian-Wu Yue
School of Information Science and Technology, Dalian Maritime University, Dalian 116026, Liaoning, China
Department of Electrical and Computer Engineering, University of Saskatchewan, 57 Campus Drive, Saskatoon SK S7N 5A9, Canada

Shuai Xu
School of Information Science and Technology, Dalian Maritime University, Dalian 116026, Liaoning, China

Ha H. Nguyen
Department of Electrical and Computer Engineering, University of Saskatchewan, 57 Campus Drive, Saskatoon SK S7N 5A9, Canada

Kui Xu, Ming Zhang, Jie Liu, Nan Sha, Wei Xie and Lihua Chen
College of Communication Engineering, Army Engineering University of PLA, No. 2, Biaoying Street, Nanjing, China

Mohammed Zahid Aslam and Yoann Corre
SIRADEL, 2 Parc de Brocéliande, St. Grégoire, France

Emil Björnson and Erik G. Larsson
Linköping University, 58183 Linköping, Sweden

Dongmei Zhang, Ximing Wang, Kui Xu and Wei Xie
Army Engineering University of PLA, No. 88, Houbiaoying Street, Nanjing, China

Yijun Yang
Unit 32020 of PLA, Wuhan, China

Saidiwaerdi Maimaiti, Gang Chuai, Weidong Gao, Kaisa Zhang, Xuewen Liu and Zhiwei Si
Key Laboratory of Universal Wireless Communications, Ministry of Education, Beijing University of Posts and Telecommunications, Beijing 100876, People's Republic of China

Abbas Akbarpour-Kasgari and Mehrdad Ardebilipour
Department of Electrical Engineering, K. N. Toosi University of Technology, Shariati St., Tehran, Iran

Hwanjin Kim and Junil Choi
School of Electrical Engineering, Korea Advanced Institute of Science and Technology, Daejeon, Korea

Hui Zhi
Key Lab of Intelligent Computing and Signal Processing (Ministry of Education), Anhui University, Hefei 230601, China

Xiaoguang Ding
QuantumCTek Co., Ltd., Hefei 230088, China

Stefan Schwarz
Christian Doppler Laboratory for Dependable Wireless Connectivity for the Society in Motion, Institute of Telecommunications, Technische Universität (TU) Wien, Vienna, Austria

In-soo Kim and Junil Choi
School of Electrical Engineering, KAIST, Daejeon, South Korea

Ju-Hong Lee
Department of Electrical Engineering, Graduate Institute of Communication Engineering, and Graduate Institute of Biomedical Electronics and Bioinformatics, National Taiwan University, No. 1, Sec. 4, Roosevelt Rd., Taipei 10617, Taiwan

Wei-En Sun
Graduate Institute of Communication Engineering, National Taiwan University, No. 1, Sec. 4, Roosevelt Rd., Taipei 10617, Taiwan

Lei Huang and Peichang Zhang
Multi-dimensional Signal Processing Lab., College of Information Engineering, Shenzhen University, Shenzhen, China

Mohamed Rihan
Multi-dimensional Signal Processing Lab., College of Information Engineering, Shenzhen University, Shenzhen, China
Department of Electronics and Communication Engineering, Faculty of Electronic Engineering, Menoufia University, Shibin Al-Kawm 21974, Egypt

Index

www.ingramcontent.com/pod-product-compliance
Lightning Source LLC
LaVergne TN
LVHW080727210926
848485LV00003B/532
9781639892617